Monitoring Biodiversity

This book is an exciting reappraisal of the role and practice of biodiversity monitoring, showing how new technologies and software applications are rapidly maturing and can both complement and maintain continuity with the best practice in traditional field skills.

Environmental monitoring is a key component in a large number of national programmes and constitutes an important aspect of understanding environmental change and supporting policy development. Taking an interdisciplinary approach, *Monitoring Biodiversity* begins by discussing monitoring as an established field and examines the various budgetary and technological challenges. It examines different methodologies, the variation between countries, and the design features relevant to understanding monitoring systems created for new policy goals or different funding situations. The huge variety of methods revealed across 18 chapters, which vary from statistical designs to remote sensing, interviews, surveys, and new ways of stacking and combining data and thematic information for visualization and modelling, underlines just how mature and multifaceted the modern practice of monitoring can be. It concludes with several problem-based chapters that discuss the design and implementation of environmental monitoring in specific scenarios such as urban and aquatic areas. All chapters include key messages, study questions, and further reading.

With a focus on Europe but with international relevance, *Monitoring Biodiversity* will be an essential resource for students at all levels of environmental monitoring, assessment, and management.

Anna Allard is researcher in the Division of Landscape Analysis, Swedish University of Agricultural Sciences. She has worked for many years in national monitoring of biodiversity on the landscape scale, in several ongoing monitoring schemes including seashores and mountains as well as national digital vegetation mapping. Her expertise is in landscape ecology and analysis of the landscape and vegetation by remote sensing.

E. Carina H. Keskitalo is professor of political science in the Department of Geography, Umeå University, and a guest researcher at the Swedish University of Agricultural Sciences. She has published widely on forest and natural resource policy and use applying qualitative methods.

Alan Brown has worked for one of the UK nature conservation agencies as the senior remote sensing manager, previously as the lead on terrestrial monitoring. His professional background is in upland ecology, vegetation survey and monitoring, computer programming, and multivariate and statistical analysis.

Monitoring Biodiversity

Combining Environmental and Social Data

Edited by Anna Allard,
E. Carina H. Keskitalo,
and Alan Brown

Routledge
Taylor & Francis Group
LONDON AND NEW YORK

earthscan
from Routledge

Designed cover image: © Travelpix Ltd/Getty

First published 2023
by Routledge
4 Park Square, Milton Park, Abingdon, Oxon OX14 4RN

and by Routledge
605 Third Avenue, New York, NY 10158

Routledge is an imprint of the Taylor & Francis Group, an informa business

British Library Cataloguing-in-Publication Data
A catalogue record for this book is available from the British Library

ISBN: 978-1-032-01593-4 (hbk)
ISBN: 978-1-032-01594-1 (pbk)
ISBN: 978-1-003-17924-5 (ebk)

DOI: 10.4324/9781003179245

Typeset in Bembo
by MPS Limited, Dehradun

For Margareta Ihse, Professor emerita in Ecological Geography, who has dedicated her working life towards the enhancement of landscape biodiversity, and for Peter Hope Jones (1935–2020), ornithologist, who pioneered monitoring in Wales.

Contents

17 Case study: reindeer husbandry plans – "Is this even monitoring?" 333
PER SANDSTRÖM, STEFAN SANDSTRÖM, ULRIKA ROOS, AND ERIK CRONVALL

18 Reflections on monitoring: conclusions and ways forward 349
E. CARINA H. KESKITALO, ALAN BROWN, AND ANNA ALLARD

Contributors

Andreas Aagaard Christensen Associate professor of landscape research and environmental planning at the Department of People and Technology, Roskilde University, Denmark. He has a background in environmental geography with research interests ranging from habitat monitoring and landscape ecology to rural sociology and planning, social–ecological modelling, nature conservation planning, environmental humanities, and environmental history.

Anna Allard Researcher in the Division of Landscape Analysis, Swedish University of Agricultural Sciences. She has worked for many years in national monitoring of biodiversity on the landscape scale, in several ongoing monitoring schemes including seashores and mountains, as well as national digital vegetation mapping. Her expertise is in landscape ecology and analysis of the landscape and vegetation by remote sensing.

Elias Andersson Associate professor of technology in the Landscape Studies Unit, Department of Forest Resource Management, Swedish University of Agricultural Sciences. He works in gender studies, amongst other areas, often applying qualitative methods.

Magnus Appelberg Professor of environmental assessment with a specialization in fish ecology. His research focuses on different aspects of environmental assessment and ecosystem-based fisheries management.

Jonas Ardö Professor of physical geography and ecosystem science at Lund University, Sweden. His focus has been on Earth observation.

Armin Benzler Research associate at the Federal Agency for Nature Conservation in Germany. His area of activity is nature conservation monitoring with special focus on High Nature Value farmland monitoring and monitoring environmental impacts of GMOs.

Alan Brown Now retired, he worked for one of the UK nature conservation agencies as the senior remote sensing manager and previously as the lead on terrestrial monitoring. His professional background is in upland ecology, vegetation survey and monitoring, computer programming, and multivariate and statistical analysis.

Anders Bryn Professor of vegetation ecology at the Natural History Museum, University of Oslo, Norway. Bryn also holds an associate position at the Norwegian Institute of Bioeconomy Research. His research focuses on ecological climatology, ecosystem mapping and distribution modelling, treeline dynamics, and citizen science.

Erik Cronvall Environmental assessment specialist and current PhD student at the Swedish University of Agricultural Sciences (SLU) in Umeå, Sweden. His PhD project focuses on improving mapping of key grazing areas and the understanding of how cumulative effects of multiple land use activities impact reindeer ranges and movements.

Andreas Eriksson Forest resource analyst at the Swedish Forest Agency and a PhD student at the Swedish University of Agricultural Sciences. He has a background in forestry and political science and works, amongst other things, on forecasts and scenario development.

Louise Eriksson Researcher in environmental psychology at Umeå University. Her research focus is on relations between people and the environment.

Anton Grafström Associate professor of mathematical statistics in the Department of Forest Resource Management, Swedish University of Agricultural Sciences. He is an expert in probability sampling and has designed several environmental monitoring programmes.

Santiago Guerrero Senior economist at the World Bank. He has more than ten years of experience conducting research and providing policy advice at the domestic and international levels in agricultural policy and environmental topics. He has held positions at the OECD, the Ministry of Environment in Uruguay, and Banco de México and has worked as a consultant for the World Bank and the Inter-American Development Bank.

Marcus Hedblom Professor of landscape architecture, especially landscape management. His research focuses, amongst others, on urban ecology and how people's health is affected by nature.

Henrik Hedenås Researcher and head of the national monitoring of landscapes in Sweden in the NILS programme at the Swedish University of Agricultural Sciences. He works mainly in conservation biology and environmental monitoring.

Jonas Hentati Sundberg Senior lecturer in the Department of Aquatic Resources, Swedish University of Agricultural Sciences. He works, amongst other areas, on the use of AI for biodiversity modelling.

Andreas Hilpold Senior researcher at the Institute for Alpine Environment at the Eurac Research Center based in Bolzano, South Tyrol, Italy, and coordinator of Biodiversity Monitoring South Tyrol. His focus is on plant communities and ecological basics of biodiversity monitoring.

Tim Hofmeester Researcher in the Department of Wildlife, Fish and Environmental Studies, Swedish University of Agricultural Sciences in Umeå. Hofmeester works with the coexistence of wildlife communities and people but is also interested in ways in which the general public and other stakeholders can contribute to wildlife monitoring.

Einar Holm Professor of geography at Umeå University, Sweden. He has long experience in research in population microsimulation and forest ownership.

Clive Hurford Research scientist and field surveyor in the Department of Geography and Earth Sciences, Aberystwyth University, Wales, UK. He has worked with monitoring in many countries across Europe and been a lead editor/author of several books on habitats and monitoring. He is the monitoring advisor and chair of the Remote Sensing Support Group for Eurosite (a Netherlands-based NGO) and regularly organizes and contributes to international conferences on biodiversity monitoring.

Christian Isendahl Professor of archaeology in the Department of Historical Studies, University of Gothenburg, Sweden. His research applies a historical ecological lens to the study of pre-Columbian socioecological systems in Latin America, with a particular focus on exploring archaeology's potential to address contemporary challenges.

E. Carina H. Keskitalo Professor of political science in the Department of Geography, Umeå University, and guest researcher at the Swedish University of Agricultural Sciences. She has published widely on forest and natural resource policy and use, amongst others, applying qualitative methods.

Gregor Levin Geographer and senior advisor in the Department of Environmental Science, Aarhus University, Denmark. His primary research focus is on mapping and assessment of land use and land cover changes and modelling of ecosystem services.

Gun Lidestav Associate professor in the Landscape Studies Unit, Department of Forest Resource Management, Swedish University of Agricultural Sciences. Her main research focus is on the dynamics of ownership and drivers of small-scale forest owners.

Torgny Lind Researcher in the Department of Forest Resource Management, Swedish University of Agricultural Sciences. His research focuses on forest decision support systems.

Urban Lindgren Professor of geography, Umeå University, Sweden. His research focuses on economic geography, including population and employment development and labour mobility.

Mats Nilsson Researcher with long-time expertise working with different aspects of remote sensing and mapping at the Swedish University of Agricultural Sciences.

Lisa Norton Senior scientist at the UK Centre for Ecology & Hydrology. Her research focuses on landscape ecology and interdisciplinary land management. Norton has been involved in the UK Countryside Survey for many years, identifying relationships between different ecosystem components and drivers of change.

Henrik Persson Researcher in forest remote sensing with a particular interest in 3D forest monitoring, using lidar, radar, and stereogrammetry. The implementation of statistical methods and feasible image processing helps him achieve his goals.

Suzanne M. Prober Senior principal research scientist with CSIRO Land and Water, Australia. Her research is centred on managing and restoring the natural diversity, ecosystem function, and resilience of vegetation communities, with a long-standing interest in balancing production and cultural and economic values with nature conservation, particularly in temperate eucalypt woodlands.

Åsa Ranlund Environmental assessment specialist in the Division of Landscape Analysis, Swedish University of Agricultural Sciences. She has a PhD in ecology and works with national environmental monitoring programmes in Sweden.

Anna E. Richards Researcher and plant and soil ecologist at CSIRO Land and Water, Australia. She has a broad interest in ecosystem management, particularly how the science of vegetation and soil dynamics, along with land management practices, can be used to better monitor, evaluate, and manage ecosystem conditions.

Ulrika Roos PhD student at the Swedish University of Agricultural Sciences (SLU) in Umeå, Sweden. Her PhD project focuses on co-planning and consultation in the context of forestry and reindeer husbandry in Sweden. She has a background in governmental work concerning land use, conservation, and rural development.

Mats Sandewall External associate in the Department of Forest Resource Management, Swedish University of Agricultural Sciences. He works as an advisor to national forestry inventories in Africa and Asia and has long-time expertise in international forestry.

Per Sandström Associate professor at the Swedish University of Agricultural Sciences (SLU) in Umeå, Sweden. He leads and participates in projects assessing multiple land use impacts and planning in relation to reindeer husbandry using multidisciplinary approaches incorporating both indigenous and scientific knowledge systems.

Stefan Sandström Researcher in the Department of Forest Resource Analysis at the Swedish University of Agricultural Sciences in Umeå, Sweden. His research revolves around landscape studies, including historical land use and multiple land use, usually focusing on reindeer husbandry as the model system.

Becky Schmidt Principal environmental scientist at CSIRO Land and Water, Australia. She specializes in science translation and synthesis, particularly for natural capital assessments and environmental–economic accounts.

Mats Söderström Senior lecturer and researcher in the Department of Soil and Environment, Swedish University of Agricultural Sciences. Mats is a pioneer in the field of precision agriculture in Sweden, including research in proximal and remote sensing of soil and crops, often with a focus on translating research into practice. He is actively involved in the development of several agricultural decision support systems, some with global coverage.

Göran Ståhl Professor of forest inventory at the Swedish University of Agricultural Sciences. He has also had administrative responsibilities for environmental monitoring at the Swedish University of Agricultural Sciences.

Julia Strobl Researcher at the Institute for Alpine Environment at the Eurac Research Center based in Bolzano, South Tyrol, Italy, responsible for the communication of Biodiversity Monitoring South Tyrol.

Ulrike Tappeiner Head of the Institute for Alpine Environment at the Eurac Research Center based in Bolzano, South Tyrol, Italy. She focuses on ecological research in mountain regions.

René Van der Wal Professor of environmental citizen science in the Department of Ecology, Swedish University of Agricultural Sciences (SLU). Much of his research focuses on society–environment relationships and the role of technology therein, often conducted in partnership with social scientists.

Veerle Van Eetvelde Associate professor of landscape science in the Department of Geography, Ghent University, Belgium. As a geographer and spatial planner, her research includes landscape ecology, landscape preference, historical geography, landscape and heritage management, and planning. She is active as president of the European chapter of the Association of Landscape Ecology (IALE-Europe) and coordinates national and European projects related to landscape and cultural heritage.

Urša Vilhar Research associate in the Department of Forest Ecology at the Slovenian Forest Institute with a PhD in forest hydrology and climatology. Her work focuses on forest water balance modelling, development of operational drought stress indices for forest vegetation, and development of tree phenology monitoring concepts and their implementation.

Luke Webber Consultant in remote sensing and data science for Metria AB, Sweden. His expertise and experience are in the field of land cover mapping, in particular vegetation and tree species classification, and the modelling of forestry parameters through machine learning.

Kerstin Westin Professor of human geography at Umeå University, Sweden. She has vast experience designing questionnaires and carrying out studies on people's attitudes, values, and behaviours.

Claire Wood Researcher working in the Land Use Group at the UK Centre for Ecology & Hydrology, Lancaster, UK. She has expertise in GIS and data management and 20 years' experience of long-term monitoring, in particular the UK Countryside Survey.

Acknowledgements

This work has developed as a result of a perceived need for an interdisciplinary and international core text, drawing upon several informal networks of European international experts in environmental monitoring. We are deeply grateful for the participation of the many authors as well as the many reviewers from widely varying fields, who have stretched their understanding by writing as well as reading in very different areas. Without you, this book would not have come into fruition! We are also grateful to the research unit for Landscape Studies at the Department of Forest Resource Management, Swedish University of Agricultural Sciences for support, in expectation of the development of future monitoring courses.

1 Monitoring biodiversity: combining environmental and social data

E. Carina H. Keskitalo, Anna Allard, and Alan Brown

Introduction and aim

Environmental or biodiversity monitoring is a key component in a large number of national programmes and constitutes an important aspect of understanding environmental change and supporting policy development. There has been a deluge of new technology, along with the increasing use of artificial intelligence (AI), deep learning, and methods of data integration – none of which were available until recent times. Users of these new concepts and methods may find many new pitfalls in the capture, correction, and storing of biodiversity data.

An additional complication to established monitoring is that we are now living in the Anthropocene, a time period in which natural systems are strongly influenced by the choices of humans. Understanding land use and environmental change in this period requires a practical knowledge of how to design systems based on multiple data inputs, both environmental and social. However, much existing literature in the field of monitoring highlights only the established ecological component, focusing on the research side of ecology around natural systems, as opposed to more explicitly including not only land areas and their content but also land management and land use data derived from social studies (e.g. Artiola et al. 2004; Randolph 2004; Lindenmayer and Likens 2018).

This book aims to respond to both the call for a broad introduction to the multiple methods relevant to monitoring and how these can be combined, along with a similar introduction to some common social methodologies. In doing so, the book aims to place monitoring in the context of the broader scientific endeavour towards understanding environmental change, including that of the social sciences and humanities. We hope to move the reader towards an increased multidisciplinarity, by acknowledging and outlining the very different disciplines needed to provide solid and valid application of methods in any one area (see Text box 1.1).

Fundamentally, we hope that you as a student or reader will be able to use the volume as an introductory handbook to understanding the design features surrounding a broader approach combining environmental and social monitoring.

Scope of the book

How monitoring is understood in this book

The book focuses on monitoring biodiversity, stocks of species, and impacts of land management in areas of land, wetland, and water. Another focus is on the wide-ranging, national

DOI: 10.4324/9781003179245-1

Text box 1.1:   The Anthropocene and changes in the natural or semi-natural systems

A separation between nature and culture has long been guiding how these systems are viewed in the research world. However, there is large agreement that we are now living in the Anthropocene – a time period in which natural systems cannot be separated from the influence of humans on them (e.g. Palsson et al. 2013). Human actions, societies, economies – and the resource use and pollution resulting from these – are major influences on natural systems.

The interactions are many and complex. For instance, increasing trade and transports at sea, on land, or by air could result in transmission of species that may be able to establish and become invasive in new areas or bring new diseases. New equilibriums may arise resulting from climate change, chaotic weather patterns, and longer periods of extreme weather. Pollution and changes in the air or water quality may have large impacts on established biodiversity such that either the species themselves or the ones that they depend on may be weakened (e.g. Matthews 2012; Stenlid and Oliva 2016; Cline 2021); see more in chapters 5 and 9.

This means that monitoring must, to an increasing degree, take both the "natural" environment and the social impacts on it into consideration to understand and help authorities at all levels be ready to respond to change, whether it is in the form of invasive species or impacts on the existing environment such as forest damage.

---

monitoring schemes covering extended geographical areas because these, in turn, will have impacts on the possibility to follow/monitor progress and/or setbacks on restoration projects – these restorations are by most definitions local or regional. Other types of monitoring will not be covered in depth here, such as monitoring the quality of water, soil, or air; the spread of harmful chemical substances; and the many other types of abiotic parameters (see Text box 1.2).

A focus on mandatory monitoring – but also support for other approaches

A fundamental assumption in much of the book is that monitoring is often undertaken for mandatory purposes; that is, as a part of supplying information that is mandated legislatively or in regulation by the state or supranational bodies. The European Union (EU) is a case in point. To meet this need, monitoring should provide often very broad information usable for different purposes and support comparisons between different national systems of regulation. Examples of mandatory requirements for different systems – from land use to water – are provided in chapter 3.

However, the book also discusses design aspects in a wide context and using a wide range of methodologies that could just as well be applied in local-scale cases and those that are not directed by mandatory monitoring.

Text box 1.2:   Monitoring biodiversity – scope of the book

Biodiversity monitoring makes repeated observations of the natural environment and the species, habitats, and ecosystems found in natural, semi-natural, and cultural landscapes, including wetlands, rivers, coastal waters, and the open seas. Its aim is to understand patterns of status and change in the past and present to make predictions about future changes and – working with modelling and experimental investigations – to help suggest possible drivers, pressures, and remedies where these changes are undesirable.

A primary focus of the book is on the types of biodiversity monitoring that are regularly required by states, unions of states, and other authorities as part of mandated data collection set out in policy instruments. The purpose is typically to make regular surveys of the natural (or, more likely, anthropogenically modified) state: for example, monitoring fish stocks in order to set fishing quotas (e.g. Weng 2007–2020; Thenkabali 2015; European Commission 2022; Francis et al. 2022).

A second focus is on *social systems* and their influence on the natural world, highlighting some specific methods of data collection and modelling for understanding both these systems and their impacts. Though not usually included in the traditional account of biodiversity monitoring, understanding social systems is important to better understand biodiversity state and change (e.g. Dick et al. 2018; Angelstam et al. 2019).

A focus on the handicraft

The book is intended to provide an overview of the "handicraft" of monitoring, setup/design, resulting data types, and reporting to higher levels. It is also intended to highlight the parts of social data that are otherwise often left out in relation to monitoring. The book is therefore oriented towards designing monitoring programmes to help identify impacts as well as responses and stressors/drivers of change, in the sense of both social and environmental systems. Chapters thus outline both common assumptions or design decisions but also aim to problematize these; in many cases, monitoring is in practice constrained by what resources are available, and design decisions may need to be made in relation to existing databases and existing monitoring systems. Our ambition is to illustrate that the "handicraft" of monitoring is not only about following a design table but about adapting monitoring – both to changing legislative demands and to cope with resource restrictions or new purposes demanded of monitoring.

Examples from different levels in the EU and European context – but illustrating broader concerns

The book draws on examples from international-, EU-, and national-level reporting, including multiple-country examples, as well as examples from fields in which there are conflicting interests. There exists a breadth of uses in relation to monitoring as treated here, ranging from micro-monitoring and research on specific sites through local-, regional-, and national-scale reporting and harmonizing of datasets to aggregations such as the EU or global constellations of ecological or land-cover data. The aim is to illustrate issues of designing (or working with) systems including components of international comparisons, national systems, and lower-level local or regional cases.

A limitation – but also an opportunity – is that examples are generally taken from the EU or European context. We hope that this will enable an understanding of how national monitoring and different systems relate to general European aims, illustrating issues that may be transferrable to other contexts. The pan-European level is especially relevant because even though it is the focal level for national and regional programmes in the EU and associated countries, it has less often been covered as a whole in the largely national literature on monitoring. The logic here is that if we can understand the issues of tailoring monitoring to one context (such as the EU), understanding what the requirements are and how these can be met, we should then be able to make similar assessments (or well-grounded assessments) in other contexts.

Areas of investigation from land to aquatic systems – but with a focus on terrestrial biodiversity

The book draws applications and examples from a broad range of environments and types of land use, including forest and mountains, wetlands and aquatic systems, cultural landscapes, agriculture and agroforestry, with a corresponding range of different users and stakeholders. Though most examples are on land, from forestry and other specific cases that the authors are familiar with, the attempt has been to illustrate which parameters these applications require and how monitoring schemes have been designed, in order to support an understanding of how other cases could be analyzed or designed.

For aquatic systems, the enormous networks of in situ devices, gathering data either autonomously or by remote navigation, to support large-scale models based on satellite data of ocean climates or to monitor the fish supply are a leading force in innovative use of new technology. These new technologies, models, and data integration methods can provide valuable lessons for the development of – at first sight – very different land-based monitoring. Monitoring aquatic systems as part of mandatory monitoring is discussed in chapter 3, and the innovative use of technology is discussed in chapter 8.

A focus on the use of knowledge from multiple disciplines – but also the need for specialists and expertise

A crucial aim of the book is to provide information to support the better integration of multiple types of environmental and social data to improve an understanding of change and implementation of policy measures. Here the main themes concern the integration of different conventional and upcoming techniques for the design of monitoring systems, the possibilities for including social data in these, and how to design monitoring systems for a variety of policy goals.

However, the book also acknowledges that even moving towards a broader inclusion of multiple disciplines (as will be discussed also in the concluding chapter, chapter 18) is a large challenge, and perhaps larger than what is commonly recognized in calls for monitoring to include interdisciplinarity. The range of different disciplines and methodologies that can support monitoring and the development and integration of data management to learn lessons for monitoring purposes is very wide indeed. It follows that the range of highly specialized expertise required to deal with these disciplines, methodologies, and variations is equally wide. The contributions included here do not pretend that any one person or team from more traditional monitoring areas could cover the range illustrated here; instead, the aim is to show what data *could* be drawn upon by including or cooperating with a range of experts. As a result, the aim is to cover the area widely by providing broad introductions, not by going deep into the technologies, because these are developing at such a rapid pace and such information will probably be outdated soon. Neither does the book cover the wider research fields of social science – for example, adaptive management, co-management, or socioecological studies – but rather illustrates how different tools can be used to build a foundation for understanding different types of management.

A focus on methodological overview

To accomplish these aims of providing overview, a large focus in the book is placed on introductory descriptions of multiple methodologies. Many of the chapters cover specific techniques, types of data, and methodologies in both environmental and social studies. Though we are acutely aware that there are numerous types of studies worthy of inclusion, the decision was to keep the book manageable to students, by covering some established environmental and social methodologies to illustrate the logic, use, and application of these. Based on these studies, designers of a new monitoring venture may be able to draw on both material and associated methodological expertise at their disposal in any one project – and why not further our knowledge, by considering new ways to integrate other types of data than only those illustrated here?

Outline

The book is structured to progress from more traditional to more novel monitoring tools, which include citizen science and well-established – but often less integrated in monitoring – social research methods.

The book discusses, firstly, monitoring as an established field and how this is presently challenged by changes to budgets as well as the need to incorporate and integrate various technologies. It discusses the different methodologies underlying and applicable in monitoring, the variation between countries, and the design features relevant to understanding – and, if the need arises, to rescale or revise – monitoring systems towards new policy goals or different funding situations.

Secondly, monitoring is placed within an understanding of the social system dynamics relevant to situate the role and conduct of monitoring, and in this way the book also extends beyond chapters on established monitoring techniques and applications. It includes foundational chapters on the design of survey questionnaires to, for instance, landowners; the use of population and property-related microdata; possibilities for modelling natural and social systems in common; and the potential use of qualitative interviews or incorporating information from existing reporting systems. The study concludes with two chapters discussing the design and considerations regarding the application of a monitoring framework for policy advice. Chapters generally also include sections on key messages, study questions, and further reading.

The sequence of chapters largely follows what a student (or other reader) might be expected to recognize, progressing from more general overviews of traditional modelling to discuss a variety of methods to explain the monitoring structure in relation to which various methods and data could be included. For these reasons, chapters on design are also provided (chapters 4, 9, and 15) to summarize some of preceding chapters and discuss design in relation to the features mentioned in these.

The next two chapters provide broad overviews of the field of monitoring and its purposes and can be read as an introduction to the book. Chapter 2, "Monitoring as a field", briefly defines monitoring within the scope of the book and is intended to serve as a map of the book at large and the components the student needs to be aware of in designing, evaluating, or participating in monitoring. Following this, chapter 3, "Demands on monitoring", provides an overview of the types of processes for reporting as well as legislation that governs how (and often why) monitoring is undertaken. It reviews legislation on levels from global to national in the areas of agricultural land, urban and forested areas, and semi-natural/natural land, water management, and aquatic systems.

Chapters 4 to 9, covering a breadth of both established and new biodiversity monitoring techniques as well as the design of monitoring systems, provide an introduction to biodiversity monitoring. Chapter 4 covers the general statistical and sampling design of monitoring systems, outlining important choices to be made with regard to sampling strategies. Chapter 5 then covers in situ data collection, from measurements in the field (covering devices and networks of measurements in ocean monitoring) to interpretation of photos or images from above. Chapter 6 discusses citizen science as a methodology for gathering large amounts of data using volunteers. Chapter 7 discusses data collection from the sky or space – remote sensing and global observation systems – as a foundation in the multiple systems that are presently becoming increasingly important for monitoring the Earth and its changes. The chapter covers the recent explosion in the availability of Earth observation data, with examples such as the Copernicus Earth Observation Programme, headed by the EU, as well as other global data portals that provide storage of data, availability of scripts and ready-made analyses, and server or computer power for computation of enormous datasets. Following this, chapter 8 discusses the addition of even more novel technologies as well as new uses of existing technology for gathering data. In some cases these are so novel that we do not yet have a good understanding of their potential uses and risks, such as drones, artificial intelligence (AI), and environmental DNA. Finally, this section is closed by a discussion in chapter 9 of how to manage the combinations of data collection from remote sensing by air or space as well as from the field, and new data types. Hybrid methods for integration that have been discussed in the previous chapters.

Chapter 10 then constitutes a bridging chapter to the focus on social methodologies and social data. It discusses what social data exist in current reporting systems, concluding that social factors are highly relevant in almost all instances of land use and management. However, so far they have not been a common component of long-term environmental monitoring, although social data and monitoring data are sometimes combined in the analysis phase linked to monitoring programmes. The following section of the book, made up of chapters 11 to 14, aims to provide the basic knowledge for the monitoring student to be able to understand some of the types of potential social data that could be relevant to increasing the knowledge of social factors.

Chapter 11 provides an overview, attempting to sketch some types of considerations necessary for understanding social studies; in particular, the fact that local studies often need to be understood in context of higher levels and that different cases vary depending on historical development. The chapter covers some of the factors that are generally relevant to designing monitoring systems on an international basis, which also have a particular emphasis in social studies.

Chapter 11 also functions as an introduction to the following chapters. Chapters 12 to 14 discuss different established social methodologies and data collection: microdata on the property level in chapter 12, survey questionnaires as methodologies for understanding different management conditions in chapter 13, and interviews with landowners or managers, for example, in chapter 14. Though qualitative studies are seldom at the forefront in monitoring, this book thus introduces qualitative methodology as a relevant way of understanding how and why land is used in different ways. This can lead to an in-depth understanding of different conditions, policies for financial subsidies, trends, or even influence evident from tradition, which in turn can illustrate the most important issues influencing land use in specific areas.

Chapter 15 then summarizes the previous chapters, revisiting the question of when a monitoring system is fit for purpose, looking especially at how monitoring schemes change over time and the importance of being flexible and able to change the method of data collection while still maintaining as much continuity as possible with existing long-term data series.

These issues of how to design an integrated social–ecological monitoring system are then problematized in the following two case study chapters, which highlight very different applications of monitoring and different social applications. Chapter 16 looks at issues around monitoring small biotopes and habitats where the current biodiversity is highly dependent on the history of cultural management. It illustrates how the authors (and others) worked in different cases and what results different approaches produced in terms of conservation or management. Chapter 17 then provides a relatively unusual case study of a system for participatory monitoring and evaluation, illustrating the breadth with which the wide array of monitoring methodologies can be applied. The case highlights the development of reindeer husbandry plans in northern Sweden, how this has been developed, and how it is undertaken, as well as the supporting features that have made the development possible, in practice, in this case. A final chapter (chapter 18) concludes the book. The chapter discusses some of the issues that will always remain, even for the student who may apply many of the lessons from this book: how completely the possible range of issues and observations can be covered, how different types of knowledge can be managed, and how different types of data can be combined.

Key messages

- In the Anthropocene, the design and understanding of ecological monitoring also need to take into account social drivers and the policy processes it is designed to inform.
- Understanding one context, such as the European one presented here, should make the reader well placed to appreciate the complexity of other, different cases in other contexts.
- Specialized, well-established methods are found in both ecological and social studies and should be applied with understanding and preferably through cooperation with experts in each field of knowledge.

Study questions

1   Why do we need to understand social systems in order to understand changes in nature?
2   What is monitoring, viewed from both narrow and broad points of view?
3   What is the Anthropocene?

Further reading

Francis, R.A., Millington, J.D.A., Perry, G.L.W. and Minor, E.S. (eds) (2022) *The Routledge Handbook of Landscape Ecology*. London: Routledge.

This book discusses many of the issues related to monitoring and integration, landscape structure and function, scale, and connectivity; landscape processes such as disturbance, flows, and fragmentation; methods such as remote sensing and mapping, fieldwork, pattern analysis, modelling, and participation and engagement in landscape planning; and emerging frontiers such as ecosystem services, landscape approaches to biodiversity conservation, and climate change.

Angelstam, P., Manton, M., Elbakidze, M., Sijtsma, F., Adamescu, M.C., Avni, N., Beja, P., Bezak, P., Zyablikova, I., Cruz, F., et al. (2019) LTSER platforms as a place-based transdisciplinary research infrastructure: learning landscape approach through evaluation, *Landscape Ecology 34*, 1461–1484. 10.1007/s10980-018-0737-6

Redman, C.L., Grove, J.M. and Kuby, L.H. (2004) Integrating social science into the long-term ecological research (LTER) network: social dimensions of ecological change and ecological dimensions of social change, *Ecosystems* 7(2), 161–171.

These papers discuss the ways in which social dimensions need to be integrated into ecological research but also some of the difficulties of integration.

Vaughan, H., Whitelaw, G., Craig, B. and Stewart, C. (2003) Linking ecological science to decision-making: delivering environmental monitoring information as societal feedback, *Environmental Monitoring and Assessment* 88(1–3), 399–408.

This paper discusses the need for integration between ecological and societal feedbacks.

References

Angelstam, P., Manton, M., Elbakidze, M., Sijtsma, F., Adamescu, M.C., Avni, N., Beja, P., Bezak, P., Zyablikova, I., Cruz, F., et al. (2019) LTSER platforms as a place-based transdisciplinary research infrastructure: learning landscape approach through evaluation, *Landscape Ecology* 34, 1461–1484. 10.1007/s10980-018-0737-6

Artiola, J., Pepper, I. and Brusseau, M. (2004) *Environmental Monitoring and Characterization*. Elsevier.

Cline, E.H. (2021) *1177 B.C. The Year Civilization Collapsed*, Revised and updated, Turning Points in Ancient History. Princeton University Press.

Dick, J., Orenstein, D.E., Holzer, J.M., Wohner, C., Achard, A.L., Andrews, C., Avriel-Avni, N., Beja, P., Blond, N., Cabello, J., et al. (2018) What is socio-ecological research delivering? A literature survey across 25 international LTSER platforms. *Science of the Total Environment* 622, 1225–1240.

European Commission. (2022) Monitoring and evaluation, https://ec.europa.eu/neighbourhood-enlargement/monitoring-and-evaluation_en (Accessed June 3, 2022).

Francis, R.A., Millington, J.D.A., Perry, G.L.W. and Minor, E.S. (eds) (2022) *The Routledge Handbook of Landscape Ecology*. London: Routledge.

Lindenmayer, D.B. and Likens, G.E. (2018) *Effective Ecological Monitoring*. CSIRO Publishing.

Matthews, J. (ed.) (2012) *The SAGE Handbook of Environmental Change. Vol. 2: Human Impacts and Responses*. Sage.

Palsson, G., Szerszynski, B., Sörlin, S., Marks, J., Avril, B., Crumley, C., Hackmann, H., Holm, P., Ingram, J., Kirman, A., et al. (2013) Reconceptualizing the "Anthropos" in the Anthropocene: integrating the social sciences and humanities in global environmental change research, *Environmental Science & Policy* 28, 3–13.

Randolph, J. (2004) *Environmental Land Use Planning and Management*. Island Press.

Stenlid, J. and Oliva, J. (2016) Phenotypic interactions between tree hosts and invasive forest pathogens in the light of globalization and climate change, *Philosophical Transactions of the Royal Society B: Biological Sciences* 371(1709), 20150455.

Thenkabali, S. (ed.) (2015) *Remote Sensing Handbook*. 3 vols. CRC Press.

Weng, Q. (ed.) (2007–2020) *Remote Sensing Applications*. 19 vols. Foremost.

2 Monitoring as a field

Anna Allard, Claire Wood, Lisa Norton, Andreas Aagard Christensen, Veerle Van Eetvelde, Alan Brown, Henrik Persson, and Louise Eriksson

Introduction

This book describes examples of biodiversity monitoring, focused on land and oceans. There are many ways of collecting data, and researchers constantly find new ways of making technology work for their purposes, influencing the development of new approaches to monitoring. Monitoring can be very important to highlight the effects of policies, especially where they may have potentially harmful impacts on nature. We encourage students to do a thorough research of current monitoring before embarking on any new monitoring schemes, in order to identify any existing monitoring, even monitoring that uses different or older methodologies.

This book aims to capture the monitoring process, starting with the need for information (often driven by policy), framing questions that could be answered by monitoring and then identifying approaches for finding answers to those questions. One important aspect of deciding on monitoring approaches is when to choose a particular type of data collection and how to combine different datasets to produce data that can help in decision making. This may include understanding the role of humans in driving change in ecological systems. In our modern world, so many different types and scales of evidence are available, from the observation of a lichen on a tree trunk to the flow of debris following a Saharan sandstorm across the world (seen in near real time by an orbit of satellites). This chapter provides a basis for understanding different monitoring approaches.

Biodiversity monitoring

The definition of *biodiversity monitoring* in this book comprises an interdisciplinary field that includes both environmental and social science. Its focus is to protect Earth's environment and human health by sustainable management and conservation or restoration of natural resources. Although we are well aware that other areas exist, such as the monitoring of atmospheric layers, weather monitoring in relation to climate change, or monitoring for security or response to emergencies, those are not included in this book. Also not included is the type of biodiversity research involving the capture of organisms, seeds, and fragments of plants swirling in the air together with insects, which constitute a lot of biodiversity dispersal on local to global scales. The rich biodiversity below ground is only touched upon in the form of environmental DNA (see chapter 8).

Our scope encompasses more than just the actual collection and evaluation of data or their potential use in forecasting, as mentioned in the European Union (EU) definition (European Environment Agency 1999). It also includes the importance of understanding

DOI: 10.4324/9781003179245-2

social and political drivers of biodiversity change. Governments tell us that biodiversity monitoring is needed to benefit society by helping to maintain *public goods* or *environmental services*, and this need comes out of a recognition of unwanted and damaging environmental change from pressures on land use, the exploitation of resources, pollution, and climate change; see chapter 3. Monitoring can show how the health of the environment is intimately linked to the health of society and can be an appropriate response to local concerns from people who will be affected by a deteriorating environment and will benefit from its restoration. In this sense, environmental protection will have many clients, including future generations (United Nations, Economic Commission for Europe 2016). In a wider setting, biodiversity and environmental monitoring often need to be a part of a programme of policy, funding, and practical measures to maintain and restore threatened environments.

Types of monitoring

As a baseline, the type of monitoring approach you need to adopt depends on the questions that you need to answer. Do you have a clear question? Are you collecting data to answer some future question? Or are you just curious and testing to see how the data you collect will enhance your knowledge associated with your interest (see Table 2.1)? Commonly, biodiversity applications can include work in the field of archaeology (see more in chapter 8). Though including broader social science is even less common, this book illustrates how to build bridges between social science and monitoring for policy and interdisciplinary research.

Curiosity driven

Curiosity-driven or passive monitoring typically has no statistical design and is not linked to answering specific questions or triggering any particular management intervention, with no requirement for reporting. The effectiveness of this type of monitoring depends on the knowledge and motivation of whoever is carrying it out. For example, a site manager might use a monitoring scheme to learn about the long-term changes on their site as a way of structuring what would otherwise be ad hoc observations or a way of gaining enough ecological understanding of change to specify a more focused hypothesis-driven scheme. To qualify as monitoring, we suggest there would have to be enough specification (what, where, and when to observe or record) so that the same scheme could be carried out by another observer to give important information about trends. Trends, for example, are possible increases or declines in populations or habitat diversity and what changes may have occurred in spatial patterns and distributions (e.g. Lindenmeyer and Likens 2018; Ten et al. 2021).

Mandated monitoring

Mandated monitoring is carried out in response to the requirements of government legislation or directives (such as the EU Habitats Directive; European Commission 1992); for example, monitoring of resources of great economic importance such as national forest inventories. There are usually some general specifications on what habitats, species, or environmental measurements to include, often leaving the details of sampling design and methods to governments or regional agencies. Here it is appropriate

Table 2.1 Some monitoring types and some of their uses

Monitoring type	Who monitors and how	Who wants the result	Who else benefits
Curiosity driven	Varied, depending on the nature of curiosity	In many cases the one who does the monitoring; for example, researchers testing a hypothesis or ecologists using old maps to investigate ruined or abandoned homesteads to find cultural biodiversity	The results of the curiosity-driven investigations often put thoughts into the minds of others (members of the public, politicians, or scientists) and become the start or pilot case of a mandated monitoring scheme.
Mandated	Typically, they are large-scale. Universities, science centres, or consultancy companies usually run the monitoring.	Authorities, for use in planning environmental management and reporting data higher up in some obligatory chain of reporting	The results are useful for settling disputes or driving opinions to be considered in political decisions in conflicts of interest; for example, land use in forests as industrial timber farming or recreation and berry picking.
Question driven	As in mandated, often done by researchers etc.; usually smaller scale. Pilot cases of regional or local scale. Looking to answer specific questions, with a scheme designed accordingly.	For quantity data, to answer questions of *what is there, how much*, and *where*. For quality data, such questions as *who owns the land in question, how do they use it* and *what are their plans for it*, or *how do they cope with current regulations?*	The planners of large-scale management, in the sense of *what* and *how much* but also in the sense of *how a current situation has developed*. Research into possible future *scenarios* and insights into how to make a desired outcome feasible
Citizen science driven	Run by researchers etc.; volunteers collect at least one data source	Commonly used in many environmental settings, especially for species monitoring. In many cases, these are question-driven and/or mandated but also occur as curiosity driven.	Being involved increases people's awareness of an environmental issue, which often influences the political willingness to preserve or protect. Reduces the cost of data collection and a larger dataset can potentially be collected.
Community driven	A specific community is carrying out the data collection, often in communication with a university or consultancy company.	The community is asking for the results.	The results are of benefit in mediating conflicts between different interests.

to be aware of the risk of setting very broad classes etc., which will make the monitoring too general in its nature and will not answer specific questions or give information on issues of concern. Funding is normally from public money or by private enterprise as a condition of exploiting public goods or carrying out activities which might be polluting or otherwise harmful to the environment. Often cycles are set for monitoring and reporting, with an expectation that the results can be merged with those from similar monitoring projects to contribute towards a national or continental assessment of, for example, water quality or habitat condition, focusing on trends over time and whether or not the objectives of policy are being met (European Commission 2022).

Question driven

Question-driven monitoring has a rigorous statistical design that is able to answer predetermined questions or hypotheses. The specification should include an effect size (specifying the degree and direction of change that would be considered ecologically significant) and specified levels of both statistical significance and the acceptable risk that a real change of this order can, by chance, go undetected as a consequence of having only a small finite sample (statistical power). Often multiple questions are asked, and in the case of habitat and land cover/land use monitoring they may be based on a conceptual model, setting out possible transitions of a habitat from a desirable current state to undesirable future states that need monitoring (Houk and van Woesik 2013; Lindenmeyer and Likens 2018).

The term *top-down* means a mandated monitoring, where, in the case of EU Directives, each member state uses the same set of requirements to specify their own schemes. In contrast, *bottom-up* projects could address specific questions closely linked to site management decisions.

Sometimes these categories appear to overlap. The monitoring designed to understand how to proceed with site management, discussed in chapter 16, is both curiosity driven and question driven, even though these direct-response surveys can often avoid needing statistical designs. Public money will often not pay for curiosity-driven monitoring, but some important historical monitoring schemes started off as such projects or have since become historical records such as the phenological observations of Gilbert White, published in 1789 (Sparks et al. 2020). In Germany, lack of knowledge about the connections between rural biodiversity and farming practices generated the long-term project Biodiversity Exploratories (Fischer et al. 2010). A problem with passive monitoring is the risk that someone who is not an expert on the landscape or site context will mistake chance changes in the sample of observations for real changes or think that the surveyed sample represents the entire population. Of course, they may – but we need to know more to have confidence in making this inference, as discussed in chapter 4.

We need, however, to know what we are looking for, so effective mandated monitoring also needs to be question-driven. The ability to detect real change can be decided a priori, with the sample size determined after a pilot study to estimate the variance, or realized only afterwards, with the statistical power dependent on how many observations we happened to get funded. If we already know the sample size is too small to answer the question, question-driven monitoring may fall at the first hurdle. Expensive monitoring projects use sophisticated designs that maximize the statistical power for a given sample size, including unequal probability sampling, ratio and regression models, and stratification and post-stratification, notably using auxiliary variables from remote sensing; see more in chapters 4, 7, 8, and 9.

Citizen science driven

Citizen science–driven monitoring has boomed in recent years and is typically run by experts setting questions and designing the survey but with the data collection done by volunteers, many of whom themselves are experts and passionate about the subject. Examples are the United Kingdom Butterfly Monitoring Scheme (UKBMS), dependent on volunteer networks for recording the datasets, to enable the assessment of trends (e.g. Macgregor et al. 2019; van Swaay et al. 2019). Volunteers, often dedicated ornithologists, collect the data for the Swedish Nesting Bird Inventory (e.g. Brlik et al. 2021; Morrison et al. 2021); see more in chapter 6.

Community driven

Community-driven monitoring, or community-based monitoring, is a type of monitoring in which a specific community is either carrying out the data collection or driving the demand for the results (Wilson et al. 2018; Khair et al. 2021), as exemplified in chapter 17, which shows what cooperation around reindeer husbandry have achieved through innovative reindeer husbandry plans.

Quantitative and qualitative data

To incorporate social science in monitoring, other approaches are often used. Mandated monitoring, coming from an authority in response to concerns, is not always straight-forward where these concerns involve people and their use of the land, culture, and ownership (both in reality and in the sense of belonging). An alternative is to start by collecting quantitative data (information about quantities, and therefore numbers) such as *research in register data*, which is the focus of chapter 12. Monitoring in terms of human evaluation of landscapes (e.g. preferences) can be done by *surveys using questionnaires*, further discussed in chapter 13.

The other type of data is *qualitative* data, which regards circumstances that can be observed or elicited but not measured, such as peoples' sense of place (e.g. Minichiello et al. 2008). Monitoring in this regard can be done by conducting *interviews*; read more about this method in chapter 14. Approaches for taking humans into consideration, when combining social data and landscape components for evaluations of how landscape and people interact, are exemplified in Text box 2.1, and some of the main differences between the two types of data are summarized in Table 2.2.

Approaches to data collection

The key information is that successful monitoring programs are based on well-defined questions, a conceptual understanding of relevant ecological processes, and a robust study design that allows for inferences to be made about ecosystem change while also remaining adaptive to new information and questions (European Commission 2022).

It is crucial to determine beforehand whether the monitoring will be used to indicate the change between two visits (long-term monitoring) or just record the status at two points in time. The statistical inferences differ significantly between these two approaches and large-scale field long-term monitoring. Monitoring works (with a few exceptions) with what is called *uncertain knowledge*, which is solved by statistical sample design. Large sample sizes

Text box 2.1:   Human evaluations of landscapes

Evaluations of landscapes may include assessments of scenic beauty – that is, the extent to which a landscape is perceived to be unattractive or beautiful – or scenic preferences, in terms of disliking or liking a particular landscape. To take this one step further, willingness to engage in different activities in diverse landscapes has been studied as a way to learn about the fit between people's activities and the features of the landscape.

An understanding of human evaluations of landscapes requires that both human and landscape components be considered. Landscape type, such as a forest or open landscape, and different characteristics including the height of trees, ground vegetation, etc., are important dimensions. In turn, people have different experiences, values, and beliefs that influence evaluations. Landscape experts (e.g., managers) but also lay people (e.g., tourists, rural and urban populations) have been studied to learn about how people assess landscapes. Overall evaluations can consider how people interact with landscape characteristics.

Elementary for an understanding of evaluations is the need to develop appropriate landscape stimuli and to use reliable measures of evaluations. Landscape character-istics need to be defined and presented in a standardized way to determine their relevance for evaluations. Photos are the most common way of eliciting landscape evaluations, but technological advances have enabled computer visualizations where landscape features can be experimentally manipulated to ensure high stimuli control. In addition, virtual reality techniques are increasingly used. Even though it is expensive to use real-world landscapes, and these enable less control over specific landscape features, such studies are characterized by higher external validity and are necessary for holistic understandings of landscape evaluations, including the role of sound and smell, for example. In addition, the scales used for evaluations require a proper conceptualization and pre-testing. Given differences between people in how they evaluate landscapes, the sample has to be carefully selected and considered when interpreting results. For an understanding of the physiological and psychological processes underlying evaluations, studies have furthermore included measures such as blood pressure and cholesterol level, as well as individuals' evaluations of the setting and the emotions a setting evokes, to enable conclusions as to why people evaluate landscapes in a certain way (Sundli Tveit et al. 2013).

Even though people evaluate landscapes differently, the methods employed to study evaluations have revealed some distinct patterns. For example, people generally evaluate forest landscapes positively if they contain many large trees of different ages and species but with sparse ground vegetation. However, there are also divergences in how experts evaluate landscapes compared to the general public (Eriksson et al. 2012). These general insights have obvious implications for policy and planning. Nevertheless, it is important to consider the specifics of a certain landscape and the people living there when using these results in practice.

Table 2.2 Some of the main differences between qualitative and quantitative data collection

Data collection	Qualitative	Quantitative
Conceptual framework	Focus on understanding human behaviour from the informant's perspective	Focus on determining facts about biodiversity or social phenomena
	Assumes a dynamic and negotiated reality	Assumes a fixed and measurable reality
Methodological framework	Data are collected through participant observation and interviews.	Data are collected through measuring, quantifying, or classifying.
	Data are analyzed by themes from descriptions by informants.	Data are analyzed through numerical comparisons and statistical inferences.
	Results are reported in the language of the informants.	Results are reported through statistical analyses.

Source: Modified after Minichiello et al. (2008).

become expensive, prohibiting detection of rare occasions; see more on design and sampling in chapters 4 and 5. Modelling the world as some part of the monitoring scheme is very common, either at the initial stage or as a stage somewhere in the chain of collection. Machine learning–based methods and, in particular, deep learning have become increasingly used in later years, and these methods require even larger amounts of annotated reference data compared to traditional model-based methods used in the past (see more in chapters 7 and 8).

There are several *scales* to consider in monitoring (e.g. Sparrow et al. 2020):

- The scale of area: from the local investigation of earthworms in a single field, a regional survey of shrub types in a county or rural patterns of housing over a cluster of counties, to a national survey of landscape classes or with an international, pan-European, or global reach.
- The scale of time: for instance, the biodiversity of pastures with different cultural histories or in stratified layers of the soil through environmental DNA, going back to prehistoric times.
- The scale of resolution: from small-scale detail, via in situ data, to the large landscape view provided by satellite data.

Types of data collection

We use different ways to collect data for biodiversity monitoring. In reality, however, most monitoring schemes use a range of different methods to achieve their goal, either as a predefined scheme or as innovative ways to complete or repeat an existing monitoring scheme and to fill in gaps of knowledge when compiling data sources. Two examples in this chapter pinpoint the wide range of different ways to collect data in the same monitoring scheme; Figure 2.1 sketches some of the ways to collect data relevant to the scope of this book.

The landscape of today is shaped by natural processes but also largely by the efforts of yesterday's inhabitants working the land. Monitoring the old landscape and the continuity

Figure 2.1 A sketch representing some of the data collection methods for biodiversity monitoring in the scope of this book, involving collection of data from many different sources and used in various ways. From social data, in registers, as understandings of human evaluations, to human use of the land, now as well as through history, and the impact it has on the ecology.

Credit: Image by Anna Allard.

of biodiversity over time is important, because the current biodiversity is highly dependent on earlier land use. Ways to monitor this include excavating the site; analysis of environmental DNA, microfossils, and pollen cores; comparing older aerial photos (often going back to the 1950s) and satellite archives (back to the 1970s); and interpreting old maps, some of which go back to the 1600s. The discipline of historical ecology and research across time using aerial photos are introduced in chapter 5.

In situ data collection

In situ data collection can be stand-alone, taking up the entire monitoring effort, but most often constitutes only a part of the whole setup, which combines the simultaneous use of traditional field observations and samples observed with innovative new technologies. Leading the way for others are the large intergovernmental monitoring networks around the oceans, which have enormous setups of in situ measurements to complement remote sensing and fill gaps in knowledge needed to provide analyses, modelling, and forecasts. Examples are the Baltic Marine Environment Protection Commission or the Helsinki Commission (HELCOM) and the OSPAR Commission, an intergovernmental cooperation to protect the marine environment of the North-East

Atlantic (HELCOM Baltic Marine Environment Protection Commission 2022; OSPAR Commission 30th 2022). Many types of devices are employed: buoys, sea vessels, autonomous floating platforms, sea floor capture, drones, trailing or drifting devices with multi-sensors, tide gauges – even sensors on marine mammals. They record such things as bio- or geochemical measurements (chlorophyll or sediments in water), nutrients, salinity, and temperature, from the surface all way down to the sea floor (e.g. Merchant et al. 2019; Sastri et al. 2019; Wang et al. 2020; Programme of the European Union 2022).

Some variables must be collected in the field, such as single species of plant or fish, investigating the finer details of soil profiles or water, or searching for archaeological and historical clues; see more on in situ monitoring in chapter 5. This type of data also makes up *ground truth*, meaning the data used in computerized classifications or models, either as training or for validation of the product afterwards (Cavender-Bares et al. 2022); see chapter 9. Field surveys can be very labour-intensive: a survey of lowland Wales (Stevens et al. 2004) took over a decade to complete and at the time was regarded as too expensive to repeat. However, remote sensing now makes it possible to adapt surveys as a baseline for monitoring, making it easier to survey both points and areas for assessment, estimates, and thematic mapping – see more on this in chapter 9.

Whatever the type of monitoring, some type of sampling design must be used, for practical reasons: working with samples and making estimates of populations is more efficient than trying to count, measure, or observe the entire population, unless we are monitoring species or habitats, with known locations to visit. The downside is that we need to follow probability-based sampling designs and use complex statistical analysis, but this is a small price to pay in the age of computers. Citizen science in some cases can make similar, representative observations based on a sound statistical design (for more on sampling design and principles, see chapters 4 and 6).

Experiments

Experiments use randomization of treatments and controls to draw conclusions about cause and effect. Some forms of monitoring and environmental impact assessment look similar to experiments, notably for stream water quality and recovery from marine pollution; for example, designs of before-after-control-impact (BACI). Though inferences from these designs can only be made with caution, they show the possibility of using "natural experiments" and controls to suggest causal mechanisms for changes in biodiversity (Hurlbert 1984; Underwood 1992; Stewart-Oaten 1996; Filazolla and Cahill 2021).

Habitat surveys

Habitat surveys are common as a way of finding out what the resource is and where. Land cover and land use surveys have a long history of answering the questions of *how much of a given resource or environmental service is there, and where do we find it?* Some of the early surveys were not designed to be repeated and often used interpretative field methods without fixed points, which could be revisited. However, the national monitoring schemes typically use layouts specifically designed to be repeatable, enabling the important possibility to add the question: *Is there any change?*

Satellite remote sensing has a different approach to data collection, aiming for frequent, complete spatial coverage, known as *wall-to-wall*, rather than observing only a scatter of points or small sample areas every few years. Remote sensing observations still have uncertainty but it is less the result of sampling choices and more to do with correction factors for atmospheric conditions and limiting spatial, spectral, and radiometric resolution. Because these sensors in effect see everything visible from above, there is also uncertainty around separating out variables of direct interest from all of the other factors that influence the way light is reflected (see more in chapter 7).

Early satellites for vegetation monitoring often covered the Earth at rather long intervals (e.g. the Landsat missions, U.S. Geological Survey 2022), the weather satellites have always made frequent overpasses but with pixel/raster sizes on the kilometre scale (e.g. National Oceanic and Atmospheric Administration 2022). Images had to be purchased from an archive or specially "tasked" and substantial funding was required to use satellite data. During the last decade, many more satellites, carrying optical instruments as well as radar and laser sensors, have covered the surface. Satellite imagery and archives of algorithms and scripts used for analysis now can be accessed freely on global cloud-based platforms together with compiled maps (e.g. phenology or grasslands), such as Google Earth Engine or the EU Copernicus Services giving access to the Sentinel satellite complex, of optical multispectral and radar data. Recent years have seen a policy shift into making maps and data as open source, downloadable for use, allowing them to be integral parts in planning or finding possible sites for selected habitats; see examples in chapter 8 or the maps and data of the UK Countryside Survey in this chapter (see Figure 2.4; UKCEH Environmental Information Data Centre 2022).

Integration of variables and scales and modelling

Monitoring often draws upon different types of data, adding images or laser and radar data from different heights, such as from drones, aeroplanes, or satellites, to the mix in order to follow different lines of enquiry. Imagery from the lower heights can be used in automated object recognition and counting but see more on drone and unmanned aerial vehicles in chapter 8. However, these images can be used in much the same fashion as field surveys, using experts to interpret them into thematic classes as a source of evidence for monitoring; see more in chapter 5. Remote sensing with active sensors (sending out and receiving answers as point clouds) such as lidar or radar makes other uses possible, including seeing the ground surface below the vegetation cover and penetrating clouds (see chapters 7 and 8).

The integration of data from soils experiments or DNA sequencing of species and the comparison with archival measurements of in situ of variables related to the ecosystems in question (e.g. from climatic or phenological data) needs to be conducted in a way that contributes towards our understanding of interactions and possible changes (e.g. Cavender-Bares et al. 2022). This is done by process modelling. Modelling is not a single method, because myriad modelling methods exist in most fields of monitoring. These include the building and analysis of mathematical models of ecological processes, including both purely biological and combined biophysical models. They can be purely analytic or used in simulations, with the aim of both understanding complex ecological processes and predicting how real ecosystems might change (Amato and Giménez 2022; Jeong et al. 2022; Priyadarshi et al. 2022).

Monitoring schemes and flexibility

Long-term monitoring might maintain unchanging classes, variables, or sampling design, which is always preferable to enable recording of the real changes happening over time, without the risk of falsely interpreting due to changes in data collection. However, many schemes have to be flexible enough to accommodate new questions asked from authorities or respond when the results indicate that the sampling has too little relevant data or the wrong variables for understanding what is happening – see more on design schemes in chapter 4.

The ways in which researchers or nations classify land and waters have often been set in tradition and are thus resistant to changes in definitions. New questions from the policy side are now changing that resistance, with the need to change the content of the classes to comply with overarching data compilations; for example, pan-European ready-made analysis layers on data portals. Because we still have to find ways to incorporate existing older data in monitoring and to compile several data sources into something new, answering new or changed questions is a task that most long-running monitoring programmes will encounter. This means that there will be gaps in knowledge, and innovative ways of filling these gaps are integral to many monitoring schemes, as you will see in several chapters and examples throughout the book. There are also legacy datasets in the form of thematic maps, dividing the area of interest into units labelled with habitats, land cover, or land use. Remote sensing imagery can also be classified into thematic maps, which may be less reliable than field surveys but are typically more repeatable and have more information about spatial and thematic variation inside each classified polygon (e.g. Congalton et al. 2014).

Regardless of the choice of classes, the data collected at different scales, in the field, near the ground by drones, or at height from aeroplanes or from space all have different possibilities. The details at ground level must be translatable to the life forms at the landscape scale of the survey from above, where we see the structure and texture and use ecological skills to translate these to vegetation associations. When monitoring comprises all of these levels, measuring them in a way that makes translations possible will greatly benefit the results; a closer look at these issues is taken in chapters 8 and 9.

Accommodating the views of different stakeholders can introduce some level of ambiguity in requirements for monitoring. For collection of data, however, one is dependent on clear, unambiguous decisions on limits and content of classes as well as exactly how to collect them. This is important for repeatability, to be able to record in a similar fashion across time, across nations, or across persons doing the collection; see Table 2.3 and example on the EU level in chapter 3. Even with clear and concise instructions, any differences of interpretation between the people making *in situ* observations must be addressed and calibrations of person-to person variations is crucial to the quality of the data, see more on that in chapter 5.

Example: monitoring biodiversity in the UKCEH Countryside Survey

The monitoring work in the Countryside Survey includes many of the elements of monitoring taken up in this book. It provides a unique and statistically robust series of datasets, consisting of an extensive set of repeated ecological measurements at a national scale. It was first undertaken in 1978 to provide a baseline for ecological and land use change monitoring in the rural environment of Great Britain, following a stratified random design, based on 1km squares. It is a national-scale long-term monitoring programme, carried out by the UK Centre for Ecology & Hydrology (UKCEH, and predecessors), investigating stock and change of habitats, landscape features, vegetation, soil, and freshwaters. Based on repeated field surveys in 1km squares in the countryside,

Table 2.3 The differences between question versus data collection – or the *what, why,* and *how* of data collection

The question *perspective on monitoring:* what *and* why	Data collection *perspective on monitoring:* how
Is tolerant of more than one opinion; can be ambiguous in what to collect and monitor	Needs variables or classes that are fixed and have clear boundaries and are unambiguous
The limit of what has to be included can change, often due to new results and/or new concerns or opinions.	Needs clear boundaries and scope of the monitoring; for example, a geographical extent, a species population, or a range of habitats
Different opinions on what should be included can make the intent ambiguous.	Needs agreement regarding exactly what is included
Sometimes has to deal with opinions based on feelings (which are not always obvious) or anecdotal information	The method is designed to be shared and taken over by another person without introducing changes using numbers, text, lines on a map.
Uncertainty can be tolerated without defining it.	Uncertainty must be formally represented by statistical methods or equivalents.
Recommendations as text descriptions and perhaps maps	Results in data and analysis, some text
Tends towards uncertainty but having more realism	Tends towards certainty and over-interpretation
Needs to maintain adaptability of interpretation	Needs to maintain continuity of recording

Policies or demands from authorities tend to be vague, minimizing conflicts at the decision stage. This leaves room for interpretation at national or lower levels regarding how to define classes and crisp borders between them. Failing to create classes that can be translatable to other systems will cause problems when the analysis and gathering of estimates for reporting starts.

the generated data and maps are directed towards policy purposes and constitute an important basis for scientific objectives, because the survey provides evidence on how multiple aspects of the environment are changing over time. Other aims are the study of ecosystem services and how changes affect the economy and well-being of humans, to estimate progress against target indicators in biodiversity strategies and to provide data for the UK Government's reporting of biodiversity. Results and analyses of status and changes over time are available at the UKCEH Countryside Survey home page (2022); see Text box 2.2.

Thus, in the context of monitoring types, the Countryside Survey is both mandated and curiosity driven (Norton et al. 2012; UKCEH Countryside Survey 2022). Since 2019, the survey has moved from an approximately decadal year-long stand-alone survey to a rolling programme, where locations are monitored over a five-year period, enabling annual updates and resilience against atypical years in terms of weather and spreading resources more evenly. The Countryside Survey programme is complemented by the UKCEH Land Cover Map (UKCEH Countryside Survey 2021), a series of satellite-derived maps representing land cover across the UK, starting in 1990 and now annually produced since 2020.

The field survey sampling strategy is based on the Institute of Terrestrial Ecology (ITE) land classification (UKCEH, Environmental Information Data Centre (2022), which divides the land and water area of Great Britain into sets of environmental strata, termed *land classes*, to be used as a basis for ecological field survey (Figure 2.2; UKCEH, Environmental Information Data Centre 2022). Originally developed by the ITE in the late 1970s, the strata were created from the multivariate analysis of 75 environmental

---

Text box 2.2: Sampling and mapping in the Countryside Survey

*The sampling included in Countryside Survey field surveys (*Norton et al. 2012):

- 1978 onwards – Vegetation sampling using large randomly placed plots (main plots, 200m^2), which sample open areas in fields, woods, heaths and moors, and targeted habitat and linear feature (4m^2 and 10 × 1m) plots (maximum number of plots, 18,466; mean plots per square, 31; Wood et al. 2018).
- 1978 onwards – Soil sampling within the main plots (maximum number, 2614), including samples from the top 15cm of the profile for physicochemical measurements and samples from the top 8cm for invertebrates and microbiology.
- 1990 onwards – Sampling of a headwater stream (Strahler order 1e3) site within the survey square, comprising a macroinvertebrate kick net sample (Murray-Bligh 1999), preserved in formalin and returned to UKCEH laboratories for enumeration; a macrophyte survey based on the Mean Trophic Rank methodology (Holmes et al. 1999) but with an extended species list; and a River Habitat Survey (Environment Agency 2003) and accompanying physicochemical data (up to 373 of the 591 squares).
- 2007 onwards – Pond sampling (one randomly located pond in each of 260 squares containing ponds) comprising a pond macrophyte survey and accompanying physicochemical and habitat data, a new survey element for 2007.
- 1978 onwards – Comprehensive repeat field mapping of landscape point, line, and area features across each 1km square (Wood et al. 2018), including detailed mapping at the polygon level, according to the Joint Nature Conservation Committee (JNCC) Broad and Priority Habitat classifications (Jackson 2000).

---

variables, including climatic data, topographic data, human geographical features, and geology data into 45 classes (Bunce et al. 1996, 2007; Barr and Wood 2011). To select 1km survey sites, originally eight random 1km squares were drawn from each of 32 environmental classes, thus comprising 256 sample squares in the 1978 survey. The number of these sites increased to 382 in 1984, 506 in 1990, 569 in 1998, and 591 in 2007. The increase in the number of survey sites reflects the incorporation of the requirement for country-level reporting (not Great Britain as a whole). An increase in sites was necessary to obtain the statistical power to report results for England (since 2007), Scotland (since 1998), and Wales (since 2007) as separate entities. This also increased the number of land classes from the original 32 to 45 classes by 2007; see Figure 2.2, Table 2.4, and Text box 2.2. Text box 2.3 provides some examples of assessment of changes over time and Figure 2.3 illustrates field surveys across time and space in the UK Countryside Survey.

Mapping the landscape is important for understanding connections in nature. The UKCEH has a long history of using satellite imagery to map land cover, from the first national land cover map of Great Britain in 1990 to the current production of annual land cover maps and land cover change data. The UKCEH land cover classes are based on the UK Biodiversity Action Plan (BAP) Broad Habitats (Jackson 2000) and are

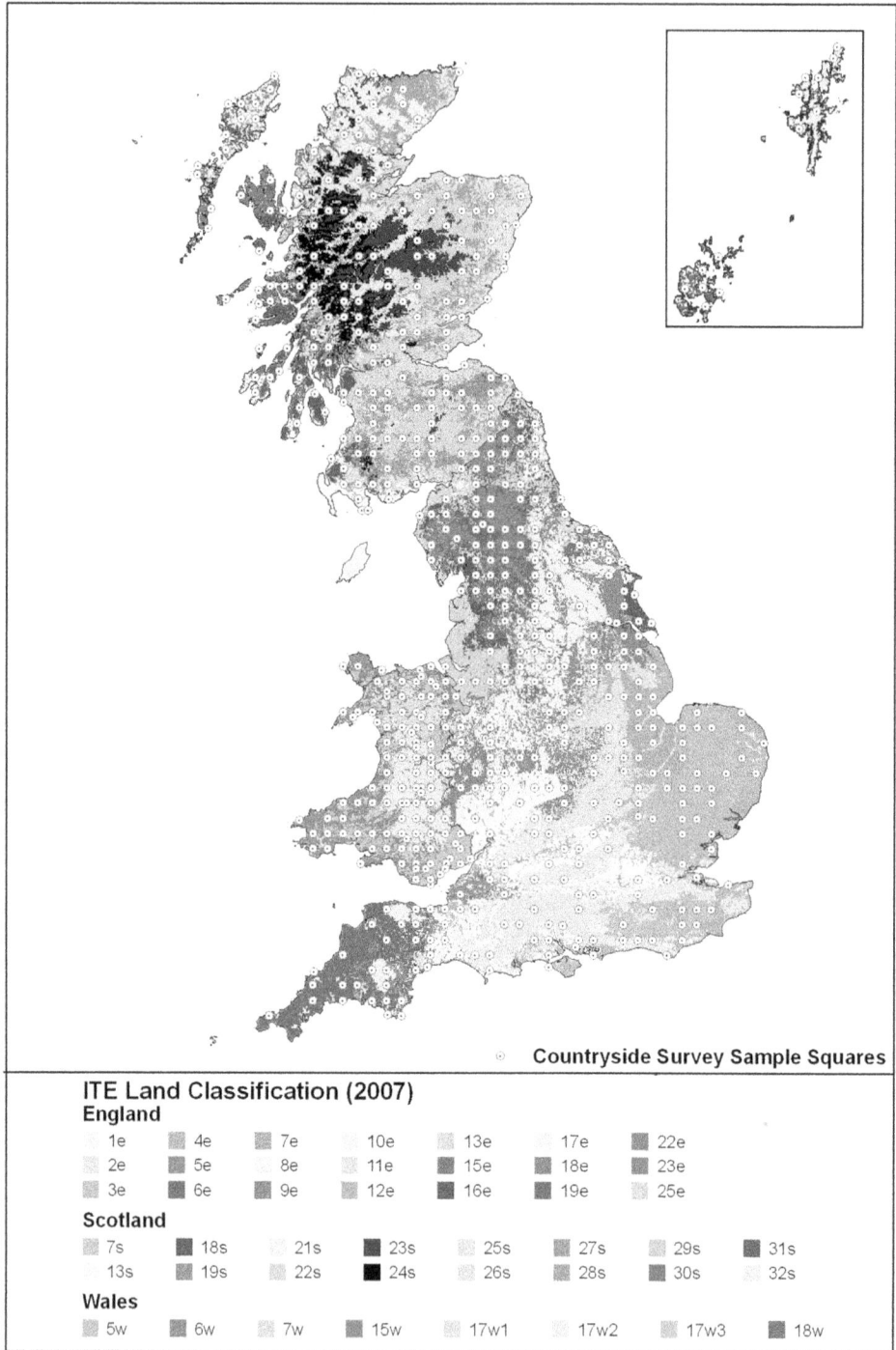

Figure 2.2 The ITE land classification and the sampling strategy for the Countryside Survey.

Table 2.4 Summary of the 45 ITE land classes, 2007 version

England

1e	Flood plains/shallow valleys, S England
2e	Low calcareous hills/variable lowlands, S England
3e	Flat/gently undulating plains, E Anglia/S England
4e	Flat coastal plains, E Anglia/S England
5e	Shallow slopes/flood plains, S-W England
6e	Complex valley systems/table lands, S-W England
7e	Sea cliffs/hard coast, England
8e	Estuarine/soft coast/tidal rivers, England/
9e	Almost flat plains, N Midlands, NE England
10e	Gently rolling/almost flat plains, NE England/N Midlands
11e	Flat plains/small river floodplains, E Midlands
12e	Large river floodplains, flat plains, margins, E Anglia
13e	Coastal plains/gently rolling low hills, NW England
15e	Flat river valleys/lower hill slopes, NW England
16e	Gently rolling low hills/flat river valleys, NW England
17e	Upland valleys/rounded hill sides, England
18e	Upland valley sides/low mountains, N England
19e	Upland valleys/plateau's, N England
22e	Intermediate mountain tops/broad ridges, N England
23e	High mountain summits/ridges, N England
25e	Flat/gently undulating river valleys, N England

Scotland

7s	Hard/mixed coasts, S-W Scotland
13s	Coastal plains/soft coasts, S-W Scotland
18s	Isolated hills/mountain summits, W Scotland
19s	Upland valleys/low mountains, S Scotland
21s	Low mountain slopes/upper river valleys, Highlands
22s	Round mountains/broad upper ridges, S Scotland/Highlands
23s	High mountain summits/ridges/valleys, Highlands
24s	Steep valley sides/intermediate mountain tops, W Highlands
25s	Undulating plains/gently sloping valleys, E Scotland
26s	Flat plains/gently sloping lowlands, central & S Scotland
27s	Low hills/undulating lowlands, Scotland except W
28s	Shallow valleys/low hill plateau's, throughout Scotland
29s	Inner rocky/mixed coasts/complex topography, W Scotland
30s	Outer rocky/mixed coasts/low hills, W Scotland/Islands
31s	Rocky/mixed coasts/low hills, N Scotland/Islands
32s	Shallow hills/complex coastlines, N Scotland/Islands

Wales

5w	Shallow slopes/flood plains, Wales
6w	Complex valley systems/table lands, Wales
7w	Sea cliffs/hard coast, Wales
15w	Flat river valleys/lower hill slopes, Wales
17w1	Low mountain ridges/valley slopes, N Wales
17w2	Rounded mountains/scarps/upper valleys, mid/S Wales
17w3	Variable landforms of hills/low mountain, Wales
18w	Upland valley sides/low mountains, Wales

Text box 2.3:   Examples of assessments of changes over time and basis for research

- Changes in the area and distribution of broad habitats including some habitat types of special interest (e.g. hedgerows, arable field margins, and upland heath).
- Changes in the condition of habitats, especially changes in biodiversity.
- Determination of how the countryside's natural resources respond to changes in land use, climate change, and government policy.
- Updating biodiversity indicators, such as UK Priority Habitats, plant diversity (specifically open habitats, woodlands, and boundary habitats), ecological impacts of air pollution (specifically areas affected by acidity and nitrogen), invasive species, and river quality (biological and chemical).
- Changes in catchment land use – the effect on ecological quality of watercourses, their biodiversity, and ecosystem function and the effect of riparian corridors on aquatic communities.
- Ecological quality – detection of differences between ecological quality in agri-environment land and the wider countryside.
- Impacts of declines in arable weeds and butterfly and bird food plants and loss of pollinators.
- Changes in types, quantity, and distribution of non-native plant species.
- The relationship between the soil microbial diversity and soil quality; the first country-level (England, Scotland, Wales) soil sampling was carried out in the 2007 survey.
- Changes in soil acidification.
- Impact of air pollution, such as nitrogen deposited from the atmosphere, as contributing to the recorded vegetation changes.

available as both raster and vector products for the whole of the UK; see Figure 2.4. The utilization of satellite data for mapping and modelling increases as they become available (Henrys and Jarvis 2019). Also, there is now the possibility of public involvement, in the shape of the interactive modelling and finding one's own niche, searching for those places that potentially have a favourite landscape type or contain habitats that include our favourite flowers (Henrys et al. 2015; Smart et al. 2019).

Classifications of ecosystems into habitats develop over time, and though the field survey habitat mapping component of the Countryside Survey and the satellite-derived UKCEH Land Cover Map both use classifications derived from the JNCC Broad Habitats (Jackson 2000), a new hierarchical classification system is now available. The UK Habitat Classification (UKHab) is designed to be compatible with all major classifications in use in the UK and Europe and also to large-scale geographic information system (GIS)-based habitat datasets, such as the UKCEH Land Cover Map, which provides a huge advantage for scoping large-scale surveys and for sharing data regionally, nationally, and internationally (UKHab 2022).

No single monitoring scheme, however, can accommodate all information needed for every level of detail, and as in most other countries, a series of bespoke monitoring schemes (rare plants, birds, butterflies, etc.) are carried out across the UK, depending on

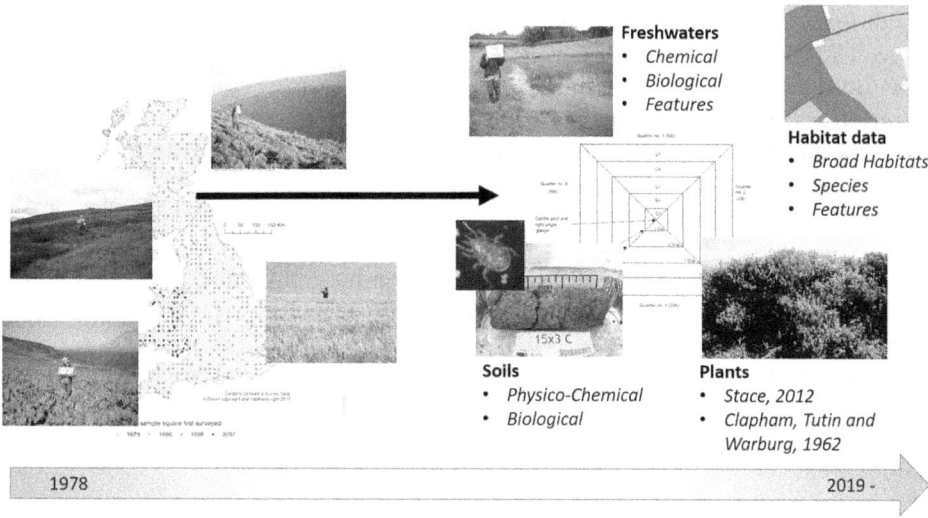

Freshwaters
- *Chemical*
- *Biological*
- *Features*

Habitat data
- *Broad Habitats*
- *Species*
- *Features*

Soils
- *Physico-Chemical*
- *Biological*

Plants
- *Stace, 2012*
- *Clapham, Tutin and Warburg, 1962*

1978   2019 -

Figure 2.3 The UKCEH Countryside Survey field survey across time and space.

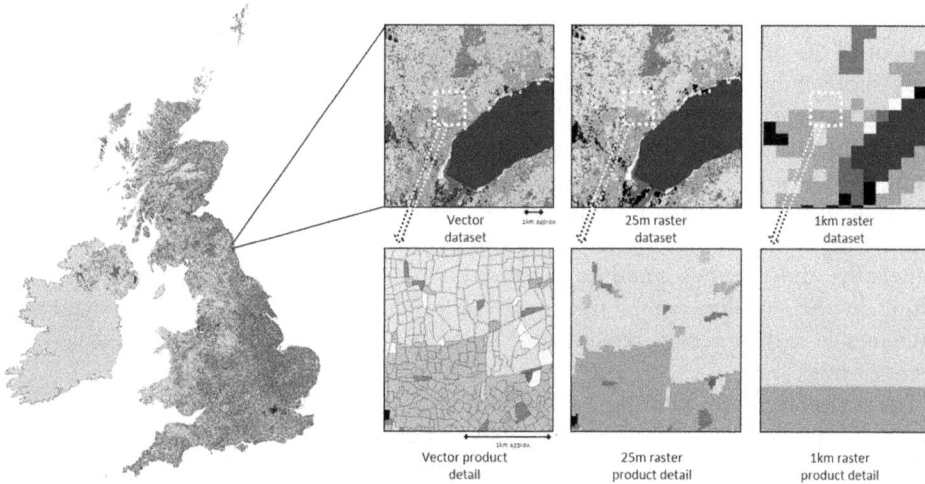

Figure 2.4 The UKCEH land cover map, 2020 (left), showing different levels of detail (right).

Source: UKCEH Countryside Survey (2021). Based on Land Cover Maps 2020 and 2007 © UKCEH 2021. Contains Ordnance Survey data © Crown Copyright 2007, Licence number 100017572.

the level of detail needed. For example, there is a complementary relationship between these general classifications used for land cover across the whole country and a second set of vegetation classes, the National Vegetation Classification (NVC), which is used by the nature conservation agencies for more detailed recording of biodiversity on protected high value sites (Rodwell 2006; JNCC 2022).

Example: high-resolution monitoring of the Belgian coast

As with many other coastal areas, the Belgian coast is vulnerable to impacts of climate change. The Climate Resilient Coast Project (CREST) ran from 2015 to 2019 (Flanders Marine Institute [VLIZ] 2022), aiming to understand local patterns of deposition and erosion of beaches as a combination of processes by waves, tides, sediment transport, wind, and human activities, using in situ measurements and related data acquisition, modelling, and monitoring techniques (Monbaliu et al. 2020).

The primary objectives were to:

- Gain a better understanding of nearshore and onshore physical processes including improved models and the validation of *grey* (not covered by vegetation) data about coastal dynamics.
- Determine the resilience of the natural coastal system (dunes and beaches) in relation to storms and wind.
- Validate calculations using state-of-the-art models, based on laboratory tests and field measurements.

To monitor aeolian dune formation and dynamics on the upper-beach, high-resolution terrestrial laser scanning techniques from permanent instrument stations, recording data continuously, were used. In this way, morphological changes could be investigated at an appropriate temporal and spatial scale, allowing the characterization of ephemeral dune dynamics. In practice, the best temporal resolution for this purpose, given the instruments and conditions, was found to be six-hourly laser scans to understand the development of the protodunes with a height ranging from 0.15 to 0.42m formed under an along-shore wind above 7 m/s (Montreuil et al. 2020); see Figure 2.6. With the terrestrial laser scanning techniques (accuracy: 5mm), point clouds were acquired every hour (using the Riegl® VZ-2000, see Figure 2.5) in subsequent survey periods each lasting 36 hours. The resulting point clouds (comprising 95,500 points per survey) were used to generate digital elevation models (DEMs) for each point in time with a cell size of 0.25cm; see Figure 2.7. Based on this time series of DEMs, the differences between consecutive time periods could be visualized individually and in a cumulative way, resulting in the visualization of microscale morphological changes alongside aggregate outcomes of such processes. Please see Montreuil et al. (2020) for a more detailed account of this.

The example of monitoring of the protodunes of the Belgian coast (Figure 2.7) illustrates a number of characteristic features of how data acquisition, sampling, storage, and analysis processes in monitoring are currently being transformed through technological development. It shows how the presence of exceedingly vast amounts of data, collected continuously at high temporal and spatial resolutions, tend to inform research processes oriented more towards data mining and filtering and how such analysis processes may link understandings of the same processes across different temporal and spatial scales. In comparison, most data acquisition was extremely costly until very recently and therefore observations were typically limited to carefully selected samples of data for specific times and places. For example, aerial imagery was typically recorded at great cost and/or only at long intervals, depending on the location (Christensen 2013). Additionally, for such imagery to be useful to monitoring, it must be interpreted and classified, which is a delicate and painstaking process when conducted manually.

Figure 2.5 (Left) Riegl® VZ-2000 permanent laser scanner, overlooking the beach (to the right).

Source: Monbaliu et al. (2020).

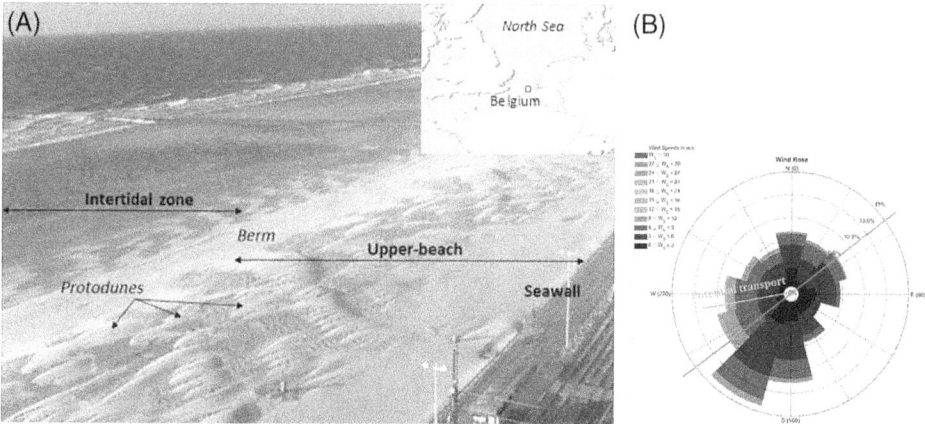

Figure 2.6 (A) A photograph with the presence of fully developed protodunes at the Mariakerke site of the project on April 26 at 18.00. The wind is blowing from the bottom left in the photo. (B) A figure showing the annual wind directions in Ostend, Belgium.

Source: After Montreuil et al. (2020).

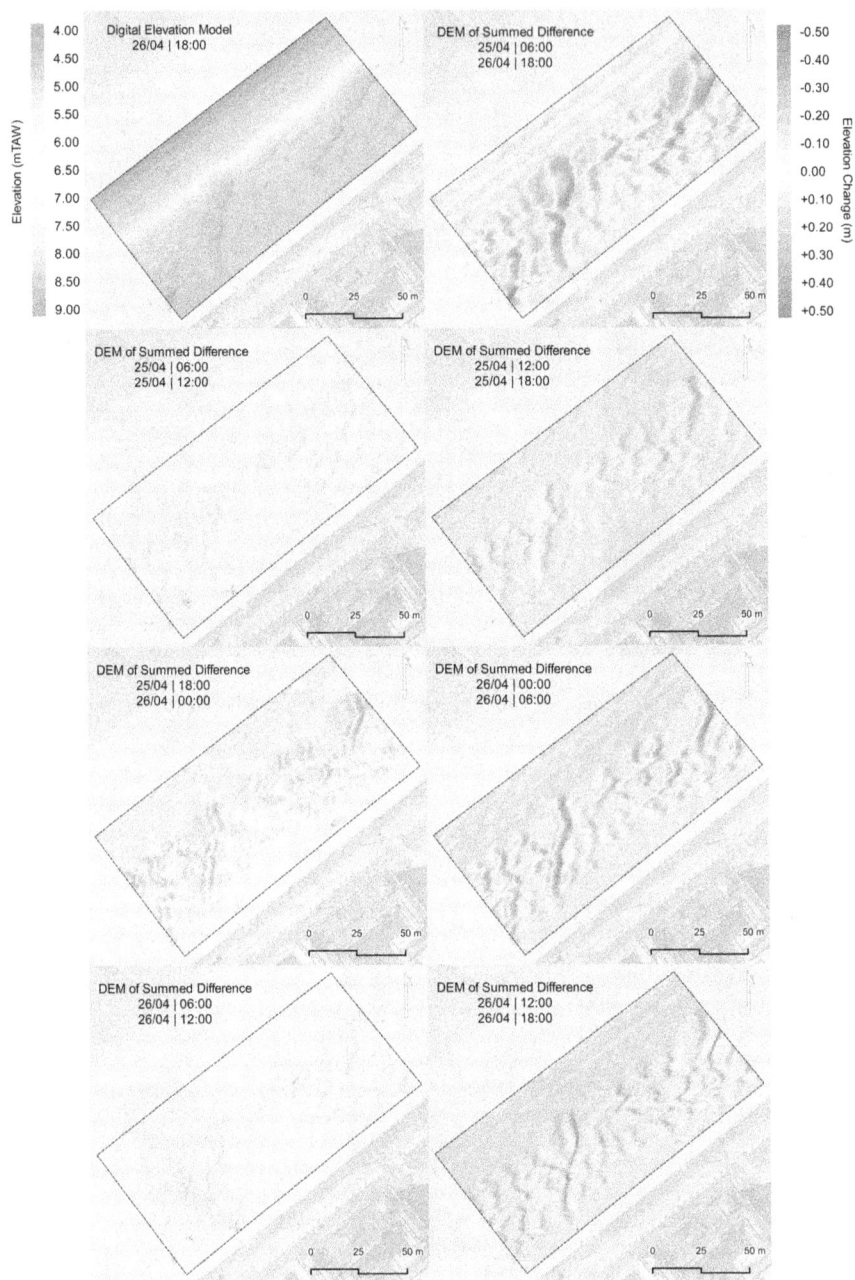

Figure 2.7 Digital elevation model (DEM) of the fully developed protodunes on April 26, 2018, at 18.00 and (top right) and DEM of differences (DoDs) showing the differences between the formation on April 25 at 6.00 and the fully developed protodunes on April 27 at 18.00 (top left). Below are six DoDs, showing summed elevation differences between consecutive surveys with a temporal resolution of observations of six hours.

Credit: Data, analysis, and cartography: Lars De Sloover, Department of Geography, Ghent University, Belgium, 2022.

This meant that until recently, only a limited area of land could be covered, and only at certain time intervals (Christensen et al. 2017). Similar limitations applied, and indeed still apply, to in situ fieldwork observations and most of the wide range of other methods used in monitoring. Therefore, sampling has traditionally been performed before acquisition and analysis of data, which then in turn has focused on deriving generally applicable insights about habitat change and persistence from relatively few observations. In contrast, current observation platforms tend to supply vast datasets that create an increasing demand for research processes where data are sampled after acquisition and where analysis to an increasing extent is unrestricted by limits imposed by temporal resolution. This is the case, for example, with respect to imagery collected as part of the Sentinel programme of the European Environment Agency (EEA), where researchers are able to develop land cover and habitat classifications directly from archives of continuously recorded imagery in combination with other types of data (Zanaga et al. 2021). Such imagery is recorded on a continuous basis at high temporal resolutions, in the same way that observations were made on the Belgian coast described above. This creates a situation where analysis efforts in monitoring tend to shift from being focused on design of observational processes to being focused on data mining, sorting, filtering, selection, and aggregation of already existing data; see example in chapter 9. In combination with deep learning algorithms and other forms of artificial intelligence used to classify data, this advent of big data processing techniques has tended to create situations of data excess, where too much data is available for it to be taken into account. This has had a range of different effects on monitoring. As such, the field is currently being transformed from a situation where research questions were formulated with respect to highly focused, specialized arrays of observation and analysis procedures to a situation where hypotheses may be tested with reference to large existing collections of observations. In this way, monitoring is becoming an increasingly flexible, explorative field of research, testing the limits of how observation may be transformed into knowledge.

Key messages

- In this chapter, we have briefly discussed some of the most important issues taken up in this book, starting with common types of monitoring and what they are for, including an understanding of who wants the results of monitoring and who could benefit from them.
- There are a large number of complementary ways to collect data, although in situ measurements are a constant factor because much of the detailed knowledge of biodiversity is hard to collect from remote sensors.
- Both imagery and data have become freely accessible in an unprecedented way, and researchers need to be flexible and incorporate both historical and current datasets potentially from different sources, making monitoring into a multidisciplinary field.
- The example of long-term national monitoring in the UK describes the issue of flexibility, with new and/or more data to incorporate into the scheme. The other example shows how the use of new technology in permanent networks of in situ devices, so common to monitoring in oceans, can be used to capture swift changes in other settings such as dune formations.

Study questions

1   What is the difference between the various types of monitoring, and to whom are they typically directed? Think about who else can benefit from the results and why.
2   How do the two types of data, quantitative and qualitative, differ, and what are they used for?
3   Read about the different types of data mentioned in this chapter, the ways of integrating them, and the amount of data needed to validate all steps. What are the main hindrances to data integration?
4   What can we learn from the near-constant changes in what is asked from monitoring, and what does that entail for the data provider who has to report the monitoring results?

Further reading

Much information on ongoing monitoring is only available on websites, and we recommend a search on sites, focusing on the type of monitoring of interest, because much can be learned from the success (as well as failures) of others.

Cavender-Bares et al. (2022) provide a recent review of monitoring biodiversity in relation to ways of collecting data and where different methods are helpful.
Lindenmayer and Likens (2018) provide a comprehensive read on ecological monitoring with common reasons for failure.

References

Amato, M.T. and Giménez, D. (2022) Quantifying root turnover in grasslands from biomass dynamics: application of the growth-maintenance respiration paradigm and re-analysis of historical data, *Ecological Modelling* 467, 109940. 10.1016/j.ecolmodel.2022.109940

Barr, C.J. and Wood, C.M. (2011) *The Sampling Strategy for Countryside Survey (Up to 2007)*, http://nora.nerc.ac.uk/19487

Brlík, V., Šilarová, E., Škorpilová, J., Alonso, H., Anton, M., Aunins, A., Benkö, Z., Biver, G., Busch, M., Chodkiewicz, T., et al. (2021) Long-term and large-scale multispecies dataset tracking population changes of common European breeding birds, *Scientific Data* 8, 21.

Bunce, R.G.H., Barr, C.J., Clarke, R.T., Howard, D.C. and Lane, A.M.J. (1996) Land classification for strategic ecological survey, *Journal of Environmental Management* 47, 37–60. doi: 10.1006/jema.1996.0034

Bunce, R.G.H., Barr, C.J., Clarke, R.T., Howard, D.C. and Scott, W.A. (2007) *ITE Land Classification of Great Britain 2007*. NERC Environmental Information Data Centre. 10.5285/5f0605e4-aa2a-4 8ab-b47c-bf5510823e8f

Cavender-Bares, J., Schneider, F.D., Santos, M.J., Armstrong, A., Carnaval, A., Dahlin, K.M., Fatoyinbo, L., Hurtt, G.C., Schimel, D., Townsend, P.A., et al. (2022) Integrating remote sensing with ecology and evolution to advance biodiversity conservation, *Nature Ecology & Evolution* 6(5), 506–519. 10.1038/s41559-022-01702-5

Christensen, A.A. (2013) Mastering the land: mapping and metrologies in Aotearoa New Zealand, in Pawson, E. and Brooking, T. (eds) *Making a New Land*. University of Otago Press, pp. 310–327.

Christensen, A.A., Brandt, J. and Svenningsen, S.R. (2017) Landscape ecology, in Richardson, D., Castree, N., Goodchild, M.F., et al. (eds) *International Encyclopedia of Geography: People, the Earth, Environment and Technology*. Oxford, UK: John Wiley & Sons, pp. 1–10.

Congalton, R.G., Gu, J., Yadav, K., Thenkabail, P. and Ozdogan, M. (2014) Global land cover mapping: a review and uncertainty analysis, *Remote Sensing* 6, 12070–12093. doi: 10.3390/rs61212070

Environment Agency. (2003) *River Habitat Survey Manual.* Bristol, UK: Environment Agency.

Eriksson, L., Nordlund, A., Olsson, O. and Westin, K. (2012) Recreation in different forest settings: a scene preference study, *Forests* 3, 923–943. doi: 10.3390/f3040923

European Commission. (1992) Council Directive 92/43/EEC of 21 May 1992, https://ec.europa.eu/environment/nature/legislation/habitatsdirective/index_en.htm (Accessed November 1, 2022).

European Commission. (2022) Monitoring and evaluation, https://ec.europa.eu/neighbourhood-enlargement/monitoring-and-evaluation_en (Accessed June 3, 2022).

European Environment Agency. (1999) Monitoring definition, https://www.eea.europa.eu/help/glossary/eea-glossary/monitoring (Accessed June 3, 2022).

Filazolla, A. and Cahill, F.J. (2021) Replication in field ecology: identifying challenges and proposing solutions, *Methods in Ecology and Evolution* 12, 1780–1792.

Fischer, M., Bossdorf, O., Gockel, S., Hänsel, F., Hemp, A., Hessenmöller, D., Korte, G., Nieschulze, J., Pfeiffer, S., Prati, D., et al. (2010) Implementing large-scale and long-term functional biodiversity research: the Biodiversity Exploratories, *Basic and Applied Ecology* 11(6), 473–485. 10.1016/j.baae.2010.07.009

Flanders Marine Institute (VLIZ). (2022) Climate Resilient Coast Project, CREST, http://www.crestproject.be/nl (Accessed June 3, 2022).

HELCOM Baltic Marine Environment Protection Commission. (2022), Home page, https://helcom.fi/ (Accessed June 3, 2022).

Henrys, P.A. and Jarvis, S.G. (2019) Integration of ground survey and remote sensing derived data: producing robust indicators of habitat extent and condition, *Ecology and Evolution* 9(14), 8104–8112. 10.1002/ece3.5376

Henrys, P.A., Smart, S.M., Rowe, E.C., Jarvis, S.G., Fang, Z., Evans, C.D., Emmett, B.A. and Butler, A. (2015) Niche models for British plants and lichens obtained using an ensemble approach, *New Journal of Botany* 5(2), 89–100. 10.1179/2042349715Y.0000000010

Holmes, N.T.H., Newman, J.R., Chadd, S., Rouen, K.J., Saint, L. and Dawson, F.H. (1999) *Mean Trophic Rank: A User's Manual.* Bristol, UK: Environment Agency.

Houk, P. and van Woesik, R. (2013) Progress and perspectives on question-driven coral-reef monitoring, *BioScience* 63(4), 297–303. 10.1525/bio.2013.63.4.10

Hurlbert, S.H. (1984) Pseudoreplication and the design of ecological field experiments, *Ecological Monographs* 54(2), 187–211.

Jackson, D. (2000) *Guidance on the Interpretation of the Biodiversity Broad Habitat Classification (Terrestrial and Freshwater Types): Definitions and the Relationship with Other Habitat Classifications.* JNCC Report, No. 307, Peterborough, UK.

Jeong, S., Jonghan Ko, J. and Yeom, J.-M. (2022) Incorporation of machine learning and deep neural network approaches into a remote sensing–integrated crop model for the simulation of rice growth, *Scientific Reports* 12, 9030. 10.1038/s41598-022-13232-y

Joint Nature Conservation Committee. (2022) National vegetation classification, https://jncc.gov.uk/our-work/nvc/ (Accessed June 3, 2022).

Khair, N.K.M., Lee, K.E. and Mokhtar, M. (2021) Community-based monitoring for environmental sustainability: a review of characteristics and the synthesis of criteria, *Journal of Environmental Management* 289, 112491. doi: 10.1016/j.jenvman.2021.112491

Lindenmayer, D. and Likens, G.E. (2018) *Effective Ecological Monitoring.* Clayton South, Australia: CSIRO Publishing.

Macgregor, C.J., Thomas, C.D., Roy, D.B., Beaumont, M.A., Bell, J.R., Brereton, T., Bridle, J.R., Dytham, C., Fox, R., Gotthard, K., et al. (2019) Climate-induced phenology shifts linked to range expansions in species with multiple reproductive cycles per year, *Nature Communications* 10, 4455.

Merchant, C.J., Minnett, P.J., Beggs, H., Corlett, G.K., Gentemann, C., Harris, A.R., Hoyer, J. and Maturi, E. (2019) Global sea surface temperature, in Hulley, G.C. and Ghent, D. (eds) *Taking the Temperature of the Earth.* Elsevier, pp. 5–55. 10.1016/B978-0-12-814458-9.00002-2

Minichiello, V., Aroni, R. and Hays, T. (2008) *In-Depth Interviewing. Principles, Techniques and Analysis.* 3rd edn. Australia: Pearson Australia Group.

Monbaliu, J., Mertens, T., Bolle, A., Verwaest, T., Rauwoens, P., Toorman, E., Troch, P. and Gruwez, V. (eds) (2020) *CREST Final Scientific Report: Take Home Messages & Project Results.* VLIZ Special Publication 85, Universiteit Gent, Belgium.

Montreuil, A.-L., Chen, M., Brand, E., De Wulf, A., De Sloover, L., Dan, S. and Verwaest, T. (2020) Early-stage aeolian dune development and dynamics on the upper-beach, *Journal of Coastal Research* (Special Issue) 95, 336–340. 10.2112/SI95-065.1

Morrison, C.A., Auniņš, A., Benkő, Z., Brotons, L., Chodkiewicz, T., Chylarecki, P., Escandell, V., Eskildsen, D., Gamero, A., Herrando, S., et al. (2021) Bird population declines and species turnover are changing the acoustic properties of spring soundscapes, *Nature Communications* 12, 6217.

Murray-Bligh, J. (1999) *Procedures for Collecting and Analysing Macro-invertebrate Samples.* Bristol, UK: Environment Agency.

National Oceanic and Atmospheric Administration. (2022) Satellites, https://www.noaa.gov/satellites (Accessed June 3, 2022).

Norton, L.R., Maskell, L.C., Smart, S.M., Dunbar, M.J., Emmett, B.A., Carey, P.D., Williams, P., Crowe, A., Chandler, K., Scott, W.A., et al. (2012) Measuring stock and change in the GB countryside for policy: key findings and developments from the Countryside Survey 2007 field survey, *Journal of Environmental Management* 113, 117–127. doi: 10.1016/j.jenvman.2012.07.030

OSPAR Commission 30th. (2022) Home page, https://www.ospar.org/ (Accessed June 2, 2022).

Priyadarshi, A., Chandra, R., Kishi, M.J., Smith, S.L. and Yamazaki, H. (2022) Understanding plankton ecosystem dynamics under realistic micro-scale variability requires modeling at least three trophic levels, *Ecological Modelling* 467, 109936. 10.1016/j.ecolmodel.2022.109936

Programme of the European Union. (2022) Copernicus Europe's eyes on Earth, Copernicus Services, https://www.copernicus.eu/en/copernicus-services (Accessed June 3, 2022).

Rodwell, J.S. (2006) *NVC Users' Handbook.* Peterborough, UK: Joint Nature Conservation Committee. All volumes are available at: https://jncc.gov.uk/our-work/nvc/

Sastri, A.R., Christian, J.R., Achterberg, E.P., Atamanchuk, D., Buck, J.J.H., Bresnahan, P., Duke, P.J., Evans, W., Gonski, S.F., Johnson, B., et al. (2019) Perspectives on in situ sensors for ocean acidification research, *Frontiers in Marine Science* 6, 653. https://www.frontiersin.org/article/10.3389/fmars.2019.00653

Smart, S.M., Jarvis, S.G., Mizunuma, T., Herrero-Jáuregui, C., Fang, Z., Butler, A., Alison, J., Wilson, M. and Marrs, R.H. (2019) Assessment of a large number of empirical plant species niche models by elicitation of knowledge from two national experts, *Ecology and Evolution* 9(22), 12858–12868. 10.1002/ece3.5766

Sparks, T., Garforth, J. and Whittle, L. (2020) A comparison of Nature's Calendar with Gilbert White's phenology, *British Wildlife* April, 271–275.

Sparrow, B.D., Edwards, W., Munroe, S.E.M., Wardle, G.M., Guerin, G.R., Bastin, J.-F., Morris, B., Christensen, R., Phinn, S. and Lowe, A.J. (2020) Effective ecosystem monitoring requires a multi-scaled approach, *Biological Reviews* 95(6), 1706–1719. 10.1111/brv.12636

Stevens, J.P., Blackstock, T.H., Howe, E.A. and Stevens, D.P. (2004) Repeatability of Phase 1 habitat survey, *Journal of Environmental Management* 73(1), 53–59.

Stewart-Oaten, A. (1996) Problems in the analysis of environmental monitoring data, in Schmitt, R.J. and Osenberg, C.W. (eds) *Detecting Ecological Impacts: Concepts and Applications in Coastal Habitats.* San Diego, CA: Academic Press, pp. 109–131.

Sundli Tveit, M., Ode Sang, Å. and Hägerhäll, C.M. (2013) Scenic beauty: visual landscape assessment and human landscape perception, in Steg, L., van den Berg, A.E. and de Groot, J.I.M. (eds) *Environmental Psychology: An Introduction.* 1st edn. Chichester, UK: BPS Blackwell.

Ten, A., Kaushik, P., Oudeyer, P.Y. and Gottlieb, J. (2021) Humans monitor learning progress in curiosity-driven exploration, *Nature Communications* 12, 5972. 10.1038/s41467-021-26196-w

UKCEH Countryside Survey. (2021) Land cover maps, https://www.ceh.ac.uk/ukceh-land-cover-maps (Accessed June 3, 2022).

UKCEH Countryside Survey. (2022) Home page, https://countrysidesurvey.org.uk/ (Accessed June 3, 2022).

UKCEH, Environmental Information Data Centre. (2022) ITE land classification of Great Britain (1990–2007), https://catalogue.ceh.ac.uk/documents/b86b3572-f480-48e1-9d31-5d3880b7f213 (Accessed May 15, 2022).

Underwood, A.J. (1992) Beyond BACI: the detection of environmental impacts on populations in the real, but variable, world, *Journal of Experimental Marine Biology and Ecology* 161, 145–178.

United Nations, Economic Commission for Europe. (2016) *Environmental Monitoring and Assessment – Guidelines for Developing National Strategies to Use Biodiversity Monitoring as an Environmental Policy Tool for Countries of Eastern Europe, the Caucasus and Central Asia, as well as Interested South-Eastern European Countries*, United Nations Publications, ECE/CEP/176. https://unece.org/sites/default/files/2020-12/1528741_E_inside.pdf

U.S. Geological Survey. (2022) The Landsat satellite missions, https://www.usgs.gov/landsat-missions/landsat-satellite-missions (Accessed May 14, 2022).

van Swaay, C.A.M., Dennis, E.B., Schmucki, R., Sevilleja, C., Balalaikins, M., Botham, M., Bourn, N., Brereton, T., Cancela, J.P., Carlisle, B., et al. (2019) *The EU Butterfly Indicator for Grassland Species: 1990–2017.* Technical Report. Butterfly Conservation Europe.

Wang, F., Zhu, J., Chen, L., Zuo, Y., Hu, X. and Yang, Y. (2020) Autonomous and in situ ocean environmental monitoring on optofluidic platform, *Micromachines* 11, 69. 10.3390/mi11010069

Wilson, N.J., Mutter, E., Inkster, J. and Satterfield, T. (2018) Community-based monitoring as the practice of Indigenous governance: a case study of Indigenous-led water quality monitoring in the Yukon River Basin, *Journal of Environmental Management* 210, 290–298.

Wood, C.M., Bunce, R.G.H., Norton, L.R., Maskell, L.C., Smart, S.M., Scott, W.A., Henrys, P.A., Howard, D.C., Wright, S.M., Brown, M.J., et al. (2018) Ecological landscape elements: long-term monitoring in Great Britain, the Countryside Survey 1978–2007 and beyond, *Earth System Science Data* 10, 745–763. 10.5194/essd-10-745-2018

Zanaga, D., Van De Kerchove, R., De Keersmaecker, W., Souverijns, N., Brockmann, C., Quast, R., Wevers, J., Grosu, A., Paccini, A., Vergnaud, S., et al. (2021) ESA WorldCover 10 m 2020 v100. 10.5281/zenodo.5571936

3 Demands on monitoring

Anna Allard, Santiago Guerrero,
Andreas Aagaard Christensen, Armin Benzler,
Magnus Appelberg, Göran Ståhl, and Mats Sandewall

Introduction

This chapter is about the policies and legislation governing much of the demand for monitoring and, by default, it leans heavily towards mandated monitoring. Many times, however, the concern about an environmental issue leading to the creation of a policy was first initiated by question-driven monitoring. In a sort of mutual development, results from mandated monitoring can trigger new and revised demands.

An advantage of policies, directives, and legislation is that they are documents open for everyone to read, and the actual concern for any of the issues becomes transparent. Anyone can follow the development over time, when issues were added or discarded, which constitutes a good explanation of what was expected to come out of these directives. By knowing what lies behind the laws or policies, we can understand the questions we need monitoring to answer.

The demands on monitoring are also shaped by the evolution of data, including the increasing availability of new types of digital data (several data portals providing both satellite data and already-analyzed high-resolution layers for Europe or other parts of the world, laser data providing models of the surface and moisture regime). This availability and the subsequent creation of data portals are also changing the demands on reporting and simultaneously forcing and enabling harmonization processes nationally and internationally. The large-scale monitoring in oceans covers both national and international territory, where monitoring has long been organized in large communities across several countries and has been at the forefront of these harmonization efforts. An example is shown later in the chapter.

In this way, the demands on monitoring can roughly be divided into three types: legislation and policy, the evolution of data, and requirements for harmonization so that data from different areas can be combined and compared. It is important to be aware of these influences on how monitoring should be conducted. There are a vast number of directives, and we give a short introduction to some of them in relation to biodiversity monitoring and within the scope of this book.

Demands on monitoring

As mentioned in chapter 2, biodiversity monitoring is needed to benefit society by helping to maintain what the European Union (EU) summarizes as public goods or ecosystem services. By that they mean to map, assess, and achieve good condition of ecosystems so they can deliver benefits such as climate regulation, water regulation, soil

DOI: 10.4324/9781003179245-3

health, and pollination and disaster prevention and protection (European Commission [EC] 2022b). This is driven by the recognition of unwanted and damaging environmental change from pressures on land use, the exploitation of resources, pollution, and climate change (United Nations, Economic Commission for Europe 2016). If we want to understand how much of a certain biodiversity resource (for humans or for the intrinsic value as a part of the whole environment) we have, whether that resource is diminishing, or what is happening to the environment, we need information. Environmental concerns may also be about perceived trends that, in future, might pose some threat to resources, whether due to changed land use or abandonment, reduced fish catches, or farmland being lost to sprawling towns and housing.

As authorities realize, perhaps from research results, that some vital resource for biodiversity or for the ecosystem that provides us with its services is likely to be lost, policies are a good way to enforce the preservation or restoration of that resource. Policies can steer the work efforts of the region of interest at local, regional, national, or international scales. Examples are preservation of biodiverse areas in wetlands, of water bodies, or on land or taking action against significant losses, such as restoring wetlands by filling in old ditches or creating new pastures on abandoned grasslands to increase biodiversity.

For example, at the EU level, The Biodiversity Strategy 2030 (EC 2022b) called on Member States to map and assess the state of ecosystems and their services in their national territory with the assistance of the European Commission. In 2020, this was expanded to assessments of the economic value of such services and integration of these values into accounting and reporting systems at both EU and national levels (Maes et al. 2018, 2020).

This means that the data collected by monitoring must be able to answer new questions that arise and show the extent of chosen classes of ecosystems according to how these are defined and understood. The data must be able to show whether ecosystems are in good health and whether a potential problem in an ecosystem is detectable and when, meaning as it develops or only when deterioration has gone too far. Data must also be able to say whether ecosystems can provide the services associated with them (Maes et al. 2020).

The social component

There is also much value in being able to assess both habitats or ecosystems and the social structures creating them from one region or country to the next to predict future shortages of habitats or other consequences of accelerating climate changes. These aspects, which follow on from both agreed-upon policies and local demand, are a core theme through the Agenda 2030 described later in this chapter (EC 2022h). Other work, for example, is directed at monitoring land use and changes in land use, like the land use, land use change, and forestry (LULCF) work stream, intended to mitigate climate change in areas of agriculture and forestry, as human activities affect changes in carbon exchange between the terrestrial ecosystem and the atmosphere (EC 2022f).

Long-term data versus new content of classes

There is a complexity in the duality of wanting to have data that fit into other classification systems or data that actually say anything about our local site. Both are important when seen from different perspectives. On one hand, long-term monitoring using specially developed classes and/or variables that best fit nature conservation sites within the national range of habitat types and species, mainly for the purpose of management decisions, creates strong local time series of biodiversity data. These series, going backwards in time, constitute a wealth of expert knowledge and observed changes, which can often be linked to changes in both cultural understanding, including policies, and with changes in the physical world, such as climate variations or chemical pollutants. Keeping a special legacy like that is, of course, important. On the other hand, these specialized datasets can be hard to combine with datasets from other countries, an important issue that has driven the recent funding to create ready-made layers of data across Europe (for example) provided by data portals. An example is Copernicus Services, which includes data layers based on monitoring data collected, analyzed, and reported to a common standard from all EU Member States; chapter 15 discusses this further.

Two directives are driving the issue towards harmonization. The EU INSPIRE Directive (EU 2022) was created to establish a pan-European spatial data infrastructure, to be able to receive data that are compatible and usable in a context of community (groups of countries) and across the borders of the different nations or states. To achieve the INSPIRE Directive, each Member State is required to annually report on how they are implementing this harmonization. Much work has been laid down over decades to create harmonized data collection to enable compilation of data across borders (e.g. EC 1999; Copernicus Services 2022). The Open Data Directive is not directly aimed at harmonization but focuses on reusing all public sector information (or governmental data), harmonized in a European data portal. In the EU, the public sector is one of the most data-intensive sectors. Public sector bodies produce, collect, and pay for vast amounts of data. Examples include geographical information, statistics, weather data, data from publicly funded research projects, and digitized books from libraries. *Open* means that public data can be readily and widely accessed and reused, sometimes under non-restrictive conditions (EC 2019). Both of these are important for the EU to provide web services for viewing and downloading data (EC 2019; Minghini et al. 2020). In this context, a monitoring scheme needs to be flexible and able to cope with new questions asked or new rules of data collection and analysis being issued from the policy side.

Tools to help compilation of data across borders

There are a number of different data portals providing the collected and compiled data as services for policymakers and researchers and for use in planning the environment on a regional, national, or international scale. Data painstakingly gathered, analyzed, and reported come back as easily accessible web services with maps and catalogues of various products with compiled information; see chapter 7. Many countries, states, and political unions worldwide have their own data portals with open data to support policy, including the European Copernicus Services catalogue (Programme of the European Union 2022b), the U.S. Government's open data (Data.Gov 2022), and the UK Data Service in the United Kingdom (UK Data Service 2022).

Aside from the already analyzed products, the free and available resource of remotely sensed data is bringing about a revolution in the possibilities for monitoring, as multi-spectral images or radar and sometimes laser point clouds are obtainable from all over the world, downloadable from server halls where they are archived (see list in chapter 7). This is in stark contrast to recent times, when they were expensive enough to hinder most from acquiring satellite data and the analysis of public data rarely was open for all to use.

By contrast, with the diversity of in situ field methods, satellite instruments make exactly the same observations wherever they are in the world, giving results that – after performing pre-processing and calibration – are comparable with common standards of recording and metadata. It is possible to order analysis-ready images, but due to resampling and simplification, they can contain subpixel registration differences, which are hidden to the user, and analysis-ready images are not recommended. In addition, companies are providing "downstream" services, such as the Google Earth Engine, where researchers and companies can access data along with data handling and visualization tools and a library of algorithms, in R-scripts or other types. The library provides help both in choosing algorithms (new ones are constantly being developed, for different purposes) and in applying the same algorithms to data from other providers for change detection (Google Earth Engine 2022).

Collection and assessment – the importance of geographical scale

Collection of monitoring data is a scale-dependent process. This means that monitoring takes place at a chosen set of spatial scales that limit and focus what is observed. Demands for monitoring from government institutions typically also focus on certain spatial scales of particular relevance to policy, planning, or regulation, whereas processes manifesting at other scales may go unrecorded and/or unmanaged.

This means that within monitoring research, scale choices are important. The same research conducted at different scales may lead to several equally valid monitoring results (Henle et al. 2014). For example, a project focusing on tree health in forest landscapes could take, as its point of departure:

- Monitoring individual trees, allowing researchers an opportunity to analyze variations in tree density and health within stands and between individual organisms.
- The primary unit of observation could instead be patches of tree vegetation, which would support analysis focusing, for example, on connectivity between patches, ecosystem size, spatial variability, and distribution of habitats within forests, etc.
- Scaling further out would allow monitoring of whole forest landscapes, supporting research into how forests are distributed, affected by geomorphological conditions, and connected through flows of genetic information, energy, and matter (Forman and Godron 1986; With 2019).

As such, choosing a scale of monitoring is important for how ecosystems can be described, and often this choice depends directly on demands for monitoring. This is because demands for monitoring reflect scales of human decision making, policy, and practice, making them equally scale dependent. Therefore, precise and policy-relevant monitoring research often depends on finding a scale of observation and analysis that makes sense both ecologically (by capturing key variation in ecosystems) and socially (by matching the scale needed for decision making; Hägerstrand 2001). It is not always possible to reconcile social and ecological units of analysis if these do not match each other spatially (Liu et al. 2008). When demands for monitoring require certain scales taken into account to ensure relevance for subsequent formulating and evaluation of policies and regulation, it can be particularly challenging to match the scales.

Example of mismatching scales: farmland monitoring in Denmark

An example of mismatching scales is the case of nitrate pollution stemming from agricultural land use in Denmark, where leaching and emission of nitrate from fertilized agricultural fields to coastal waters, aquifers, and freshwater recipients represent a major ecological problem. Therefore, various kinds of general regulation have been imposed, limiting the use of fertilizer. This kind of blanket regulation has historically been a very efficient way of lowering nitrogen loading to recipient water bodies. However, further lowering of nitrogen emissions from agriculture while maintaining high levels of productivity in agriculture will depend on more targeted, differentiated measures that focus on those fields that emit the most nitrogen due to their location, soil, and hydrology (Dalgaard et al. 2014). This means shifting the scale of observation from whole watersheds (based on water quality measurements made continuously where streams meet the sea) to individual fields (either based on numerous measurements made in drain pipe exits or based on modelled emission estimates tuned to watershed totals). Experiments with this kind of monitoring and associated regulation show promising results, because it is at the scale of individual agricultural fields where emissions of nitrogen originate. There is, in other words, a match between the social and ecological processes interacting at the field/patch scale, making it possible to trace downstream environmental impacts back to particular practices and land units (Christensen et al. 2019).

The demands for monitoring coming from political and administrative interests instead emphasize farm businesses as the preferred unit for regulating nitrogen emissions and request monitoring matching that scale. Farm businesses, though, typically consist of many fields, which change ownership through time and may be located in several watersheds. This makes it exceedingly difficult to progress with monitoring and associated policymaking, because data reflecting ecological realities do not match political demands and ambitions for policy intervention (Christensen et al. 2021).

What is at stake here is generally referred to as a lack of "fit" between social, ecological, and administrative/political systems, including the spatial units and scales that they operate within (Epstein et al. 2015). Monitoring in general depends on researchers being able to fit such demands on monitoring together, creating and maintaining monitoring frameworks that match a range of different and often contradicting demands and concerns regarding scales of observation, analysis, and reporting of results.

When policy falls short: the High Nature Value farmland indicator

This example underlines the importance of having a clearly defined objective and precise methodological framework for the successful development of monitoring, and the results coming from different countries need to be compatible. It also shows why policies need revision.

The European Commission set up a new indicator for regular reporting from Member States in 2007, as a part of the Common Agricultural Policy (CAP) of the EU (EC 2022c). The term *High Nature Value (HNV) farmland* refers to types of farmland that are important for biodiversity, and the indicator was intended to show how much of the total agricultural area had high biodiversity and track changes in the area over time.

Each Member State should mainly use existing monitoring data to do this, but the EU forgot to define exactly what was meant by three classes (Table 3.1). They also did not show how different results, stemming from different monitoring schemes, should be used

Table 3.1 Indicator types with High Nature Value in farmland

Type	Class
1	Farmland with a high proportion of semi-natural vegetation
2	Farmland with a mosaic of habitats and/or land uses
3	Farmland supporting rare species or a high proportion of European or world populations

by the member country, leading to a variety of methodologies for identifying the three classes; see Table 3.2. In addition, when trying to complete missing data to fulfil the obligation, the vague guidelines led to differences in what area types were perceived as important or what scale should be used for the monitoring (van Doorn and Elbersen 2012).

A number of Member States tried to follow the guidelines that did exist (Andersen et al. 2003) but had to back down (e.g. Zomeni et al. 2018). It was quite obvious that mismatching occurred on several levels when Paracchini et al. (2008) made an overlay analysis to estimate the distribution patterns of High Nature Value farmland in Europe based on reported biodiversity data and land cover. The European Commission will remove the indicator from the common indicator set from 2023 and onwards, due to the lack of comparability.

One of the few that did succeed was Germany, where the 16 Federal States of Germany decided to cooperate by developing a common approach to create harmonized data. The various data sources already existing in the German states were analyzed to grasp to what extent they were compatible in relevance to the three indicator classes, including biotope mapping programmes, Farm Accountancy Data Network (FADN; EC 2022a), remote sensing data, grassland monitoring, Natura 2000 monitoring, etc. The result mirrored the internal inconsistencies that other EU members had encountered, such as:

- Extensive heterogeneity existed between the relevant mapping programmes of the individual Federal States concerning spatial and temporal resolution.
- There was a limited extent and selectivity of monitoring programmes, not matching the total utilized agricultural area.
- Resampling of data was often not secured or had extended time gaps that did not fit the mandatory reporting cycle.
- Enormous costs of obtaining the high-resolution remote sensing data needed to focus on detailed biodiversity.
- Natura 2000 monitoring was done within a small random sample, not matching the total utilized agricultural area and not matching all relevant categories.
- Data availability was restricted from the FADN due to data privacy regulations.

The best solution instead proved to be the implementation of a new, strictly targeted and cost-effective monitoring programme, which was in fact more cost-effective than expanding and adapting the whole list of already running programmes. A random sampling approach was developed and biodiversity data on the common agricultural landscape were collected on a national level and in a systematic manner in Germany (Benzler and Fuchs 2016). So, finally this example illustrates how important it is to have iterative developments of both the requirements as well as the methodology used for the actual monitoring.

Table 3.2 Methods used in different EU Member States to identify High Nature Value farmland and effectiveness in identifying the three types of High Nature Value farmland

Method / Case	IRENA/ EEA	Land cover	Soil/ altitude	Management schemes	Farming systems	Species data	EU designated areas	National designated areas	Other habitat identification	Site sampling	IACS/ LPIS	HNV type 1	HNV type 2	HNV type 3
	Methods used (x=used)											*Effectiveness in identification*		
1				x							x	xx		
2							x		x			xx	x	
3							x							xx
4		x					x	x	x		x	xx		xx
5							x	x	x			x	x	xx
6	x											xx	x	
7										x		xx	xx	xx
8		x					x					xx		
9		x	x			x	x	x	x			xx	xx	xx
10							x	x				x		
11							x	x				xx		xx
12		x			x							xx	xx	xx
13							x		x			x		
14		x			x		x		x			xx		xx
15		x		x			x		x			xx	x	xx
16	x												xx	
17	x											xx	x	
18		x										xx		
19		x		x	x	x	x	x			x	xx	x	xx
20		x		x	x	x					x	xx	x	x
21		x			x							x	x	xx
22		x			x							xx	x	xx
23		x			x	x						xx		xx
24		x									x	xx		xx
Total	3	13	1	4	7	4	12	6	7	1	5	18 + 4	4 + 9	13 + 1

Source: Modified from Peppiette (2011).

XX means the indicator was effectively identified; a single X indicates partial identification.

Abbreviations: IRENA/EEA = Indicator reporting on the integration of environmental concerns into agricultural policy, a project run by European Environment Agency; IACS = The Integrated Administration and Control System (which consists of computerized databases of the subsystems); LPIS = Land Parcel Identification System; HNV = High Nature Value (HNV) farmland describes the link between extensive farming systems and their use of semi-natural land and the conservation of high biodiversity in agricultural landscapes).

Conventions and agreements for the environment on land, water, and semi-aquatic areas

Environmental monitoring may be initiated for several different reasons. One important reason is that countries need to start monitoring programmes as a direct or indirect response to international environmental agreements. Parties, usually countries, agree upon conventions as a means for promoting collaboration and developing international law and action towards specific goals that cannot be reached independently. As part of the agreements, some conventions prescribe detailed monitoring and reporting of progress, whereas others are vague and leave it to the parties to decide upon appropriate measures.

The directives are also constantly revised and added to as the Member States report on difficulties or how something was meant to be monitored but the agreed-upon methods just did not function for all states (e.g. Minghini et al. 2020). Some of the important conventions and directives are listed in Table 3.3.

In recent years there has been a change in demands from the EU, especially in the questions asked of monitoring. The enquiries have been changed into whether there are enough resources, habitats, or food supply for species or groups of species, making the analysis trend towards being more ecosystem based (accounts for ecosystem extent and condition). Another new demand is to measure the socioeconomic part of the natural capital accounting, aiming at the supply and monetary value of ecosystem services, in this way including human well-being in the reporting (Maes et al. 2020).

An example of an important driver of monitoring is the Convention on Biological Diversity (CBD), ratified by individual countries but also by the EU as a community. By being a Member State in the EU, each country is bound by EU law and to follow any prescribed efforts to reach the CBD or EU-specific targets, which often involve monitoring and reporting. Another important EU directive related to biodiversity is the Habitats Directive, which, among other things, prescribes recurrent reporting of state and change of listed species and habitat categories (EC 1992).

The Sustainable Development Goals (SDGs) are 17 objectives that Member States of the United Nations (UN) committed to achieve by 2030 and constitute the heart of Agenda 2030, aiming at peace and prosperity for the Earth as well as for people. Each goal has a number of targets that are linked to a set of indicators. Hence, achieving those goals requires measurement and monitoring of more than 240 indicators (United Nations 2015; EC 2022h).

The SDGs cover several dimensions:

- Economics (e.g. no poverty, zero hunger, decent work, and economic growth).
- Social (e.g. quality education, gender equality, reduced inequalities, peace, justice, and strong institutions).
- Environmental (e.g. climate action, clean water and sanitation, affordable and clean energy).

The directives and legislations that govern the waters of Europe went through a fitness check in 2020 to investigate the effects of the directives on water quality and current status and where the EU should alter legislation to make remedies or mitigation activities smoother for the Member States. One of the factors hindering achievement was the difficulty of establishing a governance framework that takes into account the specific conditions in each Member State. Another was the concept of "good status", which

Table 3.3 Conventions and directives relevant to monitoring, regulating the way of collecting information

International conventions	Purpose
Global scale	
Convention on Biological Diversity (CBD)	Conservation and sustainable use of biodiversity, as well as fair sharing of genetic resources. Global coverage. Entered into force in 1993.
Ramsar Convention on Wetlands of International Importance	Protection of wetlands of international importance for waterfowl. Global coverage. Entered into force in 1975.
Paris Agreement	A legally binding international treaty on climate change, reducing greenhouse gases. Adopted by 196 parties at COP 21 in Paris 2015. Entered into force in 2016.
Convention on the Conservation of Migratory Species of Wild Animals	Conservation and sustainable use of migratory species. Global coverage. Entered into force in 1983.
World Heritage Convention (WHC)	Protection of outstanding cultural and natural heritage sites. Global coverage. Entered into force in 1975.
Sustainable Development Goals (SDG)	17 economic, social, and environmental goals encompassing 169 targets to achieve by 2030. Adopted by all UN member states in 2015.
UN Gothenburg Protocol	Reduction in acidification, eutrophication, and ground-level ambient pollutant concentrations and exposure. Agreed on in 1999.
United Nations Framework Convention on Climate Change (UNFCC)	The objective is to stabilize human-induced greenhouse gas concentrations, preventing levels that could be dangerous to the climate system. Ratified by 197 countries, it binds Member States to act in the interests of human safety, even in the face of scientific uncertainty. Entered into force in 1994.
Protocol on Strategic Environmental Assessment	A legally enforced assessment procedure, aiming to ensure consideration of environmental and sustainability aspects to policies and legislation (non-mandatory) as well as plans and programmes. Europe, Caucasus and Central Asia, UN.
Convention on International Trade in Endangered Species of Wild Fauna and Flora (CITES)	An international agreement between governments to ensure that international trade in specimens of wild animals and plants does not threaten the survival of the species. Entered into force in 1975.
EU or community scale	
Bern Convention on the Conservation of European Wildlife and Natural Habitats	Protection of natural habitats and endangered species in Europe and northern Africa. Entered into force in 1982.
European Landscape Convention	Protection, management, and planning of European landscapes. Entered into force in 2004.
Convention on Long-Range Transboundary Air Pollution	Reduction of long-range transboundary air pollution. North America and Europe. Entered into force in 1983.

(Continued)

Table 3.3 (Continued)

International conventions	Purpose
Water Framework Directive (WFD)	Ensure good status for all ground and surface waters (rivers, lakes, transitional waters, and coastal waters). Including river basins and their management, as mitigation of runoff. Adopted in 2000.
EU Marine Strategy Framework Directive 2008/56/EC (MSFD)	To protect the marine environment across Europe. Member States shall take the necessary measures to achieve or maintain good environmental status in the marine environment by the year 2020 at the latest. Adopted in 2008 (EEA 2008)
The Convention for the Protection of the Marine Environment of the North-East Atlantic (OSPAR)	To protect and manage the North-East Atlantic through strategies that are legally binding on the contracting parties, recommendations, and other agreements. Entered into force in 1998. The North-East Atlantic Environment Strategy (NEAES) 2030. Adopted in 2021 (OSPAR Commission 2021).
The Convention on the Protection of the Marine Environment of the Baltic Sea Area (Helsinki Convention, HELCOM)	Covers the whole of the Baltic Sea area, including inland waters as well as the water of the sea itself and the seabed. Measures are also taken in the whole catchment area of the Baltic Sea to reduce land-based pollution. Entered into force in 1980 with updates in 2000–2013. HELCOM (2014).
Environmental Quality Standards Directive (EQSD)	Establish environmental quality standards in the field of water policy. Entered into force in 2008.
Groundwater Directive (GWD)	Have been integrated into the Water Framework Directive.
Floods Directive (FD)	To establish a framework for the assessment and management of flood risks. Reduce negative consequences on human health, economy, the environment, and cultural heritage. Entered into force in 2007.
Habitats Directive (or: Council Directive 92/43/EEC on the Conservation of natural habitats and of wild fauna and flora)	Specified areas for conservation, leading up the Natura 2000 Network to protect species and habitats. Consists of 24 articles of legislation with which all Member States must comply. Article 17 sets the terms and standards for reporting on both the habitats and species. Adopted 1992 in response to the Bern Convention.
Birds Directive (or: Council Directive 2009/147/EC on the conservation of wild birds)	The Birds Directive aims to protect all of the 500 wild bird species naturally occurring in the European Union. Adopted in 1979, amended in 2009.
The INSPIRE Directive	An infrastructure for spatial information in Europe to support community environmental policies and policies or activities that may have an impact on the environment. Entered into force in 2007.
Open Data Directive (Latest version: Directive (EU) 2019/1024)	Focussing on allowing reuse of public sector data (or government data), with open access, because they are publicly funded. Also for commercial use. Adopted in 2019.
Environmental impact assessment (EIA Directive)	All projects listed in Annex I are considered as having significant effects on the environment and require an EIA. Entered into force in 1985.

(Continued)

Table 3.3 (Continued)

International conventions	Purpose
National Emissions reduction Commitments (NEC) Directive	National emission reduction commitments for five important air pollutants. Replacing the older National Emission Ceilings Directive. Entered into force in 2016.
Regional Scale	
Alpine Convention	Protection and sustainable development of the Alps (covers 13 topics; e.g. biodiversity, water, climate, spatial planning, and green economy). Signed by eight alpine countries and the EU. Entered into force in 1995.
Strategic environmental assessment (SEA Directive)	A legally enforced assessment procedure aiming to ensure consideration on environmental and sustainability aspects implemented in plans and programmes. Regional to local scales. Was adopted in 2001.

The scale can be on several levels: global; a group of countries acting in unison, like the United Nations or the European Union; or smaller groups like the countries around the Alps or the UK. Most countries belong to some of these groups.

depends on measures to mitigate current pressures but often forgets the restoration measures needed to address pressures from the past. Finally, good status of water bodies is highly interconnected to a number of other pieces of EU legislation, such as the Nitrates Directive and the Urban Waste Water Treatment Directive, as well as better integration of water objectives in other policy areas such as agriculture, energy or transport.

There are some directives aimed specifically at water bodies, including:

- The Water Framework Directive (WFD): The policy key in the Water Framework Directive is simply that each country should work to get polluted waters clean again and ensure that clean waters are kept clean, whether marine or freshwater bodies (European Environment Agency [EEA] 2008; EC 2022i).
- The Environmental Quality Standards Directive (EQSD): This directive is a component of the WFD and regulates the monitoring and reporting of priority substances of pollution (currently 45 substances, of which 12 were added in 2013). Some of these are also hazardous substances, being either ubiquitously persistent, bioaccumulative, or toxic substances (EC 2013).
- The Groundwater Directive (GWD): The components of the Water Framework Directive dealing with groundwater cover a number of different steps for achieving good quantitative and chemical status of groundwater. The latest update was made in 2015 and includes the designation (with a view to specify such bodies that run the risk of not achieving the WFD environmental objectives) and reporting of groundwater bodies from each Member State. The EU also wants monitoring networks, established registers of protected areas, and management plans for such things as bathing areas, river basins, and bodies used for drinking water (EC 2006).

- The Floods Directive (FD): This directive deals with the assessment and management of flood risks and requires Member States to assess all watercourses and coastlines for risks of flooding and to map humans and assets at risk and to take measures to reduce those risks (EC 2007).

Specific directives also exist directly relevant to wetlands, one of the first nature conservation issues for European and international policy consideration under the 1971 Ramsar Convention on the conservation and wise use of wetlands and their resources (Ramsar 1971). Many of the objectives, targets, and actions in the following CBD and the Agenda 2030 (especially SDG 6), are directly relevant to the conservation and wise use of wetlands.

However, the Habitats and Birds directives and the WFD are the main pieces of legislation ensuring the protection of Europe's wetlands. The Natura 2000 network of protected sites and the integration of wetlands into future river basin management planning (under the WFD) provide ways to ensure their future conservation and sustainable use (EC 1992, 2022i).

Monitoring for fulfilment of policies, some examples

The Natura 2000 network

Most data, aside from curiosity-driven surveys for research, are collected to be analyzed and reported to some authority, which in turn can collate several data sources and report further on. One example of such monitoring is the national report from each Member State, due every sixth year, reporting on the status and change of a large series of protected areas, in a framework called the Natura 2000 network of protected areas by the European Commission, organized under the Habitats Directive. The network was set up to fulfil the obligation as one of the contracting parties to the Bern Convention on the Conservation of European Wildlife and Natural Habitats as the contribution to a higher level, the Pan-European Emerald Network of the Bern Convention (Council of Europe 1979; EC Directorate-General for Environment 2008).

The network celebrated 30 years in 2022 and now constitutes the largest coordinated network of protected areas in the world, covering about 18% of the EU's land area and more than 8% of its marine territory (EEA 2022; EC 2022g).

The policy behind the network was to offer a long-term haven to Europe's most valuable and threatened species and habitats, regulated in two directives, the Birds and Habitat directives. These directives have allowed the creation of a representative system of legally protected areas throughout the EU, called Sites of Community Importance (SCI; EC 1992). On the dedicated web site of the EEA, more information and maps and data are provided for the conservation of the 233 habitat types listed in Annex I of the Directive and the 900-plus species listed in Annex II (EEA 2022). Figure 3.1 shows such a site of community importance, a grassland with long-term management of grazing in southern Sweden.

Green spaces and human well-being

An example of complying with directives about human well-being and monitoring that integrates the social component of humans and how they integrate with their environment is the investigation into the health benefits of access to green areas;

Figure 3.1 An example of a Natura 2000 Site of Community Importance in the county Scania, southern Sweden. A grassland with long-term management of grazing by the seaside, with high biodiversity in plants and insects, providing good conditions for many birds.

Credit: Photo by Anna Allard.

see Figure 3.2. More than half of the population worldwide live in cities, a number that the UN (2019) expects to increase to 64% by the year 2050. This poses problems, for both planners of designers of urban areas and in the increasing demand for housing, because this massive urbanization is associated with severe environmental problems that negatively affect human well-being (Reyes-Riveros et al. 2021). Academically, the concept of human well-being is multi-faceted and can easily fit into social sciences, philosophy, or biodiversity and on the whole is multidimensional, including various aspects of life, upon which the measurements are based, aiming to study the links between ecosystem change and human well-being. The definition of what is perceived as good and beneficial to well-being is not always clear and may well differ depending on who you are and where you were born (did you grow up by the sea, in the highlands, on a farm, or in a city?). It is important to explore the different individual ways of looking at green spaces (Wood et al. 2018; Jabbar et al. 2021; Reyes-Riveros et al. 2021).

In a review of 153 research articles, Reyes-Riveros et al. (2021) found that the number of green spaces available and their percentage of vegetation cover and size improved human well-being in all aspects, especially health. The naturalness of the landscape and biodiversity were the characteristics most valued in the articles, and especially improved mental health and social relations.

Figure 3.2 A small park in a stone city provides the opportunity for a moment of tranquillity with the city noise replaced by birdsong (Visby, Sweden).

Credit: Photo by Anna Allard.

Monitoring oceans: the ICES example

A number of collaborations exist on the monitoring of the oceans, some for weather models or litter analysis and others for monitoring the resource of fish; the International Council for the Exploration of the Sea (ICES) is one example. EU fisheries management is governed by the Common Fisheries Policy (CFP), which is based on the need to ensure environmentally sustainable use of marine biological resources and long-term profitability for the fisheries sector. A priority area for the CFP is to safeguard fishery resources by adapting fishing capacity to fishing opportunities. To achieve sustainable exploitation, fish stocks are managed with the principle of maximum sustainable yield (MSY). Annually, the EU allocates fishing opportunities for most commercial species, expressed as total allowable catch (TAC). Based on scientific assessments by ICES and EU's Scientific, Technical and Economic Committee for Fisheries (STECF), the Commission presents its proposals for total allowable catch to the Council at the end of each year.

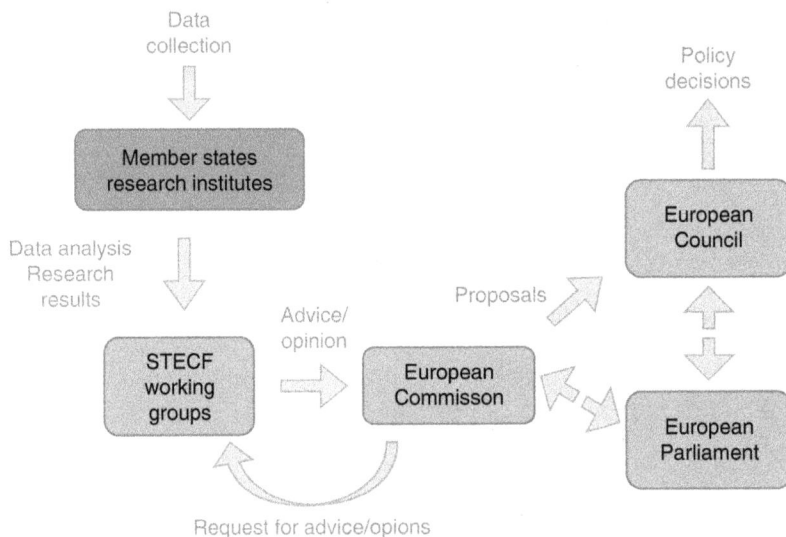

Figure 3.3 A simplified overview of STECF's role in the EU's marine policy.

Source: EC (2016).

The marine planning cycle for monitoring of aquatic resources includes several steps and authorities (EC 2016). Following the EU's data collection framework (DCF), Member States annually collect fish and fisheries data in support of CFP. The Member States data collection follows a national work plan, which is coordinated in regional coordination groups (RCGs) that can also suggest regional work plans. The data collected from the Member States are used by ICES for further analyses, and some of the data are uploaded to the EU's joint research centre (JRC) databases to fulfil the needs of the European Commission. After analysis and quality assurance, STECF working groups provide scientific opinions and recommendations in support of the Commission's policymaking (see Figure 3.3).

ICES provides advice and services on marine ecosystems for several different authorities and organizations. Among those, ICES has standing requests for recurrent advice from the European Commission. The advisory process for fishing comprises four major steps (ICES 2019; Figure 3.4). First, the request is formulated in an iterative dialogue with the requester's need for advice. Secondly, in accordance with ICES data policy (ICES 2016) and codes of conducts (ICES 2018), expert groups bring forth the best available knowledge. Thirdly, after an independent review of the data and methods used, a draft of the advice is produced and then the advice is finally approved in consensus by the ICES Advisory Committee (ACOM) as the final step. To ensure the relevance of the methods and data series used by the expert groups for ongoing advice, a benchmark process is conducted addressing the requests. This process is subjected to a peer review process. To keep the process transparent, stakeholders with observer status have the possibility to attend the workshops and drafting groups and receive approval of the ACOM.

The advice on fishing opportunities integrates ecosystem-based management (EBM) with MSY, an objective aiming to achieve highest yield over time. The basis for the advice depends on the request for the advice as well as the information and knowledge

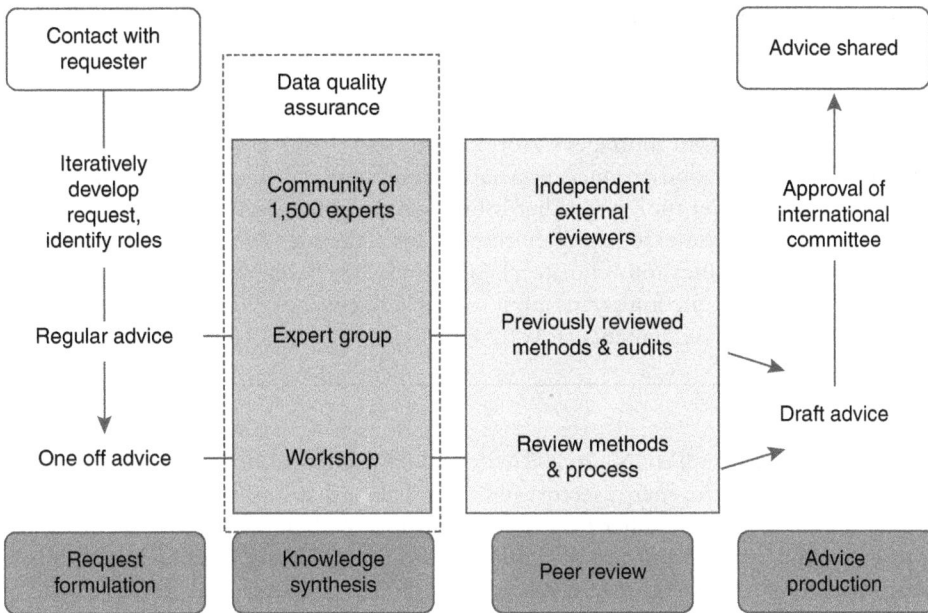

Figure 3.4 Simplified description of the ICES advisory process.
Source: ICES (2019).

available for the fish stock/stocks in question. For fish and shellfish stocks where sufficient data are available, advice based on so-called full analytical assessments and forecasts is provided. Data used are estimates on fishing mortality (F) – that is, the number of fish taken by the fisheries – and spawning stock biomass (SSB), which is the biomass of fish that would reproduce. To ensure that the spawning stock does not become too low, a number of precautionary reference points are used, both for mortality and for the biomass.

For fish stocks where data only include time series and indices on trends, advice based on proxies for MSY could be used. For other fish stocks with less data, or if knowledge of the fish stock in any sense is insufficient, ICES employs a precautionary approach (UN 1995).

Monitoring forest, assessments of stock, and estimations of greenhouse gases are global issues

Maintaining and increasing forest is a worldwide issue and hence also policy concern (see Text box 3.1). To provide answers to follow trends in forests worldwide, monitoring is important. When trying to compile data from many different countries, with various capabilities, it is also important that the rules and the framework for how to perform the monitoring be clear and concise. Forest ecosystems play a critical role in the global carbon cycle and help stabilize the climate; they regulate ecosystems, protect biodiversity, support human livelihoods, and supply goods and services that can drive sustainable growth (e.g. Mitchell et al. 2017; Food and Agriculture Organization [FAO] 2020a, 2022; Sleeter et al. 2022). The role of forest in climate change is twofold, acting as both a cause of greenhouse gas emissions and a solution for addressing it. According to the International Union for

---

Text box 3.1:   The Global Forest Resources Assessment

The need to know the status of the forest resource and how it is changing was recognized in 1948 (after World War II), when the FAO stated that "the whole world is suffering from shortages of forest products" (FAO 1948). At that time, a future shortage of wood products preventing industrial and economic growth was the major issue. Over the years, other issues linked to the role and sustainability of the forests have come to the forefront – for example, wood fuel, poverty, biodiversity, desertification, climate change, and sustainable development – and the FRA data form one important input to global discussions, climate negotiations, and UN-promoted agreements (FAO 2020b).

Conservation of Nature (IUCN), in 2021, around 25% of global emissions were from the land sector (next after the energy sector) and about half of that was from deforestation and forest degradation. On the other hand, forests are also the best solution to the problem, because an estimated one-third of the CO_2 released from burning fossil fuels is absorbed by forests every year (UNESCO, WRI, and IUCN 2021; Sleeter et al. 2022).

In most tropical areas, deforestation is commonly driven by changes in land use and agriculture, and the underlying causes relate to existing policies and strategies combined with demographic, economic, market, institutional, and technological issues. In many low-income countries, especially those with a centrally planned economy, governments assess and monitor agricultural production and other social, economic, and landscape-related parameters for purposes of governance; for example, as a basis for five-year plans and development strategies. In some countries, like Vietnam, strategies and policies firmly direct the resource management at central as well as local levels, which is further discussed in chapter 10. Those data are usually based on questionnaires and not sufficiently consistent and unbiased to reflect long-term change processes and trends. To satisfy the need for strategic landscape data, many countries develop continuous national forest assessment and monitoring systems, or NFAs (Sandewall et al. 2001; FAO 2022). NFAs are usually highly sophisticated, continuous and designed to cover a country with a representative sample over time. They involve significant costs and require a permanent organization. Although they generate high-quality data, local perspectives are not easily integrated. Monitoring can, however, be designed in many different ways, and an example of a participatory approach for setting up monitoring landscapes in a policy context is presented by Sandewall and Gebrehiwot (2015).

The FAO runs a support system for developing national forest monitoring systems (FAO 2022a). The support includes a number of tools for in situ collection of forest data, to assess greenhouse gas emissions and help to set up satellite land monitoring systems. Examples are deforestation or afforestation, where countries get help to identify and collect data for their relevant land use and its change (LULUC); forestry; and REDD+ activities (FAO 2022). REDD+ is an international framework whose name stands for Reducing Emissions from Deforestation and Forest Degradation in Developing Countries (UN, Framework Convention on Climate Change 2022).

At the global scale, the FAO coordinates a program for standardizing the forest monitoring reports from different countries. The FAO also compiles and publishes global assessments periodically to provide a consistent approach to describing the world's forests and their changes over time. Those Global Forest Resources Assessments (FRA) are based on country reports prepared by some 340 nominated national correspondents. In addition, FAO promotes the development of national forest monitoring systems (FAO 2020a; Ramirez and Morales 2021; EC 2022e) in low-income countries through local capacity building.

In addition to forests and peatlands, permanent grasslands (non-ploughed grassy fields) naturally hold large stocks of carbon, preventing its escape into the atmosphere. Assuming we know the carbon stock, we also need to follow changes in both land cover and land use to estimate how much of the carbon dioxide currently kept in the ground could potentially be released, should the land use change (e.g. Mitchell et al. 2017). Monitoring for this question reports to the policy sector of land use, land use change, and forestry, including our use of soils, trees, plants, biomass, and timber, to provide information on how we are meeting our goals (Intergovernmental Panel on Climate Change [IPCC] 2006, 2014; EC 2022e). Copernicus Services has a few products in the catalogue, among which is the Gothenburg Scenario for reduction of greenhouse emission (Programme of the European Union 2022a).

Integrated farm-level surveys

Though agriculture has the potential to provide numerous ecosystem services, the global focus is mainly on production of food, feed, and energy, providing a limited supply of other valuable ecosystem services such as carbon sequestration, preservation of habitats and biodiversity, infectious disease mediation, water quality regulation, and water flow regulation (Foley et al. 2005). Moreover, agriculture remains a significant source of environmental degradation, contributing to nearly 25% of global greenhouse gas emissions, half from farming activities and the rest from land use change (IPCC 2014). Agriculture is also one of the main sources of water pollution, causing pesticide and nutrient con-tamination; it is also one of the main drivers of biodiversity loss due to deforestation and intensive agriculture practices and the main source of ammonia emissions.

Unsurprisingly, the focus of most farm surveys is on the economic performance of the sector. For instance, the EU has the farm accountancy data network (FADN), which is a harmonized farm-level survey used to monitor farm income and business activities and to evaluate the CAP (EC 2018). The survey targets agricultural holdings, surpassing a minimum size threshold, thereby capturing the largest producers and agricultural land users. It collects information on income, labour, assets, input costs, output value, and socioeconomic characteristics of farms.

However, both the major role that agriculture plays in environmental degradation and its potential to provide ecosystem services have prompted some governments to promote initiatives to improve the environmental performance of the sector and propose farm surveys that focus on multiple dimensions (economic, social, and environmental). Recently, the EU has adopted the Farm to Fork strategy and the Biodiversity strategies. Farm to Fork aims at making the food system fair, healthy, and environmentally friendly (EC 2022d), and the Biodiversity strategy offers a plan to protect nature and reverse ecosystem degradation (EC 2022b).

In accordance with these initiatives, the EU is proposing to transform the FADN into the Farm Sustainability Data Network (FSDN) with the purpose of widening the scope of the farm surveys to cover environmental and social factors and to provide a comprehensive overview of farming activities in the EU. Integrated data collection on sustainability and economic indicators at the farm level can be done as high-precision hyperspectral remote sensing (see chapter 8) or by combined measurements and has both advantages and disadvantages (see Table 3.5). The advantages are that it facilitates evaluating the jointness and trade-offs between economic and environmental performance indicators, it permits combined policy evaluation of environmental and economic outcomes, and it can be done using existing data collection processes. For example, surveys of financial flows (e.g. input costs and sales) can be extended to collect data on physical flows related to inputs (e.g. fertilizer, pesticides, active substances, energy consumption; Vrolijk et al. 2016). Nevertheless, the integrated collection of data also comes at a cost. It increases the complexity of data collection and may require adjustment of the observation unit (e.g. commercial farms vs. small farms) and re-adjustments of sampling design, standard collection systems, and working processes.

However, providing this detailed and range of data is not always easy to do. Ireland's Agriculture and Food Development Authority (Teagasc) is one of the few government agencies in EU Member States that collect integrated information on economic, social, and environmental performance of farms. This integrated approach permits a joint economic, social, and environmental assessment of the Irish farm sector. Teagasc publishes The Teagasc National Farm Survey Sustainability Report annually, based on information from a representative sample of 900 farms in the National Farm Survey, summarizing the performance of dairy, cattle, sheep, and tillage farms. Table 3.4 shows the economic, social, and environmental indicators collected in the 2020 Sustainability Report (Buckley and Donnellan 2021). The integrated assessment of the economic, social, and environmental performance of farms has proven to be valuable for policymaking in Ireland.

Key messages

- This chapter has discussed the synergies and mutual developments between monitoring and policies and legislation. We stress the importance of creating a monitoring scheme that can be flexible enough to incorporate new demands that arise from new concerns of potential environmental problems in the ever-changing world around us.
- Sometimes it is better to simply give up trying to accommodate new demands in a cluster of older schemes and instead start another targeted scheme, which avoids the expense and labour-intensive effort to force the old ones into compliance. However, changing the whole setup of classes or variables measured also means that an older time series will be discontinued and a mismatch in the time depth is created instead.
- We also touch upon cases where the required geographical scale is not matched by the scale of monitoring or when the questions and categories framed by policy are ambiguous or fuzzy, making any questions hard to answer.
- New data covering the entire planet have become free to users (until recently, they were expensive enough to hinder most from acquiring satellite data, whether images or laser or radar data).
- The ready-made layers of analyzed data provided on many data portals around the globe constitute end products made up of a large number of different national monitoring schemes and national reporting. Examples highlight a few of the monitoring schemes being made in response to some demand. Some of the relevant directives, policies, and legislation are being addressed on global, community, and national levels.
- Lastly, we discuss a case of the most local type, where the single human observer is the data provider. Table 3.5 summarizes some advantages and disadvantages with this type of locally integrated data collection.

Table 3.4 Indicators for measuring the economic, social, and environmental performance of farms in Ireland

Dimension	Indicator	Measure	Unit
Economic	Economic return to land	Gross output per hectare	€/hectare
	Profitability	Market-based gross margin per hectare	€/hectare
	Productivity of labour	Family farm income per unpaid labour unit	€/unpaid labour unit
	Economic viability	Economic viability of farm business	1 = viable, 0 = not viable
	Market orientation	Output derived from market rather than subsidies	%
	Family farm income	Family farm income per hectare	€/hectare
Environmental	Agricultural GHG emissions per farm	Absolute GHG emissions per farm	Tonnes CO_2 equivalent/farm
	Agricultural GHG emissions per hectare	Absolute GHG emissions per hectare	Tonnes CO_2 equivalent/hectare
	Agricultural GHG emissions per kg of output	GHG emissions efficiency	kg CO_2 equivalent/kg output AND kg CO_2/€ output
	Energy GHG emissions per farm	Farm GHG energy use efficiency	kg CO_2 equivalent/kg output
	Energy emissions per kg of output	Energy GHG emissions efficiency	kg CO_2 equivalent/kg output AND kg CO_2/€ output
	NH_3 emissions per farm	Absolute NH_3 emissions per farm	Tonnes NH_3 equivalent/farm
	NH_3 emissions per hectare	Absolute NH_3 emissions per hectare	Tonnes NH_3 equivalent/hectare
	NH_3 emissions per kg of output	NH_3 emissions efficiency	kg NH_3 equivalent/kg output AND kg NH_3/€ output
	N balance	N transfer risk	kg N surplus/ha^{-1}
	N use efficiency	N retention efficiency	% N outputs/N inputs
	N surplus per kg of output	N emissions efficiency	kg N surplus/kg output
	P balance	P transfer risk	kg P surplus/ha^{-1}
	P use efficiency	P retention efficiency	% P outputs/P inputs
Social	Household vulnerability	Farm business is not viable and no off-farm employment	Binary variable: 1 = vulnerable
	Agricultural education	Formal agricultural training received	Binary variable, 1 = agricultural training received
	Isolation risk	Farmer lives alone	Binary variable, 1 = isolated
	High age profile	Farmer is over 60 years old and no members of household under 45	Binary variable: 1 = high age
	Hours worked on farm	Farm workload of farmer	Hours worked on the farm
	Total hours worked	Workload of farmer	Total hours worked on- and off-farm

Source: Buckley and Donnellan (2021)

Abbreviations: GHG = greenhouse gases; CO_2 = carbon dioxide; NH_3 = ammonia; N = nitrogen; P = phosphorus.

Table 3.5 Advantages and disadvantages with integrated collection of environmental and economic data

Advantages	Disadvantages
Jointness and trade-off between objectives/ indicators	Increased complexity of data collection
Allows for integrated policy analysis	Possible need to reconsider field of observation
Use of existing procedures and quality mechanisms	Wide variety of objectives complicates sample design
	Need to re-adjust current systems and working processes

Source: Vrolijk et al. (2016).

Study questions

1   What are the main advantages of directives and the common goals?
2   Why can the organized monitoring of oceans act as a role model to other areas of monitoring?
3   What are the 17 Sustainable Development Goals? What dimensions do they cover and why?

Further reading

Laws and legislation, with explanations and most recent updates, can be found at the web pages of the European Commission and the EU as well as the European Environment Agency. We recommend always reading up on the latest news, because new policies or health checks on the outcomes from current policies and international ratifications of agreements will, from time to time, be altered or added to. By knowing what lies behind the laws or policies, it is easier to understand the questions we need monitoring to answer.

References

Andersen, E., Baldock, D., Bennet, H., Beaufoy, G., Bignal, E., Brower, F., Elbersen, B., Eiden, G., Godeschalk, F., Jones, G., et al. (2003) *Developing a High Nature Value Farming Area Indicator.* Consultancy report to the EEA. Copenhagen: European Environment Agency.

Benzler, A. and Fuchs, D. (2016) Mapping HNV farmland in Germany 2009 to 2015, Good *Practice Workshop "HNV Farming in RDP"*, Bonn, Germany, June 7–8, 2016. http://enrd.ec.europa.eu/sites/default/files/gpw-02_4-1_germany_benzler.pdf (Accessed June 15, 2022).

Buckley, C. and Donnellan, T. (2021) *Teagasc National Farm Survey 2020 Sustainability Report.* Teagasc, Ireland: Agricultural Economics and Farm Surveys Department, Rural Economy and Development Programme.

Christensen, A.A., Andersen, P.S., Kjeldsen, C., Graversgaard, M., Andersen, E., Piil, K., Dalgaard, T., Olesen, J.E. and Vejre, H. (2021) Achieving sustainable nitrogen management in mixed farming landscapes based on collaborative planning, *Sustainability* 13, 2140. 10.3390/su13042140

Christensen, A.A., Andersen, P.S., Piil, K., Andersen, E., Vejre, H. and Graversgaard, M. (2019) Pursuing implementation solutions for targeted nitrogen management in agriculture – a novel approach to synthesize knowledge and facilitate sustainable decision making based on collaborative landscape modelling, *Journal of Environmental Management* 246, 679–686. 10.1016/j.jenvman.2019.05.107

Copernicus Services. (2022) CORINE land cover, https://land.copernicus.eu/pan-european/corine-land-cover (Accessed February 12, 2022).

Council of Europe. (1979) Convention on the Conservation of European Wildlife and Natural Habitats (ETS No. 104), https://www.coe.int/en/web/conventions/full-list?module=treaty-detail&treatynum= 104 (Accessed November 2, 2022).

Dalgaard, T., Hansen, B., Hasler, B., Hertel, O., Hutchings, N.J., Jacobsen, B.H., Jensen, L.S., Kronvang, B., Olesen, J.E., Schjørring, J.K., et al. (2014) Policies for agricultural nitrogen management – trends, challenges and prospects for improved efficiency in Denmark, *Environmental Research Letters* 9, 115002. https://iopscience.iop.org/article/10.1088/1748-9326/9/11/115002/meta

Data.Gov. (2022) The home of the U.S. Government's open data, https://data.gov/ (Accessed November 2, 2022).

Epstein, G., Pittman, J., Alexander, S.M., Berdej, S., Dyck, T., Kreitmair, U., Rathwell, K.J., Villamayor-Tomas, S., Vogt, J. and Armitage, D. (2015) Institutional fit and the sustainability of social–ecological systems, *Current Opinion in Environmental Sustainability* 14, 34–40. 10.1016/j.cosust.2015.03.005

European Commission. (1992) *Habitats Directive*, Council Directive 92/43/EEC of 21 May 1992. https://ec.europa.eu/environment/nature/legislation/habitatsdirective/index_en.htm (Accessed February 12, 2022).

European Commission. (1999) *Land Cover and Land Use Information Systems for the European Union Policy Needs, Theme 5 Agriculture and Fisheries.* Luxembourg: Eurostat. http://aei.pitt.edu/85338/1/1999.pdf

European Commission. (2006) *The Groundwater Directive 2006/118/EC*, https://ec.europa.eu/environment/water/water-framework/groundwater/framework.htm (Accessed February 12, 2022).

European Commission. (2007) *The Floods Directive 2007/60/EC*, https://ec.europa.eu/environment/water/flood_risk/index.htm (Accessed February 12, 2022).

European Commission. (2013) *Environmental Quality Standards Directive 2013/39/EU*, https://ec.europa.eu/environment/water/water-dangersub/pri_substances.htm (Accessed May 10, 2022).

European Commission. (2016) *Commission Decision of 25 February 2016 Setting Up a Scientific, Technical and Economic Committee for Fisheries*, 2016/C 74/05. https://eur-lex.europa.eu/legal-content/en/TXT/?uri=CELEX:32016D0226(01)

European Commission. (2018) Agricultural and farm economics, https://ec.europa.eu/info/sites/default/files/food-farming-fisheries/farming/documents/eu-farm-econ-overview-2018_en.pdf (Accessed February 19, 2022).

European Commission. (2019) *Open Data Directive*, https://digital-strategy.ec.europa.eu/en/policies/open-data (Accessed May 19, 2022).

European Commission. (2022a) Analyses and briefs on the economy of farms and rural areas based on the Farm Accountancy Data Network (FADN), overview, https://ec.europa.eu/info/food-farming-fisheries/farming/facts-and-figures/performance-agricultural-policy/studies-and-reports/economic-analyses-and-briefs/agricultural-and-farm-economics_en#overview (Accessed April 25, 2022).

European Commission. (2022b) Biodiversity strategy for 2030, https://ec.europa.eu/environment/strategy/biodiversity-strategy-2030_en (Accessed June 3, 2022).

European Commission. (2022c) *Common Agricultural Policies (CAP)*, https://ec.europa.eu/info/food-farming-fisheries/key-policies/common-agricultural-policy_en (Accessed May 22, 2022).

European Commission. (2022d) Farm to fork strategy, https://ec.europa.eu/food/horizontal-topics/farm-fork-strategy_en (Accessed May 27, 2022).

European Commission. (2022e) Forest and agriculture in climate action, https://ec.europa.eu/clima/eu-action/forests-and-agriculture_en (Accessed May 19, 2022).

European Commission. (2022f) Land use, land change and forestry, LULCF, https://ec.europa.eu/clima/eu-action/forests-and-agriculture/land-use-and-forestry-regulation-2021-2030_en (Accessed May 18, 2022).

European Commission. (2022g) Natura 2000 network, https://ec.europa.eu/environment/nature/natura2000/index_en.htm (Accessed April 26, 2022).

European Commission. (2022h) Nature and biodiversity law, https://ec.europa.eu/environment/nature/legislation/index_en.htm (Accessed May 19, 2022).

European Commission. (2022i) *Water Framework Directive,* https://ec.europa.eu/environment/water/water-framework/info/intro_en.htm (Accessed February 12, 2022).

European Commission, Directorate-General for Environment. (2008) *Natura 2000: Protecting Europe's Biodiversity.* European Commission. https://data.europa.eu/doi/10.2779/45963

European Environment Agency. (2008) *The Marine Strategy Framework Directive,* http://eur-lex.europa.eu/LexUriServ/LexUriServ.do?uri=CELEX:32008L0056:en:NOT (Accessed February 12, 2022).

European Environment Agency. (2022) The Natura 2000 protected areas network, https://www.eea.europa.eu/themes/biodiversity/natura-2000 (Accessed February 12, 2022).

European Union. (2022) *The INSPIRE Directive,* https://inspire.ec.europa.eu/ (Accessed February 20, 2022).

Foley, J.A., DeFries, R., Asner, G.P., Barford, C., Bonan, G., Carpenter, S.R., Chapin, F.S., Coe, M.T., Daily, G.C., Gibbs, H.K., et al. (2005) Global consequences of land use, *Science* 309(5734), 570–574. doi: 10.1126/science.1111772.

Food and Agriculture Organization. (1948) Forest resources of the world, *Unasylva* 2(4). www.fao.org/docrep/x5345e/x5345e00.htm

Food and Agriculture Organization. (2020a) *Global Forest Resources Assessment 2020: Main Report.* Rome: FAO. 10.4060/ca9825en

Food and Agriculture Organization. (2020b) *Global Forest Resources Assessment, Guidelines and Specifications FRA 2020.* Rome: FAO. https://www.fao.org/3/I8699EN/i8699en.pdf

Food and Agriculture Organization. (2022) National forest monitoring systems, https://www.fao.org/redd/areas-of-work/national-forest-monitoring-system/en/ (Accessed May 15, 2022).

Forman, R.T.T. and Godron, M. (1986) *Landscape Ecology.* New York: John Wiley & Sons.

Google Earth Engine. (2022) A planetary-scale platform for Earth science data & analysis, https://earthengine.google.com/ (Accessed June 4, 2022).

HELCOM. (2014) *Convention on the Protection of the Marine Environment of the Baltic Sea Area, 1992 (Helsinki Convention),* https://helcom.fi/media/publishingimages/Helsinki-Convention_July-2014.pdf

Henle, K., Potts, S., Kunin, W., Matsinos, Y., Simila, J., Pantis, J., Grobelnik, V., Penev, L. and Settele, J. (2014) *Scaling in Ecology and Biodiversity Conservation.* Advanced Books. 10.3897/ab.e1169

Hägerstrand, T. (2001) A look at the political geography of environmental management, in Buttimer, A. (ed.) *Sustainable Landscapes and Lifeways – Scale and Appropriateness.* Cork, Ireland: Cork University Press, pp. 35–58.

Intergovernmental Panel on Climate Change. (2006) *2006 IPCC Guidelines for National Greenhouse Gas Inventories,* prepared by the National Greenhouse Gas Inventories Programme, Eggleston, H.S., Buendia, L., Miwa, K., Ngara, T. and Tanabe, K (eds). Japan: IGES.

Intergovernmental Panel on Climate Change. (2014) *Climate Change 2014: Synthesis Report. Contribution of Working Groups I, II and III to the Fifth Assessment Report of the Intergovernmental Panel on Climate Change* [Core Writing Team, R.K. Pachauri and L.A. Meyer (eds)]. Geneva, Switzerland: IPCC.

International Council for the Exploration of the Sea. (2016) ICES data policy, 22 December 2016, https://www.ices.dk/data/guidelines-and-policy/Pages/ICES-data-policy.aspx (Accessed February 14, 2022).

International Council for the Exploration of the Sea. (2018) ICES code of conduct, https://www.ices.dk/about-ICES/how-we-work/Pages/Code-of-conduct.aspx (Accessed February 14, 2022).

International Council for the Exploration of the Sea. (2019) Advice basis, in *Report of the ICES Advisory Committee,* ICES Advice 2019, section 1.2, 10.17895/ices.advice.5757 (Accessed February 14, 2022).

Jabbar, M., Yusoff, M.M. and Shafie, A. (2021) Assessing the role of urban green spaces for human well-being: a systematic review. *GeoJournal* 87, 4405–4423. 10.1007/s10708-021-10474-7

Liu, Y., Gupta, H., Springer, E. and Wagener, T. (2008) Linking science with environmental decision making: experiences from an integrated modeling approach to supporting sustainable water resources management, *Environmental Modelling & Software* 23, 846–858. 10.1016/j.envsoft.2007.10.007

Maes, J., Teller, A., Erhard, M., Conde, S., Vallecillo Rodriguez, S., Barredo Cano, J.I., Paracchini, M., Malak, A.D., Trombetti, M., Vigiak, O., et al. (2020) *Mapping and Assessment of Ecosystems and their Services: An EU Ecosystem Assessment.* EUR 30161 EN. Luxembourg: Publications Office of the European Union. doi: 10.2760/757183.

Maes, J., Teller, A., Erhard, M., Grizzetti, B., Barredo, J.I., Paracchini, M.L., Condé, S., Somma, F., Orgiazzi, A., Jones, A., et al. (2018) *Mapping and Assessment of Ecosystems and Their Services: An Analytical Framework for Ecosystem Condition.* Luxembourg: Publications Office of the European Union. https://publications.jrc.ec.europa.eu/repository/handle/JRC120383

Minghini, M., Cetl, V., Ziemba, L.W., Tomas, R., Francioli, D., Artasensi, D., Epure, E. and Vinci, F. (2020) *Establishing a New Baseline for Monitoring the Status of EU Spatial Data Infrastructure.* EUR 30513 EN. Luxembourg: Publications Office of the European Union. https://publications.jrc.ec.europa.eu/repository/handle/JRC122351

Mitchell, A.L., Rosenqvist, A. and Mora, B. (2017) Current remote sensing approaches to monitoring forest degradation in support of countries measurement, reporting and verification (MRV) systems for REDD+, *Carbon Balance Management* 12, 9. 10.1186/s13021-017-0078-9

OSPAR Commission. (2021) Strategy of the OSPAR Commission for the protection of the marine environment of the North-East Atlantic 2030, https://www.ospar.org/documents?v=46337 (Accessed May 19, 2022).

Paracchini, M., Petersen, J., Hoogeveen, Y., Bamps, C., Burfield, I. and van Swaay, C. (2008) *High Nature Value Farmland in Europe – An Estimate of the Distribution Patterns on the Basis of Land Cover and Biodiversity Data.* EUR 23480 EN. Luxembourg: The Official Publications Office of the European Communities. https://publications.jrc.ec.europa.eu/repository/handle/JRC47063

Peppiette, M. (2011) The challenge of environmental monitoring: the example of HNV farmland, Paper presented at the *122nd European Association of Agricultural Economists (EAAE) Seminar; Evidence-Based Agricultural and Rural Policy Making,* https://ageconsearch.umn.edu/record/99586/files/peppiette.pdf

Programme of the European Union. (2022a) Copernicus Services, https://www.copernicus.eu/en/copernicus-services (Accessed May 15, 2022).

Programme of the European Union. (2022b) Copernicus Services catalogue, https://www.copernicus.eu/en/accessing-data-where-and-how/copernicus-services-catalogue (Accessed February 12, 2022).

Ramírez, C. and Morales, D. (2021) *Integrating Forest and Landscape Restoration into National Forest Monitoring Systems.* Rome: Food and Agriculture Organization. 10.4060/cb6021en

Ramsar. (1971). The Convention on Wetlands, https://www.ramsar.org/ (Accessed June 15, 2022).

Reyes-Riveros, R., Altamirano, A., De La Barrera, F., Rozas-Vásquez, D., Vieli, L. and Meli, P. (2021) Linking public urban green spaces and human well-being: a systematic review, *Urban Forestry & Urban Greening* 61, 127105. 10.1016/j.ufug.2021.127105

Sandewall, M. and Gebrehiwot, M. (2015) An approach for assessing changes of forest land use, their drivers, and their impact to society and environment, in Zlatic, M. (ed.) *Precious Forests – Precious Earth.* IntechOpen Book Series, ch. 12. 10.5772/61074

Sandewall, M., Ohlsson, B. and Sawathvong, S. (2001) Assessment of historical land-use changes for purposes of strategic planning – a case study in Laos, *AMBIO: A Journal of the Human Environment* 30(1), 55–61. 10.1579/0044-7447-30.1.55

Sleeter, B.M., Frid, L., Rayfield, B., Colin, D., Zhu, Z. and Marvin, D.C. (2022) Operational assessment tool for forest carbon dynamics for the United States: a new spatially explicit approach linking the LUCAS and CBM-CFS3 models, *Carbon Balance Management* 17, 1. 10.1186/s13021-022-00201-1

UK Data Service. (2022) Key services, https://ukdataservice.ac.uk/ (Accessed April 14, 2022).

UNESCO, WRI, IUCN. (2021) *World Heritage forests: Carbon sinks under pressure,* Paris: UNESCO, https://portals.iucn.org/library/sites/library/files/documents/2021-034-En.pdf

United Nations. (1995) *The United Nations Agreement for the Implementation of the Provisions of the United Nations Convention on the Law of the Sea of 10 December 1982 relating to the Conservation and Management of Straddling Fish Stocks and Highly Migratory Fish Stocks (in force as from 11 December 2001) Overview,* http://www.un.org/Depts/los/convention_agreements/convention_overview_fish_stocks.htm (Accessed May 23, 2022).

United Nations. (2015). Transforming our world: the 2030 Agenda for Sustainable Development, https://sdgs.un.org/sites/default/files/publications/21252030%20Agenda%20for%20Sustainable%20Development%20web.pdf

United Nations. (2019) *World Urbanization Prospects 2018: Highlights* (ST/ESA/SER.A/421). Department of Economic and Social Affairs, Population Division.

United Nations, Economic Commission for Europe. (2016) *Environmental Monitoring and Assessment – Guidelines for Developing National Strategies to Use Biodiversity Monitoring as an Environmental Policy Tool for Countries of Eastern Europe, the Caucasus and Central Asia, as well as Interested South-Eastern European Countries.* United Nations Publications, ECE/CEP/176, New York, Geneva.

United Nations, Framework Convention on Climate Change. (2022) REDD+ platform, https://redd.unfccc.int/ (Accessed April 12, 2022).

van Doorn, A. and Elbersen, B. (2012) *Implementation of High Nature Value Farmland in Agri-environmental Policies.* Alterra Report 2289. Wageningen UR.

Vrolijk, H.C.J., Poppe, K.J. and Keszthelyi, S. (2016) Collecting sustainability data in different organisational settings of FADN in Europe, *Studies in Agricultural Economics* 118, 138–144.

With, K.A. (2019) *Essentials of Landscape Ecology.* 1st edn. Oxford, UK: Oxford University Press.

Wood, E., Harsant, A., Dallimer, M., Cronin de Chavez, A., McEachan, R.R.C. and Hassall, C. (2018) Not all green space is created equal: biodiversity predicts psychological restorative benefits from urban green space, *Frontiers in Psychology* 9. https://www.frontiersin.org/article/10.3389/fpsyg.2018.02320

Zomeni, M., Martinou, A.F., Stavrinides, M.C. and Vogiatzakis, I.N. (2018) High nature value farmlands: issues in identification and interpretation using Cyprus as a case study. *Nature Conservation* 31, 53–70. 10.3897/natureconservation.31.28397

4 Designing monitoring systems

Åsa Ranlund, Anton Grafström, Alan Brown,
Henrik Hedenås, and Gregor Levin

Introduction

The design of monitoring systems should consider relevant components introduced throughout the book. This chapter outlines the most important decisions to be made before implementing a monitoring design (Figure 4.1). It is costly to realize, after years of monitoring, that the survey design or sample size was not adequate to address the objectives of monitoring. The chapter therefore also highlights the important roles of different expertise; for instance, before you collect data, consult a (design-based) statistician.

There are a multitude of different ways to design monitoring, and several are referred to in different parts of this book. In some instances, monitoring will cover an entire area, wall-to-wall, and such methods are covered in chapters 7 to 9. Sometimes, the monitoring subject is very rare but its locations well known, in which case those locations can be monitored in their entirety or with a sample. In some cases the most important objective for monitoring is to continue a long time series, irrespective of how the sample for that time series was chosen. Citizen science data are increasingly used and interpreted via statistical models (chapter 6).

Often monitoring surveys a sample instead of trying to make an inventory of all occurrences of the phenomenon that is targeted. When we measure a subset of all of the occurrences – a sample – we can allocate resources towards making more accurate measurements than if we attempt to measure the whole population, especially if we want to track changes in the population over time. A subset or sample may be selected in different ways, some subjective and some objective. To choose a sample that can be as accurate a representation of the monitored phenomena as possible, we advocate *design-based* sampling, defined in the Design-based sampling section. Many of the aspects of designing monitoring systems that we cover in this chapter are relevant irrespective of design method. The focus of the second half of the chapter is, however, design-based sampling.

The chapter includes a couple of examples on different monitoring systems and how they have been revised.

Questions

The questions formulate what we want to monitor (Lindenmayer and Likens 2010). They give the direction and focus of the monitoring scheme. The scope of the question is further developed within the objectives. Many of the questions have already been developed in the policy processes (see chapter 3), in which case we need to frame these

DOI: 10.4324/9781003179245-4

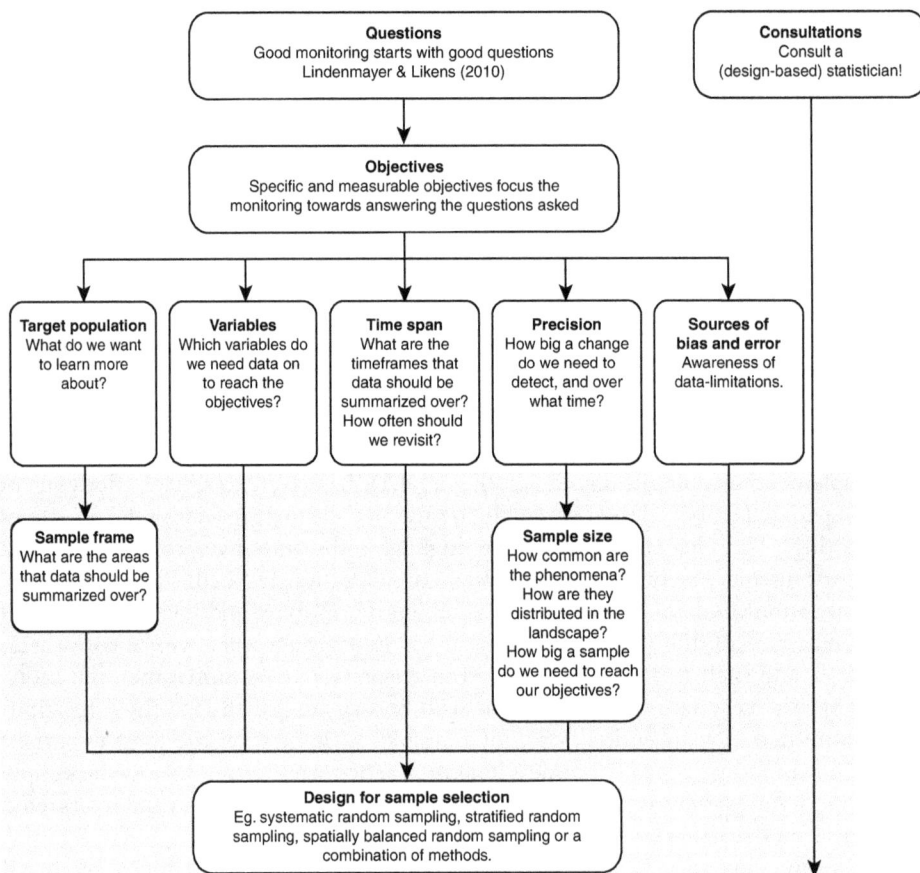

Figure 4.1 Flowchart of steps when designing monitoring systems. In the main text, we go through the different parts of the figure, aiming to clarify how one might think about designing a monitoring system based on determining questions, objectives, target population, variables, time span, precision, sources of bias and error, and design for sample selection. The grey part of the figure is relevant when selecting a sample to monitor, and we focus on design-based sampling.

as questions with a well-defined population and measurable objectives; for example, stipulating the amount and direction of change we need to be able to detect.

Measurable objectives

In the process of developing well-defined and measurable objectives, we focus the monitoring towards answering the questions asked. There is a risk that without well-defined objectives we may not select the most appropriate design or variables needed to answer the questions. To answer a question such as "What is the change in the area of broadleaf forest in Sweden?", we need specific measurable objectives. For example, objectives could include targets on precision such as "estimate the area of broadleaf forest with a relative standard deviation (RSD[1]) of less than 20%" or "detect any change in the area of broadleaf forest cover greater than 15% over a 5-year period, with a probability of at least 0.9".

The questions and objectives that the monitoring system is expected to address and the resources available set the backdrop for choosing which monitoring design will be the most appropriate. They also frame other aspects of how we should collect data: Which target population and spatial frame is relevant to reach the objectives (see Target population section)? Which variables do we need to collect data for to reach our objectives (see Variables section)? When, and for what time-period, do we need to answer the questions (see Time span section)? Are there precision requirements for the questions and objectives; that is, does it matter how certain we are in our answer or to what extent we can detect changes (see Precision section)? And, not least, which potential sources of bias and error are there (see Sources of bias and error section)?

Because all other aspects of designing a monitoring system rely on the questions and objectives, they should be considered carefully. Though monitoring programmes can be initiated with the intention for them to last a long time, over time, questions and objectives will inevitably change (Text boxes 4.1 and 4.2). When initiating long-term monitoring programmes, it can therefore be worth considering how questions and objectives might change and how the design, sampling or other, can enable the monitoring programme to continue over time even when there have been changes in questions being asked, when new environmental influences have to be taken into account, or when technical advances have been made.

Text box 4.1: National Inventories of Landscapes in Sweden: switching sampling design

The National Inventories of Landscapes in Sweden (NILS) was started by the Swedish Environmental Agency in 2003 following thorough investigations into which field variables could be collected to monitor different aspects of the status of Swedish landscapes (Ståhl et al. 2011; Hedenås et al. 2016). As a result, a large amount of field variables were collected. The monitoring system was set up as a permanent inventory of a systematically placed grid of 5km by 5km squares covering all of Sweden (the sample frame), where a random sample of 639 squares, with 12 plots field inventoried in each, was surveyed. Each year a fifth of the sample was surveyed so that after five years the entire sample was covered and a re-inventory started in year six. The first analyses of change over the entire sample could be done after ten years. For general questions about common landscape features the NILS monitoring system 2003–2020 worked well. But, whereas some parts of the collected data were rarely used, other parts came up short when demands on data shifted over the years. Specifically, data from NILS were not enough to satisfy the demands on data for reporting to the European Union (EU) Species and Habitat Directive, where data from NILS are used to report current area, and area changes, for habitats in the directive's Annex I. The design was also rigid and could not be adapted to the new requirements without violating the original statistical assumptions.

To enable higher precision for estimates of less common habitats, but continuing to collect data on general aspects of Swedish landscapes, the NILS monitoring programme was renewed. Learning from old mistakes, a new more flexible sampling design was developed to allow for future adaptations towards changes in data demand. Using one monitoring programme to survey phenomena that occur at very different frequencies in the landscape also presents a challenge. The new design relies on a two-phase sampling design where, in the first phase, a sample of tracts was chosen using random balanced sampling (see Text box 4.4). In two national inventories of deciduous forests and grasslands, each tract consists of 196 plots systematically placed within 1km by 1km squares. All plots belonging to tracts in the sample are inventoried using aerial photos and other remote sensing techniques to classify them (see Text box 5.3). Then, in the second phase, the classes from the remote sensing are used to select a sample of plots to field inventory. The remote sensing classes constitute a type of stratification. In a national inventory of alpine habitats, each tract instead consists of 1600 plots systematically placed within 1km by 1km squares (see Text box 8.1). These plots are classified through models based on remote sensing data and the classes are used in the second phase to select a sample of plots to field inventory, in the same way as for the deciduous forests and grasslands inventories. The finesse of this second phase is that we can use remote sensing information to exclude plots from an expensive field inventory when we know beforehand that they do not contain anything relating to the objectives of the monitoring programme.

Some of the habitats that the NILS programme collects data on are rare. Therefore, we collect data on different phenomena in different subsets of the sample. We use a small subset of the sample to survey common phenomena and larger subsets of the sample to survey less common phenomena. In the end, we can focus our sampling effort where it is needed and then combine data to make estimates towards, for example, reporting to the EU Species and Habitat Directive.

The Swedish Environmental Protection Agency has set targets towards detecting area change of a certain minimum magnitude for Annex I habitats for reporting within the EU's Species and Habitats Directive. The objective when deciding on sample sizes for the renewed NILS programme has been to reach these requirements for as many of the relevant Annex I habitats as possible, within the given budget. Today NILS (now National Inventories of Landscapes in Sweden) is a programme incorporating several national inventories that monitors biophysical conditions and changes in alpine habitats (Text box 8.1), deciduous forests (Text boxes 5.3 and 5.4), grasslands (Text boxes 5.3 and 5.4), and seashores (Text box 5.7) with a focus on habitats with high nature values.

Text box 4.2:   The Danish monitoring of small biotopes

Roskilde University's campaigns in the 1980s and 1990s

Agricultural land use comprises around two-thirds of the Danish terrestrial area, and habitats are often characterized by small and often spatially isolated biotopes in a matrix of intensive agriculture. In the early 1980s, Roskilde University initiated the small biotope monitoring programme to assess the extent and development of biotopes in the rural landscape. The inventory, which is based on a combination of aerial photo interpretation and field surveys, is a wall-to-wall mapping of land cover and land use with a specific focus on small remnant biotopes. The mapping is carried out for 2 × 2km areas that contain at least 75% agricultural land. The areas were sampled to reflect the variation of biophysical parameters, such as soil types and geomorphology and of agricultural production within the Danish rural landscape (Agger and Brandt 1988).

Roskilde University conducted monitoring campaigns in 1981, 1986, 1991, and 1996. The first campaign included 13 areas on Funen, Zealand, and the southern islands. Until 1991, the number of areas was gradually increased to reach 32 areas distributed over the whole country (see Figure 4.2). The inventory follows a classification hierarchy including the main land use type, geometry (area or line), proportion of woody vegetation, and soil wetness (see Table 4.1 and Figure 4.2; Brandt and Levin 2006).

Continuation within the national monitoring programme after 2000

In the early 2000s, the small biotope monitoring programme was integrated into the Danish national monitoring programme for the aquatic environment and nature (NOVANA). The Environmental Protection Agency conducted monitoring campaigns in 2007 and 2013. In addition to the registration and mapping of biotopes, botanic surveys were carried out. Within each of the 32 areas, 60 sites were laid out (see example in Figure 4.2). The sites were sampled to reflect the variation of biotope types within each area. Within a square of 0.5×0.5m a complete list of all vascular plants was elaborated. As a supplement, species within 5m from the inventory square were added. To assess the coverage of species, in 2013, the botanical survey also included a pinpoint analysis in which species that were touched by the point of a vertical needle at 16 regular locations across the site were recorded (Fredshavn et al. 2015).

Table 4.1 Biotope classification scheme

Main land use type	Geometry	Proportion woody vegetation	Soil wetness	Biotope type
Built				
Road/rail				
Land in agricultural rotation				
Land outside agricultural rotation	Area (min. 100m²)	<50%	Permanently dry	Dry grassland/heather
			Periodically wet	Bog
				Wet meadow
			Water surface	Lake/pond
		≥50%	Permanently dry	Broadleaf forest on dry soil
				Coniferous forest on dry soil
				Mixed forest on dry soil
			Periodically wet	Swamp forest
	Line (min. 100m length; 1–20m width)	<50%	Permanently dry	Field boundary
				Road verge
			Periodically wet	Ditch
			Water surface	Water course
		>50%	Permanently dry	Hedgerow

Source: Adapted from Fredshavn et al. (2015).

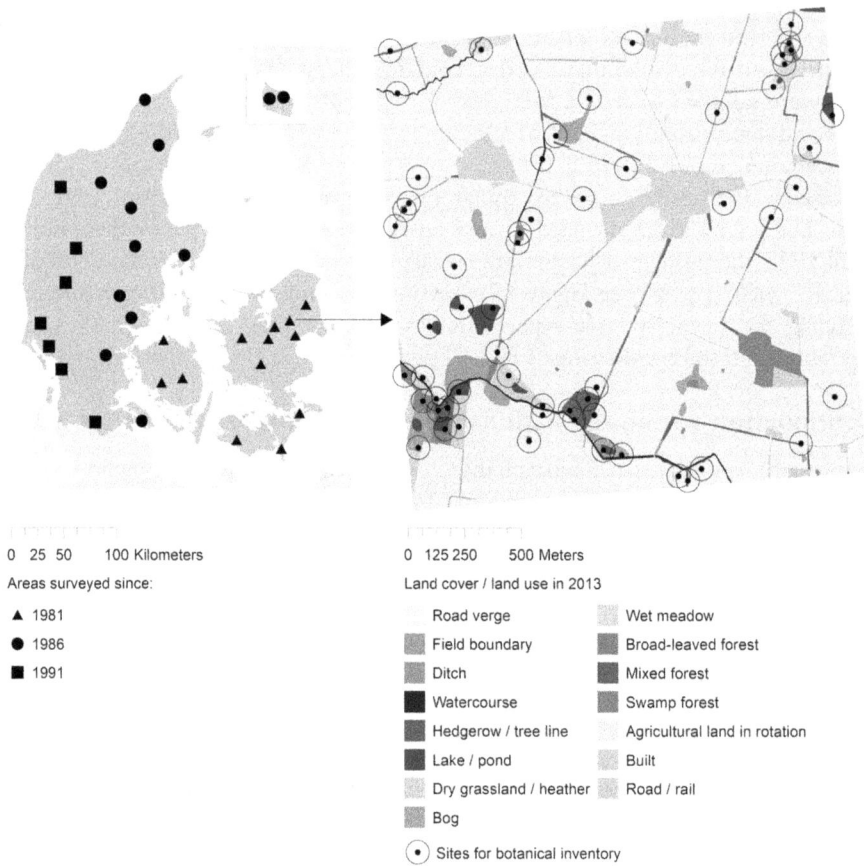

Figure 4.2 Locations of monitoring areas and an example of small biotope mapping in the Danish monitoring of small biotopes.

Lessons learned

The Danish small biotope monitoring programme was one of the first attempts to monitor habitat change in the rural landscape. The long time frame with the first campaign carried out more than 40 years ago gives a unique dataset for assessment of change in the rural landscape. Until 1996, the focus was primarily on understanding structural dynamics. After integration into the national monitoring programme, the focus shifted more towards an assessment of biological quality. Furthermore, the classification of biotope types shifted over time. Together with biases from various institutions being responsible for the different recording campaigns, this has resulted in considerable challenges when analyzing changes over time.

For an example of results from the Danish small biotope monitoring programme, see chapter 16.

Target population

The target population is what we want to learn more about through our monitoring. It can be, for example, wetlands of a certain type in a specific region or pollinators within the European Union. It is important to carefully define the target population, both because other aspects of the monitoring design relate to that definition and because the monitoring results need to be interpreted in light of it.

Sample frame

To select a sample from a population, we need a sampling frame. Choosing a sample frame is an important step because it determines where we survey the target population. It is critical that the sample frame include the entire population for the full duration of the survey. Otherwise, the survey risks becoming biased in the future if the part of the population that ends up outside the sample frame is dissimilar to the sampled population within the frame. For long-term surveys it could be necessary to make the sample frame larger than where the population is now. If, for example, the distribution of a surveyed species of butterfly might change over time, the sample frame needs to include both the areas of the current and the future distribution.

The sampling frame is a mathematical representation of a population that allows us to select a sample. In environmental surveys, the frame often represents some well-defined geographical area where the population of interest is located. Such an area frame can be considered to represent a continuous population, comprising all of the possible points in an area. It is also common that an area frame is partitioned into, and treated as, a finite number of grid cells. In probability sampling (see Design-based sampling section; probability sampling options in Appendix 2) from a finite population, every object in the population has a positive probability of being selected in the sample.

Variables

Typically, we are interested in estimating some quantity[2] that can be expressed as a population total of some variable (the sum of all values in the population) or a population mean. Some examples of population parameters are the total area of broadleaf forest, the number of butterflies, or the mean topsoil depth of an area. The basis for estimates of such quantities are the variables for which data is collected. Depending on the method of data collection, there are different aspects to consider for the variables to be useful for the current questions that the monitoring needs to answer, as well as for future flexibility and comparability. Such aspects are covered in more detail in other chapters in the book; for example, for field measurements (chapter 5), remote sensing (chapter 7), or hybrid methods (chapter 9).

Time span

Environmental monitoring including biodiversity monitoring is often initiated to describe the state of the environment and to detect environmental changes. Before choosing a design for the survey, it is important to consider when estimates of population quantities are needed, over which time period we want to detect change in them, and for how long we expect the survey to be maintained. Further, there will be a trade-off between the length of the time span over which we want to detect changes and the cost of the survey. If we want to detect a certain magnitude of change sooner, we need to survey the necessary sample size over a shorter time period. For a survey running across years, that means increasing the proportion of the sample surveyed each year, thus increasing the yearly cost.

In Sweden, two national monitoring programmes, the Swedish National Forest Inventory (NFI) and the National Inventories of Landscapes in Sweden (NILS; see Text box 4.1), both have five-year survey intervals. To reduce yearly costs and to have continuity in terms of both employment and budget, one-fifth of the total sample is surveyed every year so that the entire sample is surveyed after five years. Each sampling unit is then surveyed again every five years. With this method, changes based on the entire sample can be detected after the first ten years of sampling. For comparison, the EU stipulates in its guidelines "Reporting under Article 17 of the Habitats Directive: Explanatory Notes and Guidelines for the period 2013–2018" (DG Environment 2017) that changes should be able to be detected over a 12-year period.

Precision

When designing a survey, we want to have unbiased estimates of relevant population quantities (Figure 4.3) – for example, total area of broadleaf forest – that are as precise as possible. Estimates of variation are often presented as relative standard error (RSE), which is calculated as the ratio between the estimate of the standard deviation and the estimate of the total for the population parameter. The RSE can be multiplied by 100 to express a percentage.

There is often a trade-off between precision of estimates and survey cost. In general, larger samples give estimates with higher precision. When designing a survey, it is

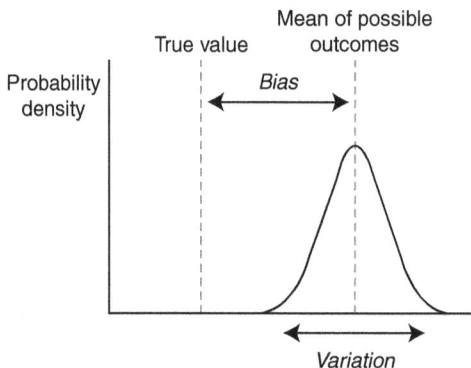

Figure 4.3 The curve shows the distribution of a statistic (e.g. mean values) for samples drawn from a population using the same method of sample selection. There is bias, a systematic error, in that method because the mean of possible estimates (e.g. the mean of sample means) is not the same as the true value of that statistic over the entire population. There is also variation, a measure of how, for example, mean values differ between different samples. Variation is the inverse of precision, so when variation is low, precision is high. Bias can also exist when monitoring aims to survey the entire population.

therefore important to consider what precision that is required of key estimates. When the requirements for precision are known, it is possible to do a power analysis (Christensen and Hedström-Ringvall 2013) to evaluate the required sample size for the survey and relate that to the resources available.

For many monitoring programmes it is important to track changes in the population quantities of interest. The ability to do that varies between monitoring designs. For example, the specific number of years required to detect changes in vertebrate populations depends on the species biology as well as the survey method (White 2019). One way to evaluate the precision required for a survey is therefore to look at what magnitude of change the survey should be able to detect, within a set time frame, for certain key population quantities (Text box 4.1).

Sample size

In general, the larger the sample, the better our resulting estimates will be. A common misunderstanding among non-statisticians is that we need to sample a large proportion of the population to get good estimates of population parameters. However, the sampled proportion is often less important than an efficient sampling design in practical applications of sampling. A small sample can be enough if it is well chosen.

An appropriate size of the sample is determined by the result of power analyses in conjunction with the amount of resources available. Resources are most obviously money and time but also specialist knowledge and field skills. Expert staff are generally the most expensive and often limiting factor in sampling, except perhaps in citizen science projects (chapter 6). When choosing the size of the sample, it is worth considering what the sampling unit is and, for field visits, how the workload at each site corresponds to the time at hand in one visit. We want to avoid adding additional plots or measurements that just tip the time needed over from one visit to two. Any technology that can make it easier and quicker to locate a plot or recording site, adds metadata, or saves time recording can make potential field visits more time efficient and facilitate a larger sample.

In any particular sampling scheme, these sorts of general principles have to be reconciled with the objectives of monitoring. In many cases, the statistical design is focussed, as in this chapter, on the most efficient way of getting an unbiased estimator of population statistics − for example, the mean habitat area − with enough precision to detect (most of the time) any increase or decrease at least as big as the limits set (the effect size). In this design, the estimate of variance is a predictor of precision (statistical power), and the sample design aims to minimize this variance.

Sources of bias and error

There are many possible sources of bias and error in monitoring (for survey samples, see Lesser and Kalsbeek 1999). For example, coverage errors can occur if monitoring (sampling frame when sampling) does not cover the entire target population, and losses or gains in the part of the population left out may be missed or exaggerated if they are atypical of the remainder. A common problem in, for example, polls is non-response error where some subjects do not respond or refuse to respond to parts of the poll. Similar non-response errors can occur also in biodiversity monitoring when, for example, a plot cannot be accessed in the field.

It has become increasingly common to use remote sensing to collect large amounts of data over vast areas, often at low cost. Such information can be very useful for monitoring systems (see chapters 7 to 9). Due to continued development of the technologies used in remote sensing it is, however, difficult to use remote sensing data alone to evaluate change without bias (see e.g. Breidenbach et al. 2022). In general, remote sensing information alone cannot replace field measures, but the methods can complement each other in design-based sampling (Text box 4.1; see also chapter 9).

Another way large amounts of data can be collected at low cost is by non-professionals in citizen science. Such methods often come with large biases that need to be considered when interpreting the data. For example, more data are often collected near densely populated areas, and there tends to be an overrepresentation of rare or conspicuous species. For more on citizen science, see chapter 6.

When selecting a sample, sampling error[3] and bias can be reduced and precision and accuracy can be increased by the way the sample is chosen and by increasing sample size. Sampling designs often assume that measurements are controlled within some level of precision and are repeatable, so that any error from them is smaller than the sampling error and therefore can be disregarded. As chapter 5 discusses in more detail, in practice these in situ measurements and observations may be subject to significant error and observer bias. This can sometimes be taken into account before the statistical analysis, just as other types of observations such as remote sensing need the calibration and correction steps mentioned in chapter 7. Though the topic of error propagation is outside the scope of this chapter, some principles are easy to see; for example, that random errors in two variables that add together can dilute the error but if one variable is divided by another the error can be increased. We need to be cautious about what inferences we can make from the results when errors are potentially high due to, for example, field methods. A common approach to compensate for errors other than sampling error in design-based sampling is to overestimate the sampling error.

Design-based sampling

When we measure a subset of a population (a sample), we can allocate resources towards making more accurate measurements than if we attempt to measure the whole population, especially if we want to track changes in the population over time. A subset or sample may be selected in different ways. Most methods for sample selection will produce bias in the estimation (see Figure 4.3). We imagine that we can use our own judgement to select a sample that represents the population, but in practice most of these samples are biased, often in favour of high values of the thing we are trying to measure. If we do not know whether or not our design is biased, we should assume that it is. This creates caveats to consider when interpreting data: in which scenarios are the results, based on those methods, reliable?

To avoid introducing bias already in the design of the monitoring programme, data can be collected using probability sampling. Probability sampling is the selection of sample units from a sampling frame using probability. In its simplest form it is like a lottery, where every unit has an equal chance of being included. A sample can also be selected using unequal inclusion probabilities; for example, varying the probability to be included between groups or strata. The main advantage of using probability sampling over alternative (subjective) methods of sample selection is that it allows us to construct statistically valid unbiased estimates of quantitative aspects of the population – based on field measurements, for example – as well as to assess the uncertainty of those estimates. Population quantities that we are interested in are referred to as *population parameters* in probability sampling. Probability sampling is used to study a population of interest, without the need to observe the entire population. Based on the observations in the sample, we can draw conclusions about the entire population.

From the random sample, we can construct an estimator. An estimator is a formula that tells us how to compute an estimate of a population parameter; for example, how to estimate the total area of broadleaf forest from field measurements in the sample. Different outcomes of the random sample give different outcomes for the estimator. In this way, an estimator is any statistic computed for the sample that is used to estimate population parameters such as the mean (average), total, or a measure of variation (e.g. variance). An estimator is a random variable and the outcome (i.e. the estimate) is a number. The magnitude of change that can be detected for a parameter depends on the variance of the estimator and the correlation between estimators of the parameter over time.

Probability sampling has the unique feature that it offers an objective way of collecting the observations needed to study a population, and it has become a universally accepted standard for information gathering. Sometimes, we can even make inferences without any assumptions about the population under study.

A sampling design is a random mechanism that specifies how the probability sample is selected. Statistical inference that requires a probability sample (by some design) is said to be design based. An example of a finite, artificial population and probability sample with unbiased estimators is given in Appendix 1.

Type of sampling design

There are many things to consider when designing a survey. However, an important starting point is to summarize what we know about the population that can be used to form a sampling design. The more information we have at hand, the better we can design the survey. Efficiency can be gained if the population can be divided (stratified) into groups (strata) that are more homogeneous with respect to our target variables, because the sum of strata variances then will be lower than the variance without stratification. Hence, we need to check whether some of the information we have, in the form of available auxiliary variables, can be used for stratification. If the strata are more homogeneous than the whole population, a small sample from each stratum might be sufficient to produce estimates of desired quality. If auxiliary variables that explain some variation in our target variables are not freely available, we need to ask whether we can afford to retrieve such variables for the whole sampling frame or a large initial sample. If we can collect auxiliary information for a large initial sample, we can use that information to select a smaller stratified sub-sample. The latter procedure is known as *two-phase sampling for stratification* and is often used in environmental surveys due to its ability to improve estimates (Text box 4.1).

Another aspect to consider when choosing a sampling design is the impact of the design on the complexity of the analysis. Evaluate who should be able to analyze data from the survey and whether it is important to keep the analysis as simple as possible. A complex design can provide a rather straightforward analysis. A *representative sample* (Text box 4.3) is often desirable because such samples provide more straightforward analysis than other approaches. Samples that are selected to be well-spread or *spatially balanced* (Text box 4.4) tend to be representative.

Text box 4.3:   Representative samples

Often, it is desirable for the sample to accurately reflect the entire population in as many ways as possible. The term *representative sample* is often used, and the common view of a representative sample is that it is a miniature version of the population. More formally, a sample is representative for a set of characteristics if the number of sampled units from every coherent subset is proportional to the size of the subset. See, for example, Grafström and Schelin (2014) for details and a stricter definition of the term representative sample. As an example, this means that if the population consists of 50% forest land, a sample representative of the population characteristic forest land should also include 50% forest land. Representative samples are needed for multipurpose surveys when several target variables are of interest. Such samples also enable estimation of parameters in subspaces and improved estimation of target variable distributions.

Text box 4.4:   Spatially balanced samples

A well-spread sample (also known as a *spatially balanced sample*) is a sample that is well-spread in some fully known population characteristics. This means that a well-spread sample is similar to a representative sample, but the term *well-spread sample* takes the sampling design into account (via the inclusion probabilities). When selecting a well-spread sample, we may deliberately overrepresent some part of the population using unequal inclusion probabilities. However, a well-spread sample selected with equal inclusion probabilities is a sample representative of the same population characteristics for which it is well-spread.

Environmental surveys commonly use some form of plot or cluster as a sampling unit. A cluster might then contain several population units that are all selected for the sample if their cluster is selected. The use of plots or clusters is considered cost-efficient, especially when the cost of travelling to different locations is high. The size of the plot or cluster can be chosen to correspond to a day's work in the field for larger national surveys, whereas regional surveys might have smaller plots. A sampling design that includes selection in phases and includes clusters and/or stratification is considered a complex sampling design. Most designs for sampling environmental populations would be classified as complex.

Design–based statistical approaches

The sampling design describes the random mechanism used to select samples. Formally, the design is defined by a probability density function on the set of possible samples. The most common design-based estimator, known as the Horvitz-Thompson estimator, weights each unit's observation by the inverse of its probability to be selected/observed.

This estimator produces unbiased scaling from the sample to the population over repeatedly drawn samples. It is in this sense the estimation is based on the design. The variation of the possible outcomes of an estimator is then also due to the sampling design. Changing the design may increase or decrease the precision of an estimator. Tillé (2020) is an up-to-date reference that covers the foundations of survey sampling theory and explains how to put this theory into practice. Sampling with an area frame often requires some extra care because the frame, the plot, and the population have geometrical representations. For that case, Gregoire and Valentine (2007) cover many different sampling techniques used in ecology, forestry, and environmental sciences.

For practical purposes, the statistical software R (R Core Team 2021) is often used to select samples from different types of frames and to perform the analysis. The R packages sampling (Tillé and Matei 2021), survey (Lumley 2021), spsurvey (Dumelle 2022), and BalancedSampling (Grafström and Lisic 2019) are examples of packages that may be of assistance in sample selection and/or analysis.

Model–assisted design–based methods

If we select the sample via a sampling design, the statistical inference can be design based, model assisted, or model based. The model is usually built using field data as training data and auxiliary data such as wall-to-wall covering data – that is, remote sensing data – as predictors. The model-assisted approach is still based on the sampling design, which means that the statistical inference is based on the randomness induced by the sampling design. Parameter and variance estimators are then derived under the concept of repeatedly drawing samples from the population with the same sampling design, and statistical modelling plays a minor role. Model-assisted estimation (Särndal et al. 1992) is used to improve the estimation in environmental surveys. The basic idea is that the bias in model-based estimation may be removed or at least reduced by adding the estimated total (or mean) of the differences between the values collected in the field and the corresponding value from the model prediction. In the model-based case, the sampling design plays a less important role because the inference is often based on an assumed model of the population (see chapter 9 for more on model types and their association to different data types).

Key messages

- The chapter illustrates that designing a monitoring system requires making a large number of choices. A key to proper functioning is to make these choices in concert with other systems that are to be integrated or harmonized (as discussed in chapter 3), just as we need well-developed cooperation between experts working with different aspects of the monitoring system, as set out in chapters 5 to 9.
- There are many different ways to collect data. In this chapter's second half we focus on designing a monitoring system using design-based methods to select a sample that we can survey in more detail to answer the questions that were the reason for initiating the monitoring. Design-based methods have the advantage of not introducing bias in the selection of the sample.
- A sample, as opposed to wall-to-wall data, facilitates focussed efforts in a survey towards a smaller part of the phenomena of interest. A sample also improves the possibility to accurately track changes.

- Different types of data can, however, inform a monitoring programme in different ways. Full-cover data can be used in the monitoring design to more efficiently select a sample or later in the analyses of already collected data. Information on social aspects can be important when initiating a monitoring system, but data on, for example, land management can also be the focus of monitoring.
- Faced with limited resources, the design of a monitoring system needs to balance the size of the sample, the effort spent at each sample unit, budget, and the importance and associated risks attached to the decisions that will be made with the results of monitoring. Leaving aside curiosity and scientific research, do we need to monitor distinctions and detect changes that make no difference to either policy of direct habitat and species management?
- On the other hand, monitoring those changes that risk the most unwelcome and – if undetected – unavoidable consequences for future states of biodiversity ought to be given the most effort. Notice how this combines not only uncertainty, which we try to control within our sample-survey design, but a measure of the utility and public preference for different outcomes in the future.

Study questions

1   A number of steps for developing monitoring systems are outlined in this chapter. How does each step relate to the results that the monitoring system will produce?
2   What are the advantages of using design-based sampling, and when might other methods of monitoring design be advantageous?
3   Suppose your power analysis shows your sample is too small to detect your threshold for minimum change. What are the risks of going ahead with monitoring/not going ahead, and what can you do to mitigate these risks?

Notes

1 See Precision section.
2 Design-based sampling makes the assumption that these population statistics are fixed, knowable quantities.
3 The sampling error describes variation between estimates from different samples. It occurs when a subset of the population is observed instead of the entire population. The sampling error is not a mistake, and for design-based sampling it can be estimated if the sample size is large enough.

Further reading

As a student, you will come across a multitude of work done by other researchers, where other solutions for sample design may have been used. We recommend that you look those sample designs up, but see Appendices 1 and 2 for a brief introduction to some of them.

Gregoire, T.G. and Valentine, H.T. (2007) *Sampling Strategies for Natural Resources and the Environment.* Chapman and Hall/CRC. 10.1201/9780203498880
Tillé, Y. (2020) *Sampling and Estimation from Finite Populations.* John Wiley & Sons.

References

Agger, P. and Brandt, J. (1988) Dynamics of small biotopes in Danish agricultural landscapes, *Landscape Ecology* 1(4), 227–240.

Brandt, J. and Levin, G. (2006) *Indikatorer for Landskabsændringer: Analyser af Komplekse Landskabsændringer på Baggrund af RUCs Småbiotopundersøgelser* [Indicators for Landscape Changes from Investigations of Small Biotopes at Roskilde University]. Roskilde Universitet.

Breidenbach, J., Ellison, D., Petersson, H., Korhonen, K.T., Henttonen, H.M., Wallerman, J., Fridman, J., Gobakken, T., Astrup, A. and Næsset, E. (2022) Harvested area did not increase abruptly – how advancements in satellite-based mapping led to erroneous conclusions, *Annals of Forest Science* 79, 2. 10.1186/s13595-022-01120-4

Christensen, P. and Hedström-Ringvall, A. (2013) Using statistical power analysis as a tool when designing a monitoring program: experience from a large-scale Swedish landscape monitoring program, *Environmental Monitoring and Assessment* 185, 7279–7293. 10.1007/s10661-013-3100-z

DG Environment. (2017) Reporting under Article 17 of the Habitats Directive: explanatory notes and guidelines for the period 2013–2018, https://circabc.europa.eu/sd/a/d0eb5cef-a216-4cad-8e77-6e4839a5471d/Reporting%20guidelines%20Article%2017%20final%20May%202017.pdf

Dumelle, M. (2022) Spatial sampling design and analysis, https://cran.r-project.org/web/packages/spsurvey/spsurvey.pdf

Fredshavn, J.F., Levin, G. and Nygaard, B. (2015) *Småbiotoper 2007 og 2013* [Small Biotopes 2007 and 2013]. NOVANA. Aarhus Universitet, DCE – Nationalt Center for Miljø og Energi, 38 s. – Videnskabelig rapport fra DCE, Nationalt Center for Miljø og Energi nr. 143, http://dce2.au.dk/pub/SR143.pdf

Grafström, A. and Lisic, J. (2019) Balanced and spatially balanced sampling, https://cran.r-project.org/web/packages/BalancedSampling/BalancedSampling.pdf

Grafström, A. and Schelin, L. (2014) How to select representative samples, *Scandinavian Journal of Statistics* 41(2), 277–290.

Gregoire, T.G. and Valentine, H.T. (2007) *Sampling Strategies for Natural Resources and the Environment.* Chapman and Hall/CRC. 10.1201/9780203498880

Hedenås, H., Christensen, P. and Svensson, J. (2016) Changes in vegetation cover and composition in the Swedish mountain region, *Environmental Monitoring and Assessment* 188, 452. 10.1007/s10661-016-5457-2

Lesser, V.M. and Kalsbeek, W.D. (1999) Nonsampling errors in environmental surveys, *Journal of Agricultural, Biological, and Environmental Statistics* 4(4), 473–488.

Lindenmayer, D.B. and Likens, G.E. (2010) The science and application of ecological monitoring, *Biological Conservation* 143, 1317–1328.

Lumley, T. (2021) Analysis of complex survey samples, https://cran.r-project.org/web/packages/survey/survey.pdf (Accessed November 2, 2022).

R Core Team. (2021) *R: A Language and Environment for Statistical Computing.* Vienna, Austria: R Foundation for Statistical Computing. https://www.R-project.org/

Ståhl, G., Allard, A., Esseen, P.-A., Glimskär, A., Ringvall, A., Svensson, J., et al. (2011) National Inventory of Landscapes in Sweden NILS – scope, design, and experiences from establishing a multiscale biodiversity monitoring system, *Environmental Monitoring and Assessment* 173, 579–595.

Särndal, C.E., Swensson, B. and Wretman, J. (2003) *Model Assisted Survey Sampling.* Springer Science & Business Media.

Tillé, Y. (2020) *Sampling and Estimation from Finite Populations.* John Wiley & Sons.

Tillé, Y. and Matei, A. (2021) Survey sampling, https://cran.r-project.org/web/packages/sampling/sampling.pdf

White, E.R. (2019) Minimum time required to detect population trends: the need for long-term monitoring programs, *BioScience* 69, 40–46. doi: 10.1093/biosci/biy144

5   Data collected in situ: unique details or integrated components of monitoring schemes

Anna Allard, Alan Brown, Clive Hurford,
Christian Isendahl, Andreas Hilpold, Ulrike Tappeiner,
Julia Strobl, and Henrik Hedenås

Introduction

This chapter takes a look at in situ monitoring, which can be a single stand-alone inventory or a series of repeat surveys. In situ data collection is also increasingly becoming a part of larger monitoring schemes, along with data collection from remote sensing (drones, planes, or satellites) or other auxiliary data sources, all of which are relevant to this chapter.

Data collected in situ take time and effort to gather, which is costly. However, it is the foundation for the other types of monitoring, because knowing the "truth" of what is there on the ground or in the water is essential to inform and validate models, classifications, and related map products. There is no getting away from the fact that we need details of the environment, such as boring into tree trunks to count tree rings, sampling water to investigate the types of algae, or identifying the dominant species in a sample plot. Knowing what or how much is growing there, the soil types, thickness of biotic and abiotic layers, the temperature or content of water bodies, and more, is necessary to give the context for understanding the ecosystems and to provide the chance to assess changes. The diversity of methods is shown graphically in Figure 2.1. The European Union (EU) monitoring services also depend on a wide variety of ground-based, seaborne, or airborne monitoring systems, as well as geospatial reference or ancillary data, collectively referred to as *in situ data*, for production and validation of the services (Programme of the European Union 2022).

Some sampling schemes are designed only to get unbiased estimators of statistics for the whole population. These might use a different set of in situ observation points on each visit, selecting them each time using probability sampling to minimize bias. More often, schemes make repeated observations at the same individual sample points to allow us to see spatial patterns of environmental phenomena, such as revisiting permanent plots, points, or transects. Permanent sample points allow in situ observations to show not only what is changing, on average, but where change is taking place. However, they can become unrepresentative of the population, and precautions must be taken not to introduce bias into estimates of population statistics.

In a statistical setting, there are also different approaches according to whether and how spatial information is used. Geostatistical analysis, originally developed by mining engineers and soil scientists, can be used to convert observations from a set of spatially indexed points into a contour map. In a more familiar setting, spatial patterns can be understood by drawing maps in the field, interpreting aerial photos, and classifying

DOI: 10.4324/9781003179245-5

wall-to-wall imagery from satellite remote sensing; see more in chapters 7 and 8. In this chapter, we stay a bit closer to the ground than satellites and consider how to make repeatable and reliable in situ observations.

In situ inventory

In situ inventory involves going to a predetermined place and making an observation, taking a sample or photograph from the ground or a drone, or carrying out some sort of recording with an instrument (which can be permanent, trailing, drifting, or carried there by the surveyor). Even though photographs will have to be viewed and analyzed later, what to photograph and where to collect a sample are also decisions that need recording. Often there will be a list of metadata – data about the circumstances of data collection. Metadata can include who was involved and where and when it took place, the type and model of instrument used (perhaps with calibration data), and anything that might influence the value or reliability of what is observed (date in the season, time, wind, temperature, turbidity of water, etc.). Experience, both in the field and in data analysis, tells us what might influence the reliability and repeatability of the field inventory and how best to record it.

Devices for data collection

Because smartphones now have screens that can be read in bright sunlight, a variety of apps have been developed for recording of geographical data by employed staff or volunteers. Monitoring data providers might develop their own script to use in a smartphone or computer pad, and companies develop and sell ready-made solutions. Examples are many of the autonomous data recorders used in freshwater or ocean environments or for recording land-based data, such as the app *Sweet for ArcGIS* developed for habitat mapping by the Countryside Survey in cooperation with ESRI (Esri UK and Ireland 2022); see also Text boxes 5.1, 5.4, and 5.7.

For water monitoring, a wide selection of devices for measurements are available in a number of networks and services for rivers, coastal waters, and oceans. Many of these networks have permanent devices; see example of permanent laser scanners in chapter 2. Other devices are on platforms that either are moored, drift, or move autonomously, carrying profiling floats, gliders, and moorings; read about an autonomous drone ship in chapter 8. Even marine mammals can be equipped with instrument tags, and commercial ships collect measurements as they sail. For measurements not yet automated, such as deep-sea measurements or sampling of water properties or ecosystem variables, research vessels are used on local to global scales. For a comprehensive summary, we recommend a visit to the website EU Ocean for observation (Eu4oceanobs 2022).

Plot location and relocation

Locating – and eventually re-locating – the place to carry out in situ observations is critical to the success of field inventory whenever we need to georeference the results with other datasets (think "ground truth" for remote sensing) or want to come back to exactly the same place in the future. Locating can be done with maps and a compass, using landmarks or aerial photographs for navigation, Global Positioning System (GPS), or apps where the current position is shown on a map or an orthophoto. GPS positions often roam around

Text box 5.1:   Example of a biodiversity monitoring program in the Alps: Biodiversity Monitoring South Tyrol

In 2019, Eurac Research, a research centre situated in the northernmost province of Italy, South Tyrol (province Bolzano/Bozen), initiated the Biodiversity Monitoring South Tyrol monitoring programme on behalf of the local government. The data are collected by employed staff in 320 terrestrial sites distributed across the region over a period of five years and in 120 aquatic sites over a period of four years (Figure 5.1). The study sites were selected using a stratified selection approach. Within each stratum, a random site selection was performed, whenever possible and logistically reasonable. The monitoring consists of field surveys, from the densely inhabited and agriculturally intensively used valley floors, along the altitudinal transect over meadows and forests up to the more extreme areas dominated by alpine grasslands, rocks, and screes (up to 3000 m.a.s.l.). Both managed and (near-) natural habitats are included.

Figure 5.1 Distribution and main categories of monitoring sites within the region (province Bolzano/Bozen, South Tyrol).

While targeting species groups that react sensitively to climate and land use changes, the survey also includes the surrounding landscape and habitat mosaic. Within the terrestrial monitoring part, vascular plants, bryophytes, birds, bats, butterflies, and grasshoppers are assessed. Additionally, samples of various invertebrate groups are collected, applying standardized sampling methods. Aquatic habitats are surveyed through the larvae of mayflies, caddisflies, and stoneflies.

All surveys are conducted in or directly around the chosen monitoring sites (Figure 5.2). The plot sizes vary depending on what is being monitored and are different depending on the surveyed taxonomic groups: The basic plot is a square of 10 × 10m, used for vascular plants and grasshoppers (31.6 × 31.6m for forests), except for settlements and lake shores, where monitoring is conducted along a transect line of 100m. In or directly around the surveyed plot invertebrate sampling via standardized methods (pitfall traps, soil extraction, beat and sweep netting) is also conducted. From this basic plot, birds and bats are assessed within a circle, using a radius of 100m for birds. Butterflies are surveyed on a transect line of 50m crossing the monitoring site. Habitats and landscape cartography are mapped within a square field of 200m in all four directions from the survey plot.

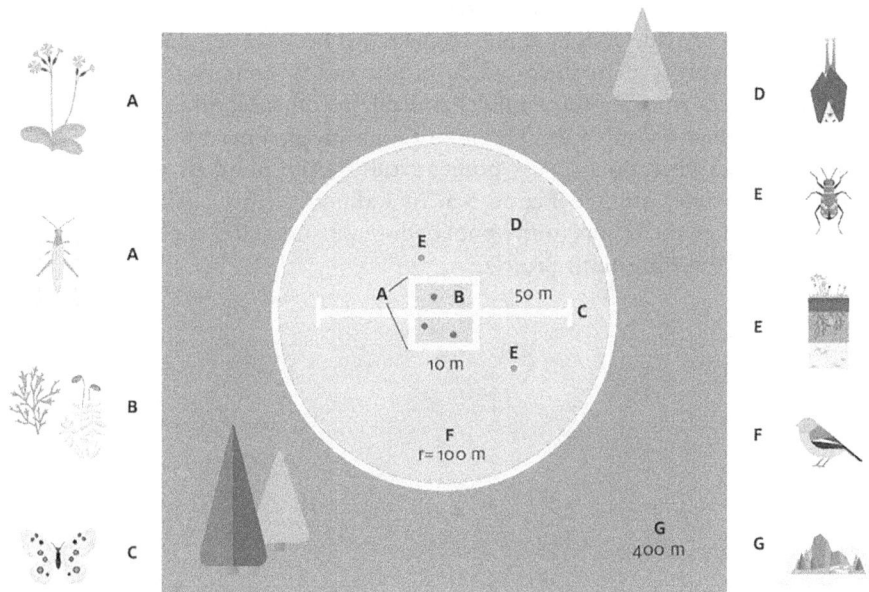

Figure 5.2 Scheme of field methodology for each organism group in Biodiversity Monitoring South Tyrol, terrestrial part.

Source: Hilpold et al. (In press).

Vascular plants are assessed once a year using coverage and a species list based on the protocol of the European Dry Grassland Group (EDGG). Samples of lichens and bryophytes are collected for later identification in the laboratory (Dengler et al. 2016, 2021). Butterfly survey is conducted four times a year (at least three times for alpine sites) between mid-May and the beginning of September, in collaboration with, and using the same methodology as, the Austrian Viel-Falter butterfly monitoring of the Alps, a combined expert and citizen science project (Österreich forscht, www.citizen-science.at 2022). Birds are monitored by acoustic and visual identification in a radius of 100m around the survey plot by an expert three times a year (twice at higher elevations), choosing the time of season to exclude migrating birds passing the areas. Standardized surveys of bats take place in all terrestrial survey points, using ultrasonic recording devices, so-called batloggers, enabling the identification of species or at least the genera by their calls. Spiders and other invertebrates are collected for later identification using pitfall traps, and soil core samples are taken for assessment of soil fauna and determination of soil types. Sweep and beat nets are used to collect and assess different groups of invertebrates. In running waters, invertebrate fauna is collected via Surber samples.

For the habitat cartography, orthophotos are used as a baseline and subsequently habitats are determined in the field. In the case of a habitat being listed in the European Habitats Directive (European Commission [EC] 1992), an evaluation of the conservation status is done. Finally, cartographic (thematic) maps are drawn up.

The first time-series data will be available when the first turn of repetition is concluded; that is, ten years after project start. However, the data from the first monitoring period are already well suited for various kinds of ecological analyses; for example, to provide insight about connections between agricultural management and biodiversity or to help public institutions in conservation issues. The project team aims to make these results available to the scientific community via publishing in scientific journals. The monitoring programme runs an interactive website, where a click on a survey point reveals species numbers that have been found at the specific site; see Figure 5.3. In addition to the scientific work, the communication of the results to the public, dissemination, and expert training are crucial parts of this long-term project.

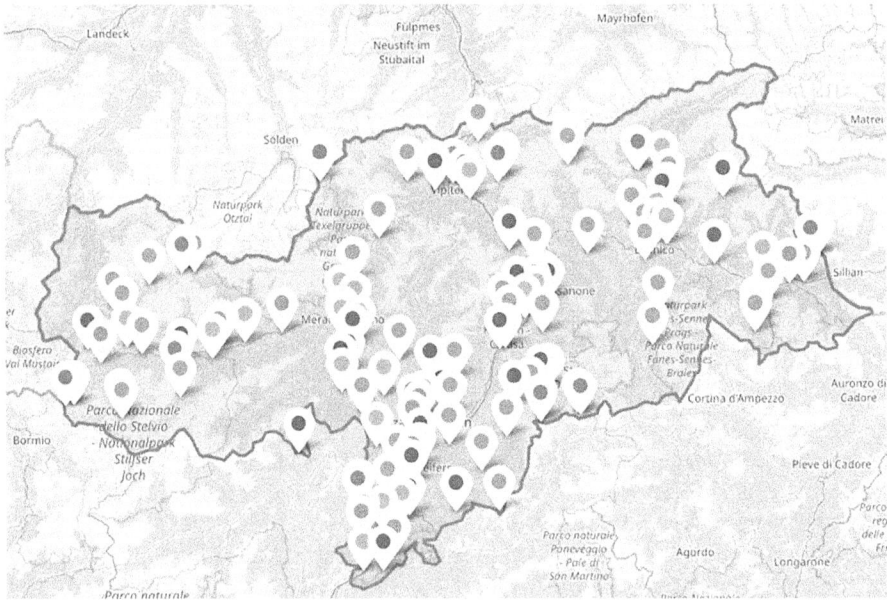

Figure 5.3 The interactive map of biodiversity monitoring in South Tyrol, containing 320 terrestrial sites. To the right is a representation of the data popping up when clicking on a site. In the future, detailed species lists will be available.

Source: From Eurac Research (2022).

slightly, due to the passing of satellites, so a mean of several GPS measurements is re-commended. Though the older methods may seem redundant, we might still need to use them to find the location of old plots in archival records, bearing in mind that old maps may not be precisely georeferenced or accurately drawn up. The precise relocation of points can be helped by leaving markers nearby that can be found with the aid of GPS, and then the actual point is measured out from the marker using a tape.

Species identification

Suppose we are in the right place. Recording habitat on land-based areas often means recording the presence, abundance, and cover of vascular plants, bryophytes, and lichens and less often fungi. The different methods of doing this have strengths and weaknesses, which are discussed below. Here, the important questions are: what do we need to record, how long does it take, and how reliable and repeatable is the method? Left unasked is whether we have the knowledge and field skills to make a correct identifi-cation, often from just a few leaves or stems of non-flowering plants; see Text box 5.3 for a solution in a national monitoring regime. The difficulty associated with finding and identifying plants can unpick the entire monitoring method, making it unreliable and prone to missing rare plants, showing changes that do not exist (called *pseudo-turnover*), and missing real trends and changes that we would consider ecologically significant. Of course, there are comparable issues of reliability and repeatability with any sort of in situ inventory, whether this is for plants, insects, fish, birds, or mammals. Online resources are increasingly helpful for plant identification, including apps that work on smartphones and can be accessed in the field. Photographing specimens and using these apps are especially important in citizen science monitoring.

Plot design

Decisions on where in the landscape to go for collection are handled in the design of the sampling scheme; see chapter 4. However, once the sample point is reached, we need a manner of collection. The plot is where we measure our list of variables, taking samples of soil or water and recording species in single plot (or a cluster of plots; square or circular areas, along transects, or a whole tract are common designs). All designs are accompanied by some assumptions about estimations of population statistics (Canfield 1941; Alberdi et al. 2010; Baraloto et al. 2013; Henttonen and Kangas 2015); see examples in Text boxes 5.1, 5.3, and 5.7. One programme may use both permanent and temporary plots (with similar or different designs) as in the long-term (100 years in 2023) National Forest Inventory (NFI) in Sweden. The two results are used either as a whole in model-assisted mapping (e.g. M. Nilsson et al. 2017) or the two parts are used separately, where the temporary plots are used as training data and the permanent plots for validation of the model; see chapter 9.

Thematic classes

Thematic classes are often assigned first using aerial photos, maps, or models and then decided on site by referring to a field classification scheme. The classification may be determined only by species (plants, insects, animals, algae) or completed by habitat or land cover criteria such as structure, the type of soil, sediment, depth of peat, or water.

The expertise and experience of the surveyors are not to be underestimated, which means that the monitoring needs highly educated staff who have attended bespoke training courses involving all of those classes encountered in the actual monitoring. If possible, a mixed team of more experienced staff together with those new to monitoring enables the experienced to guide those who are new to the classification scheme.

Another decision is how to proceed when the selected plot contains more than one class. If the design allows more than one class, how will the split be recorded or, if not, which class will be preferred by default over the other? The example in Text box 5.3 employs two types of classification, where in step 1, inventory by aerial photos is used to pinpoint potential habitats of interest to visit in the field. This allows only one class and has a hierarchy list to follow, so even if there are two habitats of interest in the circular plot in Figure 5.7C, only the highest hierarchical class is recorded. The second step of in situ visits includes the option of splitting the circular plots into three distinct areas, with clear instructions on how to record the splits, to enable revisits and to allow them to be georeferenced and relocated both in situ and in a geographic information system (GIS).

Not having clear guidance on how to record mixtures, mosaics, and transitional classes will make the surveyors uncertain and negatively affect the repeatability and efficiency of the work. Hierarchical classification schemes also allow the surveyor to use a more general class when they are unable to decide between two or more sub-classes; for example, when visiting a habitat at the wrong time of year to see those species needed to make a critical separation. These are two different types of uncertainty. In the first case, the parcel of land can be said to be intermediate between classes or have membership of more than one class. Where there is a mosaic of vegetation, this might be because the scale of the map does not allow small enough units to be defined and labelled. This is also a common problem for new technologies, where satellite imagery has a limited spatial resolution. In the second case, the real class may be very obvious when seen in the right circumstances, but during the actual field inventory we are unable to choose between classes. The first type of uncertainty is called *vagueness*, the second *ambiguity*, and these have to be handled very differently in analysis and change detection. The finer points of these terms are linguistic or semantic in nature and much discussed in many research disciplines; for example, in clinical practice (Codish and Shiffman 2005). The important point in our context is to be aware of them and to have a set of rules or protocols for dealing with what we actually find in field inventory, which will not always be limited by the assumptions made in planning the monitoring scheme.

Data collection through interpretation of images from above

Interpreting data from images taken by planes or drones is a kind of remote sensing, but the interpretation methods have more in common with in situ methods. Although used in many areas of monitoring, this section exemplifies the method by interpretation of land-based vegetation.

Aerial imagery has been around a long time and is often used as a component of monitoring and mapping projects (e.g. Allard 2017). Near-infrared imagery (first used by the military forces to differentiate between camouflaged objects and plant cover) is especially useful for vegetation monitoring, enabling us to distinguish different moisture regimes, types of vegetation, or land cover (e.g. Ihse 2007; Schott 2007; Lillesand and Kiefer 2015). The images are typically either orthophotos in a GIS, projected to constitute a photographic map with uniform scale, or (preferable for understanding the landscape) taken with overlap, creating a stereo view (3D). Early interpretation used analogue photos and instruments, but the last decades have seen a number of computerized ways to delineate in a digital environment with direct transfer of the geographical data into a GIS, and modern aerial cameras use digital sensors. Because any overlapping vertical photographs taken from the same height can form a stereo pair, the increasing use of drones for local photography gives new opportunities for exploring the use of stereo as an aid to aerial photo interpretation.

The person performing the interpretation is using the same type of skills as a person in the field deducing what is there; see Figure 5.4. The interpreter uses the whole set of ecological skills to analyze the landscape, its current context, as well as its history (why does this particular spot in the landscape look like this?). The analysis takes into account several "interpretation indicators", including the spectral properties of the vegetation,

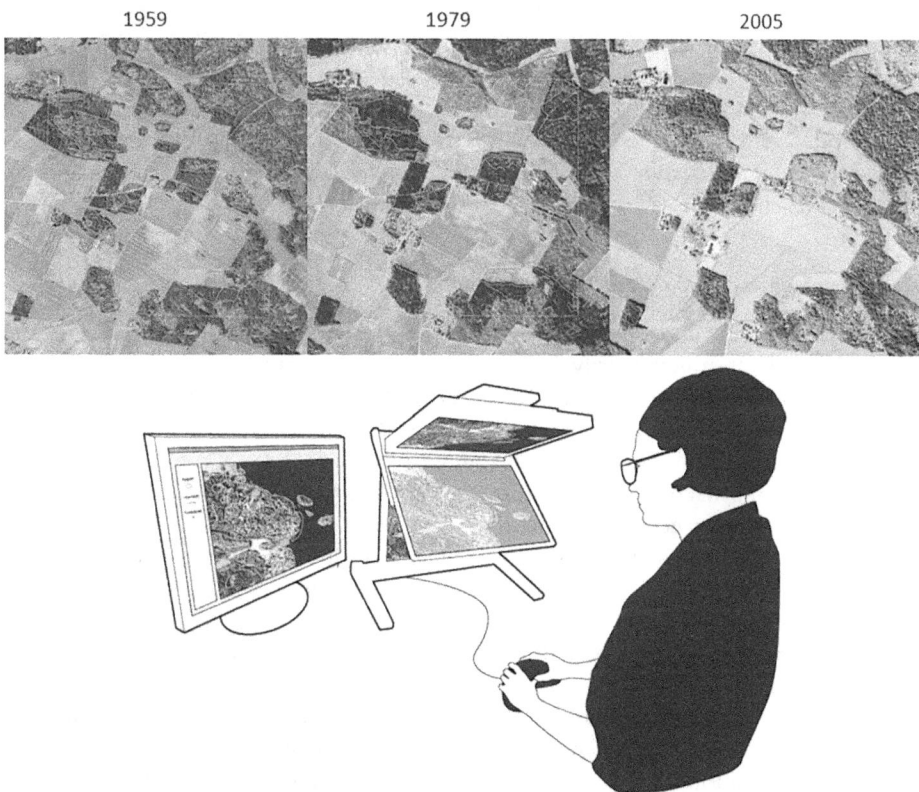

Figure 5.4 Stereo viewing of aerial images for monitoring. The upper picture set shows screenshots of the same area (1.1 × 1.1km) in 1959, 1979, and 2005, in a digital environment with scanned photos. Superimposed on the images are yellow lines delineating boundaries of land type categories (see Text box 5.2). The work process below shows images in 3D, using polarized spectacles and a 3D mouse to delineate directly onto the 3D landscape (double screen to the right). Lines, points, or polygons are simultaneously transferred to a chosen map projection in a GIS program (left), storing shapes and coordinates of all delineations, where attributes (variables and classes) are recorded into a premade database.

Credit: Figure created by Anna Allard; aerial images provided with permission by Lantmateriet, the Swedish mapping, cadastral, and land registration authority.

the time of day, the local season when the vegetation was photographed, and a range of other aspects. Most analysis from above is based more on life forms than particular species and their relative distribution in the landscape, to map the land cover or objects of interest. As with collection of data from the field, a rigorous set of rules and decision trees is applied, to record clear and unambiguous classes, and when more than one person is working on the data collection, long-term calibrations between them is essential (Ihse 2007; Allard et al. 2010; Ståhl et al. 2011; Allard 2017).

Validation of the results is more in line with other remote sensing, done by visiting a number of each class in situ to create a confusion matrix to compare the interpreted classes to reality (e.g. Chuvieco 2020). A definite advantage when working with images is that the moment of inventory can be revisited to rectify systematic mistakes. If the result is to be a map or a dataset for training a model and the main purpose is to get it right, a second interpretation, armed with the knowledge from the field visits, is recommended.

Aerial photos are a common feature in landscape ecology, an interdisciplinary science dealing with combinations of biodiversity and remotely sensed data. It combines such things as spatial patterns or functions of ecological systems in the landscape and the impact of management. Connectivity, fragmentation, and corridors are studied at multiple scales, and landscapes are studied within terrestrial, aquatic, and marine landscapes, often as changes over time (e.g. Lausch et al. 2015; With 2019; Francis et al. 2022); see Text box 5.2.

Historical ecology as a framework for examining biocultural diversity

Historical ecology is based on the truism that the present and future emerge from the past. Trans- and interdisciplinarily between ecology and anthropology (including archaeology), it is concerned with the interactions through time between societies and environments and the consequences of these interactions for understanding the formation of contemporary and past cultures and landscapes (Balée 2006; Crumley et al. 2018).

As ecological approaches aim to inform landscape management to conserve or re-establish pre-disturbance ecosystems; the anthropological approach strives towards similar goals, emphasizing past human behaviours as integral components of ecosystem dynamics (e.g., Crumley 1994; Balée 1998, 2006; Balée and Erickson 2006; Crumley et al. 2018; Isendahl and Stump 2019). Whereas applied restoration projects tend to analyze past human engagement as degrees of interference with natural ecosystems, anthropological historical ecology rejects the Cartesian separation of culture from nature to examine human behaviours' contribution to ecosystem development. Hence, historical ecology emphasizes the sociocultural complexities of human agency, decision making, problem solving, and landscape transformation (e.g., Swetnam et al. 1999; Egan and Howell 2001).

Human behaviour has impacted practically every landscape on Earth, and archaeological records provide leads to track the diversity of causal dynamics between behaviours over time. How these dynamics play out to nurture or dampen biodiversity needs to be detailed on a case-by-case basis, rather than assumed. Historical ecologists tend to focus on the landscape scale for data collection and analysis. However, the analytical perspective is multi-scalar, acknowledging that landscape-scale resource behaviours are influenced by interacting social and environmental processes and systems at work over several temporal and spatial scales, from biological processes in the soil to global socioeconomic relations. Hence, colloquial uses of the concept of historical ecology as largely interchangeable with paleoecology, paleoethnobotany, cultural ecology, or environmental archaeology are incorrect.

Text box 5.2: Example of monitoring land use and detailed landscape patterns over time

The species-rich semi-natural grasslands – that is, unfertilized, uncultivated pastures and hay meadows – are amongst the most threatened habitats in the European agricultural landscape. Many European countries have lost most of their semi-natural grasslands due to abandonment or productivity intensification during the last century (Köhler et al. 2006; European Environment Agency [EEA] 2011; Gustavsson et al. 2011; Jansson 2011; Wood et al. 2018; Ihse 2019). The biodiversity of the current semi-natural grasslands is heavily dependent on the cultural history of the patch, and continuously managed grassland shows the highest diversity in species; they also function as a sink for carbon and water catchments (Johansson et al. 2008; Reitalu et al. 2012; Lehsten et al. 2015; Boyle et al. 2021).

Monitoring changes in land use and patterns of change was done through analysis of the landscape at several points in time, using 3D aerial photos reaching back 70 years. The case study comprises 76 squares (1.1 × 1.1km, in total 97km^2), located in the mostly farmed area across the mid-central part of Sweden.

Land type categories and variables

With the intention of capturing the potential for biodiversity, the landscape categories were constructed from the existing variables in the National Inventories of Landscapes in Sweden (NILS; Ståhl et al. 2011), incorporating the notion of earlier land use in combination with the current, with a separate manual (Allard 2019); a slightly altered version is described in Glimskär and Skånes (2015). This makes it possible to characterize any patch of land from its recent history, such as a cultivated field turned into pasture. With the addition of a set of attributes (e.g. percentages of shrub encroachment, stones turning up after freeze, the appearance of pathways from cattle), the category becomes "turned long ago", thus having a slightly greater potential for recurring biodiversity after chemical fertilizers. For abandoned pastures with a regrowth of trees, the potential for biodiversity, even during the succession phase of young forest, is higher than in a formerly farmed field planted with trees or a forest replanted after clear-cutting. Also recorded were a set of line features (ditches, strips of unfertilized grass, stone walls, two-rut roads, gravel roads, and asphalt roads) and point objects (houses, stone cairns, small ponds).

Interpretation

By using several stereo photo pairs simultaneously in the digital interpretation environment (see Figure 5.4), changes in land use or objects of interest can be followed over time. The stereo pairs are coupled to each other as well as to the GIS program, and the interpreter can examine the development of every patch, which aids both the understanding and the consistency of recorded attributes (one set for each point in time). In the GIS, with auxiliary information displayed from orthophotos and digital maps, attributes are recorded in the premade geographical database, containing classes and variables.

Results

A noticeable change over this time (Figure 5.5) is the size of patches. For the category crop field in use, although the overall area has diminished, where farming is still practised, modern agriculture uses roughly the same area (crop fields in yellow), but the earlier small fields are now large. The varied former landscape, containing ditches, stone walls, or grass strips between every farmed field, has changed. Former smallholdings have turned into summer cottages or been abandoned.

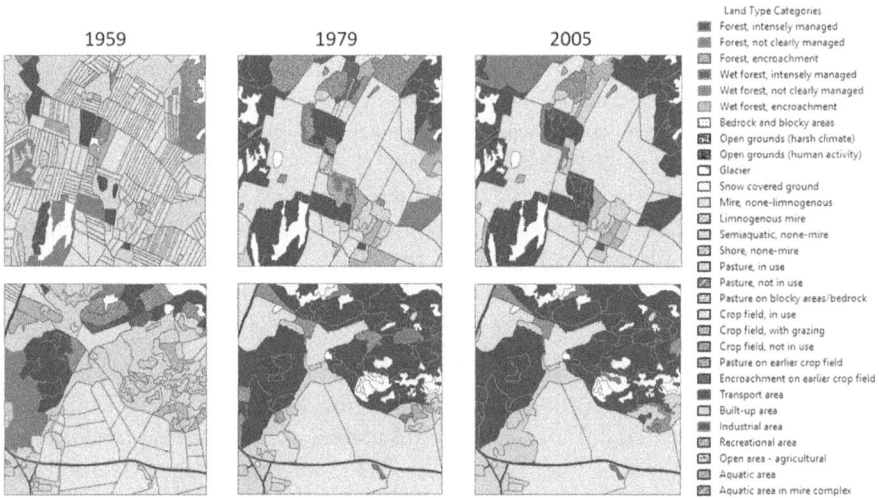

Figure 5.5 Example of two squares (1.1 × 1.1km) from the study (horizontal from left to right), from the mid-central part of Sweden. The vector polygons show the land type categories in a spatiotemporal study of the landscape, revealing the overall changes of potential for biodiversity over circa 60 to 70 years.

Peri-urban areas contain most of the abandoned crop fields; these have often been bought for development or become economically non-viable. Pastures on grassland or on dry grassland with rocky outcrops have turned into re-growth deciduous/mixed forest or have been planted with commercial monocultures of coniferous forest. The total area of changes was calculated through conversion of the vector polygons (in shapefiles) to pixels of 3.3m and snapped on top of each other to produce change trajectories (not shown here) (Allard, Hedenås et al. 2019). Table 5.1 shows the whole population for the line and point objects in the study in numbers and metres.

Table 5.1 Comparisons of the entire population of line and point objects over time in the study

Total area, 97 km²	Polygons, no.	Stone walls, metres	Paved roads, metres	Two-rut roads, metres	Ditches/ streams, metres	Biotope field islet, no.	Stone cairn in the field, no.
1959	8604	6147	129,436	94,386	264,617	383	455
1979	7168	4419	159,296	66,395	144,814	249	271
2005	6964	2929	173,575	57,121	117,837	204	195

Text box 5.3:   Aerial inventory as a component in the second phase in the monitoring of Swedish deciduous forests and grasslands by the National Inventories of Landscapes in Sweden

The National Inventories of Landscapes in Sweden (NILS) is a program that monitors biophysical conditions and changes in deciduous forests, grasslands, alpine habitats, and seashores. The sampling design of NILS has recently changed, and the background for that change is described in Text box 4.1. For the grassland and deciduous forest surveys, the sampling design consists of two phases: the first phase of selecting a sample of tracts based on balanced sampling is described in chapter 4. The second phase of selecting plots within each tract in the sample for field inventory based on remote sensing information is described here.

Step 1. Aerial inventory

Within the deciduous forests and grasslands inventories, each tract consists of a cluster of 196 circle plots. Every plot is inventoried by noting a set of classes and variables in a geographical database using aerial photo interpretation in 3D and in near-infrared (Allard, Forsman et al. 2021). Additional information such as series of orthophotos from different times, altitude, models of potential deciduous forests, broadleaf forests, and soil moisture are available to support the classification. Detailed yearly manuals are developed to facilitate consistent inventory by different people (e.g. Allard, Forsman et al. 2021; B. Nilsson et al. 2022), together with weekly calibration meetings to discuss definitions and interpretations of the landscape. Before step 2, the database is lastly checked for logical errors and corrected. In 2022 (Figure 5.6), 133,280 plots within 680 tracts were inventoried by aerial photos; see example in Figure 5.7, with a corresponding photo from the field visit shown in Figure 5.8.

Step 2: Selecting plots for in situ data collection

The classifications of each plot in step 1 are then used to select plots for field inventory. In this way we can focus the expensive field inventory on plots where we can expect to find grassland or deciduous forest relevant for the surveys. Importantly, we can exclude plots that we are certain do not contain grassland or deciduous forest within the criteria for the surveys.

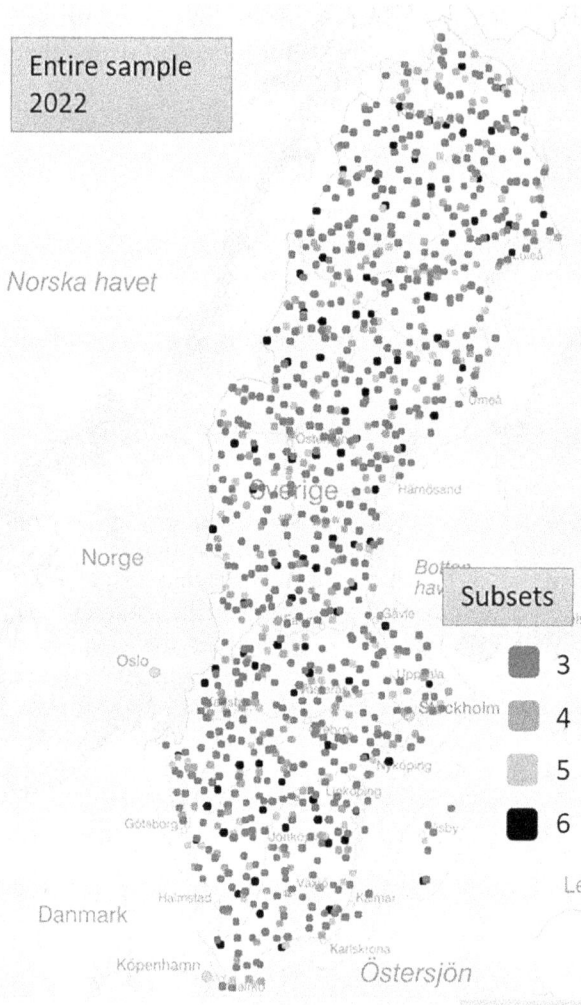

Figure 5.6 The 680 tracts (1 × 1km) inventoried in NILS 2022, each containing 196 circular plots (133,280 in total). They are divided into four subsets with different densities (from six to three as the most dense subset) where all classes are searched for in the most sparse and as the subset gets more dense, fewer classes are searched for until prioritized classes of deciduous forest and grassland types remain in subset 3.

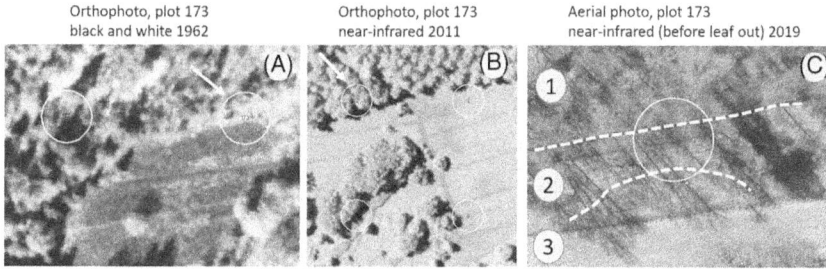

Figure 5.7 Aerial inventory in NILS. The row of images shows plot no. 173 (out of 196) in a
tract; a field photo of the plot is shown in Figure 5.8. (A), (B) Orthophotos (summer
images from 1962 and 2011, respectively); the cutouts show some of the landscape
context, and yellow arrows point to plot 173. The close-up aerial photo (C) is taken
before leaf-out, enabling the interpreter to see through the branches to the field layer
and to distinguish evergreen coniferous forest. In (C), the three different classes
occurring in the plot are delineated in white dotted lines. Area 2 of the circle falls
onto an elongated patch of pasture grassland with long continuity of grazing and is
therefore of high interest to the inventory. In the hierarchical classification, this is the
one that will be recorded in the aerial-photo inventory. Area 3 is a pasture turned
from a crop field somewhere before the 21st century and is of interest but has less
potential to contain high biodiversity. Area 1 is a young densely planted deciduous
forest less than 50 years of age and is disregarded in this inventory.

Source: Aerial images provided with permission by Lantmateriet, the Swedish mapping, cadastral, and land
registration authority.

Handling the impact of observer variation

When it comes to monitoring projects, person-to-person or observer variation is an
inconvenient reality. Not only does observer variation exist but, if not addressed, it is
probably the greatest source of error in a monitoring project, of equal or even greater
importance than sampling error. This is relevant for all types of data collection: from the
field, aerial imagery, and satellite data. All of these include steps where the people involved are making conscious decisions, and the next person deciding species, vegetation
type, or land use of the patch; the weights assigned in a model; or the steps in pre-
processing of satellite data a few years on might not make the same decision.

When the aim is to detect and report change over time, though some variation is to be
expected, the result can be an apparent *pseudo-turnover* of species between years that is not
real but is simply because different observers were involved in different years (Vittoz
et al. 2010; Morrison 2016; Filazzola and Cahill 2021). This is a constant issue to take
into consideration when performing monitoring: to always work toward calibration of
the personnel doing the collection, even if they are long-term workers, or maybe
especially with long-term workers (Galleogos Thorell and Glimskär 2009).

This issue also raises the question of how much data from the previous collection is to
be made known to the next surveyor/recorder, say five years on, when the inventory
rotation comes back to the same spot. We humans have different personalities, and some
tend to believe the first surveyor and record the same mistake again, even if we are

suspicious of the accuracy of the first, whereas others break the trend and enter a new class, species, or attribute (e.g. Furr 2009). Despite this, many long-term monitoring programmes choose to let recorders know some of the fundamental classes, often with instructions to report whether they are righting a "wrong" or whether the class has actually changed since the last visit, to alleviate the otherwise large differences that come from observer variation in general, thus making the statistical estimations more plausible, as has been done by both landscape and forest long-term monitoring in Sweden.

There are different ways to handle observer variation, depending on the scale and funding of the inventory, from single sites to the national scale. When it comes to large national inventories, one of the ways to handle observer variations is to increase the sample, which statistically ameliorates errors in estimates (Filazzola and Cahill 2021); see chapter 4. However, without training courses, calibrations and support to the staff, a large sample will not help.

There have been a number of practical reviews and studies of observer variation in species lists and cover estimates (Vittoz et al. 2010; Morrison 2016; Futschik et al. 2020; Filazzola and Cahill 2021). Typically, these compare and contrast the results from different observers looking at the same plot, covering a range of habitats and settings. They show that different observations can be influenced by both the properties of the vegetation and the physical and mental state of the observer, such as stress or mental fatigue.

The main sources of observer variation (e.g. Morrison 2016; Futschik et al. 2020), even with educated and skilled experts in monitoring, are the following:

- Overlooking or misidentification error when recording the number of species in a sample area.
- Estimation error when recording abundance for a species in a sample area.
- Recording the decision on class or habitat of the sample area or patch of land – often associated with the first two but also with personal likes and dislikes in habitat.

Because we know that many species might be overlooked, we need to consider which species are mostly likely to tell us what we need to know (given what we understand to be the main pressures driving habitat change on the site) and then assess the likelihood of these species being detected. One solution is to use multiple observers (Futschik et al. 2020; Filazzola et al. 2021). That can prove costly; however, field teams can be brought together for regular training and calibration (Seidling et al. 2020); see Text box 5.4. Observer variation can be modelled (Wright et al. 2017) and metadata collected that show the detectability – and therefore the reliability – of each species by each observer during data collection (Bornand et al. 2014; Futschik et al. 2020). What is less easy to specify is how observer variation inflates the estimated variance in sample-survey designs and how this should be taken into account in statistical power analysis of the ability to detect change (but see Mason et al. 2018). Nevertheless, it can be shown that, with care, real and significant change can be detected over the long term, despite observer differences and short-term variation in cover (Christensen and Hedström Ringvall 2013; Futschik et al. 2020).

To record changes in species cover, taking continuous photos or using remote images can help provide more reliable evidence of change than cover estimates produced by field surveyors; see Text boxes 5.3 and 5.4. It is also helpful to take stereo photos by drone or in the field (by shifting weight from one leg to the other, taking a "down"

Text box 5.4:   Field inventory of Swedish alpine habitats, deciduous forests and grasslands by the National Inventories of Landscapes in Sweden

The National Inventories of Landscapes in Sweden (NILS) is a program that monitors biophysical conditions and changes in deciduous forests, grasslands, alpine habitats, and seashores with a focus on habitats with high nature values according to the Habitats Directive (EC 1992). The sampling design of NILS has recently changed. The background for that change is described in Text box 4.1. The selection of plots within each tract within the deciduous forests and grasslands inventories is described in Text box 5.3, and the selection of plots within the inventory of alpine habitats is described in Text box 8.1. The seashore inventory is described in Text box 5.7. All surveys use the same staff and share the cost of recruitment, training courses, travel, and living costs during the field season. The main part of the field staff is seasonally employed, with a requirement of prior education in ecology and plant identification. Courses are held each year for all field staff, separated for the lowlands, alpine areas, and seashores. The office is staffed during the entire season for support on recording; for example, when a second opinion is needed for a species of plant or a habitat. For this type of support, sending in a detailed photo has proven very useful. The support also offers technical help with apps and exporting data to the databases. Manuals exist for all surveys, with a detailed description of the protocol for field survey of plots or transects (Hedenås et al. 2013; Gardfjell and Hagner 2019; Sjödin 2019; Hedenås 2021a, 2021b). Here we briefly describe the field inventory methods used in the inventories of alpine habitats, deciduous forests, and grasslands.

In situ monitoring of alpine habitats, deciduous forests, and grasslands

The sample plots used in NILS are circle plots with a 10m radius (Hedenås 2021b; Ranlund et al. 2021); see Figures 5.8 and 5.9. These are used for recording canopy cover, coverage of shrubs, diameter of trees, and amount of dead wood, for example. Each plot or sub-plot, if divided, is also classified regarding land use, land cover, type of forest, grassland, alpine habitat, and habitats listed in the Habitats Directive (EC 1992). For the decision of classes, a minimum of 0.1ha is required in most cases, but a minimum of 0.25 is required for certain forest habitats listed in the Habitat Directive. For each type of forest, grassland, or alpine habitat, we also note "quality variables" that combined will indicate the quality of the object. Further, within each sample plot are three small vegetation plots, each consisting of two concentric circle plots (Figure 5.9). One of the concentric plots is $1.0m^2$ and is used for cover estimates of field vegetation. The other concentric plots differ in size depending on which small plot it belongs to. The size of the one in the north is $0.25m^2$, the second one is $1.0m^2$, and the third is $100m^2$. These different-sized plots are used to record the presence and absence of plants. Abundance of species will be modelled based on the different-sized plots (cf. Ekström et al. 2020). The field data are recorded using smartphones and specialized apps. As each tract is inventoried, the recorded data are sent to the office. A check for systematic mistakes is done, and the field staff correct the data during the season. Errors that cannot be corrected by the field staff during the season are corrected afterwards, before compiling all field records into a database, from which the data can be extracted for estimation in the reporting phase.

(A)   (B)

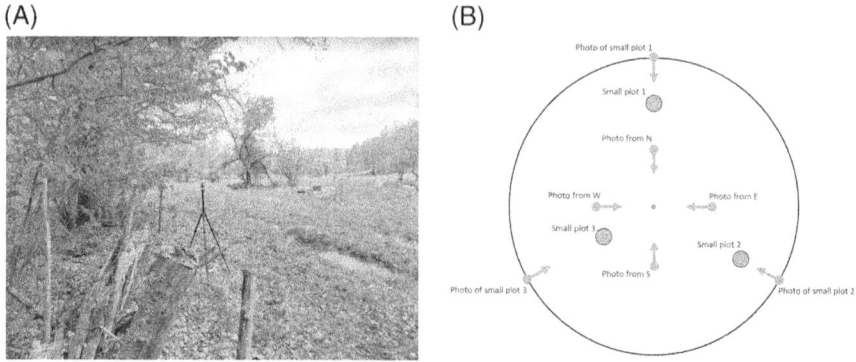

Figure 5.8 (A) Field photo from plot no. 173, inventoried in 2021 (the same plot as in Figure 5.7, showing the strip of continuously grazed grassland, between the former crop field to the left and the young forest to the right). (B) Each field plot is photographed with at least five photos that systematically reflect the plot and its surroundings (cf. Hedblom et al. 2019): four photos from the cardinal points towards the plot centre and one straight up from the canopy from the centre point. In addition, where data from the small vegetation plots are sampled, photos are taken of the small plots: one of the ground, one of the canopy, and one from the perimeter of the circle towards the small plot.

Credit: The National Inventories of Landscapes in Sweden (NILS).

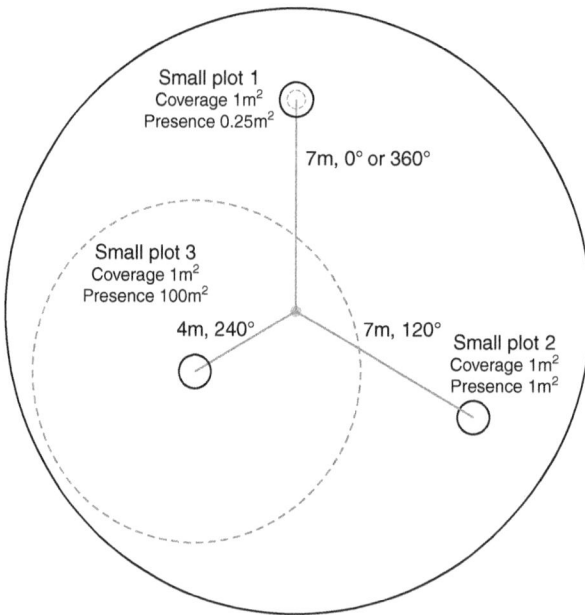

Figure 5.9 The location of the small vegetation plots in the 10m-radius circle plots used in NILS.

Source: Adapted from Hedenås (2021b).

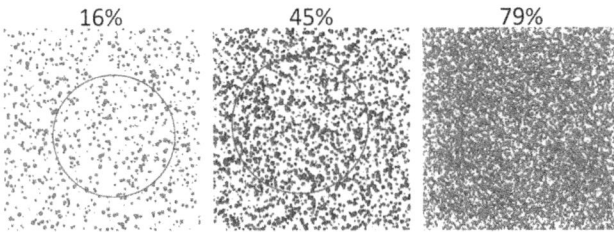

Figure 5.10 A small extract from a helping tool, developed in the NILS inventory programme. A collection of exact covers in different percentages to be consulted by the surveyor in connection to each cover decision, both decreasing time spent and increasing the reliability of cover estimates.

Source: Allard, Nilsson et al. (2021).

photo on each leg), a method used by the UK peatland survey team that worked well for short vegetation in heaths and bogs.

Lastly, the use of reference images showing different covers has proven very efficient in the aerial photo inventory of the NILS programme. NILS has developed a set of reference images of different coverages that provide a support tool to be consulted for each decision of cover in an area, which has greatly improved the cover estimates and differences between persons (Allard, Nilsson et al. 2021; Lindgren et al. 2015); see an example in Figure 5.10. Chapter 16 provides further examples of efficient and reliable monitoring projects, minimizing the impact of observer variation.

Determination of observer variation and conversion of estimates into classes

A way to determine the scale of observer variation of species is to carry out multiple sampling trials, where several experienced botanists are asked to record the same data in a set of fixed relevés (sample plots) within the same day. Opportunities for such trials arise through running in-house sampling trials in large organizations or at monitoring workshops and field trips; see examples in Text boxes 5.5 and 5.6.

Scales of transformation are sometimes used for estimations of cover or abundance, such as Domin or Braun-Blanquet (Mueller-Dombois and Ellenberg 1974; Van der Maarel 1979) scales, but trials show that the problem remains even after transformation. Sampling trials in the UK (Hurford 2006) found no difference between professional surveyors and university students. Most of the observer variation in the Domin scale occurs within the range of Domin classes 4 to 8 (see Table 5.2). This then means that at the bottom of that range (classes 4–5), we would have to ignore any changes of cover, from 4% to 33%, because there would be a 50% chance that there had been no change at all. Similarly, in the range from classes 6 to 8, we would have to ignore changes in cover from 33% to 80% for the same reason. The effects of observer error using the Braun-Blanquet scale would increase, because the cover classes are wider; the distribution of the trials converted to both scales is illustrated in Table 5.2.

The sampling trial data suggest that using the Braun-Blanquet scale, we could not know whether the change in cover had increased from 26% to 100%, decreased from 100% to 26%, or remained stable. Cover changes of this magnitude would have a dramatic impact on the biodiversity value of any habitat.

Text box 5.5:   Observer variations in species detection

English Field Unit grassland sampling trials

During the 1980s, the English Field Unit carried out a sampling trial designed to assess the scale of observer variation associated with recording species diversity and abundance in grassland vegetation (Leach and Doarks 1991). This trial involved 14 experienced grassland surveyors who were asked to independently record all of the species in two fixed quadrats situated in species–rich grassland, one 1 × 1m quadrat and one 10 × 10m quadrat.

The results from this trial showed that on average the surveyors recorded only 63% of the species; in the 1 × 1m quadrat, the most successful surveyor found 73%. As might be expected, the detection rate in the 10 × 10m quadrat was considerably lower, with the most successful surveyor recording only 63% of the species and the average falling to 55%.

These results suggest that if we are interested in recording changes in species diversity, we must be prepared to live with observer variation of ±30% or focus on recording small suites of reliable diversity indicators with high detection rates, comprising those species that we would expect to be among the first to decline or increase if the species diversity of the habitat changes. This range was corroborated by the findings of Ringvall et al. (2004). They calculated the variance in estimations for the first sample scheme in the NILS inventory (stratified random sampling, which was used from 2003 to 2020) and found that for most of the variables, a natural change of 20% to 30% was needed before the estimations were statistically significant. For some of the more rare occurrences, even this was not enough, and a greater density was necessary for a national estimation.

Countryside Council for Wales freshwater macrophyte sampling trials

In this sampling trial, the data were collected on seven dates from June 23 to September 30, 2008, all within the recommended period for macrophyte recording in the UK. Because this exercise focused primarily on species detection, as opposed to species identification, the surveyors could take samples away with them and submit their recording forms after difficult specimens had been verified (by an appropriate referee, if necessary). There was no time limit on the exercise; the surveyors carried on recording until they were satisfied that the sample was complete. The surveyors recorded data from a 100m section of river and a 500m section of river, in keeping with the requirements of the recommended methods for recording macrophytes in the UK.

The results from this sampling trial showed that no surveyor recorded more than 54% of the aquatic and emergent species in the 100m section of river, the smaller and more accessible of the two river sections. At least 48 species of aquatic and emergent plants were found to be present in the 100m stretch of river. These comprised 13 species of algae and lichen, 14 species of bryophyte, and 21 species of vascular plants. Figure 5.11 shows the distribution of the detection rates for the species during this sampling trial.

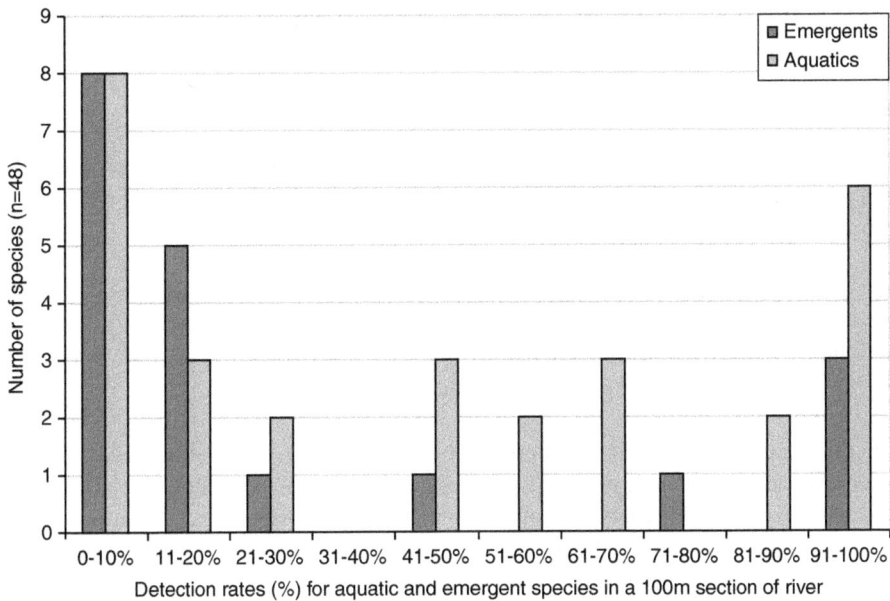

Figure 5.11 The detection rates for aquatic and emergent species in the 100m section of river. Note that half of the species have a less than 20% chance of being detected by the surveyors.

Source: Adapted from Hurford (2010).

We should not underestimate the impact of these levels of observer variation. In this case, the sampling trial results suggested that high levels of observer variation are associated with all methods currently used for monitoring river macrophytes in the UK. The key findings were that:

- Running a river classification on the surveyor data sets from the 500m section resulted in the section being classified as five different river types, with only 1 of the 11 datasets corresponding to the river type defined during the 1980s and underpinning the criteria for the conservation status of the river.
- The master dataset passed the Common Standards Monitoring macrophyte criteria for Favourable Conservation Status (Joint Nature Conservation Committee 2016), whereas the datasets of all 11 surveyors failed.
- Only the more obvious species in the two survey sections were detected by more than 75% of the surveyors, and almost half of the species were detected by less than 20% of the surveyors.

Text box 5.6:   Examples of observer variation in estimates of vegetation cover

Observer variation in estimates of vegetation cover

The main problems with recording estimates of vegetation cover arise primarily because it is a subjective measure where the estimate can be influenced by several variables, such as:

- The familiarity of the observer with the habitat or species being assessed.
- The size of the area of search.
- The complexity of the vegetation in terms of species composition.
- The structure/layering of the vegetation.

In this sampling trial, seven experienced Countryside Council for Wales habitat specialists and monitoring specialists participated in an in-house sampling trial to assess the levels of observer variation associated with recording estimates of ericoid cover in blanket bog vegetation. The ericoid species (*Calluna vulgaris*, *Erica tetralix*, and *Empetrum nigrum*) are all easy to identify and readily visible. Furthermore, the species-group could not be confused with any other species at the 1m-radius sample points. The result shows that the difference between the lowest and highest cover estimate at each sample point varied from 15% to 65%, with a mean difference of 35%. An outlier dataset from a less experienced surveyor was removed; see Figure 5.12.

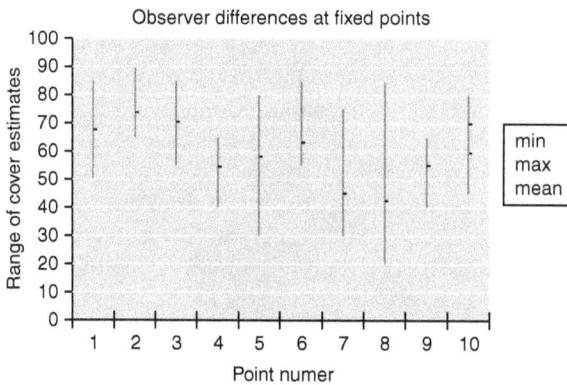

Figure 5.12 The results from a sampling trial to test the range of variation between observers estimating the percentage cover (at intervals of 5%) of ericoids at fixed points in blanket bog vegetation. The mean range of uncertainty was 36%.

Source: Adapted from Hurford (2006).

The results in Figure 5.12 are typical, with the mean range of variation at sampling points during professional multiple observer sampling trials in the UK never <30%. The covers near 10% or 100% are the easiest, and the middle percentages typically show great variation, which was also found by Gallegos Thorell and Glimskär (2009). Their study also showed that expert observers varied more than inexperienced observers primarily because non-specialist observers tend to overlook difficult species that are detected by experts.

In an online exercise during a Eurosite monitoring workshop in November 2021, 37 delegates were asked to estimate the cover of *Potentilla palustris* (to the nearest 10%); the image and results shown in Figure 5.13A-B. In keeping with previous sampling trial results, there was a wide spread of cover estimates, ranging from 20% to 90%, and though the most popular cover classes (chosen by 17 of the participants) were 70% and 80%, more participants (22) felt that the vegetation cover was outside these cover classes. Although estimating vegetation cover on photographs is not ideal, things do not get any easier in the field when structural layering is introduced.

(A)

(B)

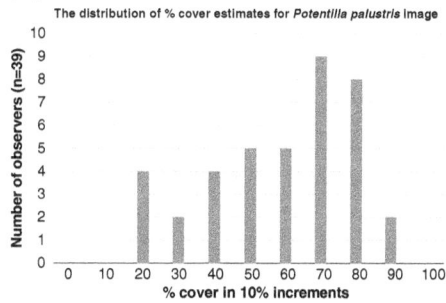

Figure 5.13 (A) The photograph used to assess the range of observer variation associated with estimates of cover. (B) The cover estimates generated by 37 delegates. Most popular estimates were 70% and 80%, but the range is between 20% and 90%, and most observers provided estimates outside the two most popular cover classes.

Credit: Photo by Clive Hurford.

Table 5.2 Even the most generous interpretation of the sampling trial results shown in Figure 5.12 results in a range of Domin and Braun-Blanquet scores that are of little practical use for informing habitat management because the analysis would need to allow one point in either direction to accommodate the effects of observer variation

Domin		Cover estimate distribution	Braun-Blanquet scale		Cover estimate distribution
+	A single individual		+	Less than 1% cover	
1	1–2 individuals		1	1%–5% cover	
2	Several individuals <1% cover		2	5%–25% cover	4
3	1%–4% cover		3	26%–50% cover	6
4	5%–10% cover		4	51%–75% cover	19
5	11%–25% cover	4	5	76%–100% cover	10
6	26%–33% cover	2			
7	34%–50% cover	4			
8	51%–75% cover	19			
9	76%–90% cover	10			
10	91%–100%cover				

The distribution of cover estimates shows the number of delegates who allocated their estimate to the corresponding cover bands on the Domin/Braun-Blanquet scales.

Different data types and uncertainty

As discussed briefly in chapter 2, the common characteristic of monitoring is working with uncertain knowledge. The discussion of field inventory and in situ recording showed how uncertainty and observer error can be a problem and how having realistic guidance and training, including standardization between field observers, can help reduce this. Chapter 4 showed how handling some types of uncertainty is not only part of the statistical design but that working with only a small sample can give unbiased estimators of population parameters that are both efficient and cost-effective. Clearly, some types of uncertainty can be tolerated to make the design more efficient and other types need to be reduced or eliminated. Understanding the details of how this can be done for a particular design is a key part of designing and running biodiversity monitoring and one that needs both theoretical expertise and experience in field inventory.

Field-based monitoring most often uses sample schemes, adopting some sort of layout that statistically represents the whole, to make estimations of entire populations.

Remote sensing from aerial imagery, radar or laser sensors, or satellites is often used to cover areas completely (*wall-to-wall coverage*); see more in chapters 7 to 9. These types of observations still have uncertainty, but it is less the result of sampling choices and more to do with correction factors for atmospheric conditions and limiting spatial, spectral, and radiometric resolution. Repeated coverages are now a reality, giving us time series for phenological profiles across the season and allowing us to know something about the habitat in seasons we did not visit. Weeding out the factors of interest among multi-spectral or hyperspectral data now occupies many monitoring schemes; for example, deducing the percentage of pine trees inside a patch of mixed coniferous trees. Here the value of good training data for classification should not be underestimated, and there is much truth in the saying (equally true of all monitoring components) "the quality of the output depends on the quality of what you put in".

Lastly, the existing maps of most countries contain a lot of valuable data but often in combinations of several factors making a whole; these are themes of classes of landscape sections, like "low to mid-height buildings with lush gardens" or "heath vegetation". Maps can also show the old landscape and, to some extent, the vegetation in it. Many countries have maps from the 1800s, and sometimes older. These are of great value for deducing the cultural heritage (e.g. Antonson 2018). In cultural landscapes where there is a good road network, online imagery from Google Maps Street View can also be helpful for checking vegetation and land cover classes (Google 2022).

These types of data include different scales: *nominal scales* (discrete classes such as land cover), *interval scales* (with even intervals between values but no fixed zero), or *ratio scales* (like interval data but with a fixed zero). Other important sources of data relevant to biodiversity monitoring can include social surveys and preferences; for example, current and preferred future land use (chapters 11–15). This information can often come in the form of rankings, known as *ordinal scale* data.

Uncertain knowledge is often referred to as *belief*, and typically – but not always – we choose to manage uncertainty for interval and ratio scale data (and sometimes rankings) using statistical probability (the measure of the likelihood, or certainty, of an event occurring). This uncertainty is referred to as *error* or *statistical error*, though the term is misleading because there is usually no implication that a mistake has been made. However, we might need other models of uncertainty such as *ambiguity* and *vagueness* when incorporating different types of information (Halpern 2003; Comber et al. 2004; Liu and Mason 2009), notably expressions of social preferences, remotely sensed data, and thematic maps; see also chapter 15.

Good practice in developing practical monitoring schemes

Most monitoring schemes contain an element of collection in situ, and there are many examples across Europe and globally; we give only a small number of examples in this book. A description of steps is useful to understand the work needed before starting a scheme, and a thorough investigation of what others have done before starting a new project is recommended. Following are some steps that are important to consider.

Manual or handbook

A manual contains concise and clear definitions of the classes; attributes; variables or species lists that are to be collected. This is harder than one might think and consumes much time and debate before some kind of consensus is reached. If the result is a map (e.g. of land cover, land use, habitat), we have to decide whether to include labels for mixtures and mosaics and whether to allow labels within the hierarchical scheme where the ambiguity cannot be resolved during the field visit. From the start, we should deal with the *edges* of the classes and create a hierarchy of where to put the types that con-stitute a mixture between two, often called *ecotones* (if not used as classes of their own) or *landscape patches* that have a degree of mosaic, making them hard to put down as any one class. Do we allow the staff to record a broader, more general class when uncertain of which one to choose?

Training and calibration of staff

In either case, a solid course at the start is highly recommended, for new recruits as well as returning staff, to calibrate how to interpret the manual and avoid what we term *schools of thinking*, meaning that, over time, veterans develop certain ways (or traditions) of recording and then teach them to new recruits (though not all of them). In the Scottish Upland monitoring (Ross and Flagmaier 2015), for example, they had:

- Joint visits with all team members on the first day on a new site, covering a range of habitats.
- Some overlap and duplicated recording between more and less experienced field staff.
- Multivariate analysis across sites to check for gaps in patterns of habitat distribution where observations might have been missed.

On-hand support from the office should be provided, where field staff can call in to discuss and send pictures to clarify the queries (habitats, species, ecotones, or mosaics). For staff working with aerial photos and already in the office, the training course is just as important. Weekly calibration meetings during the whole process of inventory and data collection, looking at examples and discussing how they are interpreted by different people, is crucial to optimize comprehensive results.

Database structure

Deciding on a database structure to store incoming data is important to avoid being swamped with digital files from a myriad of fieldworkers.

Sample

The sample is where to collect data; balanced sampling is discussed in chapter 4. Other approaches include stratified sampling (e.g. Ståhl et al. 2011), the area frame layout in modelled landscape types in Norway (Bryn et al. 2018), and the sampling inside land classes of the UK Countryside Survey (Wood et al. 2018).

Recording

The use of smartphones and apps is now the most common way of recording data, either as an existing app with some adjustments or building specialized apps to suit the collection scheme. Ideally, the recorded data are checked for mistakes by the field staff before sending them to the office, where the data are stored in the premade database. The importance of providing on-hand support from the office cannot be underestimated, to answer questions and avoid expensive stops in the work.

Repeatability

Another issue to be decided is whether the sample plots are permanent; that is, should they be possible to find again for surveillance of exactly the same plots? Or is it enough to be roughly at the same place or even visit new plots? In the Swedish NILS monitoring, the exact spots are marked, and a small map of bearings (the angle between the direction of the object to the true north) from the marked centre of the field plot to three stable landmarks is made (large boulders, telemasts, and such that hopefully will not have moved in five years). The procedure for first visit and recurring visits is described in the manual, which is updated yearly (Sjödin 2019). The time interval to the next visit is often a rotation of five years, ensuring new data for the production of reports at different levels of policymaking. All of these decisions have different statistical implications.

Long-term sample plots

Sometimes the choices have political implications, as in the long-term programmes with permanent plots, which greatly enhances the possibilities of assessing changes in the landscape, but the need then arises to keep the locations secret (as in many national monitoring schemes using permanent samples; e.g. in the UK Countryside Survey or National Forest Inventories in Sweden and Finland). The secrecy has two main reasons, both statistical in their nature. One is the tendency for inhabitants to change their way of management if they know that their land is continuously monitored, and the other is the tendency for authorities to want to "right wrongs" when they find them, as the overseers of official policies. If the aim is to monitor the "everyday" landscape, to report on how a nation is responding to chosen policies, loss of habitats, or issues like a changing climate, the statistical answer must not be that *all is right and nothing has changed*, due to a sample layout of perfectly managed landscape patches. One of the uses of wall-to-wall satellite imagery is testing whether or not the sample set of units remains representative (see discussion in chapter 4) of the wider countryside.

Text box 5.7:   Monitoring seashore habitats in Sweden

As part of monitoring for the Habitats Directive (EC 1992), Sweden carries out a separate inventory along shorelines in the Monitoring of Terrestrial Habitats (THUF) programme to detect and estimate the extent of habitats, mandated monitoring to contribute to the country´s reporting obligation and to detect significant changes over time. The European Environmental Agency produces maps from the reports. The latest map of Natura 2000 Birds and Habitats Directives of member states in the EU is available at their website (EC 2022). In Forsman et al. (2014, 2022), step 1 consists of an inventory using aerial photos in near-infrared and studied in 3D.

Although the sample design has changed to be similar to the sampling in the National Inventories of Landscapes in Sweden, the NILS monitoring (see design in chapter 4), most of the variables and classes are kept constant; see Table 5.3. The design is a sampling of 1km squares, out of all possible over the coastline and archipelagos. The sample is divided into five different subsets, where different habitats are searched for according to how common they are in nature, with the most rare (in Sweden: dune formations on and above the shore and coastal meadows) in the sample containing the most squares. The sampled squares are then clustered to save on the otherwise heavy travel costs, especially to reach islands, even though a regional year/weather bias is introduced. One-fifth, making up roughly 500 squares, of the full sample is inventoried each year.

Table 5.3 Variables and classes in the seashore monitoring

Variable	Classes and subclasses
Coast type	Mainland, island, islet, shallow
Tree/forest layer	Off mainland: treeless, single trees, forest
Shore type	Cliff, gravel/boulders, sand, coastal meadow/wetland
Reed belt (*Phragmites* spp.)	Measured in metres
Mudflat and sandflat	Yes or no (outwards along transect line)
Salt pans	Yes or no (along transect line)
Coastal lagoon	Yes or no (both sample point and transect line)
Drift line on shore	Yes or no (along transect line)
Land use at seashore	Along transect line
Land use above seashore	Within 0.1ha above transect
Sand dunes, bare + tree-covered	Above transect
Pristine forest on land upheaval	Along and above transect
Sea cliffs	Along and above transect
Shore banks dominated by gravel/stone	Along and above transect
Coastal meadow	Along transect
Exploitation (if not on map)	Jetty, house, dumping ground, road, dredging, paved
Specification into habitat	In situ only; 19 seashore habitats in Sweden
Detailed species data	In situ only; in water, at point, and along transect

Most variables are recorded both in aerial photo interpretation and in situ, with the addition of extra detailed data recorded only in situ, where the actual habitat is also determined.

When sampling is done, the inventory starts by aerial imagery in 3D, recording the variables in a database, using a hexagon grid in which points are recorded as the grid lines cross a sea shoreline. A transect is drawn from the shoreline point, adding the variables for the shore and measuring phenomena above shore (e.g. land upheaval forest and dunes with vegetation). The results become the basis for the subset of potential habitats for in situ collection, where the points are located and an in situ transect is drawn up, recording detailed vegetation data and habitats, using a smartphone with an app especially made for the inventory; see Figure 5.14 and Table 5.3 (Hedenås et al. 2013; Forsman et al. 2014, 2022).

Most variables are recorded both in aerial photo interpretation and in situ, with the addition of extra detailed data recorded only in situ, where the actual habitat is also determined.

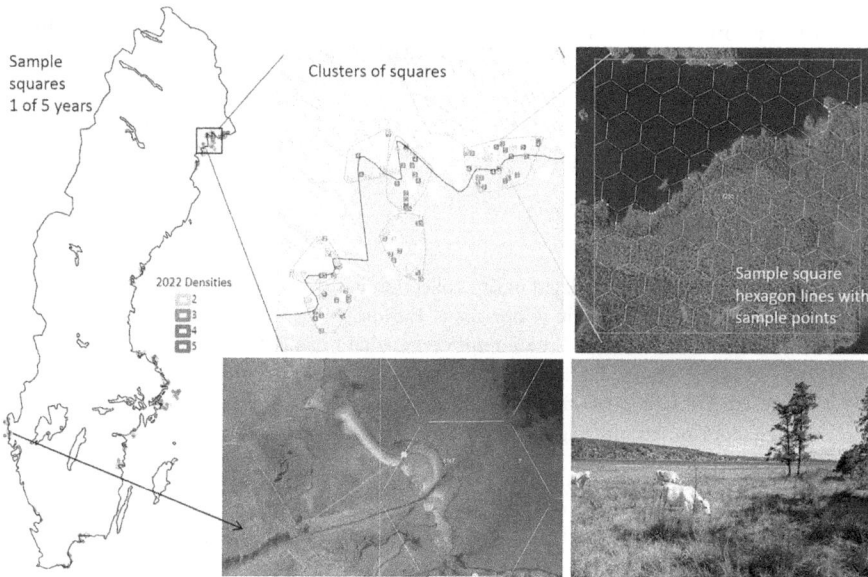

Figure 5.14 Design of monitoring habitats along the shorelines of Sweden. The map shows the samples for the year 2020, where roughly 500 sample squares are monitored (upper right photo), by 3D aerial imagery in infrared (lower left photo) and then a subset of found habitats is monitored in situ (lower right photo). The habitats at this site are coastal meadow with pasture as land use and in the water, glasswort mudflats and sandflats.

Credit: The THUF inventory; aerial images provided with permission by Lantmateriet, the Swedish mapping, cadastral, and land registration authority.

Key messages

- This chapter discusses some of the important issues to consider when performing monitoring in situ, both as the actual field inventory and from aerial inventory, where data collection has many similarities to the field. Not only is the design of the larger sample (see chapter 4) important but also the design of the actual plots visited in the field; some examples of how different programmes have solved that are shown in the text boxes.
- The importance of setting the limits and boundaries to classes and making sure that all persons involved understand them in the same way is often surprisingly time-consuming, but it is necessary.
- The handling of observer variations is an issue in all monitoring, as exemplified here, along with a discussion regarding some solutions.
- Uncertainty is something that all monitoring and data providers have to deal with and is discussed in relation to data types.
- Lastly, some steps of good practice in setting up a monitoring scheme are listed.

Study questions

1   Thinking about classifications, what are the pros and cons to using hierarchical or non-hierarchical classifications systems?
2   How would you avoid, or at least minimize, person-to-person variation, and why?
3   Read and think about the list of good practices in monitoring. How would you make that list better?

Further reading

For a comprehensive summary of updated in situ collection in the EU, we recommend the websites for The Copernicus In Situ Component (Copernicus Europe's Eyes on Earth 2022), EU Ocean for Observation (Eu4oceanobs 2022), and the intergovernmental organizations HELCOM (the Baltic Sea region; HELCOM Baltic Marine Environment Protection Commission 2022) and OSPAR (the North Atlantic region; OSPAR Commission 30th 2022).

A comprehensive work on monitoring in situ and the statistical connections to them can be found in:

Walker, K., Dines, T., Hutchinson, N. and Freeman, S. (2010) *Designing a New Plant Surveillance Scheme for the UK*, JNCC Report No. 440. Peterborough, UK: Joint Nature Conservation Committee. https://data. jncc.gov.uk/data/40e52f7b-5758-4d4a-ba19-f786fec5b718/JNCC-Report-440-FINAL-WEB.pdf

Reviews on observer variations and possible solutions can be studied in the following:

Futschik, A., Winkler, M., Steinbauer, K., Lamprecht, A., Rumpf, S.B., Barančok, P., Palaj, A., Gottfried, M., Pauli, H. and Bartha, S. (2020) Disentangling observer error and climate change effects in long-term monitoring of alpine plant species composition and cover, *Journal of Vegetation Science* 31(1), 14–25. doi: 10.1111/jvs.12822
Morrison, L.W. (2016) Observer error in vegetation survey: a review, *Journal of Plant Ecology* 9(4), 367–379. doi: 10.1093/jpe/rtv077

References

Alberdi, I., Condés, S. and Martínez-Millán, J. (2010) Review of monitoring and assessing ground vegetation biodiversity in national forest inventories, *Environmental Monitoring and Assessment* 164(1), 649–676.
Allard, A. (2017) NILS – a nationwide inventory program for monitoring the conditions and changes of the Swedish landscape, in Díaz-Delgado R., Lucas R. and Hurford C. (eds) *The Roles of Remote Sensing in Nature Conservation*. Cham, Switzerland: Springer, pp. 79–90. https://doi.org/10.1007/978-3-319-64332-8_5
Allard, A. (ed.) (2019) *Instruktion för variabeln Markslag vid Nationell Inventering av Landskapet i Sverige, NILS varv 1, Inventeringsår 2003–2007, 1970–80-Talen och 1950–60-Talen* [Instruction for the variable Land Type Category at the National Inventory of Landscapes in Sweden, NILS, Rotation 1, Inventory Years 2003–2007, the 1970s–1980s and the 1950s–1960s]. Umeå: Swedish University for Agricultural Sciences. https://www.slu.se/globalassets/ew/org/centrb/nils/publikationer/2019/ markslag_manual_version_8_2019_med-bilaga.docx.pdf (Accessed November 2, 2022).
Allard, A., Forsman, H., Granholm, A.-H., Hedenäs, H., Nilsson, B. and Ranlund, Å. (2021) *Instruktion för Nationell Inventering av Lövskogar och Gräsmarker med Hjälp av Stereoflygbilder och Ortofoton, år 2021* [Instruction for National Inventory of Deciduous Forest and Grassland Using Stereo Aerial Photos and Orthophotos, Year 2021]. Umeå: Swedish University for Agricultural Sciences. https:// www.slu.se/globalassets/ew/org/centrb/nils/publikationer/2022/nilsgraslovflygbildsmanual2021. pdf (Accessed November 2, 2022).
Allard, A., Glimskär, A., Svensson, J. and Christensen, P. (2010) Monitoring landscape and vegetation in the Swedish NILS-program, in Bryn, A., Dramstad, W. and Fjellstad, W. (eds) *Mapping and Monitoring of Nordic Landscapes, Conference Proceedings, Viten fra Skog og Landskab*. Oslo: Norwegian Institute for Forestry and Landscape, pp. 5–8.

Allard, A., Hedenås, H., Marcus Hedblom, M., Sven Adler, S. and Christensen, P. (2019) The past present and future fragmentation in the Swedish landscape, as contribution to loss of biodiversity, Presentation and abstract in Proceedings from 10th IALE World Congress, Nature and society facing the Anthropocene - Challenges and Perspectives for Landscape Ecology, Milano, July 2019. https://www.iale.de/en/international-congresses.html (Accessed December 28, 2022).

Allard, A., Nilsson, B., Pramborg, K. and Gallegos, Å. (2021) *Judging Cover of Vegetation in the NILS Programme*, Report from NILS, Swedish University for Agricultural Sciences, Umeå.

Antonson, H. (2018) Revisiting the "reading landscape backwards" approach: advantages, disadvantages, and use of the retrogressive method, *Rural Landscapes: Society, Environment, History* 5(1), 1. doi: https://doi.org/10.16993/rl.47

Balée, W. (ed.) (1998) *Advances in Historical Ecology*. New York: Columbia University Press.

Balée, W. (2006) The research program of historical ecology, *Annual Review in Anthropology* 35, 75–98.

Balée, W. and Erickson, C.L. (eds) (2006) *Time and Complexity in Historical Ecology: Studies in the Neotropical Lowlands*. New York: Columbia University Press.

Baraloto, C., Molto, Q., Rabaud, S., Hérault, B., Valencia, R., Blanc, L., Fine, P.V.A. and Thompson, J. (2013) Rapid simultaneous estimation of above ground biomass and tree diversity across neotropical forests: a comparison of field inventory methods, *Biotropica* 45(3), 288–298.

Bornand, C.N., Kéry, M., Bueche, L., Fischer, M. and Yoccoz, N. (2014) Hide-and-seek in vegetation: time-to-detection is an efficient design for estimating detectability and occurrence, *Methods in Ecology and Evolution* 5(5), 433–442. doi: 10.1111/2041-210X.12171

Boyle, D., Mamais, L. and Khabarov, N. (2021) *A Case Study – Grassland Monitoring in Estonia Sentinels Benefits Study (SeBS)*. European Association of Remote Sensing Companies. https://earsc.org/sebs/wp-content/uploads/2021/05/Grassland-Monitoring-in-Estonia_vfinal.pdf (Accessed November 2, 2022).

Bryn, A., Strand, G.-H., Angeloff, M. and Rekdal, Y. (2018) Land cover in Norway based on an area frame survey of vegetation types, *Norsk Geografisk Tidskrift* 72(3), 131–145. https://doi.org/10.1080/00291951.2018.1468356

Canfield, R.H. (1941) Application of the line interception method in sampling range vegetation, *Journal of Forestry* 39(4), 388–394.

Christensen, P. and Hedström Ringvall, A. (2013) Using statistical power analysis as a tool when designing a monitoring program: experience from a large-scale Swedish landscape monitoring program, *Environmental Monitoring and Assessment* 185(9), 7279–7293.

Chuvieco, E. (2020) *Fundamentals of Satellite Remote Sensing – An Environmental Approach*. 3rd edn. CRC Press.

Codish, S. and Shiffman, R.N. (2005) A model of ambiguity and vagueness in clinical practice guideline recommendations, *AMIA Annual Symposium Proceedings Archive* 2005, 146–150.

Comber, A.J., Fisher, P.F. and Wadsworth, R.A. (2004) Assessment of a semantic statistical approach to detecting land cover change using inconsistent data sets, *Photogrammetric Engineering and Remote Sensing* 70(8), 930–938. doi: 10.14358/PERS.70.8.931

Copernicus Europe's Eyes on Earth. (2022) The Copernicus in situ component, https://insitu.copernicus.eu/ (Accessed June 2, 2022).

Crumley, C.L. (ed.) (1994) *Historical Ecology: Cultural Knowledge and Changing Landscapes*. Santa Fe, NM: School for Advanced Research Press.

Crumley, C.L., Lennartsson, T. and Westin, A. (eds) (2018) *Issues and Concepts in Historical Ecology: The Past and Future of Landscapes and Regions*. Cambridge University Press.

Dengler, J., Biurrun, I. and Dembicz, I. (2021) Standardised EDGG methodology for sampling grassland diversity: second amendment, *Palaearctic Grasslands* 49, 22–26. https://edgg.org/sites/default/files/page/PG_49_22_26.pdf (Accessed November 2, 2022).

Dengler, J., Boch, S., Filibeck, G., Chiarucci, A., Dembicz, I., Guarino, R., Henneberg, B., Janišová, M., Marcenò, C., Naqinezhad, A., et al. (2016) Assessing plant diversity and composition in grasslands across spatial scales: the standardised EDGG sampling methodology, *Bulletin of the Eurasian Dry Grassland Group* 32, 13–30.

Egan, D. and Howell, E.A. (eds) (2001) *The Historical Ecology Handbook: A Restorationist's Guide to Reference Ecosystems*. Washington, DC: Island Press.

Ekström, M., Sandring, S., Grafström, A., Esseen, P.-A., Jonsson, B.G. and Ståhl, G. (2020) Estimating density from presence/absence data in clustered populations, *Methods in Ecology and Evolution* 11, 390–402. doi: 10.1111/2041-210X.13347

Esri UK & Ireland. (2022) *Sweet for ArcGIS*, https://resource.esriuk.com/blog/sweet-for-arcgis-highly-accurate-data-collection-with-validation/ (Accessed June 2, 2022).

EU4oceanobs, International Ocean Governance. (2022) In situ observations, https://www.eu4oceanobs.eu/eu_ocean_observing/collect-data/in-situ/ (Accessed November 3, 2022).

Eurac Research. (2022) Biodiversity Monitoring South Tyrol, https://biodiversity.eurac.edu/ (Accessed June 23, 2022).

European Commission. (1992) *The Habitats Directive*, Council Directive 92/43/EEC of 21 May 1992. https://ec.europa.eu/environment/nature/legislation/habitatsdirective/index_en.htm (Accessed May 22, 2022).

European Commission. (2022) Environment, Natura 2000 data and maps, https://ec.europa.eu/environment/nature/natura2000/data/index_en.htm (Accessed November 5, 2022).

European Environment Agency. (2011) *Landscape Fragmentation in Europe*, EEA Report, No. 2/2011. Copenhagen: EEA.

Filazzola, A. and Cahill, J.F. (2021) Replication in field ecology: identifying challenges and proposing solutions, *Methods in Ecology and Evolution* 12(10), 1780–1792. doi: 10.1111/2041-210X.13657

Forsman, H., Granholm, A.-H., Allard, A., Press, A., Gardfjell, H. and Hagner, Å. (2022) *Flygbildsinventeringen av Havsstränder i THUF, 2022 års Inventering* [Inventory of Seashores Using Aerial Photos in THUF, Year 2022]. Umeå: Swedish University for Agricultural Sciences, Section for Landscape Analysis. https://www.slu.se/globalassets/ew/org/centrb/nils/publikationer/2022/flygbildsinventeringen-av-havsstrander-i-thuf_sasong-2022.pdf

Forsman, H., Hagner, Å., Gardfjell, H. and Adler, S. (2014) *Habitat Inventory by Aerial Photo Interpretation in MOTH – Terrestrial and Seashore Inventory*. Umeå: Swedish University for Agricultural Sciences, Section for Landscape Analysis. https://www.slu.se/globalassets/ew/org/centrb/moth/publikationer/habitat-inventory-by-aerial-photo-interpretation-in-moth--terrestrial-and-seashore-inventory.pdf

Francis, R.A., Millington, J.D.A., Perry, G.L.W. and Minor, E.S. (eds) (2022) *The Routledge Handbook of Landscape Ecology*. 1st edn. Routledge.

Furr, R.M. (2009) Personality psychology as a truly behavioural science, *European Journal of Personality* 23(5), 369–401. https://doi.org/10.1002/per.724

Futschik, A., Winkler, M., Steinbauer, K., Lamprecht, A., Rumpf, S.B., Barančok, P., Palaj, A., Gottfried, M., Pauli, H. and Bartha, S. (2020) Disentangling observer error and climate change effects in long-term monitoring of alpine plant species composition and cover, *Journal of Vegetation Science* 31(1), 14–25. doi: 10.1111/jvs.12822

Gallegos Thorell, Å. and Glimskär, A. (2009) Computer-aided calibration for visual estimation of vegetation cover, *Journal of Vegetation Science* 20, 973–983. https://onlinelibrary.wiley.com/doi/full/10.1111/j.1654-1103.2009.01111.x

Gardfjell, H. and Hagner, Å. (2019) *Instruktion för Habitatinventering i NILS och THUF, 2019* [Instruction for Habitat Inventory in NILS and THUF 2019]. Umeå: Swedish University for Agricultural Sciences, Section for Landscape Analysis. https://www.slu.se/globalassets/ew/org/centrb/nils/publikationer/2019/habitatkompendium_nilsthuf_2019.pdf

Glimskär, A. and Skånes, H. (2015) Land type categories as a complement to land use and land cover attributes in landscape mapping and monitoring, in Ahlqvist, O., Janowicz, K., Varanka, D. and Fritz, S. (eds) Land Use and Land Cover Semantics – Principles, Best Practices and Prospects. Boca Raton, FL: CRC Press.

Google. (2022) Google Maps. https://www.google.com/maps/@63.8203584,20.3230125,14z (Accessed June 23, 2022).

Gustavsson, E.A., Dahlström, M., Emanuelsson, M., Lennartsson, T. and Wissman, J. (2011) Combining historical and ecological knowledge to optimise biodiversity conservation in semi-natural grasslands, in Pujol, J.L. (ed.) *The Importance of Biological Interactions in the Study of Biodiversity*. IntechOpen Book Series, pp. 173–196. https://www.intechopen.com/books/1365

Halpern, D.F. (2003) *Thought & Knowledge: An Introduction to Critical Thinking*. 4th edn. Lawrence Erlbaum.

Hedblom, M., Hedenås, H., Blicharska, M., Adler, S., Knez, I., Mikusiński, G., Svensson, J., Sandström, S., Sandström P. and Wardle D.A. (2019) Landscape perception: linking physical monitoring data to perceived landscape properties, *Landscape Research* 45(2), 179–192. doi: 10.1080/01426397.2019.1611751

Hedenås, H. (ed.) (2021a) *Fältinstruktion för Nationell Inventering av Fjällen, 2021* [Field Instructions for National Inventory of Alpine Habitats Year 2021]. Umeå: Swedish University for Agricultural Sciences, Section for Landscape Analysis. https://www.slu.se/globalassets/ew/org/centrb/nils/publikationer/2022/nilsfjallinventeringfaltmanual2021_20210720.pdf

Hedenås, H. (ed.) (2021b) *Fältinstruktion för Nationell Inventering av Gräsmarker och lövskog, År 2021* [Field Instructions for National Inventory of Grassland and Deciduous Forest, Year 2021]. Umeå: Swedish University for Agricultural Sciences, Section for Landscape Analysis. https://www.slu.se/globalassets/ew/org/centrb/nils/publikationer/2022/nilsgraslovskogsinventeringfaltmanual2021_20210610a.pdf

Hedenås, H., Gardfjell, H. and Hagner, Å. (2013) *Instruction for Seashore Inventory in MOTH, 2013*. Umeå: Swedish University for Agricultural Sciences, Section for Landscape Analysis. https://www.slu.se/globalassets/ew/org/centrb/moth/publikationer/instruction-for-seashore-inventory-of-habitats-in-moth_english-version.pdf

HELCOM Baltic Marine Environment Protection Commission. (2022) Home page, https://helcom.fi/ (Accessed June 3, 2022).

Henttonen, H.M. and Kangas, A. (2015) Optimal plot design in a multipurpose forest inventory, *Forest Ecosystems* 2(1), 1–14.

Hilpold, A., Anderle, M., Guariento, E., Marsoner, T., Paniccia, U., Plunger, J., Rüdisser, J., Scotti, A., Seeber, J., Stifter, S., et al. (In press) *Handbook Biodiversity Monitoring South Tyrol*. Bozen, Italy: Eurac Research.

Hurford, C. (2006) Minimising observer error: increasing the reliability of a monitoring project, in Hurford, C. and Schneider, M. (eds) *Monitoring Nature Conservation in Cultural Habitats: A Practical Guide and Case Studies*. Dordrecht, The Netherlands: Springer, pp. 79–92.

Hurford, C. (2010) Observer variation in river macrophyte surveys: the results of multiple-observer sampling trials on the Western Cleddau, in Hurford, C., Schneider, M. and Cowx, I. (eds) *Conservation Monitoring in Freshwater Habitats: Practical Guide and Case Studies*. Dordrecht, The Netherlands: Springer, pp. 137–146.

Ihse, M. (2007) Colour infrared aerial photography as a tool for vegetation mapping and change detection in environmental studies of Nordic ecosystems: a review, *Norsk Geografisk Tidsskrift* 61(4), 170–191. doi: 10.1080/00291950701709317

Ihse, M. (2019) Förändrade landskap och försvinnande biotoper [Changed landscapes and loss of biotopes, in Ihse, M. (ed.) *Landskap. Ett Vidsträckt Begrepp. En Antologi om Landskap* [Lanscape. A Wide Concept. An Anthology about Landscape]. Stockholm: Kungl. Skogs- och Lantbruksakademien, pp. 42–65.

Isendahl, C. and Stump, D. (eds) (2019) *The Oxford Handbook of Historical Ecology and Applied Archaeology*. Oxford: Oxford University Press.

Jansson, U. (ed.) (2011) *National Atlas of Sweden: A Cartographic Description of Agriculture and Forestry in Sweden since 1900*. The Royal Swedish Academy of Agriculture and Forestry.

Johansson, L.J., Hall, K., Prentice, H.C., Ihse, M., Reitalu, T., Sykes, M.T. and Kindström, M. (2008) Semi-natural grassland continuity, long-term land-use change and plant species richness in an agricultural landscape on Öland, Sweden, *Landscape and Urban Planning* 84, 200–211.

Joint Nature Conservation Committee. (2016) *Common Standards Monitoring Guidance for Rivers*. JNCC. https://data.jncc.gov.uk/data/1b15dd18-48e3-4479-a168-79789216bc3d/CSM-Rivers-2016-r.pdf

Köhler, R., Olschofsky, K. and Gerard, F. (eds) (2006) *Land Cover Change in Europe from the 1950'ies to 2000 – Aerial Photo Interpretation and Derived Statistics from 59 Samples Distributed across Europe*. Institute for World Forestry, University of Hamburg.

Lausch, A., Blaschke, B., Haase, D., Syrbe, R.-O., Tischendorf, L. and Watz, U. (2015) Understanding and quantifying landscape structure – a review on relevant process characteristics, data models and landscape metrics, *Ecological Modelling* 295 (2015) 31–41. doi: https://doi.org/10.1016/j.ecolmodel.2014.08.018

Leach, S.J. and Doarks, C. (1991) *Site Quality Monitoring, Methods and Approaches (with Particular Reference to Grasslands*, Project No 135. Peterborough, UK: English Field Unit, Nature Conservancy Council.

Lehsten, V., Sykes, M.T., Scott, A.V., Tzanopoulos, J., Kallimanis, A., Mazaris, A., Verburg, P.H., Schulp, E., Potts, S.G. and Vogiatzakis, J. (2015) Disentangling the effects of land-use change, climate and CO_2 on projected future European habitat types, *Global Ecology and Biogeography* 24, 653–663. eoi: 10.1111/geb.12291

Lillesand, T.M. and Kiefer, R.W. (2015) *Remote Sensing and Image Interpretation*. 7th edn. New York: Wiley.

Lindgren, N., Christensen, P., Nilsson, B., Åkerholm, M., Allard, A., Reese, H. and Olsson, H. (2015) Using optical satellite data and airborne lidar data for a nationwide sampling survey, *Remote Sensing* 7, 4253–4267. https://www.slu.se/globalassets/ew/org/centrb/nils/publikationer/2015/remotesensing-07-04253.pdf

Liu, J.G. and Mason, P.J. (2009) *Essential Image Processing and GIS for Remote Sensing*. John Wiley & Sons.

Mason, N.W.H., Holdaway, R.J., Richardson, S.J. and Chao, A. (2018) Incorporating measurement error in testing for changes in biodiversity, *Methods in Ecology and Evolution* 9, 1296–1307. https://doi.org/10.1111/2041-210X.12976

Morrison, L.W. (2016) Observer error in vegetation survey: a review, *Journal of Plant Ecology* 9(4), 367–379. doi: 10.1093/jpe/rtv077

Mueller-Dombois, D.R. and Ellenberg, H. (1974) *Aims and Methods of Vegetation Ecology*. New York: Wiley.

Nilsson, B., Granholm, A.-H., Wikander, L., Forsman, H. and Allard, A. (2022) *Instruktion för Nationell Inventering av Lövskogar och Gräsmarker via Stereoflygbilder och Ortofoton - År 2022* [Instruction for National Inventory of Deciduous Forest and Grasslands by Aerial Photos in Stereo and Orthophotos, Year 2022]. Umeå: Swedish University for Agricultural Sciences, Section for Landscape Analysis.

Nilsson, M., Nordkvist, K., Jonzén, J., Lindgren, N., Axensten, P., Wallerman, J., Egberth, M., Larsson, S., Nilsson, L., Eriksson, J., et al. (2017) A nationwide forest attribute map of Sweden predicted using airborne laser scanning data and field data from the National Forest Inventory, *Remote Sensing of Environment* 194, 447–454. https://doi.org/10.1016/j.rse.2016.10.022.

OSPAR Commission 30th. (2022) Home page, https://www.ospar.org/ (Accessed June 2, 2022).

Programme of the European Union. (2022) *The Copernicus Data and Information Access Services (DIAS)*, https://www.copernicus.eu/en/access-data/dias (Accessed February 23, 2022).

Ranlund, Å., Sjödin, M., Press, A., Gardfjell, H., Hedenås, H., Hagner, Å., Forsman, H., Christensen, P., Andersson, M. and Adler, S. (2021) *Metodbeskrivning: 2020-Års Inventeringar av Gräsmarker och Lövskogar* [Description of Methodology: Inventory of Grasslands and Deciduous Forest the Year 2020], Work report 530. Umeå: Swedish University for Agricultural Sciences, Section for Landscape Analysis.

Reitalu, T., Purschke, O., Johansson, L.J., Hall, K., Sykes, M.T. and Prentice, H.C. (2012) Responses of grassland species richness to local and landscape factors depend on spatial scale and habitat specialization, *Journal of Vegetation Science* 23, 41–51.

Ringvall, A., Ståhl, G., Löfgren. P. and Fridman, J. (2004) *Skattningar och Precisionsberäkning i NILS – Underlag för Diskussion om Lämplig Dimensionering* [Estimations and Calculations of Precision in NILS – Basis for Discussions on Appropriate Sample Layout]. Umeå: Swedish University for Agricultural Sciences, Section for Landscape Analysis. https://www.slu.se/globalassets/ew/org/centrb/nils/publikationer/publikationer2003-2009/arb_rapp_128.pdf

Ross, L.C. and Flagmeier, M. (2015) *A Resource for the Study of Scottish Upland Vegetation: Habitat and Species Data from "Plant Communities of the Scottish Highlands" and Repeat Surveys*. Scottish Natural Heritage Commissioned Report No. 880. https://www.nature.scot/doc/naturescot-commissioned-report-880-resource-study-scottish-upland-vegetation-habitat-and-species

Schott, R.S. (2007) *Remote Sensing – The Image Chain Approach*. 2nd edn. Oxford University Press.

Seidling, W., Hamberg, L., Máliš, F., Salemaa, M., Kutnar, L., Czerepko, J., Kompa, T., Buriánek, V., Dupouey, J.-L., Vodálová, A., et al. (2020) Comparing observer performance in vegetation records by efficiency graphs derived from rarefaction curves, *Ecological Indicators* 109, 105790. doi: 10.1016/j.ecolind.2019.105790

Sjödin, M. (ed.) (2019) *Fältinstruktion för Nationell Inventering av Landskapet i Sverige, NILS 2019* [Field Instructions for National Inventory of Landscapes in Sweden NILS 2019]. Umeå: Swedish University for Agricultural Sciences. https://www.slu.se/globalassets/ew/org/centrb/nils/publikationer/2019/nils_faltinstruktion_webb_ht_2019_2.pdf

Ståhl, S., Allard, A., Esseen, P.-A., Glimskär, A., Ringvall, A., Svensson, J., Sundquist, S., Christensen, P., Gallegos Torell, Å., Högström, M., et al. (2011) National Inventory of Landscapes in Sweden (NILS) – scope, design, and experiences from establishing a multi-scale biodiversity monitoring system, *Environmental Monitoring and Assessment* 173(1–4), 579–595.

Swetnam, T.W., Allen, C.D. and Betancourt, J.L. (1999) Applied historical ecology: using the past to manage for the future, *Ecological Applications* 9(4), 1189–1206.

Van der Maarel, E. (1979) Transformation of cover-abundance values in phytosociology and its effects on community similarity, *Vegetatio* 39, 97–117.

Vittoz, P., Bayfield, N., Brooker, R., Elston, D.A., Duff, E.I., Theurillat, J.-P. and Guisan, A. (2010) Reproducibility of species lists, visual cover estimates and frequency methods for recording high-mountain vegetation, *Journal of Vegetation Science* 21(6), 1035–1047.

With, K.A. (2019) *Essentials of Landscape Ecology*. Oxford Scholarship Online. doi: 10.1093/oso/97801 98838388.003.0001

Wood, C.M., Bunce, R.G.H., Norton, L.R., Maskell, L.C., Smart, S.M., Scott, A., Henrys, P., Howard, D., Wright, S., Brown, M.J., et al. (2018) Ecological landscape elements: long-term monitoring in Great Britain, the Countryside Survey 1978–2007 and beyond, *Earth System Science Data* 10, 745–763. doi: 10.745-763.10.5194/essd-10-745-2018.

Wright, W.J., Irvine, K.M., Warren, J.M., Barnett, J.K. and Yoccoz, N. (2017) Statistical design and analysis for plant cover studies with multiple sources of observation errors, *Methods in Ecology and Evolution* 8 (12), 1832–1841. doi: 10.1111/2041-210X.12825

Österreich forscht, www.citizen-science.at (2022) Projects, Viel-Falter: Butterfly Monitoring, Department of Ecology, University of Innsbruck. https://www.citizen-science.at/en/projects/viel-falter-butterfly-monitoring (Accessed November 5, 2022).

6 Citizen science: data collection by volunteers

Anders Bryn, René Van der Wal, Lisa Norton, and Tim Hofmeester

Introduction

Involvement of citizens in research is a long-lasting tradition within the natural sciences. By bringing citizens and lay people into research and monitoring, more data can be gathered over a larger geographical area. Across the last two decades, citizen science has enrolled new groups of people in an increasing and evolving set of research directions. In this section, we will explain the concept of citizen science, summarize its development, and discuss some of the motivations for involving citizens in research.

Citizen science explained

Citizen science (CS), the involvement of people from outside science in research, is a proliferating and increasingly diverse activity. One of the defining characteristics is that activities are non-paid voluntary contributions to research and associated knowledge-generation. Often, professional researchers organize citizen science activities, but this does not have to be the case (non-governmental organizations [NGOs] also frequently organize CS projects). The degree of involvement, what part of the research cycle is influenced, and how contributions to the research are made vary greatly. Therefore, there are numerous definitions, purposes, and practices of citizen science (Gura 2013; Kullenberg and Kasperowski 2016). Even within the field of environmental monitoring, citizen science can take many forms and directions.

In the most common schemes where there is a low degree of research involvement, participants contribute observations, in the form of numerical data, to research programmes organized by professionals in academic, non-governmental or private organizations (e.g. universities, conservation bodies, hunting and fishing societies). Within the framework of an existing research or monitoring programme, new entries of, for example, species, environmental conditions, or land use data can be added to a database. Research based on such contributions is sometimes called *crowdsourcing* or *contributory citizen science*, whereas the other end of the involvement gradient often is termed *participatory research*. In participatory research – sometimes split up into more specific modes such as collaborative and co-created (Miller-Rushing et al. 2012) – citizens are more often involved in the development of research goals or the methods to be used and may even take part in the analyses and dissemination of the research. However, the terminology continues to evolve and is thus used in a variety of ways to describe often very different approaches (Irwin 2018). Therefore, for the purpose of this assessment, we use citizen science as an overarching term incorporating a wide range of approaches that involve citizens taking active roles in

DOI: 10.4324/9781003179245-6

environmental monitoring and research. For a more precise and descriptive definition of citizen science, we recommend taking a closer look at the *Ten Principles of Citizen Science* (European Citizen Science Association [ECSA] 2015; Robinson et al. 2018).

The history of citizen science in monitoring biodiversity

Within the natural sciences of biodiversity mapping and monitoring, citizen science has a long tradition. In the 19th century, natural history collections became a symbol of status in industrializing countries. These collections generated societal interest for natural sciences in general, creating further appetite for sampling specimens and recording species diversity. More recently, during the 20th century, lay people have contributed millions of species records to monitoring schemes. Early on, sampling involved bringing specimens to private collections or public university museums; for example, in the form of herbarium sheets holding dried plants or stuffed bird specimens. Later on, it became more common to offer records in the form of species lists, and in the early 21st century individual species records have become the norm, currently almost exclusively submitted through mobile or web-applications. Whereas museums and universities historically took charge of collecting and archiving biodiversity records, alongside recording societies (which were often tightly linked to academic environments), recently NGOs have contributed much to increased efforts through organized sampling at regional, national, and international levels; for example, for structured bird counts or at more local levels for BioBlitzes. A BioBlitz is an organized event with the aim of identifying as many species as possible in a specific area over a short period of time (National Geographic Society 2022). Several recent citizen science approaches have emerged alongside technological development within the natural sciences and natural science societies in recent decades. Most of these new technologies and approaches offer some sort of feedback to participants. Nowadays, for example, you can contribute to land use/land cover (LULC) mapping and monitoring by combining Earth observation technologies with citizen science entries of ground-based LULC observations (LandSense 2022).

Reasons for implementing citizen science in monitoring

Reasons for using citizen science as a methodological approach are diverse. The primary reason is often to obtain more data covering much larger areas than could ever be covered by professional researchers or to get help with the processing of large amounts of data or materials. In taking this approach, a lot of research knowledge is transferred to the participants. Globally, confidence in research is challenged by fake news and alternative movements, so by involving citizens directly in the research, confidence in science can be strengthened, and the dissemination of research knowledge can reach more people (see section Data quality in citizen science projects for potential risks). In addition, participation can increase general interest in research and natural sciences, which is an important part of the shift to a green economy and improved sustainability.

Bottom-up initiatives can also generate good environmental data on species and environmental conditions. Such undertakings are often motivated by a desire for action, to prevent certain things from happening (e.g. the logging of a forest home to rare or nationally important species) or to ensure action (e.g. addressing air pollution in a city). Here, data and knowledge generation are generally driven by people outside academia, although in many cases researchers are invited to be involved.

A golden rule for citizen science projects is to make all of the data openly available for everyone to use (Principle #7, ECSA 2015). The ability to share data openly has changed fundamentally over the last decades, in accordance with the digital revolution. By making the data available, people can participate, carry out their own surveys, and interrogate data for their own use. Citizen science data are often also available and used by educational institutions and increasingly contribute to high-level policy directives, including the United Nation's (UN) Sustainable Development Goals.

Handling of citizen science participants

As stated above, citizen science covers a very broad range of activities from engaging the general public in looking at and appreciating nature to helping land owners to understand more about their land or taking part in long-term large-scale monitoring programmes that provide valuable data on species/ecosystem change. Handling participants effectively and investing in their participation is vital to ensure that these activities have value for the participants, for nature, and, where appropriate, for science (Schmeller et al. 2009; McKechnie et al. 2011; Hochachka et al. 2012; Danielsen et al. 2014).

Each specific citizen science initiative will require different handling of participants. In all cases, clear guidance, good communication, and ease of data entry are important factors contributing to successful engagement with volunteers. Though volunteers engaged in one-off events may be more motivated by the experience and what they can learn during the event, volunteers engaged in long-term monitoring are likely to be more interested in how their data are informing the detection of longer-term trends. These long-term volunteers often also become ambassadors for the project, helping with the quality control of data and the recruitment of new volunteers.

Though there may be savings to be made by using volunteers rather than paid professionals in long-term monitoring, to ensure adequate data quality and timeliness, it is important that volunteers are well managed and feel valued for their contributions, and this requires time and resources. Citizen science monitoring programmes, like the breeding bird survey run by the British Trust for Ornithology (BTO), invest heavily in supporting their volunteers.

Finding and recruiting volunteers

The recruitment of volunteers will differ according to the nature of the monitoring required and its objectives. One day, BioBlitzes or similar one-off recording exercises may be designed to elicit the interest of sections of the wider public in particular places, habitats, or species and will generally seek to recruit non-specialists to do something different and acquire new skills. Volunteers may be recruited by event advertising through organizations like Wildlife Trusts or through schools and clubs. Volunteers in these situations often have minimal skills, and it is unrealistic to expect them to develop advanced skills in a short time period; hence, recording may involve easily recognized wildlife and very simple recording forms or apps.

For citizen scientists with greater vested interests in outcomes, such as land owners or local residents, recruitment is usually place specific, and though skills may be low initially, if recording is going to occur over a period of time, there is potential for training and guidance to improve recording skills and potentially use of technology for reporting.

Volunteers involved in long-term monitoring programmes such as the BTO's breeding bird survey or the Botanical Society of Britain and Ireland's (BSBI), recording may be by highly skilled amateurs, and it is likely that such surveyors often recruit themselves to monitoring programmes rather than being actively recruited. Similarly, other less experienced volunteers with an interest in developing their skills further may well seek out opportunities with organizations that they know to carry out monitoring programmes.

Training volunteers

Training/guidelines for volunteer recorders are essential to make their recording purposeful, whether that be providing them with basic observational skills or ensuring that high-quality data are collected using identical protocols across a range of sites and volunteers.

For engagement events and one-off recording events, materials need to be simple, accessible, and attractive to encourage a wide range of non-specialists to get excited about looking out for species/habitats. Production of such material may be quite complex to avoid any misinterpretation or confusion over what is being recorded and where possible recording targets should be common, colourful, and easily identifiable. Provision of simple but effective equipment such as magnifying glasses/hand lenses or bug boxes can enhance the experience of voluntary recorders.

For monitoring that aims to collect data that will be used to evaluate state or change, it may be necessary to provide detailed training from experts as well as specific materials (and, where necessary and feasible, equipment) to ensure that volunteers are recording the correct things in a correct way. Training may be relevant to specific methodologies, like setting up and permanently marking quadrats or transects or the identification of species or habitats.

Some volunteers will be experts in their own right, particularly in species identification, where there are many amateur experts, but they may, for example, need advice and guidance in the use of specific apps to upload records as intended.

Feedback to participants/communication

Volunteer retention is often critical to ensure continuity of monitoring programmes. Finding people who are willing to collect the data that you need in a place where you need it from is not always easy, and therefore retaining volunteers is worthy of investment. Volunteers are far more likely to continue to contribute if they are well supported by staff involved in administering monitoring schemes and if they can see the benefits of their monitoring. Feedback is therefore important, whether it be through personal communication or through interactive websites that can provide updates of data submitted and information on what it has contributed to. Volunteers who send data in and receive nothing back are likely to feel that their data are not valued. It may be that volunteers do not necessarily want to see how their data have contributed to results (and this may at times be a lengthy process); an acknowledgement that the data have been received and are valued may be enough to ensure that volunteers continue to collect data (alternatively, a short resumé of the research results).

Data quality in citizen science projects

Setting clear goals and defining a protocol for how to collect, handle, store, and publish data are key aspects of any monitoring programme. These issues become even more important when many participants are involved in a program, especially if those participants are not employed within the monitoring program. Within this section, we will discuss the issues of data collection and data handling when organizing a monitoring program as a citizen science project, with a specific focus on how to make sure the data you collect are of the quality needed for the purpose of the program.

Data quality is in the eye of the beholder

What is considered good data quality is dependent on the purpose of the project or program. If your program is aimed at monitoring the geographical distribution of a specific species, you might only need verifiable records, such as photographs of the species combined with a set of coordinates. Such a project would be very suitable for a CS approach, where apps and platforms such as iNaturalist or, for example, the Swedish Species Portal can help you to simplify the protocol and reach many people. However, if you want to follow trends in the number of individuals of a species, you will need to have a more detailed protocol specifying how often a person visited a location, what transect they walked, how many individuals of the species they counted, etc. Therefore, it is important to set clear goals for your monitoring program and find a clear and simple protocol that would provide the data that you need. In particular, it is important to decide whether you need structured data (for example, systematic, area-representative, or balanced sampling) or if unstructured data are good enough (typically more ad hoc sampling, often spatially biased and unbalanced). This decision is related to the scientific goals and the analytical parts of the citizen science project/programme. Unstructured data are generally more challenging to handle in data analysis, and citizen science is not always the best solution (see discussion in Johnston et al. 2022).

It is important to realize that there is a clear trade-off between the number of participants that can partake in a citizen science project and the difficulty of the protocol. For example, when people need to be able to identify a large number of species and go out to the same place multiple times over a season to walk a specific transect and count all individuals of a group of species they see on that transect, there will be only a few enthusiastic naturalists who would be able to follow the protocol. By contrast, anyone can take a picture of a flower with their telephone and let the artificial intelligence (AI) in the iNaturalist app (iNaturalist 2022) identify the species and upload the observation to the system. Lukyanenko et al. (2016) even argued that data quality can be increased by keeping protocols very simple and projects as inclusive as possible. Furthermore, prior knowledge of the volunteers, regular feedback, and proper training can increase data quality (Lewandowski and Specht 2015). Sometimes, technology can help with making a difficult protocol simpler, by having several steps taken by the technology rather than the participant. For example, a camera trap can take pictures of local wildlife, giving verifiable vouchers of the exact species that was observed. From that same location, other instruments can register, for example, temperature or light conditions, which could be important to explain presence or activity. At the same time, the camera can store information about when it was active, so you get a record of both the presence of species or individuals and the effort spent collecting that data, with minimum input by the volunteers, who only need to set up and collect the camera trap (see section The key to quality control). In any case, it is important to provide clear and repeated feedback to participants about the data (see above), so that participants remain motivated both to continue collecting data and to better harmonize the application of the sampling protocol among participants (Kosmala et al. 2016).

The key to quality control

With the increasing number of digital platforms that can be used to obtain data collected within citizen science projects, there is an increasing potential for automatic data validation. This data validation or quality control is important to detect sources of error or bias in the data. Participants might have erroneously filled in the wrong date, location, species, number of individuals, etc. Many of these errors can be automatically checked against reference observations in a digital system. For example, the system might flag observations of migratory birds outside of certain dates for a check by administrators. Similarly, the system might check whether the reported coordinates fall within the known or expected distribution range of the species. Machine learning algorithms, experts, or other users can check observations that are accompanied by a picture or sound file to verify the identification of the species (Kosmala et al. 2016). Which of these options might be feasible to use in a project depends on the size and the aims of the project. In any case, it is important to provide feedback to the participant that entered the data where inaccuracies occurred (Van der Wal et al. 2016).

After correcting for any sources of human error, there might still be biases in the data that need to be corrected for to use the data for monitoring purposes. For example, people tend to report observations from accessible locations and of rare species rather than common ones. Similarly, unless the protocol is specifically focused on recording species absence as well as presence, data on absence are often lacking. Thus, it is important to correct for these potential sources of bias when analyzing the data. An increasing number of statistical methods, often in combination with other spatial data (remote sensing data or explanatory variables used in distribution modelling etc.) have been developed to deal with these biases (Kosmala et al. 2016; Callaghan et al. 2019; Bryn et al. 2021).

Developing technologies

The advent of the internet and subsequent digital innovation are key to the proliferation of citizen science initiatives. Such innovations have had the greatest impacts in the environmental realm where technological developments have intersected with deep-felt environmental concerns. This, in turn, has contributed to an insatiable demand for data – at ever greater spatial scales and finer temporal scales – by researchers and research institutions keen to understand more about the state of nature and consequences of ongoing changes for society. Moreover, the presence of often well-organized and skilful naturalists – people who study nature by observing plants and animals directly, in both professional and non-professional capacities – has provided fertile soil from which citizen scientists could emerge. This potent set of factors means that today's environmental citizen science is not only a child of the digital era but also an area of monitoring that will continue to be highly influenced by new technological developments. Here we will briefly capture the main families of technologies that influence, and sometimes even define, citizen science contributions to biodiversity and wider environmental monitoring.

Networks of humans and sensors

The need for more and better biodiversity data, to plug knowledge gaps without which sustainable use, management, and protection of biodiversity resources is deemed compromised, is the underpinning logic of many environmental research and conservation organisations. As a result, methods that increase scale and reliability are typically sought and adopted (Catlin-Groves 2012), state-of-the-art sensors being one of them. Citizen science in the digital age, with information and communication technologies (ICTs) allowing for the formation of communities of interest, provides the potential for scaling up approaches, but expanding the recording community from small numbers of highly skilled naturalists to much larger, widely dispersed audiences contributing high-quality data demands more. Two routes are often followed. The first is training, which can have an in-person component but increasingly involves online tools with or without built-in feedback mechanisms (Van der Wal et al. 2016), because that allows for operation at greater scales with lower costs. Using this approach, both amateur-expert naturalists and ordinary members of the public can be trained to the level where they contribute good quality data, effectively becoming "citizen sensors". A second route is to benefit from sensor technologies becoming cheaper and omnipresent, such that data quality issues can be taken out of the equation, pairing up citizens with reliable sensors to form "networks of technical sensors and human and animal (or plant and systems they are part of) participants" (Verma et al. 2016). Whereas citizen sensors approaches have been in operation for quite a while and often prove effective in terms of data generation and experiential or learning benefits, the second route, "networks of sensors, humans, and animal/plant participants", is currently a rapidly developing area, which includes sensors of all sorts.

Sensors underpinning environmental citizen science

Mobile phones have become the most important sensor in environmental citizen science, not least because this technology allows for automated provision of location data, instantaneous creation and submission of records, and recording and sharing visual or audio evidence. Indeed, it is the bringing together of so many functionalities in one device that makes mobile phones so important in environmental citizen science and the main reason why they have been embraced in monitoring so rapidly.

The most common way mobile phones are used in citizen science is as an entry point to a plethora of different programmes, most of them with their own app or website. eBird (The Cornell Lab of Ornithology 2022), for example, allows bird watchers from around the world to submit structured observations at an unprecedented scale, such that even in many data-poor countries, the accumulation of bird records is accelerating (Amano et al. 2016). Another programme, iNaturalist (2022), spurs large numbers of people into using the camera on their mobile phones to capture and share plants and animals they have spotted. Their aim is not monitoring or conducting science but connecting people to nature. Yet, more than 39 million biodiversity records from across the world are brought together through this online social network of people, connecting them and their handheld sensors to plant and animal species. The presence of such biodiversity citizen science giants, like eBird and iNaturalist, means that starting up new online citizen science projects is now difficult and increasingly reliant upon existing organizations with broad community reach (e.g. Earthwatch 2022; Vetenskap & Allmänhet 2022) or − for smaller scaled but targeted monitoring − distinct communities with a dedicated membership (e.g. Botanical Society of Britain and Ireland 2022).

Another type of sensor that is gaining traction in environmental citizen science is the camera trap, a little device designed to strap to a stake or tree and to take photographs, triggered by an animal passing by, in the absence of the person or group deploying it. Initially developed for hunters, camera traps are now increasingly deployed in citizen science in two distinct ways. First, volunteers are asked to use it for monitoring notably larger mammals (though modifications suited for mice have also been developed). Elaborate schemes exist, such as the Candid Critters project with over 500 volunteers across North Carolina, USA (Lasky et al. 2021). The latter project managed to scale up through the help of public libraries, from which camera traps could be borrowed. A second way in which camera traps have become part of environmental citizen science is through crowdsourcing projects such as Snapshot Serengeti (The Conversation 2015; the project was on the Zooniverse platform – see description below). Here, organizations deploy cameras, often in attractive locations, and images are uploaded to a website for categorization by volunteers. In doing so, the burden of work – going through countless images to detect ones with animals that then can be identified – is passed on to small armies of volunteers. This model of distributed citizen science has taken flight, leading to communities of interest and large volumes of annotated data that are or can be used for environmental monitoring. The platform that has deployed this most effectively is Zooniverse (2022), with over 50 online initiatives brought together following the same formula and 1.6 million registered users. Though not without its critics, it is fair to say that such crowdsourcing initiatives allow people to encounter special places and species and establish effective pipelines for researcher-driven monitoring involving volunteers from across the world.

More widespread deployment of other sensors is around the corner, such as sound pods for noise monitoring and detecting (audible) bird species (Open Acoustic Devices 2022). Here, excellent monitoring data could result where the material can be processed accordingly. This model of "sensor deployment + data submission" is far from new. Weather enthusiasts, for example, have deployed temperature and rainfall sensors for many decades, and such material continues to contribute to the monitoring and forecasting of weather in many countries around the world. Technological development in, for example, the Internet of Things (IoT) may be able to generate feedback loops to sensor-deploying volunteers and possibly draw more and more people into citizen science, consciously or unconsciously.

AI in environmental citizen science

Digital sensors such as camera traps or sound pods generate vast amounts of data and thus the processing of this is a key bottleneck to environmental monitoring. Photo- and sound-based observations lend themselves ideally to the application of AI routines, and these are increasingly the route through which such material is processed. This is good news for data generation, at least where automated species recognition protocols are good enough, yet they come with the risk of by-passing the volunteer, potentially limiting learning and engagement. Yet, when embedded well, such technologies can "open" certain species groups to novices, and through this route many more people could potentially contribute to environmental monitoring. Finally, it is appropriate to recall that all new (and old) technologies are accompanied by specific sources of errors and uncertainties. Depending on the technology taken in use for citizen science, specific measures should be taken to handle data quality (see section Data quality in citizen science projects).

Democratization and ethical considerations

In many countries, confidence in research and monitoring is being challenged by alternative narratives. With the emergence of social media, there is thus a growing need to strengthen trust in and knowledge about research. This can be targeted in many ways, and citizen science is one of them. By involving society in monitoring, the research process may become transparent for participants and the findings more widely available. As a result of participating in the research, commitment and insight can be created, which ultimately can contribute to increased trust (Bedessem et al. 2021).

Through participation, citizen science can also contribute to a higher focus on ecological and environmental democracy. The basic idea of ecological and environmental democracy is "ensuring environmental sustainability while safeguarding democratic values and practices" (Pickering et al. 2020). Whether these goals can be achieved through specific projects or not depends on how a citizen science project is organized: how deeply the participants are involved, what the results are used for, and how representative the participants are. However, simply by extending and increasing the amount of environmental monitoring data collected, relevant citizen science projects may contribute to improved management of nature for biodiversity. Conversely, participants may have pre-existing agendas and could manipulate results (or cause them to be non-representative); for example, through entering false positives or biased data. It is therefore obligatory in any citizen science project to make sure that ethical considerations are reflected upon at an early stage in the research process. For citizen science to live up to its democratizing potential, important issues such as equality and cultures of inclusivity need to be considered (Cooper et al. 2021).

Open platforms for registering and reading of environmental monitoring data underline an increased need for ethical considerations (and quality control, which is described in section Data quality in citizen science projects). Data privacy concerns should be carefully evaluated in citizen science projects with open databases; for example, if date and position of participants are provided. With the emergence of GDPR (General Data Protection Regulation) in 2018, the focus on data privacy concerns increased, and many European CS projects had to restructure privacy settings.

Case studies

iNaturalist and Global Biodiversity Information Facility

iNaturalist is an example of one of the most popular social networks for registering and sharing observations of species biodiversity around the world. The network is used by citizen scientists as well as professional biologists and is therefore primarily a network for crowdsourcing of biodiversity data. Through a free mobile application, including an automated species identification tool (from 2017), it is very easy for participants to contribute with new observations. Although iNaturalist is intended as "an online social network of people sharing biodiversity information to help each other learn about nature" (iNaturalist 2022), all observations can contribute to research and monitoring of biodiversity.

iNaturalist shares entries with the Global Biodiversity Information Facility (GBIF), an international organization with the aim of making biodiversity data from around the world publicly available. GBIF disseminates biodiversity data from many different sources, including a number of different citizen science projects, museum collections, and research institutions. All observations are structured according to a specific protocol (Darwin Core) and made available through a single web-based portal.

Farmer monitoring within an innovative farmer project

Farmers, as land owners and managers, have important influences on biodiversity. As part of an Innovative Farmer project in the UK, farmers have been encouraged to monitor their grassland species using standardized, repeatable approaches that enable them to measure how the species in their sward change over time and in response to their management (Innovative Farmers 2022). Training farmers in scientific approaches helps to ensure that the data collected are credible and encourages them to observe their land in ways that can benefit them and, ultimately, wider society. Helping identification through providing hands-on expertise early on, reference sources (such as identification guides), and digital tools like iNaturalist can assist with identification of different species. Specific monitoring apps, like Soilmentor from Vidacycle (2022) can help to store data for later comparison.

Älgobs (Moose observations)

Hunters have a long-standing tradition to monitor the species they hunt and provide a large amount of biodiversity monitoring data nowadays (Cretois et al. 2020). The Swedish moose (*Alces alces*) population is managed with an adaptive management system that includes rigorous monitoring. Since 1985, part of that monitoring has been done using voluntary moose observations (*älgobs*) reported in a systematic way by hunters (Singh et al. 2014). Nationwide, each hunting team reports all moose seen during the first seven days of hunting as well as the number of hours they spent outside to a website (Swedish Association for Hunting and Wildlife Management 2022). These observations are then used to calculate trends over time in the number of observations per man-hour, the number of calves per female, and the ratio of males in the population.

Natur i endring (Mapping and monitoring treelines in Norway)

Dynamic treelines are indicators of biome range shifts, tightly connected to summer temperature and length of growing season. Treelines respond to climate changes with an ongoing expansion into higher elevations in mountain regions or to higher latitudes along the boreal-arctic ecotone. However, none of the ongoing monitoring projects are representative for larger regions and variation therein. Therefore, to monitor the ongoing range shift in Norway, a free mobile app for registering of treeline locations was launched in 2018 (Nature in change 2022). The mobile app has illustrated guidelines (Figure 6.1), maps of existing entries, and a step-by-step registration procedure. The results are available on the project's web page, and participants can join seminars and look at their results in museum exhibitions.

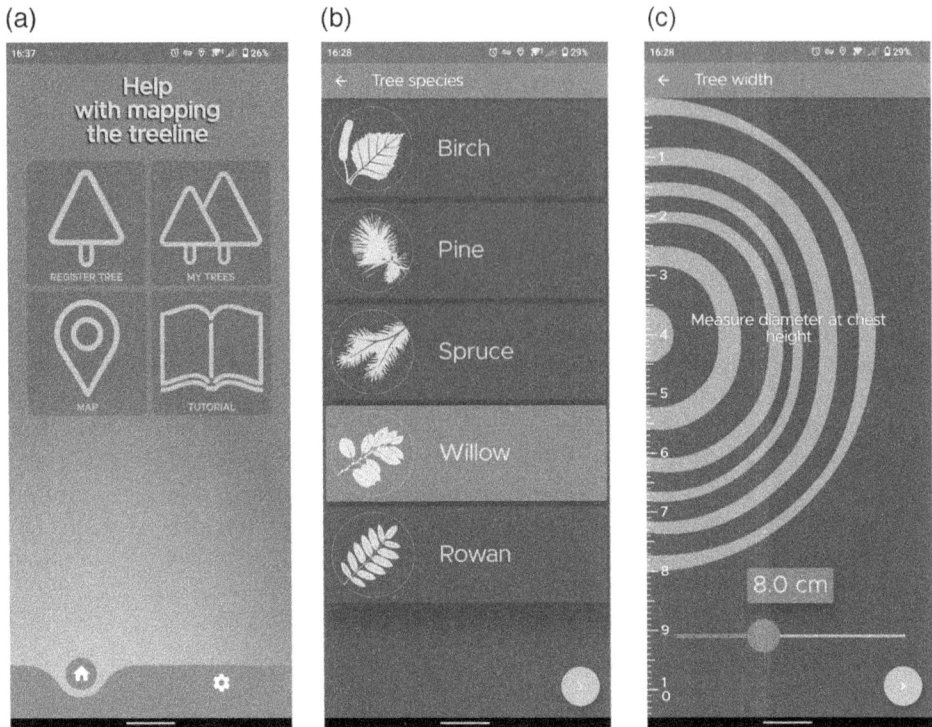

Figure 6.1 Screenshots from the mobile application used for mapping and monitoring of treelines in Norway.
Source: Nature in change (2022).

Key messages

- Citizen science is a non-paid and voluntary contribution to research usually organized by professionals.
- By involving citizens in research, knowledge can be transferred to participants and vice versa.
- Investing in recruiting and training of participants, as well as feedback whilst ongoing and afterwards, is vital to ensure valuable outcomes.
- Setting clear goals and defining a protocol for how to collect, handle, store, and publish data are key aspects of any monitoring program but are even more important when including non-professional participants.
- New technologies open opportunities for new ways of handling everything from data sampling to dissemination, but alongside come different limitations and challenges.
- Citizen science is a framework for involving society and thus a potential way of democratizing research.

Study questions

1   What are the key components of citizen science in general?
2   What are the advantages and disadvantages of using citizen science for monitoring purposes?

3   What is "spatial bias", and why is it frequently discussed within citizen science projects?
4   How can quality control of citizen science data be implemented in research or monitoring programs?
5   Which new and useful technologies for citizen science have been developed in the last decade or so?
6   How can citizen science data influence the public interest and confidence in research and monitoring?
7   How can citizen science contribute to achieving the UN's Sustainable Development Goals?

Further reading

Balestrini, M., Kotsev, A., Ponti, M., et al. (2021) Collaboration matters: capacity building, up-scaling, spreading, and sustainability in citizen-generated data projects, *Humanities and Social Sciences Communication* 8, 169. 10.1057/s41599-021-00851-5

Capineri, C., Haklay, M., Huang, H., Antoniou, V., Kettunen, J., Ostermann, F. and Pures, R. (2016) *European Handbook of Crowdsourced Geographic Information*. London: Ubiquity Press.

Hughes, A.C., Orr, M.C., Ma, K., Costello, M.J., Waller, J., Provoost, P., et al. (2021) Sampling bias shape our view of the world, *Ecography* 44, 1259–1269. 10.1111/ecog.05926

Lepczyk, A., Boyle, O.D. and Vargo, T.L.V. (2020) *Handbook of Citizen Science in Ecology and Conservation*. Oakland: University of California Press.

This book provides practical guidelines for the use of citizen science ecological research and monitoring but also some of the difficulties of this.

Robinson, L.D., Cawthray, J.L., West, S.E., Bonn, A. and Ansine, J. (2018) Ten principles of citizen science, in Hecker, S., Haklay, M., Bowser, A., Makuch, Z., Vogel, J. and Bonn, A. (eds) *Citizen Science: Innovation in Open Science, Society and Policy*. London: UCL Press, pp. 27–40.

This book discusses the main principles that define citizen science in general. The book is very useful for understanding citizen science and brings in perspectives from the research side as well as from the citizen side.

References

Amano, T., Lamming, J.D.L. and Sutherland, W.J. (2016) Spatial gaps in global biodiversity information and the role of citizen science, *BioScience* 66(5), 393–400. 10.1093/biosci/biw022

Bedessem, B., Gawrońska-Novak, B. and Lis, P. (2021) Can citizen science increase trust in research? A case study of delineating Polish metropolitan areas, *Journal of Contemporary European Research* 17(2), 304–325. 10.30950/jcer.v17i2.1185

Botanical Society of Britain and Ireland. (2022) About BSBI, https://bsbi.org/ (Accessed June 1, 2022).

Bryn, A., Bekkby, T., Rinde, E., Gundersen, H. and Halvorsen, R. (2021) Reliability in distribution modeling – a synthesis and step-by-step guidelines for improved practice, *Frontiers in Ecology and Evolution – Models in Ecology and Evolution* 9, 658713. https://www.frontiersin.org/articles/10.3389/fevo.2021.658713/full

Callaghan, C.T., Rowley, J.J.L., Cornwell, W.K., Poore, A.G.B. and Major, R.E. (2019) Improving big citizen science data: moving beyond haphazard sampling, *PLOS Biology* 17(6), e3000357. 10.1371/journal.pbio.3000357

Catlin-Groves, C.L. (2012) The citizen science landscape: from volunteers to citizen sensors and beyond, *International Journal of Zoology* 2012, 349630. 10.1155/2012/349630

The Conversation. (2015) Crowdsourcing the Serengeti: how citizen scientists classified millions of photos from home, https://theconversation.com/crowdsourcing-the-serengeti-how-citizen-scientists-classified-millions-of-photos-from-home-42930 (Accessed June 1, 2022).

Cooper, C.B., Hawn, C.L., Larson, L.R., Parrish, J.K., Bowser, G., Cavalier, D., et al. (2021) Inclusion in citizen science: the conundrum of rebranding, *Science* 25, 1386–1388. https://www.science.org/doi/10.1126/science.abi6487

The Cornell Lab of Ornithology. (2022) ebird, https://ebird.org/home (Accessed June 1, 2022).

Cretois, B., Linnell, J.D.C., Grainger, M., Nilsen, E.B. and Rød, J.K. (2020) Hunters as citizen scientists: contributions to biodiversity monitoring in Europe, *Global Ecology and Conservation* 23, e01077. 10.1016/j.gecco.2020.e01077

Danielsen, F., Pirhofer-Walzl, K., Adrian, T.P., Kapijimpanga, D.R., Burgess, N.D., Jensen, P.M., et al. (2014) Linking public participation in scientific research to the indicators and needs of international environmental agreements, *Conservation Letters* 7, 12–24. 10.1111/conl.12024

Earthwatch. (2022) Home page, https://earthwatch.org/ (Accessed June 1, 2022).

European Citizen Science Association. (2015) *Ten Principles of Citizen Science.* London: ESCA.

Gura, T. (2013) Amateur experts, *Nature* 496, 259–261. 10.1038/nj7444-259a

Hochachka, W.M., Fink, D., Hutchinson, R.A., Sheldon, D., Wong, W.-K. and Kelling, S. (2012) Data-intensive science applied to broad-scale citizen science, *Trends in Ecology & Evolution* 27, 130–137. 10.1016/j.tree.2011.11.006

iNaturalist. (2022) Home page, https://www.inaturalist.org/ (Accessed June 1, 2022).

Innovative Farmers. (2022) Home page, https://www.innovativefarmers.org/ (Accessed June 1, 2022).

Irwin, A. (2018) Citizen science comes of age, *Nature* 562, 480–482.

Johnston, A., Matechou, E. and Dennis, E.B. (2022) Outstanding challenges and future directions for biodiversity monitoring using citizen science data, *Methods in Ecology and Evolution.* 10.1111/2041-210X.13834

Kosmala, M., Wiggins, A., Swanson, A. and Simmons, B. (2016) Assessing data quality in citizen science, *Frontiers in Ecology and the Environment* 14, 551–560. 10.1002/fee.1436

Kullenberg, C. and Kasperowski, D. (2016) What is citizen science? – A scientometric meta-analysis, *PLoS ONE* 11(1), e0147152. 10.1371/journal.pone.0147152

LandSense. (2022) About LandSense, https://landsense.eu/About (Accessed November 6, 2022).

Lasky, M., Parsons, A., Schuttler, S., Mash, A., Larson, L., Norton, B., Pease, B., Boone, H., Gatens, L. and Kays, R. (2021) Candid critters: challenges and solutions in a large-scale citizen science camera trap project, *Citizen Science: Theory and Practice* 6(1), 1–17. 10.5334/cstp.343

Lewandowski, E. and Specht, H. (2015) Influence of volunteer and project characteristics on data quality of biological surveys, *Conservation Biology* 29, 713–723. 10.1111/cobi.12481

Lukyanenko, R., Parsons, J. and Wiersma, Y.F. (2016) Emerging problems of data quality in citizen science, *Conservation Biology* 30, 447–449. 10.1111/cobi.12706

McKechnie, C., Maskell, L., Norton, L. and Roy, D. (2011) The role of "Big Society" in monitoring the state of the natural environment, *Journal of Environmental Monitoring* 13, 2687–2691.

Miller-Rushing, A., Primack, R. and Bonney, R. (2012) The history of public participation in ecological research, *Frontiers in Ecology and the Environment* 10(6), 285–290.

National Geographic Society. (2022) Bioblitz and iNaturalist, counting species through citizen science, https://www.nationalgeographic.org/projects/bioblitz/ (Accessed November 5, 2022).

Nature in change. (2022) Fjell blir skog [Open mountain becomes forest], https://www.naturiendring.no/ (Accessed June 1, 2022).

Open Acoustic Devices. (2022) AudioMoth, https://www.openacousticdevices.info/ (Accessed June 1, 2022).

Pickering, J., Bäckstrand, K. and Schlosberg, D. (2020) Between environmental and ecological democracy: theory and practice at the democracy-environment nexus, *Journal of Environmental Policy & Planning* 22(1), 1–15. 10.1080/1523908X.2020.1703276

Robinson L.D., Cawthray, J.L., West, S.E., Bonn, A. and Ansine, J. (2018) Ten principles of citizen science, in Hecker, S., Haklay, M., Bowser, A., Makuch, Z., Vogel, J. and Bonn, A. (eds) *Citizen Science: Innovation in Open Science, Society and Policy.* London: UCL Press, pp. 27–40.

Schmeller, D.S., Henry, P.-Y., Julliard, R., Gruber, B., Clobert, J., Dziock, F., et al. (2009) Advantages of volunteer-based biodiversity monitoring in Europe, *Conservation Biology* 23, 307–316.

Singh, N.J., Danell, K., Edenius, L. and Ericsson, G. (2014) Tackling the motivation to monitor: success and sustainability of a participatory monitoring program, *Ecology and Society* 19(4), 7. 10.5 751/ES-06665-190407

Swedish Association for Hunting and Wildlife Management. (2022) Viltdata (Wildlife data), https:// www.viltdata.se/ (Accessed June 1, 2022).

Van der Wal, R., Sharma, N., Robinson, A., Mellish, C. and Siddharthan, A. (2016) The role of automated feedback in training and retaining biological recorders for citizen science, *Conservation Biology* 30, 550–561. 10.1111/cobi.12705

Verma, A., Van der Wal, R. and Fischer, A. (2016) Imagining wildlife: new technologies and animal censuses, maps and museums, *Geoforum* 75, 75–86. 10.1016/j.geoforum.2016.07.002

Vetenskap & Allmänhet. (2022) Mass experiment, https://v-a.se/projekt-portal/dialogaktiviteter/ forskarfredag/massexperiment-2/ (Accessed June 1, 2022).

Vidacycle. (2022) Soilmentor, https://soils.vidacycle.com/ (Accessed June 1, 2022).

The Zooniverse. (2022) People-powered research, https://www.zooniverse.org/ (Accessed June 1, 2022).

7 Remote sensing and Earth observation systems

Mats Nilsson, Jonas Ardö, Mats Söderström, Anna Allard, Alan Brown, and Luke Webber

Introduction

Following on from previous chapters covering the collection of data at ground level, this chapter discusses remote sensing and Earth observation (EO) systems that are presently becoming increasingly important for monitoring the land and its changes. Remote sensing is the technique of observing and analyzing objects from a distance without being in direct contact with them, and when studying the Earth we often use the term EO. The chapter covers the reverberations of the recent explosion in the availability of Earth observation data, with examples such as the Copernicus Earth Observation Programme, headed by the European Union (EU) and providing free data from the Sentinel satellites (optical and radar images), and other satellite products now accessible free of charge.

The chapter also discusses Google Earth Engine and similar means of rapid computation of enormous datasets, mid-level data collection using aerial remote sensing, image processing for object detection and creating 3D point clouds for surface models, and the need for manual work involving skilled ecologists; for example, stereo viewing and interpretation, supervised image classification, and creating rules-based expert systems.

Finally, the chapter discusses ways of monitoring and understanding broader systems, from otherwise harsh or inaccessible environments (e.g. high alpine or glacial areas) to the built environment, because Earth observations give access to all surfaces visible from above.

What is remote sensing?

Remote sensing is described as the science (and art) of obtaining information about an object, area, or phenomenon through the analysis of data acquired by a device that is not in contact with the object, area, or phenomenon under investigation (Lillesand and Kiefer 2015). Sensors collect the data and can be set on various platforms: acquiring data at various heights using drones or small aeroplanes, using regular aeroplanes, or sensors carried by satellites orbiting the Earth. The lower the flying height, the higher the spatial resolution, but closeness to the surface also means that we have to store a large number of images, and storage of data is often surprisingly expensive. We describe a few of the most important sensor types for monitoring biodiversity here, but there are many more, and the number of sensors are increasing rapidly. The main types of systems are usually divided into passive and active. Passive optical systems use the downwelling radiance (or incidence energy; i.e. sunlight) and collect what is reflected back from the surface of Earth, vegetation or object. Most satellite multispectral sensors are passive, as are the

DOI: 10.4324/9781003179245-7

modern digital cameras on aeroplanes (Schott 2007; National Aeronautics and Space Administration [NASA] 2022c).

The other type is active sensors, sending out pulses and measuring the return time after hitting the surface, perhaps of a leaf. Active pulses can be electromagnetic, a radio wavelength, like the radar (radio detection and ranging) systems, or the directed light of a laser – or sound in the case of underwater sonar. Read more about remote sensing and geographic information in Lillesand and Kiefer (2015), Schott (2007), or DiBiase (n.d.). A conceptual model of different types of sensors and scales with relevant monitoring examples is shown in Figure 7.1.

An advantage of remote sensing is that the same sensor produces the same type of data on every acquisition, even though the raw data that come from the sensor have to be pre-processed in various ways (called *normalization*). This is done to compensate for shadowing made by ground features (such as tall structures or mountains), for cloudiness or haze in the atmosphere, and for sun angle across sloping or undulating terrain. Satellites cover all of the Earth, including places where we find it difficult to access by boots on the ground, like glaciers, vegetation burned by wildfires, and remote mountainous areas. Using the many satellites observing the surface, we can now obtain time series, not only looking backwards in time but to study changes almost as they happen. However, poor image geometry, undetected clouds and haze, differences in scale between images, phenological differences, etc., can result in poor and biased results. It is important to check carefully that the results are valid against in situ observations before drawing any conclusions.

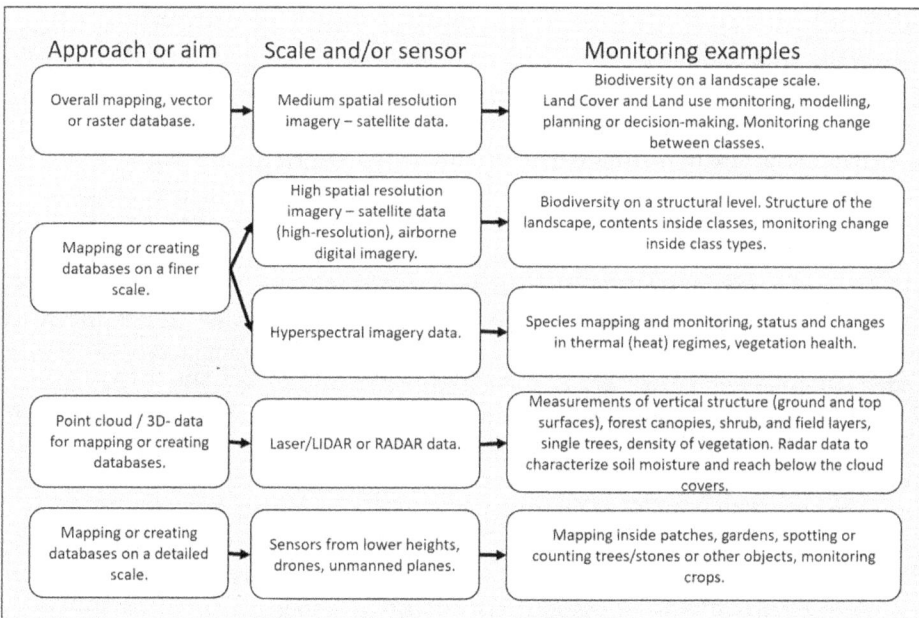

Approach or aim	Scale and/or sensor	Monitoring examples
Overall mapping, vector or raster database.	Medium spatial resolution imagery – satellite data.	Biodiversity on a landscape scale. Land Cover and Land use monitoring, modelling, planning or decision-making. Monitoring change between classes.
Mapping or creating databases on a finer scale.	High spatial resolution imagery – satellite data (high-resolution), airborne digital imagery.	Biodiversity on a structural level. Structure of the landscape, contents inside classes, monitoring change inside class types.
	Hyperspectral imagery data.	Species mapping and monitoring, status and changes in thermal (heat) regimes, vegetation health.
Point cloud / 3D- data for mapping or creating databases.	Laser/LIDAR or RADAR data.	Measurements of vertical structure (ground and top surfaces), forest canopies, shrub, and field layers, single trees, density of vegetation. Radar data to characterize soil moisture and reach below the cloud covers.
Mapping or creating databases on a detailed scale.	Sensors from lower heights, drones, unmanned planes.	Mapping inside patches, gardens, spotting or counting trees/stones or other objects, monitoring crops.

Figure 7.1 A conceptual model of different scales and sensors commonly used, depending on the aim or approach, and examples of various monitoring at these scales. In reality, a mixture of several of these are in use in biodiversity monitoring, depending on what level of scale is needed.

Source: Modified after Shahtahmassebi et al. (2021) and Lechner et al. (2020).

Gaspard–Félix Tournachon, also known as Nadar, is the first person known to have taken aerial photographs. In 1858, he took the first aerial image of Paris from a balloon. Over the years, remote sensing techniques have developed, and today we have a wide range of different techniques that can be used for various inventory and monitoring purposes. The launch of the first satellites in the 1950s was the start of a new era with satellite-borne sensors that provided data on a global scale. Landsat-1, launched in 1972 and initially known as the Earth Resources Technology Satellite (ERTS), was the first Earth observing satellite launched with the express intent to map and monitor our planet's resources. The Landsat program introduced a new sensor called "Thematic Mapper" (TM) with the launch of Landsat-4 in 1982. The improved spectral and geometric resolution of the TM sensor compared to the previous Multispectral Scanner (MSS) used on the earlier Landsat missions provided new and improved possibilities to map and monitor objects on the Earth's surface across large areas. IKONOS was launched in 1999 and was the first commercial Earth observation satellite to collect high-resolution imagery at 1m resolution.

The development of radar sensors started in the late 19th century. There was a rapid development of the technique during World War II. Synthetic-aperture radar (SAR) is a form of radar that can be used to create two-dimensional images or three-dimensional reconstructions of objects. Wavelength is an important feature to consider when working with SAR, because it determines how the radar signal interacts with the surface and how far a signal can penetrate into an object. For example, an X-band radar (wavelength: 3cm) has very little capability to penetrate into broadleaf forest because it mostly interacts with leaves at the top of the canopy. An L-band signal has a wavelength of about 23cm and thus penetrates deeper into a forest. The penetration depth into other land cover types such as soil and ice also depends on the wavelength. Interferometry (InSAR) uses SAR data to accurately measure land surface changes.

Ground penetrating radar (GPR) is widely used in archaeology where it makes it possible to survey a dig site without having to break ground to discover the likely locations of buried infrastructure (foundations of houses, trenches, waterways, cisterns, etc.), observe changes in soil structure, and identify any potential risks of damage from excavation.

Shortly after lasers were invented in the 1960s, the first lidar (light detection and ranging) systems were developed, measuring the laser returns from a surface or an object. The acronym is spelled in many ways (lidar, Lidar, LiDAR, LIDAR, LIDaR, LiDar, LiDaR, LIdar, and liDAR), where, as with radar, the lowercase lidar is the most common and recommended (Deering and Stoker 2014). When mounted on an aircraft together with a scanning device, it is commonly referred to as *airborne laser scanning* (ALS). When using lidar from any airborne carrier, it is imperative to have accurate positioning and orientation of the sensor, because the point cloud will be used to fix exact positions of trees or even parts of trees. To tie the data to a map and imagery, it is also important to know the location of an exact reference spot on the ground measured by the operator and to use global navigation systems, like Global Positioning System (GPS) and GLONASS. This is important in all collections of data from above, because we need to overlay different images, maps, and point clouds on top of each other to describe the vegetation and its changes over time. Lidar as a way of monitoring is widely used as more countries are performing nationwide scanning for various purposes, including research. Small instruments are often attached to drones for surveys of the ground, where they are especially useful when the surface below a tree canopy is of interest or the ground is hard to get to, like a glacier, but they are also used routinely for

Figure 7.2 A representation of a point cloud from airborne laser data in Abisko, in the northernmost part of the Swedish mountains. The point cloud is first draped on the national digital elevation model, and each point is coloured by the corresponding point in the orthophoto from the colour infrared aerial images. The lower corner shows a field photo taken from the area. Coloured and numbered markers show plots for in situ data collection.

Credit: Image and photo by Anna Allard.

a wide variety of monitoring: forest inventories, bathymetric surveying, archaeology, and habitat mapping. Chapter 8 takes a better look at data collection by lidar. An example of airborne laser data from the Swedish mountainous area of Abisko is shown in Figure 7.2, in which the point cloud is given the corresponding colour of the ground surface from the digital aerial image in colour infrared, with a photo from the ground set in one corner as a comparison.

The electromagnetic spectrum and sensors

In remote sensing, light rays from the sun (incident energy) bounce off objects (reflected energy) back towards the sensor in orbit. The reflected energy is what our eyes can see and what the sensor detects. In remote sensing, the electromagnetic spectrum is fundamental, where the basic unit is the photon. Photons, which are without mass, travel at the speed of light—300,000km/sec (186,000miles/sec) in the form of transverse waves analogous to the way waves propagate through the oceans; see Figure 7.3. The energy of

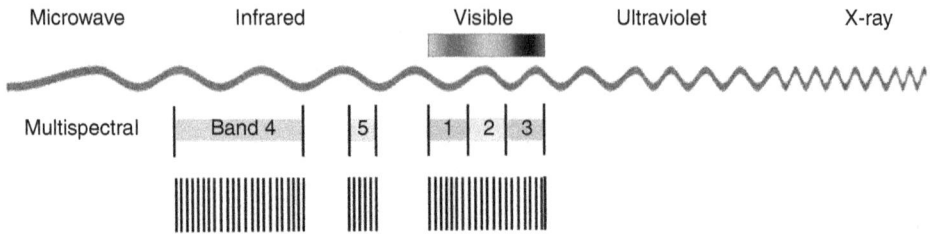

Figure 7.3 A representation of the electromagnetic spectrum, showing the range of photon wavelengths (the distance between successive crests) or frequencies (the rate of vibrations per second). Some regions have been given names, such as the visible part where human eyes can see reflected light, infrared, and ultraviolet. Sensors are often built to catch reflectance in more parts than the visible, collected in bands; in this example we have the visible reflectance in bands 1 to 3, near-infrared in band 4, and shortwave to longwave infrared in three bands (5–7, here drawn as one band).

Source: Modified after NASA (2022c) and GISGeography (2022).

a photon determines the frequency (and wavelength) of light that is associated with it. The greater the energy of the photon, the higher the frequency of light and vice versa, with all wavelengths combined in the electromagnetic (EM) spectrum (NASA 2022c).

In science, we have divided the EM spectrum up in wavelength regions, or bands, and given them names. At the very energetic (high frequency; short wavelength) end are gamma rays and x-rays, and the ultraviolet region extends from about 1nm to about 0.36μm. Longer wavelength intervals are measured in units ranging from millimetres through to metres, and the lowest frequency (longest wavelength) region, beyond 1m, is that of radio waves. The visible region (the wavelengths that our human eyes can see) is found in the middle, next to the near-infrared, medium infrared, and longwave (thermal) infrared regions. Visible light for humans is 380nm to 700nm (divided into three bands: red, green, and blue). Infrared (700nm to 1mm; shorter wavelengths – near 700nm) can be detected by special film or sensors (called *near-infrared colour images*), and at longer wavelengths thermal infrared radiation is felt as heat (NASA 2022c). Which wavelength or combination of wavelengths should be used depends on the application. It is, for example, often better to use near-infrared colour images than true-colour images for vegetation mapping.

The sensors giving us the imagery are called by different names according to how many discrete bands from the EM spectrum they can detect and whether or not the whole spectrum is covered. Multispectral sensors (the most commonly used) have between three and ten bands, typically with gaps in between where some frequencies are not included. In Figure 7.3, we show a representation of the multispectral bands red (3), green (2), and blue (1); near-infrared (4); and shortwave and longwave infrared (5–7). In a hyperspectral sensor, the same interval of wavelengths is occupied by hundreds of adjacent bands, which are referred to by their frequency interval rather than named colours, because they occupy only a small part of the colour bands. The next step is the ultraspectral sensors, having thousands of bands (GISGeography 2022; NASA 2022c). Recently, much attention has been put on the so-called red edge (referring to the region of the EM spectrum of rapid change in reflectance of vegetation on land or chlorophyll-a of algae in the oceans) as a means of detecting small differences in a diverse vegetation, leading to new standard indices used for processing remotely sensed data. The Sentinel-2 satellites carry three red edge bands, enhancing fine-tuned detection and discrimination of vegetation and chlorophyll content (Cui and Kerekes 2018; Bramich et al. 2021; European Space Agency [ESA] 2022b).

The remote sensing processing chain

Remote sensing data have to be transformed into information to be useful to the user and to allow consistent monitoring of land, water, or atmospheric properties. It is important that the remote sensing data be processed in a way that makes the results suitable for the application at hand.

Figure 7.4 shows an overview of the processing chain for optical remote sensing – the steps from data–acquisition to user applications.

Radiometric calibration converts the digital numbers (DNs) recorded by the sensor to an absolute or relative scale. The absolute calibration uses a radiative transfer model (e.g. Yang et al. 2011) to convert DNs to reflectance and to remove differences in reflectance that are caused by the absorption and scattering of wavelengths in the atmosphere. In the relative approach, DN values are normalized band by band to a reference image selected by the analyst. Geometric calibration corrects for the effects of surface relief and altitude, re-shaping the image to fit on a map projection, creating a new regular grid of pixels with values resampled from the original DNs. Atmospheric and radiometric correction, calibration, and geometric processing steps are known as *pre-processing*.

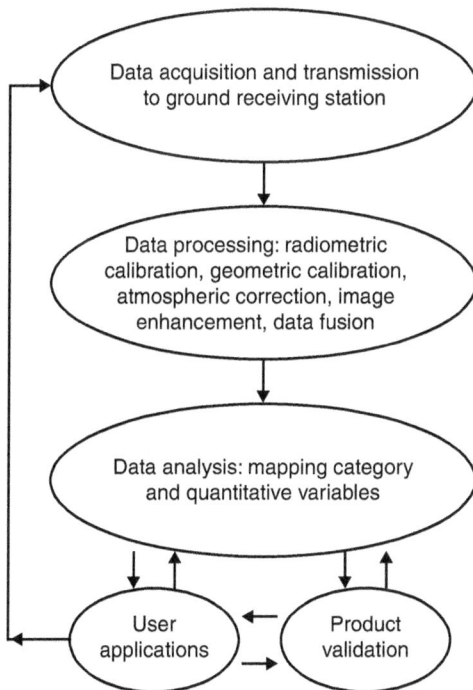

Figure 7.4 Key components of a remote sensing system.

Source: Modified after Liang and Wang (2020).

Image enhancement techniques or tools available in many remote sensing digital image processing systems are commonly used in the pre-processing of remotely sensed data to improve the performance of, for example, image classifiers or to assist visual interpretation. It is also common to fuse or merge remote sensing data from various sensors and resolutions.

After processing, the remote sensing data can be used for mapping and monitoring. Multiple channels can be combined into indexes and ratios, including vegetation indexes and fractional images from spectral unmixing. Images can be co-registered to create an image stack; for example, with images from different satellite instruments collected in a narrow time window or a time series over a season or successive years from the same instrument on different acquisition dates. Analysis – for example, creating a land cover map – can be made at the level of pixels, or pixels can be combined into objects using segmentation algorithms, and the pixels or objects are then classified. All outputs should be validated and an assessment made of the accuracy of the classification or change product using in situ observations.

Though both data pre-processing and processing were originally carried out by users and in many cases still are, in order to have control of the process, standard pre-processed imagery is now readily available from multiple vendors in the form of analysis-ready data (ARD).

The processing of synthetic aperture radar (SAR) is different in detail but has similar steps for calibration, ortho-correction and image registration so that radar images can be registered and stacked with optical images and maps.

The data revolution

Since Landsat-1 was launched in July 1972, about 50 years ago, the remote sensing community has experienced a data revolution. This revolution includes a massive increase in data availability, driven by the rise of missions and platforms launched, the diversity of sensor systems, and related increases in spatial, temporal, and geometric resolutions of launched sensors. In short, EO data have gone from being rare and expensive, with commercial systems having to be tasked in advance, to being frequent, mainly free (Wulder and Coops 2014), and often available in near real time. Currently the – Landsat TM collections (U.S. Geological Survey 2022c) provide >1 million downloaded images per month, and the long time series with consistent radiometric and geometric data properties supports long-term monitoring of land and water resources (Wulder et al. 2019). U.S.-based data providers dominated the early era of Earth observation, but today the EU, India, China, Japan, and other countries are important data providers as well. NASA, the National Oceanic and Atmospheric Administration (NOAA), and the U.S. Geological Survey are important U.S.-based data providers of large and diverse EO datasets. Copernicus, EU's EO programme, supplies both satellite and reference data on the ground (in situ) data to downstream service providers, public authorities, and other international organizations (see Text box 7.1 and section Copernicus Land Monitoring Service). In addition to free data from these large organizations, there has been an increase in commercial providers of high-spatial-resolution (<4m) satellite data such as IKONOS, WorldView, Quickbird, Planet, and the Pleiades.

Aerial platforms, unmanned aerial vehicles (UAVs; see chapter 8), and ground-based remote sensing provide increasing amounts of data as costs decrease and technical performance increases. These tools are important for local monitoring and data collection

Text box 7.1:   Global data

NASA Open data portal is NASA's clearinghouse site for open data. It provides an extensive and searchable catalogue including a range of EO data as well as many other types of data. Most of the data are stored on other sites, but the metadata are available and searchable (NASA 2022a).

The U.S. Geological Survey (USGS) maintains Earth Explorer, which includes a range of EO data (aerial imagery, Landsat, AVHRR, Sentinel-2) and other data (digital elevation data, land cover data, and more, all available through a user-friendly graphical user interface; USGS 2022b).

The USGS Science Data Catalog provides seamless access to USGS research data and monitoring data. This data catalogue holds a range of EO data and geospatial data including atmosphere and climate, environmental issues, life science, and much more (USGS 2022d).

The Global Forest Watch (GFW) is a thematic data portal designed for forest monitoring. It provides >100 global and local datasets and online alerts of tropical deforestation and fire. It also provides global and country statistics related to forests and deforestation (GFW 2022).

The World Meteorological Organization (WMO) Catalogue for Climate Data provide a set of quality-controlled data on climate (temperature, precipitation, ice, and more; WMO 2022).

The Global Earth Observation System of Systems (GEOSS) portal is a coordinated, independent EO information and processing system. It offers a single access point for users seeking data or imagery as well as analytical software packages (GEOSS 2022).

The Alaska Satellite Facility (ASF) portal downlinks, processes, archives, and distributes remote sensing data to scientific users around the world. ASF operates the NASA archive of synthetic aperture radar (SAR) data from a variety of satellites and aircraft, providing these data and associated specialty support services to researchers in support of NASA's Earth Science Data and Information System (ESDIS) project (ASF 2022).

---

due to low cost and high flexibility, and their imagery can also be used as calibration and validation data for both airborne and satellite-based EO studies and other monitoring efforts.

Information derived from remote sensing

Remote sensing data have to be transformed to information to be useful from a user perspective and to allow consistent monitoring of land, water, or atmospheric properties. Information derived has developed from being dominated by general and static land cover/vegetation classes/categories to time series of more specific physical and structural properties of soil, vegetation (exemplified in Text box 7.4), water, and the atmosphere and object identification (Arvor et al. 2013; Lang et al. 2019). In the earlier era of remote sensing, the time span between data collection and information extraction could be long, sometimes several years, whereas today several services provide real-time or near-real-time data and information. This has opened up additional means of using data; for example, in meteorology, fire detection and fire monitoring, oil spill detection (Al-Rouzouq et al. 2020), as well as early warning systems for food shortages; for example, the UN World Food Programme (WFP), The Food and Agriculture Organization of the United nations (FAO), or The Famine Early Warning Systems Network (FEWS-NET; Funk et al. 2019).

Data portals/providers often provide information, not only data. This information is commonly in the form of pre-defined and validated products (see Text box 7.1 and section Copernicus Land Monitoring Service), often well documented and evaluated. For example, the Copernicus Services catalogue (Programme of the European Union 2022c) gives access to a comprehensive and searchable list of information products as a good starting point for spatial data retrieval. The catalogue is complementary to Copernicus conventional data access hubs including providers such as the ESA (providing EO data from the Sentinels) and EUMETSAT (European operational satellite agency for monitoring weather, climate, and the environment). NASA's Earth Explorer and Application for Extracting and Exploring Analysis-Ready Samples (AppEEARS) are two U.S.-based data access points.

Processing levels tell users to what extent a dataset has been pre-processed (Weaver 2014), where level 0 corresponds to raw data, level 1 the acquired data in physical units, and level 2 is derived geophysical variables at the same resolution and location as level 1. At level 3, the original sensor grid is re-sampled to provide variables mapped to uniform, georeferenced, space–time grid scales in a consistent and gap-filled form. Level 4 are geophysical outputs or results from analyses of lower-level data (NASA 2022b; ESA 2022a). The normal monitoring user commonly uses the standardized, georeferenced level 3 and 4 products because data quality, documentation, and quantification of errors increase with higher processing level.

Sources or portals for Earth observation data

Sources providing EO data are many and diverse, with a fast development of platforms, interfaces, and access methods. Some are thematic, some are sensor specific or for certain regions, some are applied and concern certain topics, and so on. There are many sources or portals for EO data, and only a few are taken up here; see Text box 7.1.

Copernicus Land Monitoring Service

Copernicus is the European Union's Earth observation programme. It includes a range of services and data on the atmosphere, marine environments, land, climate change, security, and emergency (Programme of the European Union 2022b).

The Copernicus Land Monitoring Service provides EO products and is divided into four main components including global, pan-European, local, and imagery and reference data. The Global Land Service systematically produces a time series of bio-geophysical products related to the land surface at mid to low spatial resolution globally. The products monitor vegetation, water, energy, and the terrestrial cryosphere. These products are relevant for global land monitoring of vegetation and soil properties, revealing changes due to both natural and anthropogenic causes. Some products based on SPOT

Vegetation data are available from 1999 onwards, whereas others based on Sentinel-3 and PROBA-V are available from 2014 onwards.

The pan-European component is coordinated by the European Environment Agency (EEA) and produces CORINE land cover datasets (1990 onwards), high-resolution layers (imperviousness, forests, grasslands, water, and small woody features), and biophysical parameters (high-resolution snow as ice products and high-resolution phenology and productivity). The European Ground Motion Service (EMGS) aims to provide consistent, regular, standardized, and reliable information regarding natural and anthropogenic ground motion phenomena such as landslides and subsidence, as well as deformation of infrastructure. EMGS started in November 2016 and is produced from Sentinel-1 radar data with the aim to provide millimetre accuracy and to be complementary to national services, global navigation satellite systems (GNSS), and in situ observations.

The local component aims to provide specific and more detailed information that is complementary to the information obtained through the pan-European component. It focuses on different hotspots in the form of areas with specific environmental challenges and is based on high-resolution imagery (2.5 × 2.5m spatial resolution) in combination with other data. The component includes the Urban Atlas, data on riparian and coastal zones, as well as Natura 2000 areas. Natura 2000 sites are important for nature conservation where detailed land cover/land use mapping are provided at several time intervals from 2006 onward as well as mapped changes.

The imagery and reference data section provides European and global satellite image mosaics, digital elevation data, and hydrological data. The Land Use and Coverage Area frame Survey (LUCAS) focuses on the state and the dynamics of changes in land use and cover and has been repeated every three years since 2006.

Access to data and metadata is provided through direct download when browsing the datasets as well as through web map services (Web Map Service [WMS] and Web Coverage Service [WCS]). Complete and INSPIRE (Infrastructure for Spatial Information in Europe; European Commission 2022) compliant metadata of the core Copernicus land products are stored in a metadata catalogue.

Further information for all Copernicus land products is available in algorithm theoretical basis documents, product user manuals, and validation reports (Programme of the European Union 2022c). It should be noted that there are a range of similar, national, and international data services and data portals providing complementary data.

Text box 7.2:   Phenology

Phenology is the study of climate-dependent periodical phenomena of plants and animals (Abbe 1905) with a focus on the timing, causes, and inter-relations of these phenomena (Leith 1974). Plant phenology influences carbon, water, and energy exchanges between the biosphere and the atmosphere as it changes leaf area index, photosynthesis, albedo, and canopy evapotranspiration (Jin 2015). Climatic changes influence plant phenology through increasing temperatures and altered precipitation patterns, resulting in altered vegetation seasonality in terms of start, end, and length of season.

These changes can be studied through analysis of time series of Earth observation data, typically with vegetation indices such as the Normalized

Difference Vegetation Index (NDVI), the Enhanced Vegetation Index (EVI), or the recent Plant Phenology Index (PPI), especially developed for phenology studies (Jin and Eklundh 2014). These indices can be studied using a coarse resolution such as the MODIS satellite (Figure 7.5) or a finer resolution such as the European Sentinel satellites (Figure 7.6). PPI, a physically based vegetation index, has been shown to scale linearly with leaf area index (LAI), whereas NDVI and EVI show strong saturation at LAI >3. PPI is also less sensitive to snow influence compared to EVI and NDVI, a crucial quality for EO-based monitoring of evergreen needle-leaf forest phenology at northern latitudes (Jin and Eklundh 2014).

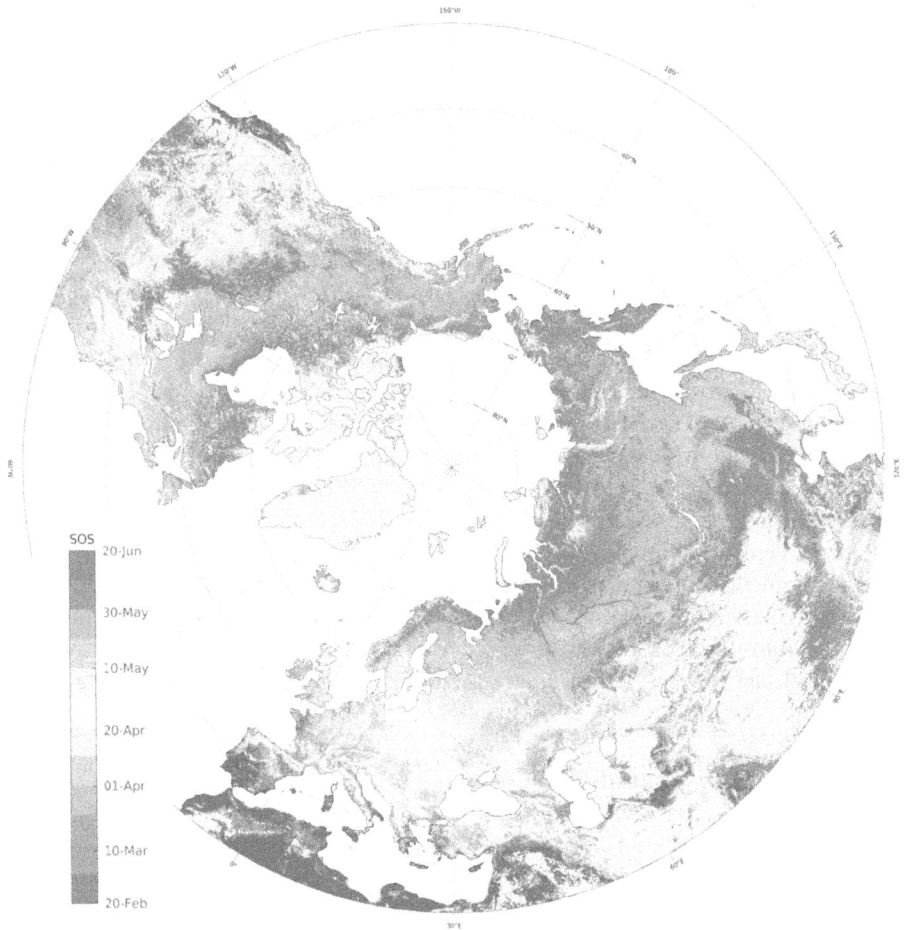

Figure 7.5 MODIS. Average (2000–2018) start of the vegetation season (SOS) for the Northern Hemisphere derived from a time series of PPI calculated on MODIS NBAR data at 0.05° of spatial resolution.

Credit: Illustration and processing by Honxjiao Jin, Lund University, Sweden (Jin et al. 2019). The start, end, and length of the vegetation season influence the carbon cycle and several other ecosystem services and changes over time in vegetation seasonality can be interpreted as a response to climate change.

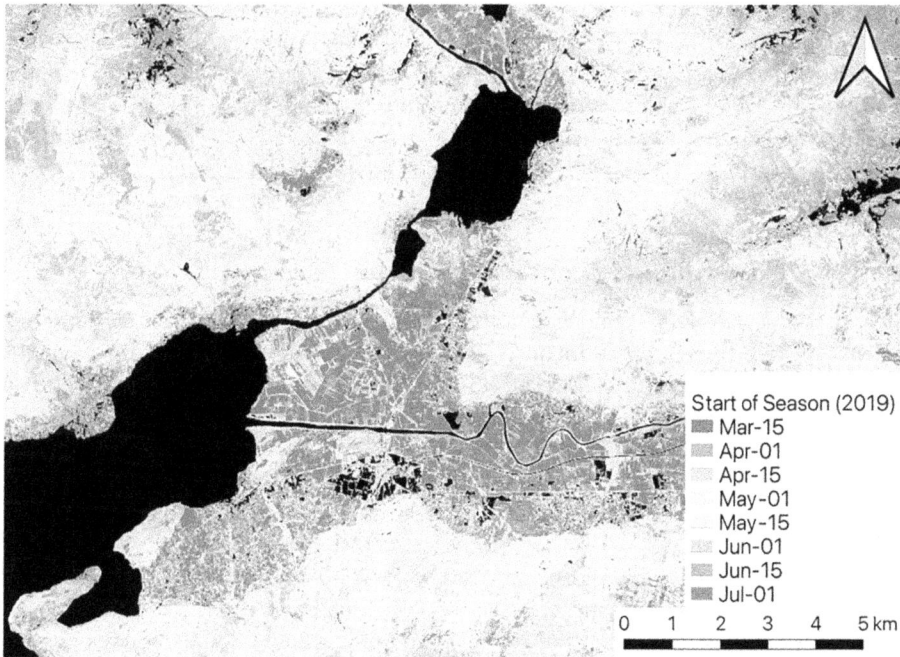

Figure 7.6 Sentinel-2. Start of the vegetation season (SOS) 2019 around Mezzola Lake in
northern Italy derived from Sentinel-2 data PPI and TIMESAT.

Credit: Illustration by Zhanzhang Cai, Department of Physical Geography and Ecosystem Science, Lund
University, Sweden. The differences in SOS are due to differences in vegetation type and altitude.

In 2021, Copernicus released a European-wide dataset called *High Resolution
Vegetation Phenology and Productivity* (HR-VPP; Copernicus Land Monitoring
Service 2022) based on Sentinel-2 data and with a 10m spatial resolution. This
product includes three groups of products; (1) The raw vegetation indices (VIs)
provide the vegetation vigour for each pixel for the respective observation day for
four biophysical variables: PPI, NDVI, fraction of absorbed photosynthetically
active radiation (FAPAR), and LAI. (2) The seasonal trajectories (STs) products are
provided yearly after the end of the vegetation growing season. These are derived
as a regular time series with data every ten days created by fitting a smoothing and
gap filling function to the raw PPI. (3) The vegetation phenology parameters
(VPPs) are derived from the seasonal trajectories of PPI on a yearly basis
(Figure 7.6). The metrics are provided for up to two growing seasons per year
(normally one) and include start of the season, end of season, length of season,
seasonal productivity, and several other metrics. The products are further described
and evaluated in Tian et al. (2021) and in the product documentation (Copernicus
Land Monitoring Service 2022). A similar dataset based on MODIS (500m spatial
resolution) for start of season (SOS), length of season (LOS), and vegetation
productivity for the 2000–2016 period is available from the EEA (EEA 2022; e.g.
Figure 7.5).

EO-based phenology studies have reported significant trends in spring phenology based on a time series of MODIS PPI data for Northern Europe (Jin and Eklundh 2014). Figure 7.5 shows the 2000–2018 average date of the start of the vegetation season for the Northern Hemisphere. The current HR–VPP dataset provides possibilities for high-spatial-resolution assessment of phenology and vegetation productivity in Europe.

Tools for processing Earth observation data

Software tools available for general handling, visualization, and analysis of remote sensing/EO data include free/public domain (QGIS, SNAP, SAGA GIS, etc.) and commercial (IDL/ENVI, ERDAS, ARCGIS, IDRISI) geographic information science (GIS) and image processing packages as well as a range of open source software libraries/code (see, for example, a collection at Github 2022). Computational and statistical software bundles such as MATLAB and R as well as languages such as Python have extensive libraries related to remote sensing, and image processing software can also be used and typically includes tools for handling EO data.

With an increasing amount of EO data, online or cloud-based solutions for accessing and processing data are becoming more and more attractive. Users can access and process data in the cloud and just download the results of the processing. Cloud solutions also allow users to collaborate in the cloud and to share processing chains, algorithms, and results.

Online/cloud computing services are an alternative, allowing processing of large datasets without downloading them. Examples of cloud computing services used in remote sensing include Amazon web service (AWS), Google Earth Engine (GEE), We knowledge Earth Observation (WEkEO), and Data and Information Access Services (DIAS). DIAS and WEkEO are designed to give single point access to all Copernicus data along with cloud-based processing resources, tools, and additional relevant data. GEE includes an extensive ready-to-use data catalogue (Google Earth Engine 2022) with multi-peta-bytes of EO, climate-related, and geophysical data. GEE's application programming interface (API) is available in Python and JavaScript and includes a set of standard tools and methods as well as facilities to share applications and code. The massive growth of EO data has triggered the development of cloud-based tools as a solution to handle the ever-increasing pool of big data collected by EO satellites (Wang et al. 2018, 2020). Flexible data portals such as AρρEEARS (USGS 2022a) provide user-oriented and flexible data delivery through web-based pre-processing, hence saving time and processing resources for the end user.

Methodologies for analysis of EO data are diverse and include a range of statistical (descriptive, classifications, regressions, time series analysis, etc.) and artificial intelligence (AI)-based methods such as machine learning, deep learning, and various types of neural network methods (Zhu et al. 2017; Reichstein et al. 2019). Some of these are described further in chapter 8.

Application examples

The development of new sensors and platforms as well as the availability of free and open EO data have resulted in an increased use of remote sensing in many applications. EO data, or products derived from EO data, suitable for a wide variety of applications are available at different scales, ranging from local to global.

Satellite data from sensors, such as Landsat TM and the Multispectral Instrument (MSI) onboard Sentinel-2, are commonly used in various mapping and monitor applications. One example is the Global Forest Watch (GFW), which provides information about forest changes around the entire world based on Landsat data. Another example is the European CORINE Land Cover (CLC) program that was initiated in 1985. In CLC, the landscape is separated into 44 land cover classes and it was updated in 2000, 2006, 2012, and 2018. There are also national mapping and monitoring programs such as the Swedish National Land Cover Mapping (NMD) that provide a more detailed classification of the landscape than CLC.

Satellite images have been used in agriculture to map and monitor crops since the launch of Landsat-1. Another important application is precision agriculture (PA), which has been enabled by the advent of global navigation satellite systems (GNSS) and the availability of spectral data from satellites or drones.

The development of airborne lidar systems provides excellent possibilities to map and monitor forests. Countries like Finland and Sweden regularly produce nationwide maps with estimates of forest variables like tree height and standing volume based on lidar data. Lidar datasets are also powerful tools in archaeology where they are used to detect and map archaeological features and structures. Geo-radar is another important tool in archaeology that is used to detect and map underground structures.

Mapping and monitoring of the atmosphere

Satellite-based monitoring of the atmosphere allows for global-scale near-real-time monitoring of the air quality and pollution at regional-scale resolutions. Measurements of solar radiation, sources and sinks of greenhouse gases, and radiative forcing are important inputs to models for tracking and predicting climate change.

Atmospheric satellites typically use optical spectrometers to measure the presence of gases and aerosols. Examples of meteorological satellites that carry these instruments include ESA's MetOp satellites that measure ozone gases, methane, carbon monoxide, nitrogen dioxide, and carbon dioxide and NASA's Aura satellite, which extends these measurements to include chlorofluorocarbons and water vapour content. The Sentinel-5P satellite has a larger spectral range than previous satellites that extends to ultraviolet wavelengths, making it possible to detect formaldehyde, sulphur dioxide, UV-B radiation, and carbon monoxide as well. The Sentinel-5 sensor will also be flown aboard future ESA MetOp satellites.

An example of how these measurements of the atmosphere have been utilized comes from ESA's Copernicus Atmosphere Monitoring Service (CAMS; Programme of the European Union 2022a), operational since 2015. It builds upon existing research and products to provide global information and forecasts on air pollution and health, greenhouse gases and climate forcing, and solar energy. Data provided by CAMS are used in policymaking at national, regional, and local scales to limit exposure to and reduce atmospheric pollutants. Emissions from volcanoes and forest fires are tracked and used to provide advanced information on air quality. Measurements of solar radiation from CAMS are also used to model the efficiency of different solar energy installations (Peuch et al. 2018).

Mapping and monitoring in glacial environments

Glaciers are sensitive to climate change and, as such, the monitoring of glaciers can provide critical information on current warming trends. As vast stores of freshwater, the accelerated melting and retreat of glaciers can have severe impacts on both the local environment, such as effects on water supply, surging glaciers, glacial outburst floods, and localized desalination of the water column, and the global environment due to rising sea levels and changes to global climatic patterns. Remote sensing provides an effective means for the mapping and monitoring of glaciers.

Data from optical sensors, both spaced-based such as Landsat and Sentinel-2, and from aerial photography can be used for the purposes of glacial monitoring. However, spaceborne synthetic aperture radar (SAR) sensors, such as ERS, Envisat ASAR, and Sentinel-1, are not blocked by clouds, unlike optical (colour) sensors, which are, and can be used for large scale mapping of remote glacial environments where aerial photography is not feasible. The polarization and scattering properties of radar, the interaction of radar with water due to its dielectric properties, and the interferometric coherence of multiple radar images, can be used to estimate and map a wide variety of glacial properties; some are listed in Text box 7.3.

Text box 7.3:   Glacial mapping

Glacial facies (glacial zones defined by their density and fluid percolation properties, including dry-snow, wet snow, firn, and bare ice) have been mapped over Svalbard (Barzycka et al. 2019) using the radar backscatter coefficient (strength of the returned SAR signal) and by polarimetric decomposition (classification based on the ratio of the returned signal in different polarizations).

Snow cover can be mapped through the use of interferometric coherence (coherence measures the degree which two radar signals share the same phase), because the coherence of snow degrades at a different rate between two radar images compared to the surrounding environment in the image; this can be exploited for mapping snow cover (Kumar and Venkataraman 2011). Snow cover can also be mapped based on the scattering and polarization of the radar signal (Muhari et al. 2017).

Snow wetness and snow density are important for estimating the total volume of water held in the glacial snowpack and are often used as an input for hydrological models for monitoring water supply. Using physical models of radar scattering as a function of snowpack properties, the physical snow parameters can be retrieved from the radar signal by inverting these models (Shi and Dozier 1995).

Glacial mass balance can increase through snow accumulation or decrease through melting and is an important indicator of glacier health and climate change. Radar-based interferometry is used to create surface elevation models, and changes over time in the surface elevation can be used to calculate the gain or loss of glacial mass (Lin et al. 2017).

Glacial surface velocity forms a part of the mass balance calculations but is also useful to characterize the stability or instability of glaciers. Hazards such as surging glaciers and glacial outburst floods can be potentially identified in advance from trends in surface velocity (Jiskoot et al. 2003). Radar-based differential interferometry and multiple aperture interferometry can be used to retrieve three-dimensional vectors of surface velocity (Jung et al. 2009).

Mapping and monitoring in urban areas

The ability to visit urban green spaces (UGS) affects the quality of life of local residents through a wide range of ecosystem services (Millennium Ecosystem Assessment 2005), and green areas and trees also help to cool down the air temperature in cities, otherwise dominated by stone, concrete, and asphalt (urban grey areas). To ensure the benefits of a diverse flora and fauna as well as the health benefits of residents of having greenery nearby, policymakers and planners of infrastructure should aim for integration of green systems into the urban fabric and sociospatial structures, sometimes referred to as *liveable cities* (Banzhaf et al. 2020; Mathey et al. 2021; Shahtahmassebi et al. 2021). Although the study of UGSs has become rather common, the main focus has often been on larger patches, and studies of smaller patches such as street trees, urban gardens, and public parks are still few, even though collectively such patches can cover substantial areas (Shahtahmassebi et al. 2021).

Using the Swedish National Land Cover Database (see Text box 8.4), Statistics Sweden has been able to map the green structures in urban areas nationwide for the first time. The findings were that, on average, 94% of the urban population had access to at least one green area within 200m of their home. Proximity to green areas differed only slightly between large and small localities (Moström 2019). Figure 7.7 shows an urban area in the southern part of Sweden with different green areas (grassy areas in beige/orange, depending on the wetness regime, and trees or forest in different greens) from the production of the second version of the Land Cover Mapping.

Mapping is a useful investigation tool in places that are hard to get to, together with subsequent analysis in a GIS to get statistics of eventual causes for concern or the need for action. In urban areas with high buildings or inaccessible backyards, the use of remote

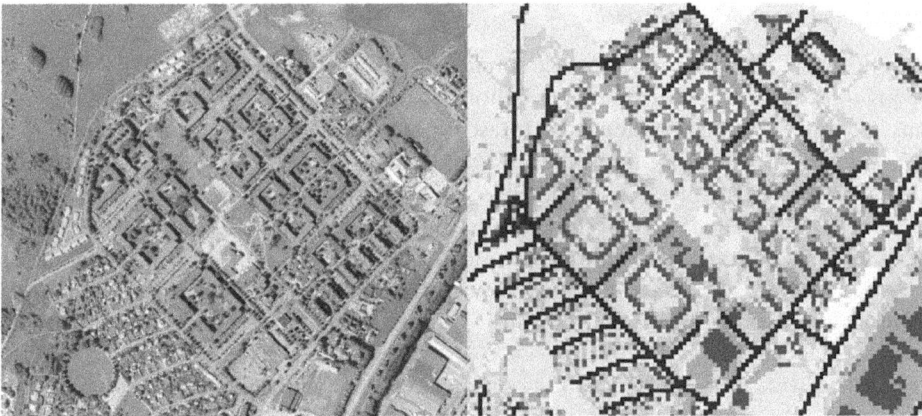

Figure 7.7 Example of urban green areas from the Swedish National Land Cover Mapping (NMD), where the land cover is integrated with moisture regime, showing the grassy areas in three different hues of beige/orange, depending on the degree of moisture, and trees and shrubs are differentiated into coniferous, deciduous, or hardwood deciduous and are shown in green hues.

Credit: Figure by Anna Allard; aerial image provided with permission by Lantmateriet, the Swedish mapping, cadastral, and land registration authority. The Land Cover database is available as open-source data (Nilsson et al. 2021).

sensing in various spatial and temporal scales gives the bird's-eye view, making the monitoring methodology extra useful. An example of mapping UGSs by Sentinel-2A of two cities in Bulgaria and Slovakia is provided by Vatseva et al. (2016). Classification into UGSs is dependent on the spectral (different sensors and lidar) as well as spatial scale – the finer the spatial scale (drones or UAVs), the finer the resolution of detail in the map (Shahtahmassebi et al. 2021).

Cities are becoming smart in a variety of ways that enable us to monitor, understand, analyze, and plan the city, with the aim of enhancing resource efficiency, equity among citizens, and quality of life in real time and provide better foundations for sustainable and healthy living. Banzhaf et al. (2020) mapped the city of Leipzig, Germany, as classes of urban land use and land cover (uLULC), a concept that refers not only to the green but also to the built elements of a city. The integration of both urban grey and green structures at the scale of local districts helps to understand the urban structure and residential needs for urban ecosystem services.

To map and analyze the heterogeneous fabric of Leipzig, a number of data sources were laid on top of each other. The layers consisted of satellite images in very high resolution, an NDVI, calculated from the red and near-infrared bands of digital ortho-imagery and lidar derivatives to differentiate between the ground surface and various vegetation structures according to their height. The next step was an object-based image analysis (the programme eCognition) that differentiates grey and green structures with high precision and at refined scale. Banzhaf et al. (2020) were able to classify the urban area according to different types of housing as well as green lawn areas, shrubs/young trees, and older/mature trees.

Spatial information linked with allocated population and health-related indicators could then identify built-up types with highest population densities and local districts with deficits in the provision of different green structures (Banzhaf et al. 2020).

Mapping and monitoring in forest

Remote sensing techniques are well suited for mapping and monitoring forest, for reasons of ecology, management, or economy, and have long been widely used, with many examples of classification into forest types. With the extra dimension of lidar data, mapping the distribution of forest ecosystems and characterizing the three-dimensional structure of forests is possible, by such variables as height of the canopy and inferred age of the trees and some knowledge of what type of ground can be found under the trees. Others use remote sensing for national assessment and reporting of forest degradation, a global phenomenon and an important indicator and precursor to further forest losses and carbon emissions (Mitchell et al. 2017; Atzberger et al. 2020; Lechner et al. 2020).

For assessments from authorities and policymakers and planners of management of the status of the forests around the world, Mitchell et al. (2017) created a global map, classifying all forested areas into four classes: intact, fragmented/managed, degraded, or deforested areas. At the EU scale, the Copernicus initiative provides a pan-European map of forests, and Atzberger et al. (2020), in their report, narrowed the issues of concern down a number priority targets and provided recommendations on the remote sensing techniques that would be of most use for the different monitoring targets.

The prioritized targets for forest monitoring in 2020 are:

* Wildfire monitoring
* Pest infestation
* Storm damages
* Drought
* Phenology
* Illegal logging

When assessing the national forest resource, updated and almost interactive mapping can be highly useful, which both the owners of forest and the local, regional, and national authorities can consult for different purposes such as clear-cutting or to preserve older forest in the day-to-day management (see Text box 7.4). Also, these show where there are known objects of cultural heritage, high nature values, already protected forest, and so on.

Text box 7.4:   Nationwide mapping of forest attributes derived using lidar data

Sweden is a forested land to a large degree, and information about the forest landscape is needed for many monitoring purposes, when planning legislation, conservation actions, or forest activities. The first national raster maps with predicted forest variables such as standing volume and tree height were produced in the early 2000s, then based on Landsat TM data.

Due to the arctic location, Sweden was pressed down by a very thick continental ice sheet during the last Ice Age, and still the land mass is slowly rising from the sea by land upheaval, making the older elevation data more and more inaccurate. With the aim of creating a new national digital elevation model (DEM), a nationwide scanning by lidar was done (between 2009 and 2018) by the Swedish mapping, cadastral, and land registration authority.

Soon researchers, forest managers, and many others realized the very high value of being able to freely download these data for a variety of additional uses; for example, for integration with in situ data from the National Forest Inventory and satellite data and for estimations of several types of forestry variables. As a result, the next generation of forestry estimations was based on lidar data (Figure 7.8). Funded by the government, the project was carried out in cooperation between the Swedish Forest Agency and the Swedish University of Agricultural Sciences (SLU).

The lidar dataset has a point density of 0.5–1.0 points/m^2 and it was used in combination with data from approximately 11,500 field plots from the Swedish National Forest Inventory to predict stem volume, aboveground tree biomass, basal area, Lorey's mean tree height, and mean stem diameter (Nilsson et al. 2017; Hultgren et al. 2020).

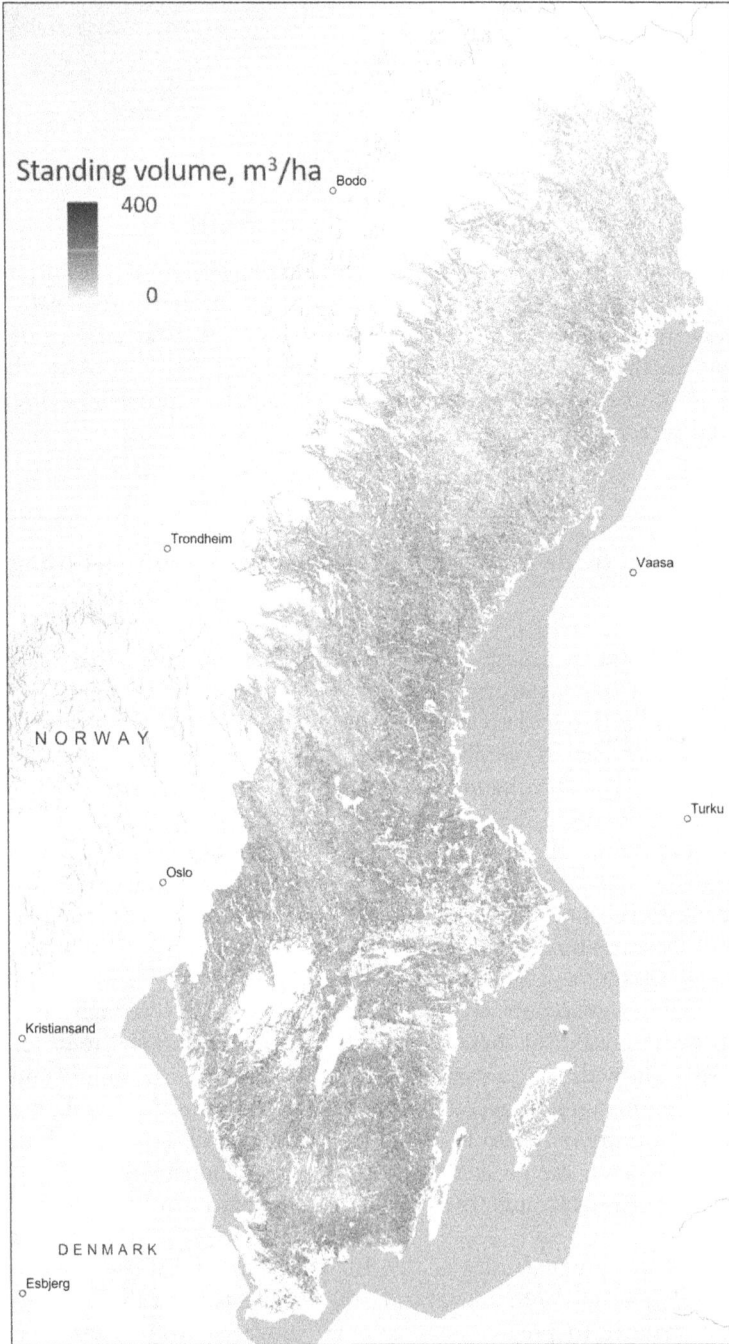

Figure 7.8 The forest map is provided as open data and has been much appreciated by the Swedish forest sector, authorities, researchers, and many other users.

Source: Swedish Forest Agency (2022).

Remote sensing in precision agriculture

According to a generally recognized definition by the International Society for Precision Agriculture (ISPA), *precision agriculture* (PA) is "a management strategy that gathers, processes and analyzes temporal, spatial and individual data and combines it with other information to support management decisions according to estimated variability for improved resource use efficiency, productivity, quality, profitability and sustainability of agricultural production". As a concept that was possible to apply in practical agriculture, PA emerged in the mid-1990s and has since evolved through new methods and techniques. It is now one of the solutions brought forward to meet global challenges in food production (European Parliament 2019). An example of precision agriculture is given in Text box 7.5.

Text box 7.5:   Example of remote sensing data collected from different platforms in precision agriculture

Remote sensing in practical use and access to satellite data in decision support systems

Remote sensing from satellites and proximal sensing from tractor-mounted sensors are currently the main techniques used in different practical PA solutions. Farmers access Sentinel-2 images through decision support systems (DSS), in which the images can be converted to variable-rate application files suitable for controlling agricultural equipment such as fertilizer spreaders (Mulla 2021). One example usage of satellite images in a DSS is to generate variable-rate nitrogen (N) maps during the growing season to adapt the amount of N applied according to the need (Figure 7.7). The technique (applied to e.g. wheat, barley, or maize) builds upon relationships between biomass and N uptake in the crop and vegetation indices within the visible-near-infrared region of the electromagnetic spectrum (e.g. Söderström et al. 2017). The goal in practice is often not only to increase the yield but also to reach a certain protein content that is needed for various types of uses and may increase the value of the harvested grain. The farmer's perspective may be to find the economical optimum N rate.

On-the-go sensing

Another approach is data collection by multispectral sensors mounted on tractors (Taylor et al. 2021). This technique has the advantage that it can be used regardless of cloud cover, and it is possible to develop what are called "on-the-go" applications. Here, the sensor can communicate with a computer in the tractor, and the amount of N to apply can be estimated directly without the need for mapping. The estimated requirement is transmitted to the spreader, which adjusts the rate accordingly (Figure 7.9). This technique has been available since the late 1990s. Tractor-based sensors may be passive and rely on the incoming sunlight but also active sensors are available (Taylor et al. 2021). The latter makes them very flexible in terms of usage, and the equipment may be used at any time-of-day.

Sensing from drones – new possibilities

More recently, remote sensing from drones (UAVs) has become an alternative to satellite-based and tractor-based sensors in PA (Maes and Steppe 2019). Crop sensing using multispectral cameras on UAVs can generate similar information as from the other sensing platforms (Figure 7.7), but it can also provide information that is different: centimetre spatial resolution, far better than what is currently possible from satellite and tractor sensors. This makes it possible to develop other types of applications; for example, detecting individual plants for weed management (Christensen et al. 2021). Through artificial intelligence applications, it is possible to automatically locate and map each weed plant and even determine the species. Instead of uniform application of herbicides, targeted spraying can substantially reduce the amount of chemicals used.

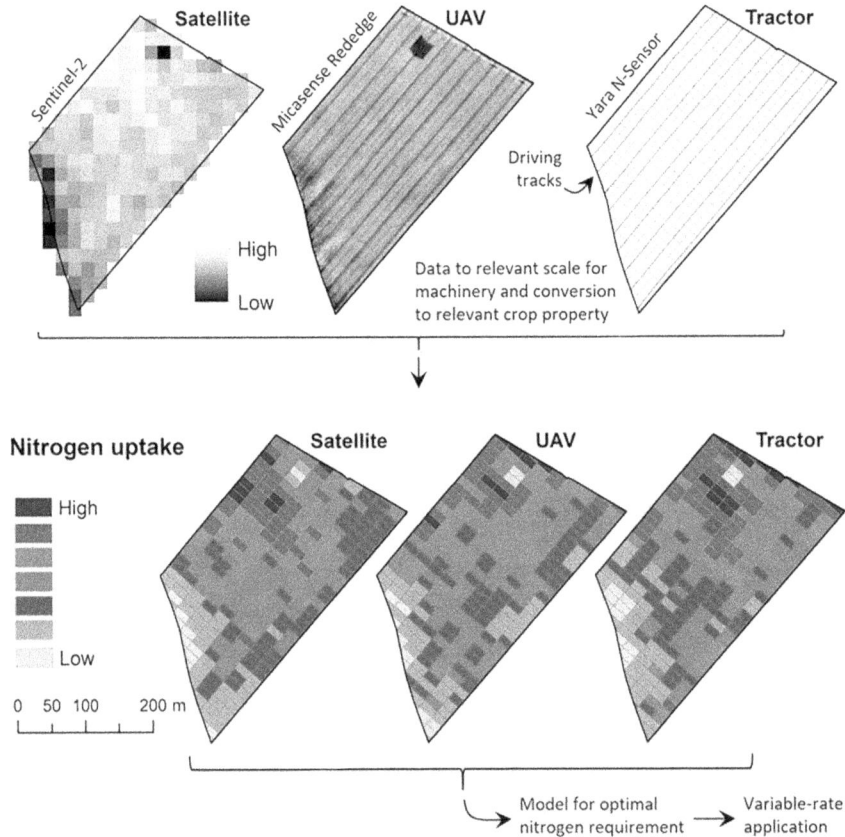

Figure 7.9 This example shows a 7.7ha wheat field. In this case, the fertilizer spreader has a 24m working width (shown by the distance between driving tracks). This determines the spatial resolution of the output data. Despite the very different sources of input data (top row: maps of a vegetation index, collected within five days, typically used for nitrogen uptake modelling; Gitelson et al. 2003; Wolters et al. 2021), the resulting maps of nitrogen uptake (bottom row) show a similar overall pattern, although there are differences between individual plots. Through agronomic models, an output map is converted to nitrogen requirements in a variable-rate application file, which is used to adjust the fertilizer spreader. In case of a sensor mounted on a tractor, the whole process of sensing and conversion to variable-rate adjustment of the spreader is done on the go, as opposed to the satellite and UAV approach.

Credit: Illustration by Mats Söderström.

Key messages

- This chapter has introduced remote sensing methods for a wide range of applications and in environments ranging from glaciers and forests to urban settings. Though air photography has been a mature technology since World War I, this is now improved by new digital cameras and new airborne instruments such as lidar.
- Over the last three decades, satellite remote sensing of the natural environment has developed from a highly specialized activity, typically involving a few expensive images with poor spatial detail, to something almost any agency or individual can do using an abundance of free online analysis-ready data and processing tools.
- This has put the emphasis of analysis firmly back where it should be: taking care of the technical aspects of pre-processing and image analysis but most influenced and shaped by our understanding of context, ecology, and landscape. This is helped by the increasing quality of satellite imagery, with more useful red edge channels for vegetation monitoring, increasingly high spatial detail able to show the small elements of the landscape, and an important step up from 7- or 8-bit encoding of DNs to 12-bit, which allows optical instruments to record far more detail from shadowed slopes and in winter months at higher latitudes. Though rather technical on paper, the net effect of innovation in instruments is to make remote sensing less limited and more informative about what we want to observe.
- There are, however, some dangers in going from bespoke restaurant catering to the remote sensing equivalent of fast food: the ease of consumption can be deceptive and the sheer availability of data can be too much to digest unless we are discriminating. We need to be careful to check that the results are valid against in situ observations – for example, to avoid automatically treating the differences between images or classifications as if they are real – and track sources of bias, error, and error propagation through the processing chain before firming up conclusions.
- If we are careful in this way, and especially if we have good calibration and a statistically competent set of ground observations to train and test the analysis, remote sensing can – and will – revolutionize our ability to monitor biodiversity.

Study questions

1 Go to the online resources of the European Space Agency/Copernicus and watch the videos to learn more about satellite remote sensing.

2 Look in the literature for examples of real processing chains and see if you can identify the different steps and how each step in pre-processing influences the quality of the analysis.

3 Find out the differences between *spectral space*, *image space*, and *feature space*. What type of visualization is enabled by these different ways of looking at a satellite image? How does feature space resemble multivariate data – for example, records of multiple species in multiple plots – and what might be the similarities in the way we combine data into a smaller number of dimensions?

4 Look online at an image from Landsat-7, Landsat-8, and Sentinel-2 for an area you know. What are the differences between them? Landsat-7 is 8 bits per channel, and Landsat-8 is 12 bits (stored as 16 bits). How does this affect what can be recorded for landscapes with low light or shadowed slopes, and why is this important? If you can,

download copies and look at these in a GIS package that allows you to "stretch" the image to see different parts of the range of digital numbers. Why can't we see all this detail at once on a (hint!) computer screen showing 8 bits per channel?

References

Abbe, C. (1905) *A First Report on the Relations between Climates and Crops*. Washington, DC: Government Printing Office.

Alaska Satellite Facility. (2022) Data portal, https://asf.alaska.edu/ (Accessed June 22, 2022).

Al-Ruzouq, R., Gibril, M.B.A., Shanableh, A., Kais, A., Hamed, O., Al-Mansoori, S. and Khalil, M.A. (2020) Sensors, features, and machine learning for oil spill detection and monitoring: a review, *Remote Sensing* 12(20), 3338.

Arvor, D., Durieux, L., Andrés, S. and Laporte, M.-A. (2013) Advances in geographic object-based image analysis with ontologies: a review of main contributions and limitations from a remote sensing perspective, *ISPRS Journal of Photogrammetry and Remote Sensing* 82, 125–137.

Atzberger, C., Zeug, G., Defourny, P., Aragão, L., Hammarström, L. and Immitzer, M. (2020) *Monitoring Forests through Remote Sensing*, Final Report. European Union. https://ec.europa.eu/environment/forests/pdf/report_monitoring_forests_through_remote_sensing.pdf

Banzhaf, E., Kollai, H. and Kindler, A. (2020) Mapping urban grey and green structures for liveable cities using a 3D enhanced OBIA approach and vital statistics, *Geocarto International* 35(6), 623–640. doi: 10.1080/10106049.2018.1524514

Barzycka, B., Błaszczyk, M., Grabiec, M. and Jania, J. (2019) Glacier facies of Vestfonna (Svalbard) based on SAR images and GPR measurements, *Remote Sensing of Environment* 221, 373–385.

Bramich, J., Bolch, C.J.S. and Fischer, A. (2021) Improved red-edge chlorophyll-a detection for Sentinel 2, *Ecological Indicators* 120, 106876. https://doi.org/10.1016/j.ecolind.2020.106876

Christensen, S., Dyrmann, M., Laursen, M.S., Jørgensen, R.N. and Rasmussen, J. (2021) Sensing for weed detection, in Kerry, R. and Escolá, A. (eds) *Sensing Approaches for Precision Agriculture*. Cham, Switzerland: Springer Nature, pp. 275–300. https://doi.org/10.1007/978-3-030-78431-7

Cui, Z. and Kerekes, J.P. (2018) Potential of red edge spectral bands in future Landsat satellites on agroecosystem canopy green leaf area index retrieval, *Remote Sensing* 10(9), 1458. https://doi.org/10.3390/rs10091458

Deering, C.A. and Stoker, J.M. (2014) Let's agree on the casing of lidar, *LiDAR News Magazine* 4(6). https://lidarmag.com/wp-content/uploads/PDF/LiDARNewsMagazine_DeeringStoker-CasingOfLiDAR_Vol4No6.pdf

DiBiase, D. (n.d.). *Nature of Geographic Information*. The Pennsylvania State University, B.C. Open Textbook Project. https://opentextbc.ca/natureofgeographicinformation/

European Commission. (2022) Inspire knowledge base, infrastructure for spatial information in Europe, https://inspire.ec.europa.eu/ (Accessed November 8, 2022).

European Environment Agency. (2022) Vegetation productivity 2000–2016, https://www.eea.europa.eu/data-and-maps/data/annual-above-ground-vegetation-productivity-1 (Accessed November 6, 2022).

European Parliament, Directorate-General for Parliamentary Research Services, Daheim, C., Poppe, K. and Schrijver, R. (2019) Precision agriculture and the future of farming in Europe: scientific foresight study, https://data.europa.eu/doi/10.2861/020809

European Space Agency. (2022a) Processing levels, https://sentinel.esa.int/web/sentinel/user-guides/sentinel-2-msi/processing-levels (Accessed November 8, 2022).

European Space Agency. (2022b) Radiometric resolutions, https://sentinels.copernicus.eu/web/sentinel/user-guides/sentinel-2-msi/resolutions/radiometric (Accessed November 8, 2022).

Funk, C., Shukla, S., Thiaw, W.M., Rowland, J., Hoell, A., McNally, A., Husak, G., Novella, N., Budde, M., Peters-Lidard, C., et al. (2019) Recognizing the Famine Early Warning Systems Network: over 30 years of drought early warning science advances and partnerships promoting global food security, *Bulletin of the American Meteorological Society* 100(6), 1011–1027.

GISGeography. (2022) Multispectral vs. hyperspectral imagery explained, https://gisgeography.com/multispectral-vs-hyperspectral-imagery-explained/ (Accessed November 8, 2022).

Gitelson, A.A., Gritz, Y. and Merzlyak, M.N. (2003) Relationships between leaf chlorophyll content and spectral reflectance and algorithms for non-destructive chlorophyll assessment in higher plant leaves, *Journal of Plant Physiology* 160(3), 271–282. https://doi.org/10.1078/0176-1617-00887

Github. (2022) Awesome-EarthObservation-Code, https://github.com/acgeospatial/awesome-eartho bservation-code/blob/master/README.md

Global Earth Observation System of Systems. (2022) Data portal, https://www.geoportal.org (Accessed June 22, 2022).

Global Forest Watch. (2022) Forest monitoring designed for action, https://www.globalforestwatch. org/ (Accessed November 8, 2022).

Google Earth Engine. (2022) A planetary-scale platform for Earth science data & analysis, https://earthengine.google.com/ (Accessed November 8, 2022).

Hultgren, B., Nilsson, L. and Nilsson, M. (2020) *Statusrapport - Uppdatering av Skogliga Grunddata Från Laserskanning 2018–2019* [Status Report – Update of National Forest Data from Laser Scanning 2018–2019]. Swedish Forest Agency. https://www.skogsstyrelsen.se/globalassets/om-oss/regeringsuppdrag/uppdatering-av-skogliga-grunddata/rapport---uppdatering-av-skogliga-grunddata-fran-laserskanning-2018-2019.pdf

Jin, H.X. (2015) *Remote Sensing Phenology at European Northern Latitudes. From Ground Spectral Towers to Satellites.* PhD Thesis, Lund University.

Jin, H.X. and Eklundh, L. (2014) A physically based vegetation index for improved monitoring of plant phenology, *Remote Sensing of Environment* 152, 512–525.

Jin, H.X., Jönsson, A.M., Olsson, C., Lindström, J., Jönsson, P. and Eklundh, L. (2019) New satellite-based estimates show significant trends in spring phenology and complex sensitivities to temperature and precipitation at Northern European latitudes, *International Journal of Biometeorology* 63(6), 763–775.

Jiskoot, H., Murray, T. and Luckman, A. (2003) Surge potential and drainage-basin characteristics in East Greenland, *Annals of Glaciology* 36, 142–148.

Jung, H.S., Won, J.S. and Kim, S.W. (2009) An improvement of the performance of multiple-aperture SAR interferometry (MAI), *IEEE Transactions on Geoscience and Remote Sensing* 47(8), 2859–2869.

Kumar, V. and Venkataraman, G. (2011) SAR interferometric coherence analysis for snow cover mapping in the western Himalayan region, *International Journal of Digital Earth* 4(1), 78–90.

Lang, S.G., Hay, J., Baraldi, A., Tiede, D. and Blaschke, T. (2019) GEOBIA achievements and spatial opportunities in the era of Big Earth Observation Data, *ISPRS International Journal of Geo-Information* 8(11), 474.

Lechner, A.M., Foody, G.M. and Boyd, D.S. (2020) Applications in remote sensing to forest ecology and management, *One Earth* 2(5), 405–412. https://doi.org/10.1016/j.oneear.2020.05.001

Leith, H. (1974) *Phenology and Seasonality Modelling.* Springer.

Liang, S. and Wang, J. (2020) *Advanced Remote Sensing.* 2nd edn. Academic Press. https://doi.org/10.1016/B978-0-12-815826-5.00001-5.

Lillesand, T.M. and Kiefer, R.W. (2015) *Remote Sensing and Image Interpretation.* 7th edn. New York: Wiley.

Lin, H., Li, G., Cuo, L., Hooper, A. and Ye, Q. (2017) A decreasing glacier mass balance gradient from the edge of the Upper Tarim Basin to the Karakoram during 2000–2014, *Scientific Reports* 7(1), 1–9.

Maes, W.H. and Steppe, K. (2019) Perspectives for remote sensing with unmanned aerial vehicles in precision agriculture, *Trends in Plant Science* 24(2), 152–164. https://doi.org/10.1016/j.tplants.2018.11.007

Mathey, J., Hennersdorf, J., Lehmann, I. and Wende, W. (2021) Qualifying the urban structure type approach for urban green space analysis – a case study of Dresden, Germany, *Ecological Indicators* 125, 107519. https://doi.org/10.1016/j.ecolind.2021.107519

Millennium Ecosystem Assessment. (2005) *Ecosystems and Human Well-Being: Synthesis.* Washington, DC: Island Press.

Mitchell, A.L., Rosenqvist, A. and Mora, B. (2017) Current remote sensing approaches to monitoring forest degradation in support of countries measurement, reporting and verification (MRV) systems for REDD+, *Carbon Balance Manage* 12(9), 1–22. https://doi.org/10.1186/s13021-017-0078-9

Moström, J. (2019) *Green Space and Green Areas in Urban Areas 2015.* Statistics Sweden (in Swedish).

Muhuri, A., Manickam, S. and Bhattacharya, A. (2017) Scattering mechanism based snow cover mapping using RADARSAT-2 C-Band polarimetric SAR data, *IEEE Journal of Selected Topics in Applied Earth Observations and Remote Sensing* 10(7), 3213–3224.

Mulla, D. (2021) Satellite remote sensing for precision agriculture, in Kerry, R. and Escolá, A. (eds) *Sensing Approaches for Precision Agriculture.* Cham, Switzerland: Springer Nature, pp. 19–58. https://doi.org/10.1007/978-3-030-78431-7

National Aeronautics and Space Administration (2022a) Data.NASA.Gov, A catalog of publicly available NASA datasets, https://data.nasa.gov/ (Accessed November 8, 2022).

National Aeronautics and Space Administration. (2022b) NASA Earth data, open access for open science, https://www.earthdata.nasa.gov/engage/open-data-services-and-software/data-information-policy/data-levels Accessed November 8, 2022).

National Aeronautics and Space Administration. (2022c) NASA science, tour of the electromagnetic spectrum, https://science.nasa.gov/ems/01_intro (Accessed November 8, 2022).

Nilsson, M., Allard, A., Ahlkrona, E., Jönsson, C., Petra Odentun, P., Berlin, B., Karlsson, L. and Olsson, B. (2021) *Agenda för Landskapet AP 7 Validering* [Agenda for the Landscape – Statistical Evaluation]. Umeå: Swedish University for Agricultural Science.

Nilsson, M., Nordkvist, K., Jonzén, J., Lindgren, N., Axensten, P., Wallerman, J., Egberth, M., Larsson, S., Nilsson, L., Eriksson, J., et al. (2017) A nationwide forest attribute map of Sweden predicted using airborne laser scanning data and field data from the National Forest Inventory, *Remote Sensing of Environment* 194, 447–454. https://doi.org/10.1016/j.rse.2016.10.022

Peuch, V.H., Engelen, R., Ades, M., Barré, J., Inness, A., Flemming, J., Kipling, Z., Agusti-Panareda, A., Parrington, M., Ribas, R., et al. (2018) The use of satellite data in the Copernicus Atmosphere Monitoring Service (CAMS), in IGARSS 2018–2018 IEEE International Geoscience and Remote Sensing Symposium. IEEE, pp. 1594–1596. https://doi.org/10.1109/IGARSS.2018.8518698

Programme of the European Union. (2022a) Copernicus Europe's Eyes on Earth, Atmosphere Monitoring Service, https://atmosphere.copernicus.eu/ (Accessed November 8, 2022).

Programme of the European Union. (2022b) Copernicus Europe's Eyes on Earth, Copernicus Services, https://www.copernicus.eu/en/copernicus-services (Accessed November 8, 2022).

Programme of the European Union. (2022c) Copernicus Europe's Eyes on Earth, Copernicus Services catalogue, https://www.copernicus.eu/en/accessing-data-where-and-how/copernicus-services-catalogue (Accessed November 8, 2022).

Reichstein, M., Camps-Valls, G., Stevens, B., Jung, M., Denzler, J., Carvalhais, N. and Prabhat. (2019) Deep learning and process understanding for data-driven Earth system science, *Nature* 566(7743), 195–204.

Schott, R.S. (2007) *Remote Sensing – The Image Chain Approach.* 2nd edn. Oxford University Press.

Shahtahmassebi, A.R., Li, C., Fan, Y., Wu, Y., Lin, Y., Gan, M., Wang, K., Malik, A. and Blackburn, G.H. (2021) Remote sensing of urban green spaces: a review, *Urban Forestry & Urban Greening* 57, 126946. https://doi.org/10.1016/j.ufug.2020.126946

Shi, J. and Dozier, J. (1995) Inferring snow wetness using C-band data from SIR-C's polarimetric synthetic aperture radar, *IEEE Transactions on Geoscience and Remote Sensing* 33(4), 905–914.

Swedish Forest Agency. (2022) Forest map, https://www.skogsstyrelsen.se/skogligagrunddata/ (Accessed May 13, 2022).

Söderström, M., Piikki, K., Stenberg, M., Stadig, H. and Martinsson, J. (2017) Producing nitrogen (N) uptake maps in winter wheat by combining proximal crop measurements with Sentinel-2 and DMC satellite images in a decision support system for farmers, *Acta Agriculturae Scandinavica, Section B – Soil and Plant Science* 67(7), 637–650. https://doi.org/10.1080/09064710.2017.1324044

Taylor, J.A., Anastasiou, E., Fountas, S., Tisseyre, B., Molin, J.P., Trevisan, R.G., Chen, H. and Travers, M. (2021) Applications of optical sensing of crop health and vigour, in Kerry, R. and Escolá, A. (eds) *Sensing Approaches for Precision Agriculture.* Cham, Switzerland: Springer Nature, pp. 333–368. https://doi.org/10.1007/978-3-030-78431-7

Tian, F., Cai, Z., Jin, H., Hufkens, K., Scheifinger, H., Tagesson, T., Smets, B., van Hoolst, B., Bonte, K., Ivits, E., et al. (2021) Calibrating vegetation phenology from Sentinel-2 using eddy covariance, PhenoCam, and PEP725 networks across Europe, *Remote Sensing of Environment* 260, 112456.

U.S. Geological Survey. (2022a) The application for extracting and exploring analysis ready samples (AppEEARS), https://lpdaac.usgs.gov/tools/appeears/ (Accessed November 8, 2022).

U.S. Geological Survey. (2022b) Earth Explorer, https://earthexplorer.usgs.gov/ (Accessed November 8, 2022).

U.S. Geological Survey. (2022c) Landsat series of Earth Observation, https://www.usgs.gov/landsat-missions/landsat-collections (Accessed November 8, 2022).

U.S. Geological Survey. (2022d) Science data catalogue, https://data.usgs.gov/datacatalog/ (Accessed November 8, 2022).

Vatseva, R., Kopecka, M., Otahel, J., Rosina, K., Kitev, A. and Genchev, S. (2016) Mapping urban green spaces based on remote sensing data: case studies in Bulgaria and Slovakia, in Bandrova, T. and Konecny, M. (eds) *Proceedings, 6th International Conference on Cartography and GIS, 13–17 June 2016, Albena, Bulgaria.*

Wang, L., Diao, C., Xian, G., Yin, D., Lu, Y., Zou, S. and Erickson, T.A. (2020) A summary of the special issue on remote sensing of land change science with Google Earth engine, *Remote Sensing of Environment* 248, 112002.

Wang, L., Ma, Y., Yan, J., Chang, V. and Zomaya, A.Y. (2018) pipsCloud: high performance cloud computing for remote sensing big data management and processing, *Future Generation Computer Systems* 78, 353–368.

Weaver, R. (2014) Processing levels, in Njoku, E.G. (ed.) *Encyclopedia of Remote Sensing.* New York: Springer, pp. 517–520.

Wolters, S., Söderström, M., Piikki, K., Reese, H. and Stenberg, M. (2021) Upscaling proximal sensor N-uptake predictions in winter wheat (*Triticum aestivum* L.) with Sentinel-2 satellite data for use in a decision support system, *Precision Agriculture* 22(4), 1263–1283. https://doi.org/10.1007/s11119-020-09783-7

World Meteorological Organization. (2022) WMO catalogue for climate data, https://climatedata-catalogue.wmo.int/ (Accessed November 8, 2022).

Wulder, M.A. and Coops, N.C. (2014) Satellites: make Earth observations open access, *Nature* 513(7516), 30–31.

Wulder, M.A., Loveland, T.R., Roy, D.P., Crawford, C.J., Masek, J.G., Woodcock, C.E., Allen, R.G., Anderson, M.C., Belward, A.S., Cohen, W.B., et al. (2019) Current status of Landsat program, science, and applications, *Remote Sensing of Environment* 225, 127–147.

Yang, G., Chunjiang, Z., Qiang, L., Wenjiang, H. and Jihua, W. (2011) Inversion of a radiative transfer model for estimating forest LAI from multisource and multiangular optical remote sensing data, *IEEE Transactions on Geoscience and Remote Sensing* 49(3), 988–1000. doi: 10.1109/TGRS.2010.2071416

Zhu, X.X., Tuia, D., Mou, L., Xia, G.S., Zhang, L., Xu, F. and Fraundorfer, F. (2017) Deep learning in remote sensing: a comprehensive review and list of resources, *IEEE Geoscience and Remote Sensing Magazine* 5(4), 8–36.

8 New and changing use of technologies in monitoring: drones, artificial intelligence, and environmental DNA

Anna Allard, Luke Webber, Jonas Hentati Sundberg, and Alan Brown

Introduction

Following on from chapter 7 on remote sensing and Earth observation systems, this chapter discusses the emerging use of some of the newer technologies that may be so novel that we do not yet have a good understanding of their potential. Some technologies might not be new, but as development proceeds, so do a multitude of innovative uses of them, and some examples from different areas of monitoring will be discussed here. The networks of in situ data collection in marine monitoring employ and develop many and varied types of technologies.

We discuss artificial intelligence (AI) or deep learning for purposes of modelling and classification and recent developments in the processing of DNA that have made it possible to use in biodiversity monitoring in the past or present.

Drones offer the possibility of collecting remote data in high detail, providing swift real-time monitoring – for example, of outbreaks of insects damaging forest or crops or wildfires in the landscape – to use in combination with other data, either satellite derived or from in situ measurements. These auxiliary data sources are becoming increasingly used in a large number of disciplines and monitoring efforts as the technology develops and the cost of devices decreases.

Emerging uses of technology

The new technologies of data collection from drones and other airborne systems, as well as the emerging availability of national laser scanning programmes and satellites bearing radar scanning devices, have changed the scope of monitoring in recent decades.

Using collected data to feed into a model to better understand what is happening in the environment constitutes a major part of most monitoring, exemplified here by how it is done in oceans. The foundations of ocean monitoring are the numerical models and the process of data assimilation. Observational data on different aspects of the ocean are collected from many platforms, including satellites and in situ data from ground stations, many with innovative use of technology or devices for collection. Examples are airborne or seaborne sensors, on buoys, ocean research vessels, and autonomous underwater floats and gliders. Compiled data are then fed into numerical ocean models to describe the state of the ocean (Copernicus Europe's Eyes on Earth 2022; EC 2022a; Mack et al. 2020). With remote observation, the upper water layer is analyzed with different water quality parameters, giving a more comprehensive picture of environmental conditions related to

DOI: 10.4324/9781003179245-8

biodiversity, eutrophication, and hydrographical conditions (e.g. Anttila et al. 2018; Mack et al. 2020). Many factors interact that we need to be aware of and take into account while running a model – for example, between ocean and the atmosphere – thus, information on wind, temperature, pressure, and heat fluxes is important to include in a given ocean numerical model. Colour is a major factor in remote sensing due to chemical contents and suspended materials in the water and content of algae, all of which give different colours and reflectance, and over decades much work has been put into solving these issues (e.g. Vandermeulen et al. 2020). Although other parameters are involved in other types of monitoring, the framework is similar to most of them, requiring new skill sets for the staff. All of these new datasets have led to multiple hybrid technologies combining widely different datasets, to be used as support in the making of models or as reference data (training computerized algorithms for classification systems and validating the classifications when done); see more on this topic in chapter 9.

Even when the cost of acquiring a device is relatively low, this type of data collection, analysis, and manipulation often generates large datasets, which in turn generate high costs for storage and hosting web services. It is desirable to minimize time in the field, usually the costliest part of a monitoring scheme, though auxiliary information can be used to shorten periods spent in situ by gathering only the necessary elements. Of course, we have to get out there to run the devices, optimizing the time spent on each site, tract, or plot to collect detailed data. Some countries have accessible sets of air photo imagery, as orthophotos or 3D stereo possibilities. Where these are missing or, more important, when we want to gather auxiliary data from the same date and environmental or atmospheric conditions as in situ measurements, collection of images and lidar or radar data (see chapter 7) by different types of drone is helpful; see examples in Text boxes 8.1 and 8.2.

High-resolution lidar scanners are also developing fast and can be found in a variety of places, such as covering fields of view in cars and in some smartphones, allowing instant 3D scans; see an example of dune monitoring using high-resolution lidar scanners in chapter 2. In disciplines such as archaeology, drones or aeroplanes are often equipped with lidar scanners, at times together with cameras, providing the almost magical analysis of the ground area without trees and most of the vegetation. Revealing otherwise hidden structures, the technique has generated a large increase in archaeological discoveries in recent years. Archaeologists have also developed a novel range of viewing methods including artificial lighting of a surface model to reveal new objects by shadowing and the creation of stereo mates that allow 3D viewing of an orthophoto mosaic.

The most detailed data are, of course, on the molecular level, where the novelty is in the rapid laboratory processes, enabling more data to be processed at lower costs. The use of environmental DNA and ancient DNA, even on the level of isotopes, as an auxiliary data source in investigations of diversity has become an important part of the composition; for example, in determination of species composition in underwater sediments.

Drones and unmanned systems in monitoring

Monitoring biodiversity by drone is an increasingly common application, used both to increase the spatial or temporal resolution in remote sensing and as ground truth/in situ collection of detailed data, to assess land-based as well as water biodiversity and several other variables. Lightweight drones equipped with digital cameras or other sensors are not exactly new, but they have emerged as a remote sensing survey tool of choice for ecologists, conservation practitioners, environmental scientists, archaeologists, and

Text box 8.1:   Example: introducing drone data from alpine areas into a national inventory programme

The National Inventories of Landscapes in Sweden (NILS) programme collects vegetation data, mandated by the authorities, with the purpose of reporting to the Swedish government for policymaking and planning of the national environmental quality objective of the mountainous areas and to fulfil reporting obligations to the EU. Field data from the inventory in the alpine area are also used as ground truth in national land cover mapping. The sampling design of NILS has recently changed, and the background for that change is described in Text box 4.1.

The Swedish mountain region, almost a third of the country, has been monitored in situ by the NILS programme since 2003. However, new questions are now being asked of the monitoring data, forcing adaptations to both sample design and innovations in data collection. The sample size had to be increased (to catch rare or semi-rare occurrences and to increase the information on plant species; e.g. Christensen and Hedström Ringvall 2013). The new sample is selected by a method of balanced sampling (see more on sampling designs in chapter 4). Described here is only the inventory of alpine habitats within the NILS programme.

Remote sensing lends itself very well to this type of landscape because most of what grows there owes its appearance to natural conditions, and to a high degree. This might be the underlying bedrock, which in its weathered form provides a soil rich or poor in nutrients, or geomorphology, surface undulations, and wind exposure, where areas that are swept free of snow in winter or exposed to constant winds in summer develop very specialized vegetation types. Even though human influence is still present, from tourism, grazing, and the like, it is rather small in comparison to those natural controls that make vegetation models rather accurate.

As satellite data have become more available (allowing, for example, analyses of seasonal time series), the modelling for a balanced sample of areas to visit has become accurate enough to reduce the overall work by eliminating the step of inventory by aerial photos (to exclude areas containing no vegetation of interest to the inventory). In the lowlands, where the vegetation is mostly a product of human management, a phase of inventory by aerial photos in 3D will reduce the overall work in the field by avoiding those sample tracts without vegetation of interest (e.g. no deciduous forest or grassland), saving travel time and time spent in the field; see more in chapter 5.

The reduced phase of aerial inventory consists of scrutinizing national orthophotos in a geographic information system (GIS) on the tract level, without collecting data to exclude any tract where there is no open land above the treeline of interest to the inventory.

The next development in the NILS data collection is to introduce drones to see what can be collected by these images (Figure 8.1). Drone imagery might reduce the amount of variables collected in situ (there are many in NILS, which take time to collect and are therefore expensive) and is able to cover a larger area, putting the plot in its context. A little extra time is, of course, introduced, because field staff are the ones flying the drones. The drone imagery in 3D re-introduces the inventory step on the single plot level done by the aerial inventory staff, but this time the work is done after the field inventory, giving the extra confidence of the data and field photos taken in situ. Another benefit is that some variables might be measured more accurately in the drone images; for example, shrub cover and bare soil/stone cover. In the future, tests will be made on semi-automatic or automatic recognition, but as we have seen with other tests, these methodologies need more development to be viable.

Figure 8.1 Flying drones in the Swedish mountains, summer 2021.

Credit: Photo by Sven Adler.

The newly applied scheme for amplifying the number of sample plots and a test using drones to lessen the number of variables, especially coverage of species and/ or groups of species, is outlined in Figure 8.2.

The entire alpine area is divided into 1km × 1km tracts (each one being a sample), and every tract consists of 1600 field plots, all possible choices for field visits. The squares are randomly divided into layouts of different densities, from a large number of squares, where only a few types of really interesting data are monitored, to a low number of squares where all data are collected. A model is constructed of vegetation types, including bare substrate (boulder fields, partially vegetation-covered areas due to intense exposure to wind, or high alpine areas above the limit of vegetation), using multiple input data, including Sentinel-2 satellite data in time series, national cover of laser data, a moisture regime from another model, and in situ data from the almost 20 years of NILS monitoring in the mountains. The model used was a generalized additive model (GAM), a common type for vegetation prediction (e.g. Kosicki 2020). All plots containing vegetation types not of interest are taken at this stage (lakes, permanent snowbeds, glaciers, and steep cliffs – which actually are of interest but are too dangerous for field staff to visit without climbing equipment and special skills). The rest

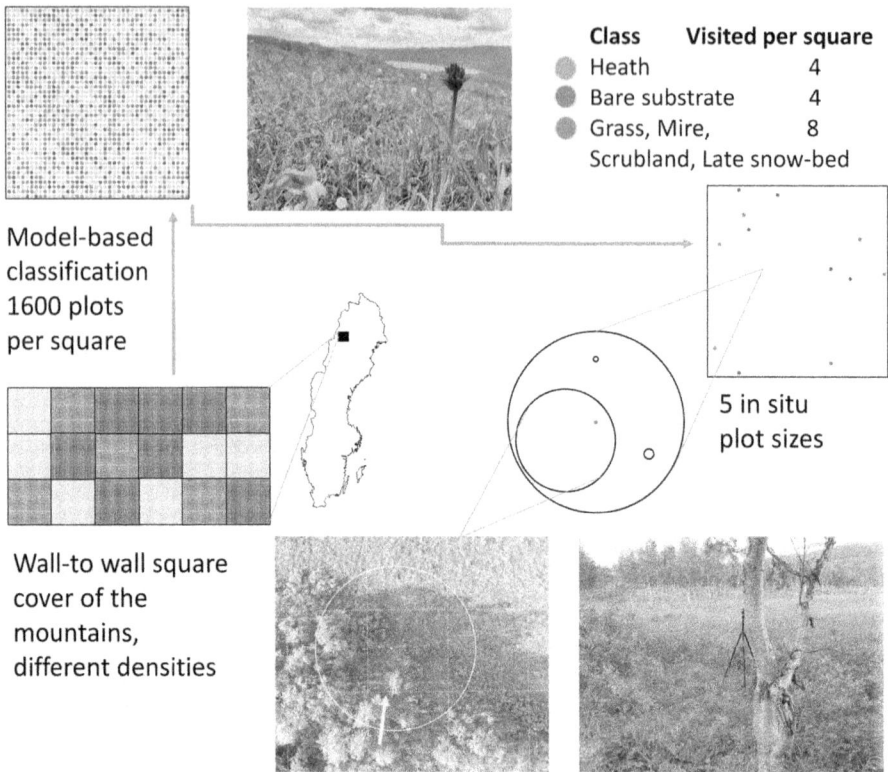

Model-based
classification
1600 plots
per square

Class	Visited per square
Heath	4
Bare substrate	4
Grass, Mire, Scrubland, Late snow-bed	8

5 in situ
plot sizes

Wall-to wall square
cover of the
mountains,
different densities

Figure 8.2 Introducing balanced sampling into the monitoring scheme of the NILS programme
to amplify the number of sample plots and drone technology to possibly decrease the
number of field variables, thereby decreasing the time spent at each plot.

Source: Adapted with permission from Adler et al. (2022); the drone image was cleared for publication by
Lantmateriet, the Swedish mapping, cadastral, and land registration authority.

potentially consist of vegetation to monitor and are included in the balanced
sampling.

- Heath (four types of heath, depending on exposure and available moisture).
- Bare substrate (below the high alpine areas).
- Green areas, a large group with grass, scrubland (*Salix* spp.), late snowbeds
  (areas that have lingering snow in the spring but quickly become green later in
  the season), and mire (often sloping mires that contain very little water and
  appear grass-like in remote sensing; all mountain mires are in this group).

From these (coloured dots in red, violet, and green in Figure 8.2), a sample of
four each of heaths and bare substrate and eight from the green areas per square is
taken. A maximum of 12 plots per square is set, which can be weighted towards
"the most wanted" group. The sampling is balanced by the Normalized Difference
Vegetation Index (NDVI; which quantifies vegetation by measuring the difference
between near-infrared light, which is reflected by vegetation, and red light, which

is absorbed by vegetation), moisture, and coordinates and performed independently for each square and also for each group in the square.

After the customary starting course in the alpine area, field staff are sent out to collect the data, which with a rolling rotation of five years will amount to 2000 to 3000 plots in the mountains, and they may also fly drones on all of them in the future.

The collection of data is threefold, each part using extensive manuals describing all classes and variables to determine and all collected in apps. The parts include habitat according the Habitats Directive (EC 1992), vegetation type according to the vegetation maps of the Swedish Mountain Area (Rafstedt 1984), and detailed lists of species and vegetation structure of the NILS monitoring programme (Ståhl et al. 2011).

The drone imagery, in 3D, was then taken to the office staff of interpretation specialists, who developed a manual (because this was new and no existing manual was available) and collected variables in a GIS database. The special interest is to see whether coverage of vegetation, divided into species groups, will be possible and to what extent, because coverage is easier and often much more accurate from above. A small number of 16 plots were tested at the end of 2021, with promising results. This step will have to develop further and a strict and clear manual, setting limits and definitions, will have to be prepared, in addition to the interpretation staff learning new skills, if or when the practice becomes a feature of the NILS programme.

Text box 8.2:   Example: marine monitoring using drones

Various types of marine drones have become commercially available during the last ten years and can be used in monitoring aquatic environments. Three main types can be distinguished: gliders for which movements in depth are regulated through the platform's buoyancy; autonomous underwater vehicle (AUVs) that operate underwater, often in close proximity to a piloting vessel; and unmanned surface vessels (USVs), which are small, unmanned boats. The focus here is on USVs, hereafter referred to as surface drones or drones.

A vast number of different types of surface drones have been and are being developed by research institutes and private enterprises. The main differences are with regards to propulsion mechanisms (propeller, wind, waves), size (0.5–10m overall length), and level of autonomy (from more or less self-navigating to actively piloted). The available payload (amount of equipment a drone can carry) varies with size and determines what type of scientific sensors can be used.

A benefit of using drones is the low impact on the environment compared to regular research vessels. The small size allows for environmentally friendly propulsion technologies such as wind and wave energy, and drones are often equipped with solar panels to charge batteries for sensors and steering mechanisms. Many drones can operate almost silently, which is a benefit when monitoring biota, because many animals are sensitive to underwater sound (Duarte et al. 2021).

One type of sensor that has been frequently used with marine drones is the echo sounder (Swart et al. 2016; Goulon et al. 2021). Echo sounders are typically mounted in the hull of vessels and transmit signals downwards. The strongest echo received is usually from the sea floor, but weaker echoes of fish, plankton, or gas bubbles can also be detected. The Saildrone, a U.S.-manufactured sailing drone, has been used over several seasons to monitor populations of Alaska pollock (de Robertis et al. 2019).

In the Baltic Sea, researchers from the Department of Aquatic Resources at the Swedish University of Agricultural Sciences have run a Norway-manufactured Sailbuoy (Ghani et al. 2014) since 2019 around the largest seabird colony in the Baltic Sea on the island of Stora Karlsö. With the goal of investigating the amount of fish available to fish-eating seabirds, the drone has been used continuously during three months each season, collecting data from the echo sounder and a range of other sensors (salinity, temperature, chlorophyll, phycocyanin, turbidity). Summarized data are transmitted at pre-determined intervals (often 10 minutes) over the Iridium satellite, and raw data are retrieved when the drone is retrieved, typically every two to four weeks. Despite its slow speed (average, 1knot), it can cover vast distances over time and thereby contribute to data with high spatiotemporal coverage (Figure 8.3).

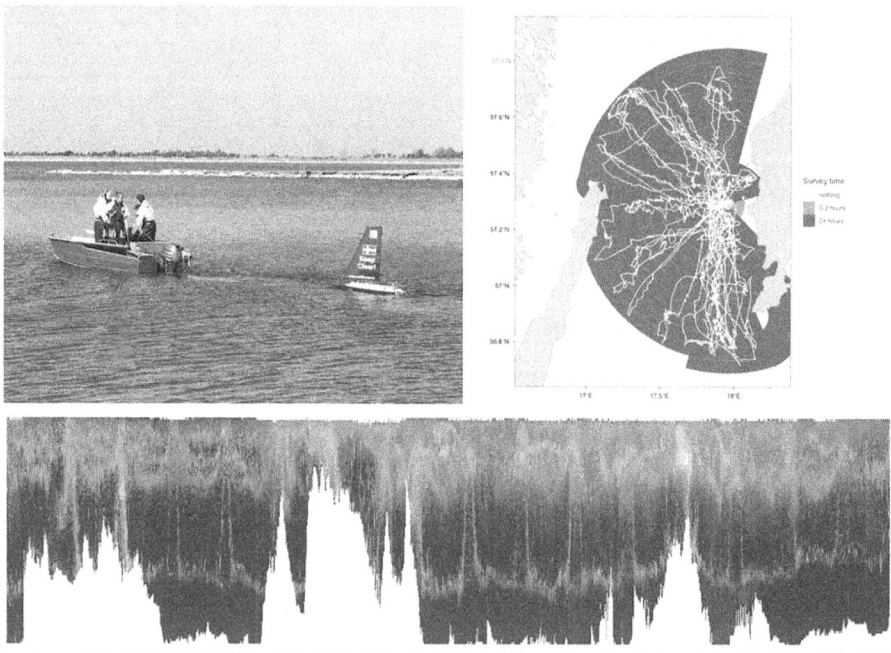

Figure 8.3 (A) SLU Aqua Sailor (Sailbuoy) launched on mission on eastern Gotland, Baltic Sea, Sweden. (B) Sailing route April to July 2020 between the islands Öland and Gotland (in the Baltic Sea). (C) Echogram showing fish abundance (green to yellow colours) and sea bottom topography (white) during two weeks in June to July 2020. The Y-axis show depth from the sea surface to 100m.

Credit: Figure and photo by Jonas Hentati Sundberg.

others. Many of the drones in ocean monitoring are carried on floating platforms or attached to trawlers, ferries, or fishing boats, monitoring the impact on the ecosystem below. Other terms used for the airborne platforms are *unmanned aerial system* (UAS) or *unmanned aerial vehicle* (UAV), including rotary drones as well as small fixed-wing aeroplanes kitted out with cameras or sensors. The possibility of choosing the exact spots for monitoring in a variety of environments has led to site-specific and user-specific data collection. Drones allow us to choose just where we want to create stereo models (imagery in 3D) and when in the phenological season, without having to rely on expensive, pre-planned programmes of air photography, see example in Text box 8.1.

This, in turn, creates some methodological challenges. Drone operators must consult the legislation regulating drone operations in the country of intended use. In the European Union (EU), the European Aviation Safety Agency (EASA) also provides some common rules and regulations (EASA 2022). Other considerations include flight planning: if we want to detect or follow plants or vegetation structures, the ecological response of the current season is an important determinant of the stage of development the vegetation has reached and what flowers are in bloom or whether dieback and desiccation has started – which impacts on whether they will be detected or not. Flight planning also means the knowledge of site-specific conditions, weather, and local environment conditions (e.g. wind profiles), as well as thinking of sun angles and shadow effects (e.g. Duffy et al. 2017; Ancin-Murguzur et al. 2020; Reckling et al. 2021). Shadows in the images from topography, houses, or vegetation structure (e.g. shrubs and trees) are troublesome during digital analysis due to high contrast, but the contrast can be reduced by combining images from different dates and applying various indices (e.g. Rahman et al. 2019; Zhou et al. 2021). For a summary of the different types of sensors and resolutions, with examples of biodiversity monitoring, see chapter 7.

Monitoring, directed at area management and planning of tourist impacts, was tested in Norway by using drones, validated by in situ measurements. The measurements in the drone images were made at semi-detailed level (life forms instead of plant species) and proved successful. The variables for surveying patterns of visitor use included trail condition (width, depth, and proliferation), vegetation structure (and disturbance in the structure), trampling, and trash along the trails (Ancin-Murguzur et al. 2019).

Fish monitoring using broadband and multi-frequency echo sounders

Broadband and multi-frequency echo sounders are also relevant new applications that have so far been used in particular for fish monitoring. Echo sounders are sensors that transmit high-frequency sounds (typically in the kilohertz spectrum) and base analyses on the echoes received. The echo sounder was developed for navigational and military applications, but it was found in the 1960s that fish and plankton in the water mass could be observed, which has had fundamental implications for commercial and recreational fisheries as well as for aquatic monitoring and research, see an example in Text box 8.2.

There are two main types of scientific analyses of biological echoes: (1) those based on identifying and counting individual echoes of fish and plankton (echo counting) and (2) those based on integration of total echo strength across distances and/or depth layers to receive quantitative measurements (echo integration). One of the main challenges with underwater acoustics is how to distinguish between organisms and species. Traditionally, trawl fishing has been used to complement quantitative measurements of abundance with species determination. However, in recent years, it has been discovered that the combination of several echo sounders (multi-frequency acoustics) and the use of broadband technology can be utilized to distinguish between species. By looking at the relative echo strength for a target at different frequencies (i.e. an acoustic signature), one can get a good idea of what type of organism was hit by the sound wave.

Size, body shape, and body composition are factors that have been found to influence the acoustic signature. For example, fish with swim bladders generally create strong echoes, especially at lower frequencies (38–70 kHz), whereas fish without swim bladders such as Atlantic mackerel have generally weaker echoes that increase in strength at higher frequencies.

Multi-frequency acoustics is routinely used to distinguish between different species of fish in Norwegian acoustic surveys (Korneliussen et al. 2016). This creates better, more accurate abundance estimates, and the need for validation through trawling (in which fish mortality is high) is lower than when using only one frequency. The combination of underwater footage and several echo sounders has been used to distinguish between different fish species and krill in the California current ecosystem. The three species present (krill, hake, and anchovy) were easily distinguishable by their acoustic signatures (Benoit-Bird and Waluk 2020).

Semi-automated recognition

An approach to developing a method of semi-automatic search for rare plants, confirming their presence, assessing their health, and verifying population trends by drone was tried in the Blue Mountains of California (Reckling et al. 2021). When monitoring sensitive ecosystems, using drones has the benefit of not impacting the actual plants, and staff do not have to climb to measure or observe in situ. To reduce time in the field and avoid overcollection of images, a model was used to plan the flights, in this case a Maxent machine learning predictive model of potential sites for the endemic plant *Geum radiatum* Michx. The model was developed with ten environmental layers as predictors and known plant locations as training data. By image interpretation of the images, they found the known 33 plants plus four new sites. However, the trial of semi-automated detection of the plants, using a neural network object detector, was not successful from the drone images (Reckling et al. 2021). A potentially useful method of automated recognition in marine monitoring has been put forward, using technology from, for example, factories or mail sorting; that is, letting a computer system automatically identify and count organisms by image recognition, large or small, and training them by machine learning algorithms (Osterloff et al. 2019).

Drone-derived data as part of arctic and alpine monitoring

Mountainous areas and alpine tundra ecosystems above the treeline are sensitive to changes in climate, because they often occur at the very limit of what they can tolerate. However, the changes are often at the scale of subtle shifts within a vegetation type rather than a complete switch from one type to another (Allard 2003).

Vegetation cover and species composition are variables that do change and are most pronounced in the upper limit of shrubs and trees. The reasons for changes are hard to interpret, because these ecosystems are impacted by chaotic trends in onset of spring and fall, often following climatic change or variations in grazing practices. Ground-truth data remain indispensable for their interpretation, as with indexes for phenology studies or Arctic greening (new vegetation on earlier barren land), which are sensitive to the spatial and temporal scales over which they are observed (e.g. Červená et al. 2020; Heijmans et al. 2022). Expansion of shrub vegetation is by far the most reported field-observed vegetation change in the Arctic tundra region, contributing to field-observed and satellite-observed Arctic greening in a review of changes by Heijmans et al. (2022). Monitoring possible impacts of changes in climate and the resulting impacts on the seasonal vegetation length is one of the drivers of the phenological services from Copernicus; see chapter 7.

Monitoring the southernmost relict area of the arctic–alpine tundra in Europe (the Krkonoše Mountains of the Czech Republic) is a part of the International Network for Terrestrial Research and Monitoring in the Arctic (INTERACT; Červená et al. 2020). Drones have been introduced for data collection, comparing data from two hyperspectral sensors to data from Sentinel-2, with the aim of classifying tundra vegetation cover. Their approach was to test three methods:

- Satellite imagery: maximum likelihood classification (pixels)
- Drone imagery: support vector machine
- Drone imagery: neural net and object-based approaches

Červená et al. (2020) achieved good results from both tested sensors, with an overall accuracy 84.3%. To get comparable results for Sentinel-2A, the classification legend had to be simplified to reach 77.7%.

Monitoring in archaeology using remote sensing, drones, and lidar

Another area of monitoring in which the use of investigations from the space or air have become increasingly applied is archaeology. In Egypt, a time of political unrest after 2011 created a vacuum in protection of archaeological sites around the River Nile, and from visual interpretation of satellite images, thousands of holes made by grave looters digging for archaeological artefacts to sell in the open market were detected and mapped. Monitoring such an enormous area would have required a long time but was made possible by enlisting university students to volunteer (Parcack 2019). Sarah Parcack went on to co-found a crowdsourced company, using citizen science monitoring at a global scale. The first project was to explore satellite images, broken up into tiles, to search for undiscovered archaeological sites in Peru. Training, education, and a version of peer review (where six persons had to agree that the tile contained potential sites of interest) were provided to the 10,000 persons enlisted worldwide. The potential sites were then turned over to the professionals for a thorough check and following field investigations or digs (Parcack 2019); see more on citizen science in chapter 6.

Lidar has been used in archaeology since the 1970s to map European castles and structures in open areas and fields. Advances in technology have made lidar available for use in areas of dense vegetation, by creating surface models only showing the ground surface beneath the vegetation; see more in chapter 7. The starting point was a survey made in Caracol, Belize, in 2009, where Chase et al. (2011) succeeded in penetrating the heavy tree cover to reveal many previously unknown structures and causeways and thousands of agricultural terraces in a fraction of the time a manual survey would have taken. Since then, the complexity and infrastructure of the Mayan population have been mapped in large areas using airborne lidar and field investigations to reveal the landscape of cities, causeways, and agriculture (Canuto et al. 2018). The area surrounding the 12th-century temples at Angkor Wat in Cambodia is being investigated in much the same manner, and lidar has revealed large tracts of urban structures around the temple area as well as roads leading to a network of cities. To sustain not only all inhabitants but also the foundations of the temple structures, the Khmer culture had developed complex water management with dam structures (e.g. Chevance et al. 2019).

Environmental DNA and ancient DNA for assessment of biodiversity

Environmental DNA (eDNA) in monitoring is another emerging method used to assess biodiversity, especially in water, sediment, or soil but also for air quality; see Figure 8.4. DNA is extracted and then amplified using primers (i.e. bits that are coded, such as enzymes or ribosomal DNA) in a chain reaction to generate thousands or millions of the DNA pieces in the samples (Ruppert et al. 2019). The information is rather commonly used in studies of ocean or freshwater and fed into models of, for example, metapopulation dynamics (Martel et al. 2020).

Metabarcoding is the most common focus, meaning that the aim is to determine species composition within a sample, rather than focus on just one organism. From eDNA metabarcoding we can detect species or establish lists of complete biological communities; for example, from pollinating insects, where the method is used to determine what flowers they visit, or from stool samples from wild animals to reveal their diet. DNA is resistant to degradation, surviving in the soil for a long time as an archive. eDNA will then help us monitor aspects of biodiversity both now and in the past, migrations of species, and new invasions of non-indigenous species (Jeunen et al. 2019; Lin et al. 2021).

The technology also has the potential to overcome the taxonomic issues of invertebrates in interdisciplinary methods; for example, to retrieve mammal DNA in dense tropical forests from the mosquitos who have sucked their blood. A sampling scheme of insect traps can be determined and overall biodiversity assessed. Linking the distribution

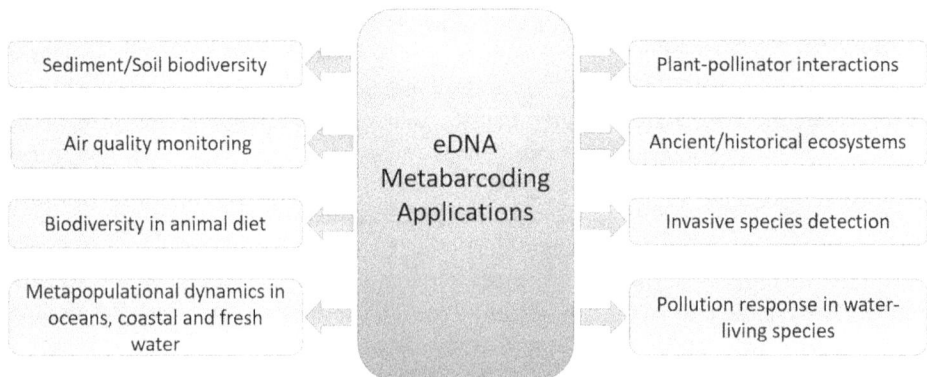

Figure 8.4 Applications of eDNA metabarcoding relevant to biodiversity monitoring in aquatic and terrestrial ecosystems.

to maps, produced by remote sensing, brings together traditional field-based ecology with in-depth molecular methods and advanced computational tools (Mena et al. 2021).

A review by Ruppert et al. (2019) goes through the basic methodology, benefits, and concerns regarding eDNA in disciplines using the technique, including biodiversity monitoring across all habitats and taxonomic groups, ancient ecosystem reconstruction, plant–pollinator interactions, diet analysis, invasive species detection, pollution responses, and air quality monitoring; see Figure 8.4. As an emerging monitoring method, there are many pitfalls and roadblocks to be considered and avoided; the methodologies often differ but may still have the ability to revolutionize modern biodiversity surveys, both past and present, in the molecular era (e.g. Thomsen and Willerslev 2015; Hering et al. 2018). When we go really far back in time, the DNA research is called *ancient DNA* or *aDNA*. This can be used for tracing the movement of species across spaces, whether plant, animal, or human. Tracing the impacts of historical climate changes and their sometimes profound impacts on civilizations and the possibilities to survive on what can grow in the environment is exemplified in Text box 8.3.

Monitoring biodiversity and bioculture using isotopes

A method using even smaller parts than DNA is stable isotope analysis (SIA), which is fairly common and may not be new, but as processing of the isotopes develops and becomes more available, the technique might be applied more widely. In marine monitoring it is still young but promising for monitoring food webs and assessments of eutrophication by identifying anthropogenic inputs of nitrogen and carbon (Briant et al. 2018; Mack et al. 2020). The methodology allows the researcher to follow where the individual started life and then follow the movements throughout its life span, whether animal or human, and thus create maps of movement, and because there are few places on the globe not impacted in some way by human culture, these studies have relevance to studies of biodiversity (Cavender-Bares et al. 2022). Hence, it is widely used for migration studies, such as the movements of people and culture in mediaeval Finland or in Greece in the neolithic era (Whelton et al. 2018; Lahtinen et al. 2021). The herding strategies of sheep in Bronze Age Italy were mapped using strontium isotope analysis, revealing insights of a centre for intense herding and wool manufacture as well as trade (Sabatini et al. 2022). In 873 AD, the Great Viking Army battled English forces at Repton, situated far inland in Mercia, according to the Anglo-Saxon Chronicle (a compilation of accounts, surviving in several versions, of the history of England from 60 BC to the 12th century; see e.g. Downham 2013). Archaeological digging found many warriors at the battlefield; two men were buried in distinctly Scandinavian style. Carbon dating indicated several hundred years prior to 873, which puzzled the researchers. Using strontium isotope analysis, Cat Jarman (2021) showed that a significant amount of the diet of the individuals was marine. A speciality of the marine environment is that much of the isotopes available to fish are very old, and the isotopes incorporated in the bodies of these two men were already about 200 years old, making the dating of the skeletons to the right age of circa 870 AD.

Artificial intelligence and deep learning in monitoring

Deep learning has brought about major advances in the field of biodiversity monitoring, with efficiency improvements in processing massive datasets and large gains in predictive

Text box 8.3:   Ancient DNA in monitoring, migration in response to changes in climate and biodiversity

Ancient DNA (aDNA), meaning DNA obtained from remains of past organisms, is a technique that has been available since the beginning of the second millennium. Since then, the development of rapid ways of sequencing DNA from ancient remains, only a decade ago, has made it possible to map and monitor the movements of our ancient relatives all over the world. By following different mutations in DNA to pinpoint when and where groups of people were living, at which points their DNA became mixed with another group, it is possible to deduce their movements and migrations, connecting them to the people living in the area today. Thus, more knowledge has been gained regarding how modern humans, the Neanderthals, and the Denisovans (who once inhabited Eurasia) have spread and also mated, leaving part of their DNA in the genome of the modern people, outside of Africa (Liu et al. 2021). For further insights, we recommend Liu et al. (2021), which provides a comprehensive summary of the developments of mapping the movements of our human ancestors around the globe, or Stephens et al. (2019) for an impressive compilation of knowledge from more than 250 archaeologists around the world to create a reconstruction of land use history; this example concentrates on Europe.

In Europe, around 4900 years ago, an environmental change took place that brought about much harsher conditions. Evidence of this climatic event has been found, for example, around the steppe of Eurasia where the human cultures of the time seem to have collapsed, driving a major migration of the steppe people towards Central Europe (Liu et al. 2021; Tarasov et al. 2012). The steppe people (themselves a mixture of at least two hunter-gatherer groups, from present day Russia and the Caucasus region) admixed with the cultures already living in Central Europe, leaving traces of their DNA. The study of ancient genomes has revealed that present-day European ancestry consists of three major genetic components – European hunter-gatherer ancestry, early farmer ancestry and steppe ancestry – in varying proportions across Europe (Haak et al. 2015; Liu et al. 2021). A few hundred years later (circa 4600 years ago), the steppe people reached the UK, mixing with the older population, and Brück (2021) used aDNA to understand more about the social connections of the people living in Bronze Age UK from the way they were buried. She found, for example, that a child buried together with adults was not necessarily related to them and speculated that kinship relations were not always determined by biogenetic links and that affective and enduring relationships were created through cultural links.

Evidence of a severe climate change to a long period of drier conditions circa 3200 years ago has been found at many places around the Mediterranean area, causing a series of events, with famine, migration, and strife, and in a very short period the formerly thriving Bronze Age culture collapsed (Cline 2021). Examples of this are found in the study of ancient biodiversity through pollen in the layered sediments of the Dead Sea, showing an almost complete loss of tree pollen (Langgut et al. 2014) or through information derived from stalagmites (with layers functioning is the same way as tree rings) in Greece (Finné et al. 2017).

accuracy and modelling capability (Ball et al. 2017). Deep learning as a subset of artificial intelligence is replacing traditional machine learning methods. Deep learning has been applied by researchers to images (Willi et al. 2019), video (Schofield et al. 2019), and audio classification (Stowell et al. 2019) and for species identification (Wäldchen and Mäder 2018), individual counting (Sylvain et al. 2019), and landscape-level mapping of the environment (Kussul et al. 2017). Primarily these new applications all rely on variations of the same deep learning model framework, convolutional neural networks, which have excelled at image classification and image recognition problems. In cases where changes over time are of interest for identifying changes in biodiversity, recurrent convolutional neural networks have been used. When the data do not come from an image-based source such as satellites, drone video, or aerial photography, representing the data as imagery, utilizing the spectrogram of an audio recording has shown equal success using the same deep learning models and techniques (Bergler et al. 2019).

Camera trapping and deep learning

Camera trapping of wildlife to obtain detailed data on the location and behaviour of animals for monitoring and conservation of ecosystems is not a new technology, but deep learning to automatically detect data on species is an emerging use (Høye et al. 2021). Camera traps using infrared and triggered by a combination of body heat and movement, which avoids responding to vegetation moving in air currents, enable inexpensive and unobtrusive collection of wildlife pictures; see Figure 8.5 for an example in Wales. Analyzing the pictures, especially in larger schemes, has been a time-consuming and manual task. Norouzzadeh et al. (2018) trained deep convolutional neural networks to identify, count, and describe the behaviours of 48 species in the Serengeti in a dataset comprising 3.2 million images, reaching an accuracy of 93.8%.

Training data for classification using deep learning

Deep learning uses artificial neural networks to perform sophisticated computations on large amounts of data. The idea of a self-teaching machine, using the structure and function of the human brain, is older, but the recent surge in availability of data and computational power has put neural networks and their variants into the spotlight. These models learn to perform tasks (like regression or classification) by considering relationships between input variables and the target variable without specifying any predefined rules. There are tutorials and courses online, but the basic steps are the layers, each consisting of a series of nodes (called *neurons*). In the input layer (in classification, the pixels), there is a series of hidden layers where each neuron in the layer processes the inputs, multiplying them by their respective weights and adding a bias. In the output layer, a predicted output is assigned according to the weights added in the hidden layers (pre-chosen, based on the task by the experts) to create a predicted output. The prediction is often not anywhere near the real target class, and, similar to human brains needing training to develop expertise, the output goes through a training process called *backpropagation*, where the weights are adjusted to reduce the error. The network is then iteratively trained with the new weights until the error is as reduced as much as possible. Deep learning systems require powerful hardware because of the large amount of data being processed.

Another challenge for researchers is that as model complexity has increased, it has become harder to fine-tune classifications when compared to traditional machine

Figure 8.5 Camera trapping in Wales. The infrared camera is triggered by a combination of body heat and movement, to avoid responding to vegetation moving in air currents. The badger (*Meles meles*) is seen marking (top row) and digging for worms (bottom row). The camera tells us when this takes place and how often, shows light levels and air temperature, and records both sound and video. As well as directly counting numbers with the camera, we can inspect the signs and prints left – and look for these elsewhere – and by understanding habitat use, we can know where suitable habitat might be found elsewhere. The cameras are cheap and use technology developed for security cameras.

Credit: Image by Alan Brown.

learning, which has meant a shift in focus from modelling towards the collection and generation of labelled training data. Collecting an adequate number of data to train a model is well known in remote sensing; in classical machine learning, collecting training data in amounts of hundreds to a few thousands is time-consuming. However, deep learning models are data hungry, and tens of thousands to hundreds of thousands of labelled samples can be required to adequately train a model – the more difficult the classification problem the more samples needed. Obtaining enough samples of high enough quality can be difficult because it generally requires researchers to manually create each sample; see an example in Text box 8.4 and more in chapter 9.

Advances in unsupervised and semi-supervised deep learning models can reduce the total number of samples needed but may introduce unintentional biases into the prediction results. Even when training datasets have been produced manually by researchers, mis-labelling or under/over-representation of certain classes can lead to inaccurate predictions or poor performance for specific classes. Generating the necessary training data is typically the most time-consuming component of building deep learning models and is typically where the largest gains in accuracy can be made by collecting more representative samples. Once these datasets have been created and models trained, unclassified datasets can then be classified with relative ease and speed. In addition, as new training samples are collected,

Text box 8.4:   Development and production of Sweden's National Land Cover Database (NMD) version 2, 2024

From 2002 onwards, Sweden has employed remote sensing to produce a national-scale mapping of land cover classes. As more satellite imagery has become accessible over time, from Landsat to SPOT, and now with the European Space Agency's Sentinel satellites, the methods employed to make land cover classifications have advanced in both complexity and accuracy of results.

The most recent national land cover map, known as *Nationella Marktäckedata* (NMD), can be traced back in its origins to the continuous mapping of nature types in protected areas (*Kontinuerlig Naturtypskartering av Skyddade områden,* KNAS) project, first released in 2002. After six updates to KNAS over the following years, *CadasterENV*, the next land cover map, was released in 2012. NMD version 1 was released in 2019; see Figure 8.6. The following year, work to improve the data and methods used had already begun, and the improvements made are expected to be included in the production of NMD v2 to be released around 2024. The same data used to train models in KNAS are still in use today as part of NMD, and with each completed project the amount of available training data has increased and has been crucial in bringing about continual improvements in accuracy. Having such a dataset is central for long-term monitoring, where differences between products need to be attributed to either actual land cover change or improvements in classification accuracy.

Each new land cover mapping has brought about a gradual improvement; between NMD v1 and the upcoming NMD v2, the changes are much more significant. Up to and including NMD v1, most land cover classifications were produced using a maximum likelihood model based on either one or two cloud-free satellite images. Although maximum likelihood models are simple to apply and easy to explain, they are inefficient on large datasets and less effective at modelling complex problems.

For NMD v2, the project goals included improving upon earlier classifications as well as implementing automation where possible. Both goals are complementary, moving from maximum likelihood models to deep learning models where not one or two images are used but hundreds of images in sequence, requiring a degree of automation to handle the much larger datasets. Moving from manual selection of cloud-free images to programmatic access of the satellite image catalogue and accompanying cloud mask data allowed for the implementation of deep learning models to achieve higher classification accuracy. These efficiency improvements in handling satellite data programmatically and the increase in satellite data have also meant that land cover classifications can be produced with much greater frequency. Crucial for monitoring applications, major revisions can be performed yearly rather than every five years, as was the case for previous mapping projects in Sweden. Where the land cover is highly sensitive to climatic changes over time – for example, around semi-permanent water bodies or snow coverage within mountainous zones – these new techniques and data have allowed for monthly updates and monitoring.

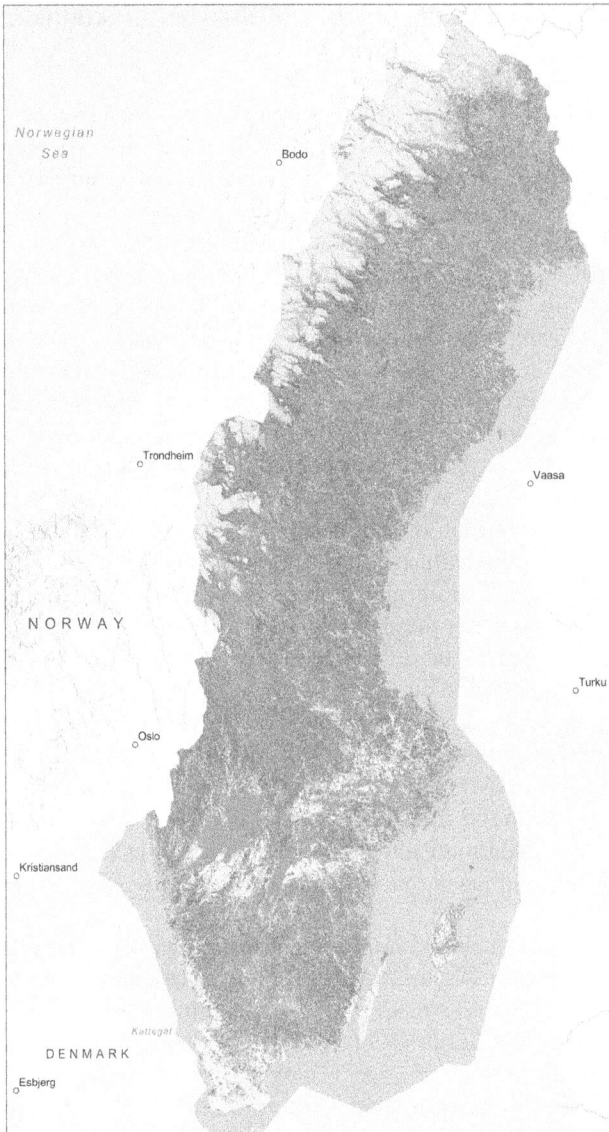

Figure 8.6 The Land Cover Database of Sweden, NMD, version 1.

Credit: The map is open source data from the Swedish Environmental Protection Agency.

Depending on the land cover class within NMD, different machine learning models have been employed, including gradient boosted trees, unsupervised clustering algorithms, and many variations of neural networks, including convolutional neural networks and recurrent neural networks. Advances in model complexity allowed for the inclusion of multi-sensor datasets within deep learning models; NMD primarily utilizes Sentinel-2 optical data. However, for certain classes it was necessary to include, for example, Sentinel-1 radar imagery when trying to increase class separation and accuracy over wetlands or include lidar data when trying to distinguish between low-lying vegetation and bushes.

The choice of model is dependent on each land cover class and is driven by how difficult the class is to correctly identify in the satellite data and the amount of available training data. More complex problems require a more complex model, but as the model complexity increases, so does the need for training data. The choice of whether to model all classes together in a single model as was done for the mountain vegetation classification or whether to produce a model for each class as was done for the forest classification where each tree species had its own model is also driven by how complex the classification problem is. Producing a multi-class model is more efficient, but it is also far harder to adjust after training or to use to explain the relationship between the satellite data and resultant land cover classification. This is a problem inherent to all deep learning models – with thousands to millions of parameters, they are intuitively harder to interrogate and understand why they produce the classifications they do. The shift from maximum likelihood models and manual selection of images to deep learning models that are fed hundreds of images from a database has resulted in increased classification accuracy but at the cost of being able to correct models through human intervention. For the purposes of land cover classification and monitoring, this has resulted in a shift in priorities: for NMD v1, land cover classification models were endlessly tweaked until they produced the correct result, but as the models in use grow in complexity, the collection and processing of sample data take priority, see Figure 8.7.

Land cover classes	User Accuracy (%)		Producer Accuracy (%)	
	NMD v1	NMD v2	NMD v1	NMD v2
Wetland	93.8	**94.5**	91.7	**96.6**
Arable land	77.2	**83.6**	92.1	**93.1**
Open ground without vegetation	51.4	**62.5**	73.1	**92.6**
Open ground with vegetation	70.1	**84.8**	72.5	**85.1**
Buildings	98.6	**98.7**	78.9	**82.2**
Urban	93.3	**95.2**	64.8	**78.7**
Roads/Railways	89.8	**92.4**	84.1	**96.8**
Lakes and waterways	94.8	**97.9**	94.8	**95.4**
Pine dominant forest	47.9	**83.6**	67.6	**75.0**
Spruce dominant forest	78.7	**92.5**	64.2	**72.6**
Mixed coniferous forest	50.0	**57.1**	34.2	**77.1**
Mixed deciduous and coniferous forest	32.3	**43.1**	44.2	**54.1**
Deciduous dominant forests (aspen, birch, willow, rowan)	35.0	**66.3**	34.1	**76.4**
Hardwood deciduous dominant forests (elm, ash, hombeam, beech, oak, linden, maple)	52.7	**83.8**	77.7	**83.1**
Mixed deciduous forest	23.1	**67.9**	2.3	**68.4**
Clearcut forest	56.1	**74.7**	55.3	**53.2**

Figure 8.7 Improvements in classification accuracy (highlighted in bold) between NMD v1 and NMD v2. Accuracy assessment made over a 100km^2 test area in southern Sweden using aerial photography and sample data from Sweden's Forest Inventory (Riksskogstaxeringen).

existing models can be updated and their accuracy improved without the need to train a new model from scratch, although with caution, because we need some knowledge of what was used in the original training.

A new reality in a twin world

A *digital twin*, a concept with origins in engineering, is a virtual representation that serves as a real-time digital counterpart of a physical object. An ambitious new development in modelling is the EC virtual model of the globe, called Destination Earth or DestinE, again changes the scope of monitoring, because the success of the digital replicas of our planet's complex Earth system will require substantial amounts of detailed monitoring data. The aim is to build a highly accurate replica of Earth to model and simulate natural phenomena, including the influence of human activities visualized as videos. Thematic categorisations from the different domains of Earth science will be used in the building process, such as extreme natural disasters, climate change adaptation, oceans, and biodiversity (Bauer et al. 2021; EC 2022b).

In 2021, 28 countries committed to accelerating the use of green digital technologies (a Green Deal) for the benefit of the environment by setting up digital twins to help monitor climate change and environmental degradation (EC 2022c). A key element of the Green Deal is its dependence on an openly accessible and interoperable European dataspace as a central hub for informed decision making (Bauer et al. 2021).

Blair (2021) summarized some of the challenges such an undertaking will face:

- Integration: the environmental asset (including data and modelling) must be brought together in one logical place. There is also a need to understand the potentially synergistic relationship between data and process understanding and to derive software architectures where the associated models can work together.
- Interoperability: to allow different components to work together as part of a larger digital twin architecture.
- Scalability: the sizes of the datasets are potentially very large and the necessary storage and processing capacity must be available.
- Considerations of complexity: environmental systems, by their very nature, are highly complex and exhibit unexpected emergent behaviour. There is a need for modelling systems to better capture such complexity, including interactions, couplings, feedbacks, and dynamics in the system, and a subsequent need to look at complexity through new lenses.
- Data science and AI techniques need to be tailored for the natural environment. Environmental systems, by their very nature, are highly complex and exhibit unexpected emergent behaviour. The models need to accommodate complex and heterogeneous data from different temporal and spatial scales.
- The uncertainty across the pathways from data collection to decision making (on action or policy) must be addressed.

DestinE will be a cloud-based platform with the potential of digital modelling of the Earth system, focusing on the effects of climate change on water and marine environments, polar areas, the cryosphere, biodiversity, or extreme weather events, together with possible adaptation and mitigation strategies. The hope is that the research done in this virtual copy of nature will help to predict, for example, major environmental degradation and disasters with fidelity and reliability in a new way. DestinE will use a

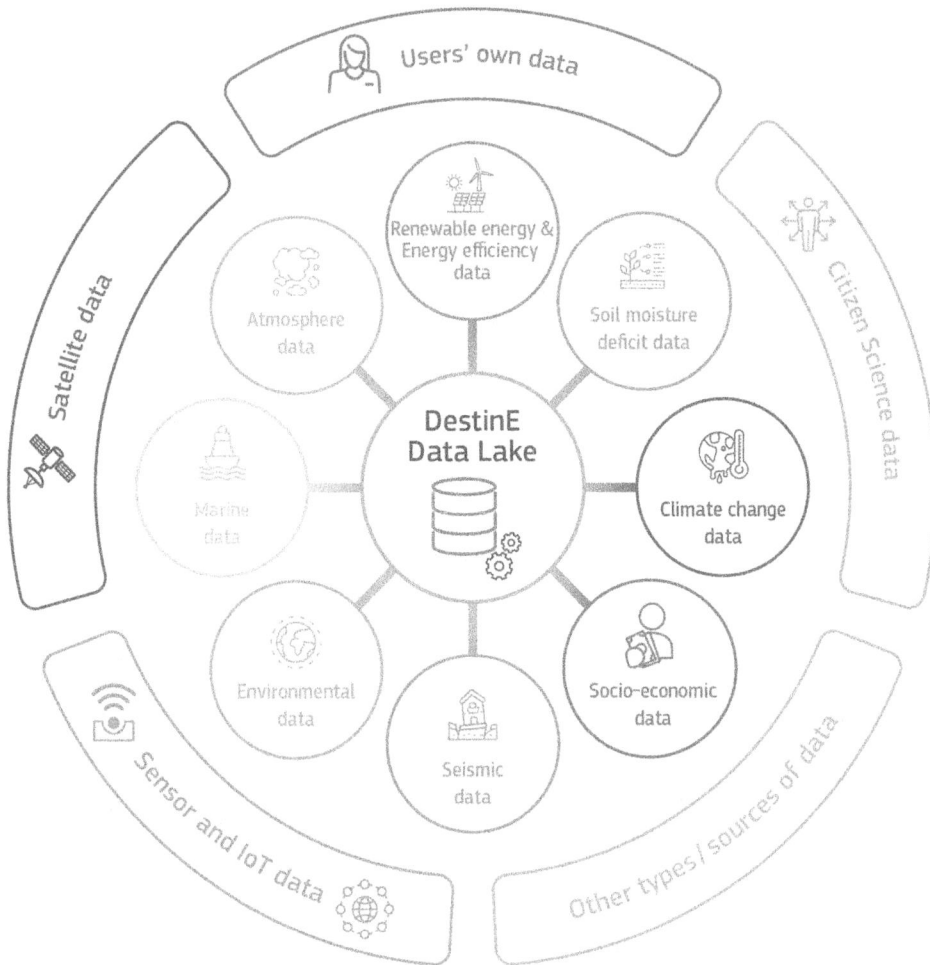

Figure 8.8 The DestinE data lake and all types of data destined to be collected by the Member States to build the model.

Source: EC (2022b).

dedicated *data lake*: a pool of data building on a federation of distributed data sources (EC 2022b); see Figure 8.8.

The support to policymakers might include the following:

* Monitor and simulate the Earth's system developments (land, marine, atmosphere, biosphere) and human interventions.
* Anticipate environmental disasters and resultant socioeconomic crises to save lives and avoid large economic downturns.
* Enable the development and testing of scenarios for more sustainable development.

The German Federal Agency for Cartography and Geodesy (BKG) has begun creating a digital twin with the intention to address various societal problems using data from a multitude of sources (Hopfstock et al. 2022). In Sweden, the Digital Twin Cities Centre at Chalmers University has been focusing on urban areas and research in noise pollution and crowd movement. The Global Biodiversity Information Facility (GBIF) is an international network and data infrastructure funded by the world's governments that aims to provide anyone, anywhere, open access to data about all types of life on Earth to which many countries link their open-source data (GBIF 2022).

Key messages

- In recent years, there has been a deluge of published papers showing the growing range and depth of results involving digital data users from platforms such as Google Earth Engine and Copernicus Services (see chapter 7 for more on these).
- New access to enormous server halls makes it possible for a wider range of researchers, including providers of monitoring data and maps, to carry out intricate analyses without having to own the computers able to perform massive computations. These might involve layers of digital images from multiple satellites or digital maps, compiling ready-made models, and carrying out other computer-heavy processes. The resulting flood of work makes new demands on what is to be reported by data providers, for example, to the EU.
- This chapter has illustrated some examples of these developments in relation to current technology – with many more likely to be developed in the coming years, leading to a need for increasing sophistication in monitoring. It also shows how scientists find uses for technology borrowed across different disciplines, honing the levels of detail in existing projects and incubating completely new datasets. This can create new needs and ways to solve environmental concerns, and, in that sense, the scientists themselves can be regarded as fertile pools of community-driven monitoring.

Study questions

1   Compare the way in which drones acquire images using a camera to take a whole frame with the "pushbroom" scanners used by the MSI instrument in Sentinel-2, which add one line to the image at a time. Why should these work differently: what are the advantages and disadvantages? Why do the frequencies of the red, green, and blue channels in a camera overlap but those of the satellite instrument do not, and does this matter when deriving vegetation indices (see Text box 7.2)?
2   We say the more difficult the classification problem, the more samples we need: does this mean there is no limit to the number of classes we can separate with remote sensing data? What happens if we have the same number of samples but more channels? Do we understand this sort of hyper-dimensional data? Read chapter 15 in Parker (2015) and see what you think. In which case, how can we understand how AI and machine learning work in tens or hundreds of dimensions?
3   What is *overfitting* in the context of machine learning? Why isn't this a good thing?
4   Read examples of habitat mapping from satellite imagery working with pixels and working with segmentation and objects (look for the software package eCognition). What are the advantages and disadvantages of these approaches? Which makes the best use of our ecological understanding of landscapes?
5   How can we use eDNA so that satellite images can predict biodiversity?

References

Adler, S., Hedenås, H., Hagner, Å., Ranlund, Å. and Christensen, P. (2022) *Utvärdering av NILS Fjällinventering* [Evaluation of NILS Inventory of Alpine Habitats], Work Report. Umeå: Swedish University for Agricultural Sciences.

Allard, A. (2003) *Vegetation Changes in Mountainous Areas, a Monitoring Methodology Based on Aerial Photographs, High-Resolution Satellite Images and Field Investigations*. Doctoral Thesis in Geography with emphasis on Physical Geography, No. 27, Department of Physical Geography and Quaternary Geology, Stockholm University.

Ancin-Murguzur, F.J., Munoz, L., Monz, C. and Hausner, V.H. (2020) Drones as a tool to monitor human impacts and vegetation changes in parks and protected areas, *Remote Sensing in Ecology and Conservation* 6(3), 105–113. https://doi.org/10.1002/rse2.127

Anttila, S., Fleming-Lehtinen, V., Attila, J., Junttila, S., Alasalmi, H., Hällfors, H., et al. (2018) A novel earth observation based ecological indicator for cyanobacterial blooms, *International Journal of Applied Earth Observation and Geoinformation* 64, 145–155. doi: 10.1016/j.jag.2017.09.007

Ball, J.E., Anderson, D.T. and Chan, C.S. (2017) A comprehensive survey of deep learning in remote sensing: theories, tools and challenges for the community, *Journal of Applied Remote Sensing* 11(4), 042601.

Bauer, P., Stevens, B. and Hazeleger, W.A. (2021) Digital twin of Earth for the green transition, *Nature Climate Change* 11, 80–83. https://doi.org/10.1038/s41558-021-00986-y

Benoit-Bird, K.J. and Waluk, C.M. (2020) Exploring the promise of broadband fisheries echosounders for species discrimination with quantitative assessment of data processing effects, *The Journal of the Acoustical Society of America* 147(1), 411–427.

Bergler, C., Schröter, H., Cheng, X.R., Barth, V., Weber, M., Nöth, E., Hofer, H. and Maier, A. (2019) ORCA-SPOT: an automatic killer whale sound detection toolkit using deep learning, *Scientific Reports* 9(1), 1–17.

Blair, G. (2021) Digital twins of the natural environment, *Patterns* 2(10), 100359. https://doi.org/10.101 6/j.patter.2021.100359

Briant, N., Savoye, N., Chouvelon, T., David, V.R., Rodriguez, S., Charlier, K., et al. (2018). Carbon and nitrogen elemental and isotopic ratios of filter-feeding bivalves along the French coasts: an assessment of specific, geographic, seasonal and multi-decadal variations, *Science of the Total Environment* 61, 196–207. doi: 10.1016/j.scitotenv.2017.08.281

Brück, J. (2021) Ancient DNA, kinship and relational identities in Bronze Age Britain, *Antiquity* 95(379), 228–237. https://doi.org/10.15184/aqy.2020.216

Canuto, M.A., Estrada-Belli, F., Garrison T.G., Houston, S.D., Acuña, M.J., Kováč, M., Marken, D., Nondédéo, P., Auld-Thomas, L., Castanet, C., et al. (2018) Ancient lowland Maya complexity as revealed by airborne laser scanning of northern Guatemala, *Science* 361, 6409. doi: 10.1126/science.aau0137

Cavender-Bares, J., Schneider, F.D., Santos, M.J., Armstrong, A., Carnaval, A., Dahlin, K.M., Fatoyinbo, L., Hurtt, G.C., Schimel, D., Townsend, P.A., et al. (2022) Integrating remote sensing with ecology and evolution to advance biodiversity conservation, *Nature Ecology & Evolution* 6, 506–519. https://doi.org/10.1038/s41559-022-01702-5

Červená, L. Kupková, L., Potůčková, M. and Lysák, J. (2020) Seasonal spectral separability of selected grasses: case study from the Krkonoše mts. tundra ecosystem, *The International Archives of the Photogrammetry, Remote Sensing and Spatial Information Sciences* XLIII-B3-2020, 2020 XXIV ISPRS Congress. https://www.int-arch-photogramm-remote-sens-spatial-inf-sci.net/XLIII-B3-2020/371/2020/isprs-archives-XLIII-B3-2020-371-2020.pdf

Chase, A.F., Chase, D.F., Weishampel, J.F., Drake, J.B., Shrestha, R.L., Slatton, K.C., Awe, J.J. and Carter, W.E. (2011) Airborne lidar, archaeology, and the ancient Maya landscape at Caracol, Belize, *Journal of Archaeological Science* 38(2), 387–398. https://doi.org/10.1016/j.jas.2010.09.018

Chevance, J.-B., Evans, D., Hofer, N., Sakhouen, S. and Chean, R. (2019) Mahendraparvata: an early Angkor-period capital defined through airborne laser scanning at Phnom Kulen, *Antiquity* 93(371), 1303–1321. doi: 10.15184/aqy.2019.133

Christensen, P. and Hedström Ringvall, A. (2013) Using statistical power analysis as a tool when designing a monitoring program: experience from a large-scale Swedish landscape monitoring program, *Environmental Monitoring and Assessment* 185(9), 7279–7293.

Cline, E.H. (2021) *1177 B.C. The Year Civilization Collapsed*, Revised and updated. Turning points in ancient history. Princeton, NJ: Princeton University Press.

Copernicus Europe's Eyes on Earth. (2022) Copernicus Marine Service, https://marine.copernicus.eu/ (Accessed November 11, 2022).

de Robertis, A., Lawrence-Slavas, N., Jenkins, R., Wangen, I., Mordy, C.W., Meinig, C., Levine, M., Peacock, D. and Tabisola, H. (2019) Long-term measurements of fish backscatter from Saildrone unmanned surface vehicles and comparison with observations from a noise-reduced research vessel, *ICES Journal of Marine Science* 76(7), 2459–2470.

Downham, C. (2013) *Celtic, Anglo-Saxon, and Scandinavian Studies: Vol. 1. No Horns on Their Helmets? Essays on the Insular Viking-Age.* Aberdeen, Scotland: Centre for Celtic Studies, University of Aberdeen.

Duarte, C.M., Chapuis, L., Collin, S.P., Costa, D.P., Devassy, R.P., Eguiluz, V.M., Erbe, C., Gordon, T.A.C., Halpern, B.S., Harding, H.R., et al. (2021) The soundscape of the Anthropocene ocean, *Science* 371(6529), eaba4658.

Duffy, J.P., Cunliffe, A.M., DeBell, L., Sandbrook, C., Wich, S.A., Shutler, J.D., Myers-Smith, I.H., Varela, M.R. and Anderson, K. (2017) Location, location, location: considerations when using lightweight drones in challenging environments, *Remote Sensing in Ecology and Conservation* 4(1), 7–19. https://doi.org/10.1002/rse2.58

European Aviation Safety Agency. (2022) Easy access regulations for unmanned aircraft systems, https://www.easa.europa.eu/document-library/easy-access-rules/easy-access-rules-unmanned-aircraft-systems-regulation-eu (Accessed November 9, 2022).

European Commission. (1992) The Habitats Directive, Council Directive 92/43/EEC of 21 May 1992, https://ec.europa.eu/environment/nature/legislation/habitatsdirective/index_en.htm (Accessed November 11, 2022).

European Commission. (2022a) Monitoring and evaluation, https://ec.europa.eu/neighbourhood-enlargement/monitoring-and-evaluation_en (Accessed November 11, 2022).

European Commission. (2022b) Shaping Europe's digital future, Destination Earth, https://digital-strategy.ec.europa.eu/en/library/destination-earth (Accessed November 13, 2022).

European Commission. (2022c) Shaping Europe's digital future, EU countries commit to leading the green digital transformation, https://digital-strategy.ec.europa.eu/it/node/9585 (Accessed November 13, 2022).

Finné, M., Holmgren, K., Shen, C.-C., Hu, H.-M., Boyd, M. and Stocker, S. (2017) Late Bronze Age climate change and the destruction of the Mycenaean Palace of Nestor at Pylos, *PLoS ONE* 12(12), e0189447. https://doi.org/10.1371/journal.pone.0189447

GBIF, The Global Biodiversity Information Facility. (2022) Free and open access to biodiversity data, https://www.gbif.org/ (Accessed November 13, 2022).

Ghani, M.H., Hole, L.R., Fer, I., Kourafalou, V.H., Wienders, N., Kang, H., Drushka, K. and Peddie, D. (2014) The SailBuoy remotely-controlled unmanned vessel: measurements of near surface temperature, salinity and oxygen concentration in the Northern Gulf of Mexico. *Methods in Oceanography* 10, 104–121.

Goulon, C., Le Meaux, O., Vincent-Falquet, R. and Guillard, J. (2021) Hydroacoustic Autonomous Boat for Remote Fish Detection in LakE (HARLE), an unmanned autonomous surface vehicle to monitor fish populations in lakes, *Limnology and Oceanography: Methods* 19(4), 280–292.

Haak, W., Lazaridis, I., Patterson, N., Rohland, N., Mallick, S., Llamas, B., Brandt, G., Nordenfelt, S., Harney, E., Stewardson, K., et al. (2015) Massive migration from the steppe was a source for Indo-European languages in Europe, *Nature* 522, 207–211. https://doi.org/10.1038/nature14317

Heijmans, M.M.P.D., Magnússon, R., Lara, M., Frost, G., Myers-Smith, I., Van Huissteden, K., Jorgenson, M., Fedorov, A., Epstein, H., Lawrence, D., et al. (2022) Tundra vegetation change and impacts on permafrost, *Nature Reviews Earth & Environment*, 3, 68–84. doi: 10.1038/s43017-021-00233-0

Hering, D., Borja, A., Jones, J.I., Pont, D., Boets, P., Bouchez, A., Bruce, K., Drakare, S., Hänfling, B., Kahlert, M., et al. (2018) Implementation options for DNA-based identification into ecological status assessment under the European Water Framework Directive, *Water Research* 138, 192–205.

Hopfstock, A., Hovenbitzer, M., Lindl, F. and Knöfel, P. (2022) Building a digital twin for Germany – using large-scale, high-resolution lidar to support policymakers, *GIM International* [online]. https://www.gim-international.com/ (Accessed November 11, 2022).

Høye, T.T., Ärje, J., Bjerge, K., Hansen, O.L.P., Iosifidis, A., Leese, F., Mann, H.M.R., Meissner, K., Melvad, C. and Raitoharju, J. (2021) Deep learning and computer vision will transform entomology, *PNAS* 118(2), e2002545117. https://doi.org/10.1073/pnas.2002545117

Jarman, C. (2021) *River Kings, a New History of the Vikings from Scandinavia to the Silk Roads*, London: William Collins.

Jeunen, G.J., Knapp, M., Spencer, H.G., Lamare, M.D., Taylor, H.R., Stat, M., Bunce, M. and Gemmell, N.J. (2019) Environmental DNA (eDNA) metabarcoding reveals strong discrimination among diverse marine habitats connected by water movement, Molecular Ecology Resources 19, 426–438. doi: 10.1111/1755-0998.12982

Korneliussen, R.J., Heggelund, Y., Macaulay, G.J., Patel, D., Johnsen, E. and Eliassen, I.K. (2016) Acoustic identification of marine species using a feature library, *Methods in Oceanography* 17, 187–205. http://dx.doi.org/10.1016/j.mio.2016.09.002.

Kosicki, J.Z. (2020) Generalised additive models and random forest approach as effective methods for predictive species density and functional species richness, *Environmental and Ecological Statistics* 27, 273–292. https://doi.org/10.1007/s10651-020-00445-5

Kussul, N., Lavreniuk, M., Skakun, S. and Shelestov, A. (2017) Deep learning classification of land cover and crop types using remote sensing data, *IEEE Geoscience and Remote Sensing Letters* 14(5), 778–782.

Lahtinen, M., Arppe, L. and Nowell, G. (2021) Source of strontium in archaeological mobility studies – marine diet contribution to the isotopic composition, *Archaeological and Anthropological Sciences* 13, 1. https://doi.org/10.1007/s12520-020-01240-w

Langgut, D., Neumann, F.H., Stein, M., Wagner, A., Kagan, E.J., Boaretto, E. and Finkelstein, I. (2014) Dead Sea pollen record and history of human activity in the Judean Highlands (Israel) from the Intermediate Bronze into the Iron Ages (~2500–500 BCE), *Palynology* 38(2), 280–302. doi: 10.1080/01916122.2014.906001

Lin, M., Simons, A.L., Harrigan, R.J., Curd, E.E., Schneider, F.D., Ruiz-Ramos, D.V., Gold, Z., Osborne, M.G., Shirazi, S., Schweizer, T.M., et al. (2021) Landscape analyses using eDNA metabarcoding and Earth observation predict community biodiversity in California, *Ecological Applications*, 31(6), e02379. https://doi.org/10.1002/eap.2379

Liu, Y., Mao, X., Krause, J. and Fu, Q. (2021) Insights into human history from the first decade of ancient human genomics, *Science* 373(6562), 1479–1484. https://doi.org/10.1126/science.abi8202

Mack, L., Attila, J., Aylagas, E., Beermann, A., Borja, A., Hering, D., Kahlert, M., Leese, F., Lenz, R., Lehtiniemi, M., et al. (2020) Synthesis of marine monitoring methods with the potential to enhance the status assessment of the Baltic Sea, *Frontiers in Marine Science* 7, 552047. https://www.frontiersin.org/article/10.3389/fmars.2020.552047

Martel, C.M., Sutter, M., Dorazio, R.M. and Kinziger, A.P. (2020) Using environmental DNA and occupancy modelling to estimate rangewide metapopulation dynamics, *Molecular Ecology* 30(13), 3340–3354. https://doi.org/10.1111/mec.15693

Mena, J.L., Yagui, H., Tejeda, V., Bonifaz, E., Bellemain, E., Valentini, A., Tobler, M.W., Sánchez-Vendizú, P. and Lyet, A. (2021) Environmental DNA metabarcoding as a useful tool for evaluating terrestrial mammal diversity in tropical forests, *Ecological Applications* 31(5), e02335. https://doi.org/10.1002/eap.2335

Norouzzadeh, M.S., Nguyen, A., Kosmala, M., Swanson, A., Palmer, M.S., Packer, C. and Clune, J. (2018) Automatically identifying, counting, and describing wild animals in camera – trap images with deep learning, *PNAS* 115(25), E5716–E5725. https://doi.org/10.1073/pnas.1719367115

Osterloff, J., Nilssen, I., Järnegren, J., van Engeland, T., Buhl-Mortensen, P. and Nattkemper, T.W. (2019) Computer vision enables short- and long-term analysis of *Lophelia pertusa* polyp behaviour and colour from an underwater observatory, *Scientific Reports* 9, 6578. doi: 10.1038/s41598-019-41275-1

Parcack, S. (2019) *Archaeology from Space: How the Future Shapes Our Past*. New York: Henry Holt and Company.

Parker, M. (2015) *Things to Make and Do in the Fourth Dimension*. Penguin Books.

Rafstedt, T. (1984) *Vegetation of the Swedish Mountain Area, Jämtlands County. A Survey on the Basis of Vegetation Mapping and Assessment of Natural Values*. Department of Physical Geography, Stockholm University, Swedish Environmental Protection Agency, Stockholm (in Swedish, with summary in English).

Rahman, M.M., McDermid, G.J., Mckeeman, T. and Lovitt, J.A. (2019) Workflow to minimize shadows in UAV-based orthomosaics, *Journal of Unmanned Vehicle Systems* 7(2), 107–117.

Reckling, W., Mitasova, H., Wegmann, K., Kauffman, G. and Reid, R. (2021) Efficient drone-based rare plant monitoring using a species distribution model and AI-based object detection, *Drones* 5(4), 110. https://doi.org/10.3390/drones5040110

Ruppert, K.M., Kline, R.J. and Rahman, M.S. (2019) Past, present, and future perspectives of environmental DNA (eDNA) metabarcoding: a systematic review in methods, monitoring, and applications of global eDNA, *Global Ecology and Conservation* 17, e00574.

Sabatini, S., Frei, K.M., De Grossi Mazzorin, J., Cardarelli, A., Pellacani, G. and Frei, R. (2022) Investigating sheep mobility at Montale, Italy, through strontium isotope analyses, *Journal of Archaeological Science: Reports* 41, 103298. https://doi.org/10.1016/j.jasrep.2021.103298

Schofield, D., Nagrani, A., Zisserman, A., Hayashi, M., Matsuzawa, T., Biro, D. and Carvalho, S. (2019) Chimpanzee face recognition from videos in the wild using deep learning, *Science Advances* 5(9), eaaw0736.

Stephens, L., Fuller, D., Boivin, N., Rick, T., Gauhtier, N., Kay, A., Marwick, B., et al. (2019) Archaeological assessment reveals Earth's early transformation through land use, *Science* 365(6456), 897–902. doi: 10.1126/science.aax1192

Stowell, D., Wood, M.D., Pamuła, H., Stylianou, Y. and Glotin, H. (2019) Automatic acoustic detection of birds through deep learning: the first bird audio detection challenge, *Methods in Ecology and Evolution* 10(3), 368–380.

Ståhl, S., Allard, A., Esseen, P.-A., Glimskär, A., Ringvall, A., Svensson, J., Sundquist, S., Christensen, P., Gallegos Torell, Å., Högström, M., et al. (2011) National Inventory of Landscapes in Sweden (NILS) – scope, design, and experiences from establishing a multi-scale biodiversity monitoring system, *Environmental Monitoring and Assessment* 173(1–4), 579–595.

Swart, S., Zietsman, J.J., Coetzee, J.C., Goslett, D.G., Hoek, A., Needham, D. and Monteiro, P.M.S. (2016) Ocean robotics in support of fisheries research and management, *African Journal of Marine Science* 38(4), 525–538.

Sylvain, C., Hervet, É. and Lecomte, N. (2019) Applications for deep learning in ecology, *Methods in Ecology and Evolution* 10(10), 1632–1644. https://doi.org/10.1111/2041-210X.13256

Tarasov, P.E., Williams, J.W., Kaplan, J.O., Österle, H., Kuznetsova, T.V. and Wagner, M. (2012) Environmental change in the temperate grasslands and steppe, in Matthews, J. (ed.) *The SAGE Handbook of Environmental Change*. Vol. 2, Chapter 34, 215–244, Sage Publications Ltd., Online, https://doi.org/10.4135/9781446253052

Thomsen, P.F. and Willerslev, E. (2015) Environmental DNA – an emerging tool in conservation for monitoring past and present biodiversity, *Biological Conservation* 183, 4–18. https://doi.org/10.1016/j.biocon.2014.11.019

Vandermeulen, R.A., Manninob, A., Craig, S.E. and Werdell, J.P. (2020) 150 shades of green: Using the full spectrum of remote sensing reflectance to elucidate color shifts in the ocean. *Remote Sensing of Environment* 247, 111892. https://doi.org/10.1016/j.rse.2020.111892

Whelton, H.L., Lewis, J., Halstead, P., Isaakidou, V., Triantaphyllou, S., Tzevelekidi, V. Kotsakis, K. and Evershed, R.P. (2018) Strontium isotope evidence for human mobility in the Neolithic of northern Greece, *Journal of Archaeological Science: Reports* 20, 768–774. https://doi.org/10.1016/j.jasrep.2018.06.020

Willi, M., Pitman, R.T., Cardoso, A.W., Locke, C., Swanson, A., Boyer, A., Veldthuis, M. and Fortson, L. (2019) Identifying animal species in camera trap images using deep learning and citizen science, *Methods in Ecology and Evolution* 10(1), 80–91.

Wäldchen, J. and Mäder, P. (2018) Machine learning for image based species identification, *Methods in Ecology and Evolution* 9(11), 2216–2225.

Zhou, T., Fu, H., Sun, C. and Wang, S. (2021) Shadow detection and compensation from remote sensing images under complex urban conditions, *Remote Sensing* 13(4), 699. https://doi.org/10.3390/rs13040699

9 Managing hybrid methods for integration and combination of data

Anna Allard, Andreas Aagaard Christensen, Alan Brown, and Veerle Van Eetvelde

Introduction

This chapter concludes and reflects on a series of chapters discussing both established methods that are widespread within monitoring (chapters 4 and 5 and Appendix 1) and innovative new methods at the cutting edge of the field (chapters 6–8). Here we consider how we can use these types of data together to form integrated, diverse data collections with the potential to support analytical tasks linking social and environmental data and push the boundaries of data collected through different methods. This involves linking in situ methods, air photo interpretation, satellite remote sensing, and machine learning, as well as survey data, interviews, and demographic and register data, among others. In chapter 15, we discuss how to use hybrid approaches and adaptive monitoring in combination with models. In this context, it is important to understand what data are used as inputs to models, how they can be characterized, what quality criteria we can use to estimate their usefulness, and how they can be classified and compared. That is the topic of this chapter, where we discuss the characteristics of data, including issues relating to classes and hierarchies, biases and conditions stemming from the original purpose of data collection associated with each layer or dataset, and the units used in data collection.

Any specific characteristics of data, including spatial resolution and thematic detail of the different data inputs, often affect analysis and reporting in the way they limit what can be mapped. A good example of this is the way spatially explicit monitoring data (map data) indicate how the extent of habitats and/or species distributions is affected by decisions about classification and observation paradigms taken before or during map production. In map data, classes are defined to support multiple interpretations. For example, an oak forest includes much more than just oaks, even though it could feasibly be represented as a single data object in habitat and land cover maps. As we incorporate the notion of the forest into the class and include glens, roads, tracks, and fragments of open areas in the same class, a more comprehensive understanding of an oak forest develops, reflecting internal heterogeneity and patterns. In this way, single data objects/observations can have detailed information, stored in both the classification and associated variables. Alternatively, the same object (an oak forest patch) can be described at finer scales to account for its internal patterns of species distribution and structure. For example, classifying individual pixels in a raster data model results in a salt-and-pepper look where pixels represent various land covers or habitat types within the forest boundary. This is often more accurate and certainly delivers more information for data users to analyze, but some information about the extent and characteristics of patterns in the data is missing when compared with the previously mentioned data model where the

DOI: 10.4324/9781003179245-9

forest was mapped in larger units combining pixels, based on its internal heterogeneity. Instead, each user will have to interpret the scatter of classified pixels as understandable units of landscape on their own account, meaning that, for example, the extent of the oak forest in question may differ from one analysis to another because a larger share of data interpretation tasks has been distributed to the data user. As can be seen, both of the approaches outlined here have gaps in knowledge, reflecting certain concerns, constraints, and decisions involved in data production. This raises a number of questions, including how we cope with gaps in datasets, at what levels of scale data can be combined, and how uncertainties and specific characteristics of each dataset can be assessed and taken into account. We provide some examples of systems for this and datasets in different combinations in monitoring designs.

Combinations of multiple layers: an overview

There are many and varied ways of combining data sources. In monitoring, this is often done by analyzing and assessing datasets as layers – that is, as overlapping map sheets referenced to a common coordinate system – which are analyzed spatially by overlaying them in a geographical information system (GIS). As such, representing data as layers is a particular type of analysis relevant when shared geographical extent, location, and variation are the primary ordering principles linking datasets together, which is most often the case with respect to monitoring data. However, it often takes quite a lot of work to fit datasets together as layers within a common geographical reference system, both spatially (all coordinates align in the stack of layers) and thematically (variables and classes are compatible between layers and can be interpreted in the same context). How this can be done varies with what we use as input layers and, of course, with expectations about the results (e.g. maps in raster or vector formats, estimates of occurrences or cover, or for use as input for modelling). Where the combined data form the basis for some further step in a larger assessment or analysis scheme, this may influence how data should be combined and represented.

An example of the process of combination

An illustrative example of the process is the planned analysis and data production framework of the second version of the Swedish land cover database (Nationella Marktäckedata, NMD) to be released in 2024. Within this framework, existing data will be used (including monitoring data, maps, statistics, agricultural data, wetland surveys, national lidar data, satellite data, etc.) to create a series of new layers and models. These are then used in different ways in the combination scheme for a final unified and singular classification of up to 48 classes of vegetation, including moisture regime (e.g. dry, mesic, or wet grassland). Even with a relatively simple classification system, the number of tasks to perform when combining such a multitude of data within a single framework of interpretation and analysis is great.

Table 9.1 lists all of the data inputs (at least 49), including basic information layers (raw/not pre-processed images from Sentinel-1 and -2 satellite sensors, mosaics from the SPOT satellite, etc.) followed by the supporting information layers (soil types, maps, vectorized layers, borders, and catchment areas). Listed are also available data layers for training and validation (called *reference data*) and planned for future collection. In the second version, extra training and validation data will be collected to accommodate all 48 classes (Nilsson et al. 2021).

Table 9.1 Input, support, and reference data to be used to create version 2 of the Swedish Land Cover
Database

Provider	Basic information
European Space Agency Service	Sentinel-1, Sentinel-2
SACCESS Service	SPOT mosaics
Lantmateriet	National: lidar data, vectorized buildings and water
Board of Agriculture	Vectorized farmed and non-farmed fields
Statistics Sweden	Vectorized roads and railroads

Provider	Supporting information
Lantmateriet	DEM (2m), maps: cadastral; terrain and road, hydrographic network, mountain vegetation map
Forest Agency	Clear-cut forest areas
Geological Survey	Soil types, soil depth
Statistics Sweden	Urban borders, county borders and infrastructure objects (six layers)
Agency for Marine and Water Management	Coastline infrastructure objects
Meteorological and Hydrological Institute	River catchment areas
Maritime Administration	Territorial border and maritime economic border
University for Agricultural Sciences	Forest digital map
Environmental Protection Agency	Nature types map (KNAS), continuous forest map, and Swedish land cover data
European Environmental Agency	CLC 2018 layer

Provider	Reference data
County board administrations	County separate inventories
Forest Agency	Inventories of key biotopes, forest type, High Nature Value
Environmental Protection Agency	Inventories of protected natural areas, Natura 2000 areas, protected areas (DOS NVR)
University for Agricultural Sciences	Inventory data: National Inventories of Landscapes in Sweden, National Forest Inventory, Tree Portal
Board of Agriculture	National inventory of meadows and pastures
Auxiliary data collected	From aerial photos, satellite images, Google Maps

At least 49 input layers consisting of basic and supporting information as well as training data for classifications are
included. It is anticipated that additional layers might be used, depending on availability and needs encountered at
the production stage.

All layers are processed and aligned (so that each pixel is geometrically on top of every
other corresponding pixel in other layers), followed by the next steps:

- The raw satellite data are normalized (atmospheric and geometrical corrections, manual masking out of clouds, etc.) and aligned on top of each other in stacks of data. The process is done to create a single satellite image, where the best/most representative data are taken from several points in time for the final classification. Another purpose is to perform analyses of time series.
- Point clouds from radar and laser are converted, where the laser is made into a series of 10m raster layers, to be used for the new wetness index etc.
- The latest map data are prepared by GIS analysis and converted to raster data.

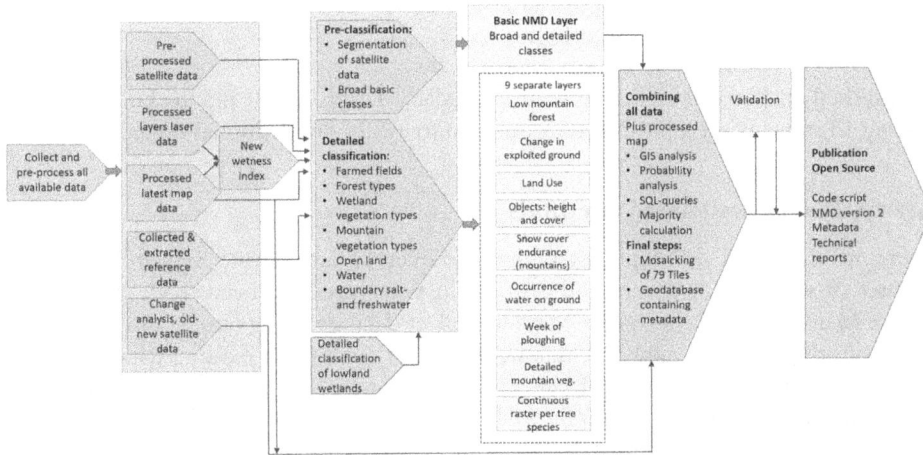

Figure 9.1 A simplified outline of the step-by-step process involved in classifying land cover in Sweden using existing digital layers of data in combinations. The input data (Table 9.1) consist of at least 49 digital layers as the point of departure. All steps (boxes) and tasks (bullets within boxes) indicate some degree of work and adjustment of data layers involved in the transformation of multiple datasets into a unified and validated map.

Source: After Swedish Environmental Protection Agency (2022).

Figure 9.1 Illustrates the following sequence of analysis and data processing steps involved in the combinations, starting with the processed layers from Table 9.1:

- Map and laser (lidar) derivatives, together with support data (e.g. soils, depth and type) are used to create a wetness index, which will function as input data in the classification process.
- A layer of detailed wetland classification of the Swedish lowland, developed by a consultancy company in cooperation with the NMD working group, is combined into the classification process (Hahn et al. 2021).

Two classifications are made, one using fewer, broad classes, which will function as the basic layer within which the fine-tuning into detailed classes will take place. A broad class of "open vegetated land" might be fine-tuned into three narrower classes dominated by grasses, shrubs, or dwarf shrubs, each further divided into three moisture classes.

- Nine extra separate layers are produced that can explain different phenomena. For example, the last time a crop field was tilled, minimum extent of snow patches (snowbeds) in the mountains, or maximum surface water around lakes and streams, or intermittently flooded terrain (presented as minimum and maximum or frequency layers).
- From laser data, a layer of heights and coverage of objects is produced, in which objects of interest are extracted (houses, trees and shrubs; above 0.5 m) to be used, for example, by planners to see cover of trees in grazed lands or for analysis of fluctuations in the mountainous treeline. Finally, all prepared and developed data are layered together to classify the final digital map, comprising 48 classes.

- The result goes through a validation process, using existing data when possible or by collecting extra data (see Text box 9.3).

The finished map database, along with metadata, scripts used, and technical reports detailing all steps, is then published as open-source data on a national digital platform.

Data types and associated methods

Recent advances and development within biodiversity monitoring indicate that rapid processes of scientific discovery and changes in perceived data needs have been set in motion. Many new and innovative ways to collect data are being tested, and this means that monitoring will inevitably include new ways of combining existing data with new types of data. This is driven by both the availability of new forms of data and the urgency of being able to predict (and avoid) future losses of biodiversity; that is, by opportunity as well as motivation. Some of the data types and associated methods of combining them are exemplified in Table 9.2.

An effective means often used in monitoring, which here refers to repeated observations of biodiversity, is modelling. Typically, models lean heavily on robust sets of biodiversity data derived from in situ observation, because they need data to be fitted or validated. However, models can also help assess data representativeness (e.g. by highlighting any bias), support proper data collection (e.g. covering the relevant gradients), or be used to make more effective use of biodiversity observations (Honrado et al. 2016; Ferrier et al. 2017). Models often form the primary basis for interpreting and assessing the meaning or content of other types of data than those used to develop the model in question. For example, models based on in situ observations may be used in the context of remotely sensed data that capture similar variables to predict habitat suitability and characteristics for much larger areas than those visited in person.

Design-based models can be valuable for improving existing programmes, by contributing to identification of gaps, removing bias, and fine-tuning spatial and temporal coverage as the first data are collected and analyzed or defining priorities for local densification of observation networks (see examples in chapters 4 and 8). Models are also helpful for testing hypotheses from monitoring data by supporting stratified sampling strategies along gradients of expected biodiversity drivers or considering the goals of related management programmes (e.g. Honrado et al. 2016). Sensitivity or uncertainty analyses can also be used to define expected variation at each observation site, allowing the differentiation of real trends from background variation while accounting for uncertainty in projections (e.g. Naujokaitis-Lewis et al. 2013).

Predictive models of species distributions provide insights on the drivers of biodiversity across scales, including interactions between these drivers. Such models can be used to develop spatially explicit forecasts of biodiversity responses to environmental pressures, such as invasion by non-native species and changes in climate or land use change (Honrado et al. 2016). To better understand the intrinsic complexity of ecosystems and different drivers of change within them, the method of logic and counterfactual reasoning offers helpful insights, where predicted, or feared, future outcomes can be investigated through constructing opposite scenarios (i.e. predicting likely outcomes for hypothetical but possible scenarios under different conditions than those observed). If such scenarios are developed using data on actual conditions and situations from earlier times, the predictions

Table 9.2 An overview of widespread methods used to combine data

Method	Application context
Design-based	In a statistical workspace; e.g. where survey-sampling designs use remote sensing data for stratification and/or predictive modelling (e.g. Honrado et al. 2016)
Model-based	Explanatory modelling and geostatistical methods are added to existing data, including the retrospective use of remote sensing and GIS data to improve the performance and spatial detail of an existing scheme across space and/or time (e.g. Ferrier et al. 2017).
Co-registration	Stacking different data layers in a GIS workspace, where imagery (raster), object maps (vector polygon), and, for example, lidar and radar (vector point cloud) layers are stacked and combined using algorithms (e.g. Stumpf et al. 2018)
Co-registration using expert systems	Similar to the above but using an expert system such as eCognition, where, for example, aerial photo interpretation is used to extract thresholds for classification steps in a CART (classification and regression tree) rule-based system for object-based image analysis (OBIA) classification of stacks of GIS layers and satellite imagery (e.g. Lourenço et al. 2021)
Statistical classification	Using methods such as machine learning or deep learning to classify image stacks of pixels or objects
Geographic information systems (GIS)	Using spatially explicit information processing platforms to co-model data, including editing and constructing thematic classes. This is often done in software with a wide range of functions, including probability estimation, often in combination with models (e.g. Sarzynski et al. 2020; Vila-Viçosa et al. 2020).
Predictive models	Statistical techniques using machine learning and data mining to predict and forecast likely future outcomes across space and/or time. The process involves using known results/outcomes to create, process, and validate models (e.g. Ferrier et al. 2017).
Logic and counterfactual reasoning	Using logical arguments and contextual information to build alternative (counterfactual) yet possible scenarios, of the type "What if?" This is used to combine existing evidence and is especially important when reasoning about cause and effect (e.g. Grace et al. 2021).

can be checked against actual outcomes and can be used to tune models. A good description of how this works is provided in Grace et al. (2021).

Data types and classifications (discussed in the sections Data types and conversions and Achieving thematic accuracy in classifications based on combinations of varied datasets) from different sources are typically combined in models to understand what factors influence the environment. These may include archival data from earlier surveys, maps, inferred elements of biodiversity in other types of inventories (e.g. an inferred landscape type based on nesting preference of birds), and a wide range of other data types. Some such combinations of data sources contain the building blocks of what we want to know, but often we will have to complete the data in some way to fill in the gaps. This can be done by adding variables and/or spatial reference points; for example, by collecting extra field data from the present or the past, sending drones to collect photos or laser data, studying older maps to try to glean the data we want, or constructing time series of imagery for analysis. Co-registration is a widely used method to do this. It consists of processes for stacking layers of imagery or point clouds in a GIS with the help of algorithms. It is a necessary pre-condition for this that a common geographical reference system can be established. This is done by accurately pinpointing each pixel or point in one chosen coordinate system, which can prove quite challenging if there are inconsistencies in the georeferencing of one or more of the layers, especially when combining point clouds to images (Sarzynski et al. 2020). Stumpf et al. (2018) exemplifies a process chain of co-registration between images of Landsat-8 and Sentinel-2, involving corrections of displacement and striping (the differences between bands/swaths of observed Earth as the satellite passes over the surface) along the track and across them to correlate images. Co-registration can also involve the employment of experts to search for objects of interest or automated search and/or segmentation and classification approaches such as within object-based image analysis (OBIA). Various types of software can be used; one of the most common is eCognition (Hidayat et al. 2018; Lourenço et al. 2021). There are many websites to draw information from, including educational sites of universities and dedicated GIS websites, as well as a plethora of articles testing different methods in relation to vegetation and mapping studies.

Data types and conversions

The methods used for different data types are developing fast, and we recommend going through the latest literature when choosing methodology for working with and analyzing data. Many websites provide information on data types and common workflows to pre-process and combine data sources. Here we outline some key data types with a view to discussing how they can be combined when forming part of multi-layer analysis workflows for biodiversity monitoring. Often this involves converting between data types. It should be noted that the categories of data and associated methods defined here are non-exclusive. They partly overlap and represent a vocabulary of selected concepts, which is useful when working with data combinations, rather than a strict nomenclature.

Spatial data

Spatial data is held in a GIS using annotated *coordinate systems*. Location is fundamental to monitoring, and every object has its own unique coordinates (location and/or extent). Coordinate systems and map projections used, including underlying geoids (models of the Earth's surface shape), may be different depending on the country or location, but most GIS have functions to translate between them. In the field, species data are most often collected at points, plots, circles, or squares or along lines or belts. This is true also when using interpretations from drones or other sensor-derived data. Feature Manipulation Engine (FME), a type of batch-processing GIS, or similar tools are often used as data integration platforms to streamline the translation of spatial data between geometric and digital formats, intended for use in software like GIS, computer-aided design (CAD), and raster graphics.

Raster data

Raster data are data held in the pixel-based data model used by sensors in remote sensing, sometimes called *imagery*, *grid cell data*, or *grids*. These are commonly square but can be other shapes depending on how data are recorded, processed, and represented. (The pixels seen on a computer screen should not be confused with the grid of measurements made by the remote sensing instrument, which are diffuse, overlapping ovals with more reflected light collected from the centre.) In an interpreted image, each pixel typically has its own value and class. Classes can represent many things, either land cover or height above sea level or rainfall, depending on what has been measured by the sensor. Common spatial resolutions for vegetation studies (the resolution here is the pixel size when projected onto the ground surface) from modern satellite instruments are 10m × 10m (e.g. from the Sentinel-2 satellite's MSI), and from the Landsat satellite Thematic Mapper the size is 30m × 30m. When using aeroplanes, drones, or other unmanned aerial vehicles (UAVs) as observation platforms, the pixel size varies with flying height and instrument configuration. In discrete rasters, every cell is completely filled with a single class in distinct categories (or themes) and usually consists of integers to represent classes. For example, the value 1 might represent grass; the value 2, open sand areas; and so on. In contrast, continuous rasters contain data modelled based on gradients; for example, in surface elevation models where gradual changes in height over the surface reference point are modelled using numerical float variables.

Vector data

Vector data are discrete geometrical instances or objects in the form of points (or vertices) made up of *X* and *Y* coordinates, joined by lines between the points to make up an enclosure called a *polygon* (see examples of polygons in Text box 5.2). Vector data in a GIS are governed by topology, defining rules for data representation in support of associated analysis and data transformation logics. For example, topologies may define rules for self-enclosure of objects, gaps, shape complexity, overlap, similarity, and logical consistency.

Comparing and combining raster and vector data

To combine vector and raster data, it is often useful to convert the vectors into a raster format, matching the pixel size of the raster data. On this basis, it is then possible to lay data layers on top of each other and compare or synthesize them using a process called *map algebra*. However, unless the pixel resolution of the raster involved is very small compared to the scale of mapping used in the vector data, it is unlikely that all of the edges of vector objects will lie along the grid where adjacent pixels meet, so many pixels will include both a polygon and a piece of its neighbour, a phenomenon called *mixels*. One way of solving this is to distribute the mixels as evenly as possible between the two classes (taking extra care at points where more than two polygons might be represented, in narrow pointy ends, for example). When converting between data models and formats in this way, simplicity is compromised but the ability to compare geographies of diverse phenomena is gained.

Raster data can also be converted to vector data, based on sets of rules for how to categorize data and define geometries from classified pixels. The simplest way is, of course, to cluster pixels that have the same values (grass with grass, for example), but often small-scale variations (for example, in mosaic landscapes) in the real world lead to the formation of very small polygons of often only one pixel cell. Therefore, approaches that are more complex are often needed, involving segmentation of data into polygons based on distributions and patterns of pixel combinations. These conversions are commonly done by segmentation algorithms, which can be fine-tuned to create objects of the required range of sizes and shapes.

Lidar data

Light detection and ranging, or *lidar*, is a remote-sensing technology that uses pulsed laser energy (light) to measure ranges (distance), producing point clouds with information on observed reflection intensity and location, often sampled very densely (creating large datasets). Lidar technology can produce higher quality results than traditional photogrammetric techniques for lower cost, and its use has exploded in recent years. Working with point clouds involves a few layers of technology: a lidar scanner, a place to store the point cloud data it collects, and a data integration platform (e.g. FME, GIS) to process and analyze the data.

The data come in a range of formats, where LAS, short for laser, represents the industry standard format for lidar. Once intended for airborne applications, it is now commonly used for terrestrial and mobile purposes. Nourbakhshbeidokhti et al. (2019) have outlined a useful workflow for processing and analyzing lidar data. In biodiversity monitoring, classified lidar points (e.g. coloured according to height or into any corresponding image by combination techniques) is useful for producing "bare earth" digital elevation models (DEMs), where structures and vegetation are stripped away, or to develop a digital surface model (DSM), which can be combined into normalized surface models (nDSMs) to measure only the heights of objects of interest. These processes involve careful understanding of laser data and instrument returns; for example, in forests, where one pulse can hit several branches and more than one return is registered from a pulse. Dense laser datasets are also beneficial for capturing the detail of a rough or complex topography or creating a decent bare earth model for an area covered by forest. Analysis typically involves calculating statistics on a point cloud (for example, to find the minimum and maximum values of some component, as well as variations and distributions) or testing the data for certain criteria using an expression.

Radar data

Radar, which stands for radio detection and ranging, is a detection system that uses radio waves to determine the distance (range), angle, and radial velocity of objects relative to a site of observation. High-tech radar systems are associated with digital signal processing and machine learning and are capable of extracting useful information from very high levels of noise (i.e. random, usually unwanted signals). Often analysis tasks are conducted within some script-based programme (such as R; Dokter et al. 2018). Radar datasets are of two basic types: imaging (represented as a map-like image in e.g. weather radar and military air surveillance) and non-imaging (represented as points with numerical values). Modern uses of radar are highly diverse, including air and terrestrial traffic control, radar astronomy, defence systems, marine radars, and self-driving cars). In biodiversity monitoring, it is used in ocean surveillance systems, meteorological precipitation monitoring, surface modelling (because it can penetrate through clouds), and surveillance of migratory birds (e.g. Becciu et al. 2019). Ground penetrating radar is used for geological and archaeological observations, and sounding radar data are used for monitoring ice sheets (Tang et al. 2022).

Objects

An *object* is anything that we want to distinguish in the real world; for example, a house, a copse of trees, a road, or a field of grass. In GIS, the same word is used to refer to pixels, points, lines, and polygons, and in image processing, object is also used as a specialized term to refer to groups of pixels that are combined into a single larger unit in an object-based image analysis (OBIA or GEOBIA). In OBIA, objects are created by combining neighbouring pixels using a segmentation algorithm.

Thematic labels

Thematic labels contain classifications and/or interpretation of GIS objects. The themes of geodata can be anything, really, and geodata are represented in various data formats where thematic labels may be applied in various ways, including raster, vector, geographical databases, and multitemporal data or time series (data representing the same empirical phenomena over a period of time). Common ways of grouping data together using thematic labels are as follows:

- Cultural, such as administrative boundaries, cities, or planning data.
- Socioeconomic, such as demographic data, crime and other practice data, and transport routes by road, rail, or air.
- Environmental, such as vegetation data, soils, or phenology, and hydrographic data about lakes, rivers, and oceans, as well as data for weather, climate, elevation, etc.

A key task in integrated monitoring is to cut across these groups and combine data from different categories in new ways.

Resampling

When combining different sizes of pixels, it is common to use transformations to downsize larger pixels to match smaller ones, or vice versa, in a process called *resampling* (see Figure 9.2). Notice how this, again, can introduce mixels if each of the larger pixels does not correspond to a whole number of smaller pixels. Re-projecting data onto a new map projection or moving two images in coordinate space to exactly overlay one another (so-called image-to-image registration) also requires resampling. In any conversion between different sizes or between different coordinate systems or geoids – where the centres of the pixel cells will not match – we need to specify the output grid and an algorithm to combine pixel values, including thematic data. The four most common ways to resample raster grids in a GIS are the following:

- Nearest neighbour – This technique takes the cell centre from the input raster dataset to determine the closest cell centre of the output raster. This means that it does not alter any values in the output raster dataset, and it is used for categorical, nominal, and ordinal data, such as land cover classification, buildings, and soil types that have distinct boundaries and discrete limits.
- Majority resampling – This is similar to nearest neighbour, but instead of taking the class from the single cell with the created overlap to the new pixel, the algorithm uses the majority class of neighbouring cells. So, if the majority class is pavement, any other classes (e.g. grass) will be ignored and the whole cell will be labelled pavement. This is commonly used in land cover applications.

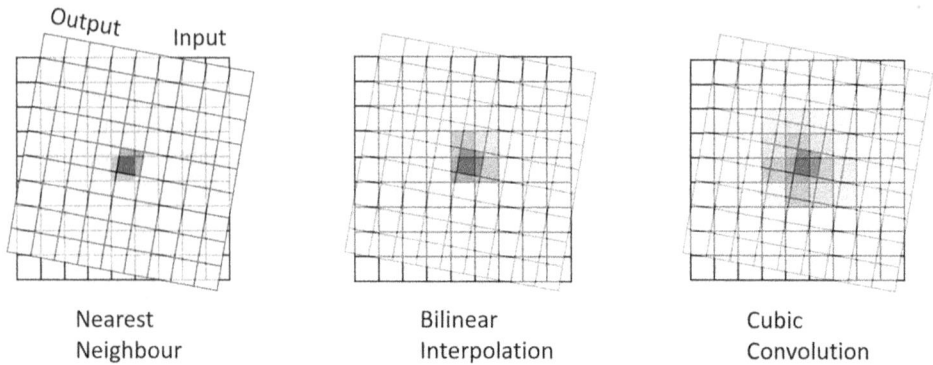

Figure 9.2 The principle of resampling raster data. The value or classification of the output cell is influenced by one, four or 16 grid cells on the input image using one of the available resampling techniques in a GIS.

- Bilinear interpolation – This technique calculates values of a grid location based on four nearby grid cells. It assigns the output cell value by taking the weighted average of the four neighbouring cells in an image to generate new values. The output raster grid is somewhat smoothed and is useful for continuous datasets without distinct limits; for example, digital elevation models or temperature gradients.
- Cubic convolution – This also uses an average of surrounding cells but takes the nearest 16 into account. The result is a smoother output that is useful for continuous surfaces where we want to smooth noise in the data. *Noise* is what we call unwanted pixel values, classified from something that is present but does not add to the result; for example, a scatter of small clouds obstructing underlying information or a boulder-rich area where the surface beneath the boulders is the important issue. Radar images typically contain quite a lot of noise, and the cubic convolution method is a good choice for these.

Classifications and hierarchies

When constructing a theme, we use classification systems as a way to summarize our knowledge of biodiversity and associated patterns in the environment. Classification systems come in two basic formats, hierarchical and non-hierarchical.

Hierarchical classification

The most common type is the hierarchically structured type of classification system, which offers better consistency owing to its ability to accommodate different levels or nested hierarchies of information, starting with structured broad-level classes, which allow further systematic subdivision into more detailed sub-classes. At each level, the defined classes are mutually exclusive. The lower down in the levels, the more criteria are added to increase information density; for example, three levels of forest:

- First level: 1 – Deciduous forests
- Second level: 1.1 – Beech forests
- Third level: 1.1.1 – Luzulo-Fagetum beech forests

Using this approach, it is possible to iteratively and strategically select a more informative class when enough information is available to do so and return to or re-aggregate data to less informative classes when less information is available. This supports combined analysis of datasets with differing levels of information.

Non-hierarchical classification

In non-hierarchical classification systems, it is necessary to choose between specific classes from the beginning, with little opportunity to disaggregate or aggregate classes during analysis and processing. This demands harder work in the way of calibration and education of staff to ensure exact consistency, because there is no retreat backwards. However, non-hierarchical systems have advantages when it comes to analysis, because we do not have to deal with potentially missing data; for example, where staff have chosen a higher hierarchical level when recording observations. This eliminates confusion at the analysis stage when otherwise we would be unsure whether broader classes are chosen due to uncertainty in the field (for methodological reasons) or because they are true (for empirical reasons). In comparison, analysis using hierarchical classification will often have to cluster the detailed levels into a common, broader level post factum, to avoid uncertainty in the analysis.

Text box 9.1:   Example of combinations of classes: the Earth Observation Data for Habitat Monitoring system

The differences in classes of different layers require careful handling in hybrid systems if we are to preserve the information content of any layers being combined. An example of a system that utilizes a whole chain of input data and translations between them is the Earth Observation Data for Habitat Monitoring (EODHaM) system. This system has adopted the Land Cover Classification System (LCCS; di Gregorio and Janssen 2005) taxonomy, which is a hierarchical system of the type described above developed by the United Nations Food and agricultural Organization (FAO; Lucas et al. 2015). A second translation is applied using a system called *general habitat categories* (GHC; Bunce et al. 2008), also a hierarchical system encompassing classes extending from single species or crops up to types of landscapes.

To achieve this, the EODHaM uses a combination of pixel- and object-based procedures, by using Earth observation data with expert knowledge to generate classes according to the LCCS taxonomy. The system comprises the following steps:

- Data input involving preparation and pre-processing, including orthorectification, radiometric, atmospheric, and/or topographic correction.
- Spectral feature extraction, segmentation, and classification to LCCS Level 2 (first stage).
- Classification to Level 3 of LCCS and beyond (second stage), which involves interpretation using expert knowledge.
- Translation of these classes to a system called general habitat categories (Bunce et al. 2008) and Annex I Classes (third stage) of conservation importance (European Commission 1992).

- A module focussing on change detection and validation of outputs, which include maps of land cover, habitats, and changes in these.
- Output products subsequently feed into modules that perform ecological modelling at the landscape level, biodiversity indicator extraction, and biodiversity indicator change detection.

Physiognomic classification

Physiognomic or physiographic classification is based on expert choices of a set of functional and morphological attributes of dominant plants in the community and is useful to describe the vegetation of large areas. The units or formations can be arranged in a hierarchical system. To characterize the structure of plant communities, it is often important to use both the vertical (i.e. stratification) and horizontal (i.e. open/closed canopy or age tiers in forests) dimensions (Vigo 2005; International Association of Vegetation Classification [IAVS] 2022). Many forest inventories are examples of this type of classification (e.g. Fridman et al. 2014).

Environmental classification

Environmental classifications are related to, in addition to vegetation, soil conditions and climate, because they have an important effect on the resulting structure and composition of plant communities (Vigo 2005; IAVS 2022). Examples of this type are the landscape monitoring programmes in Norway and the classification system from the UK Institute of Terrestrial Ecology (ITE), called the ITE land classes (Bunce et al. 2007; Bryn et al. 2018).

Physiognomic-environmental classifications are a common mixture, combining the physiognomy of plant communities with their ecology (mainly climate, soil, and biogeography). An example is the *International Classification and Mapping of Vegetation* adopted by UNESCO (1973).

Floristic classification

Floristic classifications are based on the taxonomic identity of the plants and incorporate both historical and biogeographical information, because each plant species has its own geographic distribution and particular population and metapopulation history. This type of classification is especially useful to describe habitats for conservation purposes. Classifications are made in vegetation plots along with an estimation of abundance. They either define a set of plants living under the same ecological conditions or record all of the plants in tiers. The Swedish National Inventories of Landscapes in Sweden (NILS) field inventory record of a specified list of plants (Ståhl et al. 2011) and the UK National Vegetation Classification (Rodwell 2008) are examples.

Socioecological classifications

Socioecological classifications are based on previously determined socioecological groups, defined as groups of plants that have similar ecological requirements. Each socioecological group indicates either a specific environmental condition or a range of

Table 9.3 Basic quality characteristics of vegetation classification approaches for vegetation

Characteristic	Meaning
Comprehensiveness	Classification systems should include vegetation types that encompass, as well as possible, the full range of vegetation variation within their spatial, temporal, and ecological extents. This includes the need to appropriately summarize transitional and rare plant species assemblages.
Consistency	A similar set of concepts and procedures should be consistently used for the definition of vegetation types. Because broad-scale classification projects may address the classification of vegetation with strikingly different features or be intended to satisfy many potential users, it is useful to explicitly define different units.
Robustness	Minor changes in the input data (e.g. adding or deleting some plot records) should not considerably alter the result of plot-based class definition procedures.
Simplicity	A vegetation classification may be difficult to understand and to apply by potential users when vegetation types do not have simple definitions or when assignment rules (or naming rules) are complex. This should be avoided.
Distinctiveness of units	Vegetation types should be distinct with respect to the values of the primary vegetation attributes. Distinctiveness may sometimes be artificially increased by the choice of class definition procedures (e.g. sampling design).
Identifiability of units	Vegetation types should be easy to identify in the landscape. This requires clear, reliable, and simple assignment rules that may complement possibly more complex consistent assignment rules.
Indication of context	Vegetation types should preferably reflect and be predictive with respect to its context, such as soil conditions, climatic factors, management practices, or biogeographic history.
Compatibility	Vegetation types of a given classification system may be required to have clear relationships with the vegetation types of other classification systems (whether of vegetation or not) because this facilitates transferring information from one classification system to another.

Source: Modified after De Cáceres et al. (2015).

environmental conditions (Vigo 2005; IAVS 2022). An example of socioecological classification can be found in Duvigneaud (1974).

All classifications have some characteristics in common; a set of basic characteristics to be considered in classification approaches is listed in Table 9.3.

Achieving thematic accuracy in classifications based on combinations of varied datasets

Monitoring data consist of *observations*, which can take many forms and be produced in a multitude of ways, as illustrated above. Therefore, it is a major concern when doing monitoring to secure the highest possible *thematic accuracy* of data while ensuring comparability with earlier datasets to allow accurate assessments of change and persistence (Jepsen and Levin 2013). Thematic accuracy reflects how well and with what degree of nuance recorded observations describe a range of empirical objects. In the context of biodiversity monitoring, such objects are typically land units characterized by their land cover, which, under sustained influence of various factors including human land uses, function as habitats for species assemblages.

How such objects (land units and habitats located on them) are described in monitoring data varies considerably because empirical reality allows a broad range of possible observations to be made, even for the same objects located in the same space, and because different interests, agendas, and needs are expressed in the way monitoring and observation procedures are defined (Ellwanger et al. 2018). Therefore, it is of the greatest importance to ensuring successful monitoring results that observation methods are adapted carefully to the needs of analysis as well as to empirical conditions. In practice, this means that semantic choices concerning what variables to collect and how areas/habitats are defined and classified form a cornerstone of research into monitoring design.

Most often, such choices aim to find the best compromise between two competing concerns: (1) how to achieve the highest possible degree of comparability between datasets and within datasets covering large areas of diverse landscapes and (2) recording as accurate and relevant an account of each habitat type and landscape as possible. Often, tough choices have to be made with respect to how these opposing needs in monitoring are reconciled in practice, because of the great variety of landscapes, habitats, and land units that need to be encompassed by any given monitoring framework. As an example of these types of variation, we can compare how monitoring takes place in two different landscape contexts: the dry *montado* and *dehesa* landscapes of Portugal and Spain and the rainfed former open field landscapes of Denmark, the Netherlands, and Germany.

In *montado* and *dehesa* landscapes, ecosystem functionality is affected deeply by shade from stands of oak interacting through numerous feedback loops with understories of shrubs, herbaceous vegetation covers, and grasses grazed by cattle husbandry, producing diverse, multifunctional patterns of habitats integrated with, and coupled to, human land use practices (Godinho et al. 2016). In such landscapes, accurate monitoring of biodiversity must take vegetation cover and human management actions across multiple vertical storeys (tessera) into account, including how these interact. As such, single tree canopies, clusters or stands of trees, and patterns of underlying vegetation correlated with canopy cover as well as interactions with grazing practices are key phenomena being mapped and characterized as part of monitoring efforts (Plieninger 2006). In line with this, monitoring methods have been designed to accommodate a high degree of vertical thematic precision (i.e. concerning how objects are defined and described) and a high degree of integration between information about land use practices and land cover in the way records are stored and linked (i.e. how relationships between objects are defined and observed).

In comparison, the rainfed agricultural landscapes of Atlantic Northern Europe comprise a range of landscape systems where a majority of the surface area is covered by fields with rotational, semi-permanent and permanent crops with low levels of in-field biodiversity and high rates of vegetation change due primarily to human land use practices (Stoate et al. 2009; Renes 2010). In such landscapes, most biodiversity is located either within large corridors and core areas intersecting the farmed landscape or within interstitial habitats (often referred to as *small biotopes*; see chapters 4 and 16), which are areas carrying permanent vegetation embedded within the matrix of production surfaces (Bunce et al. 2005; Levin 2006). These include hedges, ponds, tree stands, grass strips, road verges, streams, and small wetlands. In such landscapes, where a majority of the area is inhospitable, the amount or share of land area taken up by interstitial habitats is a critical factor for biodiversity, as well as the connectivity and diversity of the habitats. Here habitats often only cover a few metres in width and a few hundred square metres in area.

Therefore, accurate monitoring depends primarily on achieving a sufficiently fine-grained spatial resolution (i.e. how objects are defined spatially), making it possible to capture minute changes in habitat area, in combination with variables describing impacts of human land use on habitat suitability (i.e. how objects are described thematically and what relationships they have to surrounding areas; Martin et al. 2019).

As these examples illustrate, monitoring biodiversity in two different landscape settings can lead to the definition of equally different semantic frameworks. The semantics and methods used in *montado* and *dehesa* landscapes would not be relevant in Northern Europe's former open field landscapes and vice versa. But often it is necessary to compare, aggregate, and synthesize results across such frameworks. Such research relies on the ability of researchers to assess exactly how datasets are different, including how semantic decisions are reflected in the data compiled and compared. This is what makes it possible to take into account effects of differing methods, observation techniques, sampling strategies, classification frameworks, and other contextual factors that need to be isolated from those aspects of a given dataset that represent features of the empirical reality being monitored. In this context, it is worthwhile to consider what factors to take into account when comparing and analyzing datasets. As we have seen above, these include the thematic characteristics of data (what phenomena the data represent), temporal and geometrical reference points (where and when the phenomena were observed), and the intended use relative to other data, policy processes, and institutions (for what purpose the data were created). In Text box 9.2, we show six parameters that indicate the range of data characteristics taken into account when assessing compatibility, comparability, and data integrity in integrated monitoring projects.

Combining social and environmental factors in datasets across disciplinary boundaries

The six characteristics of data outlined in Text box 9.2 provide an introduction to the kind of considerations that have to be taken into account when monitoring environments in the context of people and their societies. As can be seen, it is a challenging interdisciplinary task to find ways of combining observations pertaining to social and ecological phenomena in monitoring. It is also a task that historically has been neglected and that has only recently been given sufficient emphasis. This reflects a long history of dualistic thinking in the Western world, whereby environmental and social phenomena have been studied in isolation, even though they have existed together and to a large degree can be seen to co-constitute each other in modern landscapes (Petrosillo et al. 2015). As Lesley Head has noted, this is evident in the way that "dominant metaphors – cultural landscapes, social-ecological systems, human impacts, human interaction with the environment, anthropogenic climate change – all contain within them a dualistic construction of humans and the non-human world" (Head 2012). Overcoming this distinction is arguably necessary for successful monitoring of environmental change and persistence, at a time when human-dominated landscapes are the most prevalent on the planet, taking up an estimated 75% of the ice-free terrestrial surface area in the year 2000. Landscapes have thus been transformed historically "into predominantly anthropogenic ecological patterns combining lands used for agriculture and urban settlements and their legacy; the remnant, recovering and other managed novel ecosystems embedded within anthromes" (Ellis 2011). As such, in a very real sense, for any subsection of the planetary surface there is only a single set of phenomena – a nature including humans and a society including natures. In this view, distinctions between social and ecological realities are

Text box 9.2:   Characteristics of data in integrated monitoring

1   **Thematic characteristics of data**
    The way the data reference phenomena and their characteristics. This
    includes choices regarding what types or classes of objects/processes to
    include and exclude in observation procedures. Empirical reality is
    complex and multiform; therefore, only a subset of objects and processes
    present in any empirical context can be observed, while the rest go
    unnoticed. Questions to consider here include how objects are defined,
    classified, and distinguished from each other; which objects are included;
    how their characteristics are represented by variables; and with what
    techniques the variables are observed. In general, the breadth and diversity
    of variables collected tends to co-determine subsequent options for the
    classification of objects and analysis of flows of change or transformation
    affecting them.

2   **Temporal characteristics of data**
    The way the data reference points or periods in time at which phenomena
    were observed or inferred to exist. This includes the temporal resolution and
    density of observations, as well as information on what temporal reference
    points the data are made relative to (a specific time, a cycle, a long-term trend,
    etc.), in addition to choices such as for what duration of time observations
    need to persist and how data are sampled (either at equal time intervals of
    following a specific strategy), both affecting the temporal variability of the
    resulting data. Questions to consider here include how temporally variable the
    phenomena are, whether they are cyclical and/or reversible, and how this
    affects monitoring.

3   **Geometric characteristics of data**
    The way the data reference spaces or locations at which the phenomena were
    observed or inferred to exist. This includes questions concerning how
    observations are located on the Earth's surface, including the spatial scale
    and resolution of the data, how and at what scale shape complexity is
    observed, decisions on minimal mapping units employed, and geographical
    reference systems used. Questions to consider here include how large the
    observed phenomena can be expected to be, how large an area they exist in, as
    well as how much detail is needed with respect to recording the size, shape,
    density, and distribution patterns of the phenomena.

4   **Relational characteristics of data**
    The way the data represent relationships between the phenomena studied as
    well as with other phenomena. This includes how the phenomena form
    assemblages, clusters, and complexes of interacting components; how they
    are related processually; and any functional effects of their spatial and
    temporal configuration. Questions to consider here include in what way the
    phenomena under study interact with other components of the environ-
    ment, what the effects of these interactions are, and under what conditions
    they occur.

5   **Societal characteristics of the data**
    The status and role of the data, relative to those human societies that created it. This includes how the data are declared, described, and presented in the context of societies where the data are ascribed a specific authority, domain of relevance, and/or truth-value when they are published and used. Questions to consider here include in what way characteristics of the data and information about conditions for correct data use are reported and declared, how misuse can be avoided, etc. It also includes questions of how the data are made suitable to fit into, support, challenge, and co-create policymaking processes, control and reporting schemes, democratic deliberation processes, and decision-making flows.

6   **Performative characteristics of the data**
    The way in which the data organize social practice and orchestrate behaviour. This includes how the data are able to perform in society; how they co-construct data users who have access to it; how that access informs and frames actions, interventions, and land use practices in society, as well as how they support viewpoints raised in debates and advance political agendas affecting the socioecological systems from where they were derived. Such processes can drive complex feedback loops from empirical realities through observational practices back to the environments studied. Questions to consider here include in what way social groups and institutions are involved in data collection and use, as well as how ownership, access, authorship, editing rights, and use rights of the data are defined.

likely to obscure empirical observation and analysis rather than support it, and this is the underlying reason why it is relevant to build hybrid datasets that include a broader view of the relationship between societies and ecologies coinciding in time and space.

Bridging traditional methods and new technologies

When we want to bridge gaps between datasets and make something else or more out of combinations of what we have, we can collect extra or auxiliary data or we can transform the data we do have using various methods. One such approach is *segmentation*, a technique that creates digitized land-based objects (a whole river, a house, or a forest) from the raster cell grid. These objects can then be classified using their shape, size, and spatial and spectral properties, typically using a rule-based approach (a set of rules programmed into an expert system). The rules can be used to create a thematic map of vegetation, habitat, or land cover or in urban areas for high-spatial-resolution mapping of houses, gardens, and other green spaces.

The human brain is very good at seeing patterns. Working with aerial photographs, satellite imagery, or other spatial data, we can use this basic landscape ecological skill to draw (vector) polygons or group pixels into raster objects and then label them according to a chosen classification system, but this is time-consuming. Segmentation algorithms are fast and automated and can take into account multiple data layers – far more than we can see at once with our own eyes. A good hybrid method is to use automatic methods but use our interpretation skills to fine-tune the segmentation. There are a number of choices to make: how big you want your objects to be, which layers from the sensors you want to use, and what weight you will put on each layer. Based on a segmented map, analysis tasks can proceed to classification of the segments, and new choices have to be made, based on geometry, area, colour, shape, texture, adjacency, etc. For example, what defines a house, a forest, or a lake? Here we might need expert advice.

Interpretation of aerial imagery (from aeroplanes, drones, or other unmanned vehicles) is often used as a bridge between space-borne remote sensing and in situ data. The methodology of interpretation, as in the spatial resolution, lies somewhere between field and space. With respect to sampling, aerial image data show commonality to other forms of remote sensing, in that we want to space out the samples as much as we can to not be biased by place (i.e. similar cover or use of the land, due to the areas lying adjacent in the landscape; e.g. Lillesand and Kiefer 2015; Liu and Mason 2016). When interpreting aerial imagery, however, the process is more similar to field data collection (see more on interpretation of aerial photos in chapter 5). We use most of the skills of an ecologist, although not at the species level, instead analyzing the structure, the texture, and the ecological context of a larger part of the landscape.

Accuracy assessments

Accuracy assessment is an important part of any classification project and compares the classified map or image to a set of data that we consider correct, often called *ground-truth*, *reference*, or *validation data*. Either there is a complete dataset to use or we must collect data to fill in the gaps, from the field, from interpreting high-resolution imagery, from existing classified imagery, or from GIS data layers (see Text box 9.3).

Typically, we assess the accuracy of data by collecting in situ (ground-truth) observations at a set of random points; this is once again a sampling problem but across the output classification, in which we compare the ground and the image-analysis classes and set them up in a confusion matrix. Often, ground-truth observations have already been split into two sets, one of which is used to train a classifier, and the other (referred to as *holdouts* or *test data*) is used for validation. There are several ways of sampling the validation points: they can be randomly or systematically placed all over the map or randomly placed within a grid so they are more evenly spread out, or they can be stratified so that a minimum number of points are placed in each class or category, or they can be clustered around placed centroids. In an error matrix, we then measure how many sample points were misclassified, according to the validation data, by each class and as an overall accuracy of the entire classified image (e.g. Congalton and Green 1999; Liu and Mason 2016).

If instead we want to validate the classification inside an area (polygon or segment), this can be rather easily done by laying out transect lines (coordinates for start and stop) through polygons, thereby producing a selection of sub-areas. An often-used method is analyzing $0.25m^2$ quadrats randomly laid along these lines. From experience, the number per polygon needs to be at least 30 quadrats (Swedish Environmental Protection Agency 1987).

An advantage using remote sensing and images from above is that when we know where things went wrong in classification, we can go back to the exact time of the first inventory and redo it, armed with the new knowledge (depending on the purpose of the inventory and, of course, the amount of samples), without having to deal with issues like changes in season or weather, cutting of hay, or grazing interfering with the renewed collection (Ihse 2007; Allard 2017).

Text box 9.3:   Aerial photo interpretation as a bridge between classification and accuracy assessment of space-borne data and in situ data

Developing and updating the National Land Cover Database is a joint work between a number of authorities and stakeholders (Swedish Environmental Protection Agency 2022) The test phase for version 2 involved a trial of increasing the classes from 24 to 48 to include wishes from various stakeholders to better suit environmental planning. The results from these tests have made a forward plan possible (see Text box 9.1).

Trials included dividing an existing broad class of open vegetated land (wetland and other types) into narrower classes dominated by grasses, dwarf shrubs, and shrubs and then further into three wetness classes and non-vegetated land into classes of exploited and natural land. Forested land was already divided into acceptable classes using experience from earlier mapping but needed better separation of wet deciduous and hardwood deciduous forest from other deciduous forest. Three very different landscapes, together covering all six of Sweden's biogeographical vegetation zones (Wastenson et al. 1996), were tested:

- Alpine and northern boreal vegetation zones in the mountains.
- Middle and southern boreal vegetation zones in the forest areas near the east coast.
- Boreonemoral and nemoral vegetation zones in the south.

Two of the test areas had enough available training data, but the third (southern area), where open land is a dominating feature, was lacking and is the one described here; see Figure 9.3.

Training data

The National Inventories of Landscapes in Sweden (NILS) programme has been a part of this rolling scheme of land cover database/map since the start of development in 2012 through production as well as further development and was organizing the search for quality digital data layers (e.g. from authorities or universities) for use as training data in the classifications. For many of the common forest types, the inventory data of the Swedish National Forest Inventory (NFI) provided adequate training data, and mire and wetland division and classification were ready-made (Hahn et al. 2021). However, not enough data were available for the divisions of open vegetated land or for separation of hardwood deciduous forest, and the decision was to collect extra training data (Allard and Adler 2020).

Using the NILS square grid of Sweden, 100 1km^2 squares, each of 196 10m circular plots were sampled; 38 were for collecting training data and the rest for collecting validation data, with both sets plotted out as widespread as possible. The aerial photos in near-infrared were interpreted in 3D, using the latest near-infrared photos from the rolling scheme of Lantmateriet, the Swedish mapping, cadastral, and land registration authority (that is, one to two years old in the south of Sweden); see Figure 9.3. The collection of data had triple purposes, besides the training/validation of the classification algorithm; these data would also act as training for the models of deciduous forests used in the NILS programme. Lastly, the collected data and knowledge were to be used in training courses for the NILS interpretation staff, for step 1 in the NILS inventory (see chapter 5).

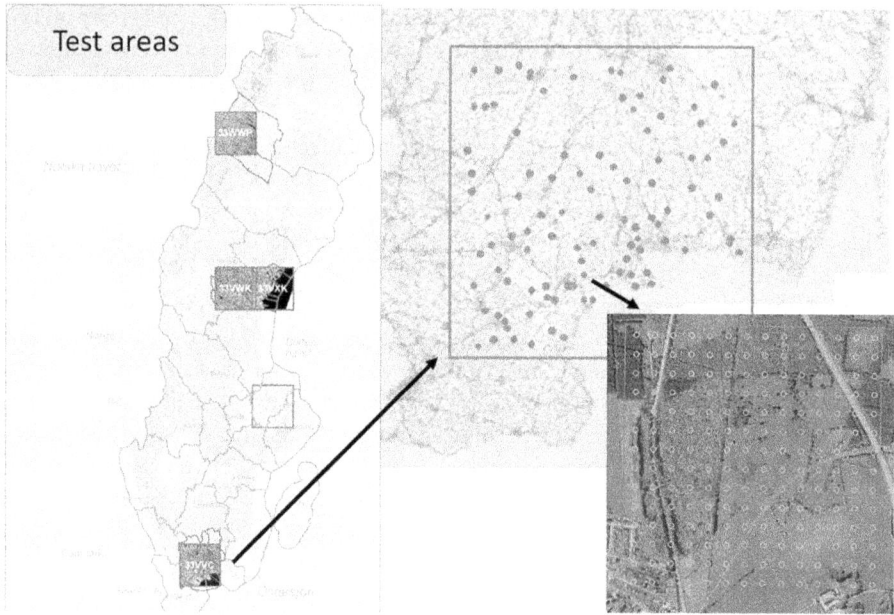

Figure 9.3 Three test areas for classification, based on the sub-areas (granules of 100km × 100km) along the swaths from Sentinel-2 overpasses. In the southern area, 100 squares of 1km^2 (red squares of the close-up map) each containing 196 circular plots (yellow circles on the near-infrared aerial photo) were sampled for collection of data.

Source: Maps and aerial photos are provided by Lantmateriet, the Swedish mapping, cadastral, and land registration authority.

The 38 squares used for training data were interpreted before field validation using a classification scheme especially developed to suit the classification of NMD; see Table 9.4. Dominating and subordinate land cover were recorded as well as mixtures, with attributes of intrinsic history such as abandoned management (encroachment) and texture (homogenous or mosaic patches). In addition, extra attributes such as roads or stone walls passing through the circular plot (all of which affect the reflectance of the patch) were recorded.

Field check of interpreted data

Each square was visited but because the goal was training data that were "right" or "homogeneous class vegetation" rather than for use as statistical estimates, the field visits targeted such areas and as many as possible field photos per square were taken during a four-week field trip; see Figure 9.4. The learning curve then was to see how to recognize the different wetness classes, how to recognize beech–oak forest from beech forest or other deciduous forest types, how to recognize pastures from the grazed farm fields and whether they were situated on sandy soil or richer mesic soil, and so on. Trees in towns and villages, especially in gardens, are often of exotic origin, and the distinction between hardwood deciduous and the more general class other deciduous becomes quite impossible, and they were all classed as other deciduous.

Table 9.4 Classification system, developed especially for training data of deciduous forest and open land, vegetated and non-vegetated

Class/variable	Dense	Sparse	Encroach-ment	Homo-genous	Mosaic structure	Shrub-dominated	Subordinate class
Forest non-wetland							
1 Deciduous	0/1	0/1	0/1	0/1	0/1	0/1	0/1
2 Mixed deciduous	0/1	0/1	0/1	0/1	0/1	0/1	0/1
3 Hardwood deciduous	0/1	0/1	0/1	0/1	0/1	0/1	0/1
4 Unsure deciduous	0/1	0/1	0/1	0/1	0/1	0/1	0/1
5 Clear-cut/young deciduous	0/1	0/1	0/1	0/1	0/1	0/1	0/1
6 Deciduous – not dominating	0/1	0/1	0/1	0/1	0/1	0/1	0/1
Forest on wetland							
1 Deciduous	0/1	0/1	0/1	0/1	0/1	0/1	0/1
2 Mixed deciduous	0/1	0/1	0/1	0/1	0/1	0/1	0/1
3 Hardwood deciduous	0/1	0/1	0/1	0/1	0/1	0/1	0/1
4 Unsure deciduous	0/1	0/1	0/1	0/1	0/1	0/1	0/1
5 Clear-cut/young deciduous	0/1	0/1	0/1	0/1	0/1	0/1	0/1
6 Deciduous – not dominating	0/1	0/1	0/1	0/1	0/1	0/1	0/1
Field layer on open ground, including visible fragments							
7 Grass – meadow like	0/1	0/1	0/1	0/1	0/1	0/1	0/1
8 Grass – lawn	0/1	0/1	0/1	0/1	0/1	0/1	0/1
9 Grassland wet	0/1	0/1	0/1	0/1	0/1	0/1	0/1
10 Dense reed	0/1	0/1	0/1	0/1	0/1	0/1	0/1
11 Dwarf shrub	0/1	0/1	0/1	0/1	0/1	0/1	0/1
12 Substrate – gravel/block	0/1	0/1	0/1	0/1	0/1	0/1	0/1
13 Sand	0/1	0/1	0/1	0/1	0/1	0/1	0/1
14 Rocky outcrop	0/1	0/1	0/1	0/1	0/1	0/1	0/1
15 Artificial – asphalt	0/1	0/1	0/1	0/1	0/1	0/1	0/1
16 Artificial – crop field	0/1	0/1	0/1	0/1	0/1	0/1	0/1

Reclassification after field

After fieldwork, the 38 squares were re-visited by aerial interpretation, using the newly gained knowledge, thereby delivering data that were as accurate as possible (Allard and Adler 2020). All of the knowledge gained became the basis for a course for training the rest of the NILS interpretation staff. It was also the basis for trusting the interpretation for validation data later on.

Validation data

When validation started the year after, the aim was to validate at least 70 raster pixels per class for construction of confusion matrices. As many forest classes as possible were taken from the NFI, but some of the uncommonly occurring classes (e.g. mixed coniferous/deciduous forest and all deciduous classes) had to be completed by extra collection, as was the case with open lands, mires, and wetlands. The remaining squares did not suffice for all of these, and about 40 extra squares were added from aerial photos in 3D, within which classified pixels were randomly placed for validation.

Figure 9.4 Aerial photo in near-infrared and corresponding field photo (indicated by an arrow) of a deciduous forest, where both interpretation in 3D and the field gave a mix of hardwood and other deciduous.

Source: Aerial photo provided by Lantmateriet, the Swedish mapping, cadastral, and land registration authority.

Credit: Field photo by Anna Allard.

Classification and uncertainty in validation data

We used the same classification system as the map but with the possibility of recording uncertainty, as in "this could also be" (for example, a pixel was classified into mire with dominance of dwarf shrub, but it could also be a mire with a dominance of shrub). In this way, it is possible to keep the crispness of a non-hierarchical classification system while still making it possible to develop a sort of fuzzy validation to gain a better sense of how right (or wrong) we are.

For the mountain area, we had used all of the available data as training and a complete set of validation data was necessary to collect. Overall, the interpretation of nearly 5000 pixels was needed to close the data gap (Nilsson et al. 2021).

Continuity of methods versus innovation

It is not a perfect world, but we use and appreciate the multitude of digital layers available, despite all of the differences introduced by people. We are in the midst of a revolution in innovation and will in all likelihood face both new technologies and new datasets to combine in the context of continuous monitoring schemes. Increasingly large and complex layers of data can be handled and recorded with increasingly diverse platforms and instruments. GIS resources for analysis and interpretation have rapidly become more user-friendly and complex at the same time, and ready-made data analysis tools are becoming increasingly widespread (Smith et al. 2021). As such, new data types and approaches need to be combined with existing data. Long-term monitoring of vegetation and biodiversity will have protocols and time series of data, going back through the years, and that continuity is hard to let go of, sometimes hindering innovations while supporting rich long-term analysis processes. The solution to that challenge is to find points of intersection where new and old data can meet by modelling and extracting the parts of older data that are compatible and useful, and thus keep long-term knowledge in place while still being able to use new ways of data collection and analysis. When possible, the data that are missing – the gaps – can also be bridged by extra collections. By using hybrid methods, we can get the best of two worlds. If we do not adapt, we cannot go forward. Yet if we do not preserve long-term analysis options, new data have little to be compared with and their relevance is diminished. Successful strategies for handling this conundrum have been those that exhibit a high degree of analytical pragmatism and where researchers use what they can get in terms of combining diverse types of data and then add data iteratively to mitigate issues with respect to gaps and linkages between elements.

Key messages

- In this chapter, we have provided an overview of how data types and ways of modelling within monitoring can be combined, as well as how these can be used to add information together to form coherent, comprehensive, and integrative datasets and analytical results.
- We have also indicated a range of characteristics and quality criteria of data and analysis approaches to take into consideration, with an emphasis on aspects of these that are relevant to hybrid data and methods.
- In addition, the chapter illustrates the importance of studying landscapes and their management as a continuous process, illustrating that continued emphasis on time series data and an increasing interdisciplinary emphasis on socioecological data hybrids may be crucial to the field of monitoring.

Study questions

1. Find out about raster data, vector data, and objects (in an OBIA environment). What are the advantages and disadvantages of each format; for example, in how they cope with spatial detail and represent continuously variable factors such as soil moisture? How is information lost when we convert from raster to vector or vector to raster or when we resample a raster image as part of changing the projection? Are the losses the same across the whole image? (Hint: they might not be.)
2. When combining maps and data from different sources, it is crucial to know what was initially meant when classification/labelling was conducted (for example, what does the word *forest* indicate and mean exactly?). Read up on different definitions of forest in European countries through the last decades, and think about what happens if we just combine two or more of these maps of forests.
3. What are the strengths and weaknesses of hierarchical and non-hierarchical habitat classifications? Is the answer different for field use than for use in later analysis?

4   How might we overcome dualistic human world versus natural world thinking in biodiversity monitoring? Is this different to simply combining traditional scientific methods with an analysis focussed on understanding needs and preferences of society?

Further reading

Many websites provide information on data types and modelling and the different ways of doing those, such as Gisgeography.com, ESRI, or sites for radar data.

An easy introduction to the complexities is found in Liu and Mason's *Image Processing and GIS for Remote Sensing: Techniques and Applications* 2nd ed. (2016), and some solutions of the complexities are also provided in chapter 15.

De Caceres et al. (2015) include a comprehensive review about vegetation classification in all its forms; much of that is also found on the website of the International Association of Vegetation Classification.

References

Allard, A. (2017) NILS – a nationwide inventory program for monitoring the conditions and changes of the Swedish landscape, in Díaz-Delgado, R., Lucas, R. and Hurford, C. (eds) *The Roles of Remote Sensing in Nature Conservation*. Cham, Switzerland: Springer, pp. 79–90. https://doi.org/10.1007/978-3-319-64332-8_5

Allard, A. and Adler, S. (2020) *Insamling av Kvalitetshöjande Vegetationsdata från NILS för Modellering och Klassificering* [Collection of Quality Vegetation Data from NILS for Modelling and Classification]. Report from NILS, Swedish University for Agricultural Sciences, Umeå. https://www.slu.se/globalassets/ew/org/centrb/nils/publikationer/2021/judging_cover_of_vegetation_nils_programme_2021.pdf

Becciu, P., Menz, M.H.M., Aurbach, A., Cabrera-Cruz, S.A., Wainwright, C.E., Scacco, M., Ciach, M., Pettersson, L.B., Maggini, I., Arroyo, G.M., et al. (2019) Environmental effects on flying migrants revealed by radar, *Ecography* 42(5), 942–955. https://doi.org/10.1111/ecog.03995

Bryn, A., Strand, G.-H., Angeloff, M. and Rekdal, Y. (2018) Land cover in Norway based on an area frame survey of vegetation types, *Norsk Geografisk Tidsskrift* 72(3), 131–145. https://doi.org/10.1080/00291951.2018.1468356

Bunce, R.G.H., Barr, C.J., Clarke, R.T., Howard, D.C. and Scott, W.A. (2007). *ITE Land Classification of Great Britain 2007*. Lancaster: NERC Environmental Information Data Centre. https://doi.org/10.5285/5f0605e4-aa2a-48ab-b47c-bf5510823e8f

Bunce, R.G.H., Groom, G.B., Jongman, R.H.G. and Padoa-Schippoa, E. (2005) *Handbook for Surveillance and Monitoring of European Habitats*, Alterrareport, No. 1219. Wageningen, The Netherlands: Alterra, Wageningen University.

Bunce, R.H.G., Metzger, M.J., Jongman, R.H.G., Brandt, J., de Blust, G., Rossello, R.E., Groom, G.B., Halada, L., Hofer, G., Howard, D.C., et al. (2008) A standardized procedure for surveillance and monitoring European habitats and provision of spatial data, *Landscape Ecology* 23, 11–25.

Congalton, R.G. and Green, K. (1999) *Assessing the Accuracy of Remotely Sensed Data Principles and Practices*. Boca Raton, FL: Lewis Publishers.

De Cáceres, M., Chytrý, M., Agrillo, E., Attorre, F., Botta-Dukát, Z., Capelo, J., Czúcz, B., Dengler, J., Ewald, J., Faber-Langendoen, D., et al. (2015) A comparative framework for broad-scale plot-based vegetation classification, *Applied Vegetation Science*, 18(4), 543–560. doi: 10.1111/avsc.12179

di Gregorio, A. and Jansen, L.J.M. (2005) *Land Cover Classification System (LCCS): Classification Concepts and User Manual for Software*. Version 2, Technical Report 8. FAO Environment and Natural Resources Service Series, Rome.

Dokter, A.M., Desmet, P., Spaaks, J.H., van Hoey, S., Veen, L., Verlinden, L., Nilsson, C., Haase, G., Leijnse, H., Farnsworth, A., et al. (2018) bioRad: biological analysis and visualization of weather radar data, *Ecography* 42(5), 852–860. https://doi.org/10.1111/ecog.04028

Duvigneaud, P. (1974) *La synthèse écologique* [The ecological synthesis]. Paris: Doin.

Ellis, E.C. (2011) Anthropogenic transformation of the terrestrial biosphere, *Philosophical Transactions of the Royal Society A: Mathematical, Physical and Engineering Sciences* 369, 1010–1035. https://doi.org/10.1098/rsta.2010.0331

Ellwanger, G., Runge, S., Wagner, M., Ackermann, W., Neukirchen, M., Frederking, W., Müller, C., Ssymank, A. and Sukopp, U. (2018) Current status of habitat monitoring in the European Union according to Article 17 of the Habitats Directive, with an emphasis on habitat structure and functions and on Germany, *Nature Conservation* 29, 57–78. https://doi.org/10.3897/natureconservation.29.27273

European Commission. (1992) The Habitats Directive, Council Directive 92/43/EEC of 21 May 1992, https://ec.europa.eu/environment/nature/legislation/habitatsdirective/index_en.htm (Accessed November 11, 2022).

Ferrier, S., Jetz, W. and Scharlemann, J. (2017) Biodiversity modelling as part of an observation system, in Walters, M. and Scholes, R. (eds) *The GEO Handbook on Biodiversity Observation Networks*. Cham, Switzerland: Springer, pp. 239–257. https://doi.org/10.1007/978-3-319-27288-7_10

Fridman, J., Holm, S., Nilsson, M., Nilsson, P., Ringvall, A.H. and Ståhl, G. (2014) Adapting national forest inventories to changing requirements – the case of the Swedish National Forest Inventory at the turn of the 20th century. *Silva Fennica* 48(3), 1095. https://doi.org/10.14214/sf.1095

Godinho, S., Guiomar, N., Machado, R., Santos, P., Sá-Sousa, P., Fernandes, J.P., Neves, N. and Pinto-Correia, T. (2016) Assessment of environment, land management, and spatial variables on recent changes in *montado* land cover in southern Portugal, *Agroforestry Systtems* 90, 177–192. https://doi.org/10.1007/s10457-014-9757-7

Grace, M.K., Akçakaya, H.R., Bull, J.W., Carrero, C., Davies, K., Hedges, S., Hoffmann, M., Long, B., Nic Lughadha, E.M., Martin, G.M., et al. (2021) Building robust, practicable counterfactuals and scenarios to evaluate the impact of species conservation interventions using inferential approaches, *Biological Conservation* 261, 109259. https://doi.org/10.1016/j.biocon.2021.109259

Hahn, N., Wester, K. and Gunnarsson, U. (2021) *Satellitbaserad Övervakning av Våtmarker – Nationell Slutrapport Första Omdrevet* [Satellite Based Monitoring of Wetlands – National Final Report from First Rotation]. Stockholm: Naturvårdsverket, Rapport 6950. http://www.myrar.se/ or https://www.naturvardsverket.se/978-91-620-6950-6

Head, L. (2012) Conceptualising the human in cultural landscapes and resilience thinking, in Bieling, C. and Plieninger, T. (eds) *Resilience and the Cultural Landscape: Understanding and Managing Change in Human-Shaped Environments*. Cambridge: Cambridge University Press, pp. 65–79. https://doi.org/10.1017/CBO9781139107778.006

Hidayat, F., Rudiastuti, A.W. and Purwono, N. (2018) GEOBIA an (geographic) object-based image analysis for coastal mapping in Indonesia: a review, *IOP Conference Series: Earth and Environmental Science* 162, 012039.

Honrado, J.P., Pereira, H.M. and Guisan, A. (2016) Fostering integration between biodiversity monitoring and modelling, *Journal of Applied Ecology* 53(5), 1299–1304. https://doi.org/10.1111/1365-2664.12777

Ihse, M. (2007) Colour infrared aerial photography as a tool for vegetation mapping and change detection in environmental studies of Nordic ecosystems: a review, *Norsk Geografisk Tidsskrift* 61(4), 170–191. doi: 10.1080/00291950701709317

International Association of Vegetation Classification. (2022) Vegetation classification, https://sites.google.com/site/vegclassmethods/home (Accessed May 24, 2022).

Jepsen, M.R. and Levin, G. (2013) Semantically based reclassification of Danish land-use and land-cover information, *International Journal of Geographical Information Science* 27, 2375–2390. https://doi.org/10.1080/13658816.2013.803555

Levin, G. (2006) Farm size and landscape composition in relation to landscape changes in Denmark, *Geografisk Tidsskrift* 106, 45–59. https://doi.org/10.1080/00167223.2006.10649556

Lillesand, T.M. and Kiefer, R.W. (2015) *Remote Sensing and Image Interpretation*. 7th edn. New York: Wiley.

Liu, J.G. and Mason, P.J. (2016) *Image Processing and GIS for Remote Sensing: Techniques and Applications*. 2nd edn. Wiley.

Lourenço, P., Teodoro, A.C., Gonçalves, J.A., Honrado, J.P., Cunha, M. and Sillero, N. (2021) Assessing the performance of different OBIA software approaches for mapping invasive alien plants along roads with remote sensing data, *International Journal of Applied Earth Observation and Geoinformation* 95, 102263. https://doi.org/10.1016/j.jag.2020.102263

Lucas, R., Blonda, P., Bunting, P., Jones, G., Inglada, J., Arias, M., Kosmidou, V., Petrou, Z.I., Manakos, I., Adamo, M., et al. (2015) The Earth Observation Data for Habitat Monitoring (EODHaM) system, *International Journal of Applied Earth Observation and Geoinformation* 37, 17–28. https://doi.org/10.1016/j.jag.2014.10.011

Martin, E.A., Dainese, M., Clough, Y., Báldi, A., Bommarco, R., Gagic, V., Garratt, M.P.D., Holzschuh, A., Kleijn, D., Kovács-Hostyánszki, A., et al. (2019) The interplay of landscape composition and configuration: new pathways to manage functional biodiversity and agroecosystem services across Europe, *Ecology Letters* 22, 1083–1094. https://doi.org/10.1111/ele.13265

Naujokaitis-Lewis, I.R., Curtis, J.M.R., Tischendorf, L., Badzinski, D., Lindsay, K. and Fortin, M.-J. (2013) Uncertainties in coupled species distribution–metapopulation dynamics models for risk assessments under climate change, *Diversity and Distributions* 19, 541–554. https://doi.org/10.1111/ddi.12063

Nilsson, M., Allard, A., Ahlkrona, E., Jönsson, C., Petra Odentun, P., Berlin, B., Karlsson, L. and Olsson, B. (2021) *Agenda för Landskapet AP 7 Validering* [Agenda for the Landscape – Statistical Evaluation]. Umeå: Swedish University for Agricultural Science.

Nourbakhshbeidokhti, S., Kinoshita, A., Chin, A. and Florsheim, J. (2019) A workflow to estimate topographic and volumetric changes and errors in channel sedimentation after disturbance, *Remote Sensing* 11, 586. 10.3390/rs11050586.

Petrosillo, I., Aretano, R. and Zurlini, G. (2015) Socioecological systems, in Fath, B. (ed.), *Encyclopedia of Ecology*. 2nd edn. Oxford: Elsevier, pp. 419–425. https://doi.org/10.1016/B978-0-12-409548-9.09518-X

Plieninger, T. (2006) Habitat loss, fragmentation, and alteration – quantifying the impact of land-use changes on a Spanish *dehesa* landscape by use of aerial photography and GIS, *Landscape Ecology* 21, 91–105. https://doi.org/10.1007/s10980-005-8294-1

Renes, H. (2010) Grainlands. The landscape of open fields in a European perspective, *Landscape History* 31, 37–70. https://doi.org/10.1080/01433768.2010.10594621

Rodwell, J.S. (2008) *British Plant Communities*. 2nd edn. Vols 1–4. Cambridge University Press.

Sarzynski, T., Giam, X., Carrasco, L. and Lee, J.S.H. (2020) Combining radar and optical imagery to map oil palm plantations in Sumatra, Indonesia, using the Google Earth Engine, *Remote Sensing* 12, 1220. https://doi.org/10.3390/rs12071220

Smith, G., Kleeschulte, S., Soukup, T., Garcia, R., Banko, G. and Combal, B. (2021) An operational service for monitoring grassland dominated Natura2000 sites with Copernicus data, *IEEE International Geoscience and Remote Sensing Symposium IGARSS, Brussels, Belgium, July 11–16, 2021*. doi: 10.1109/IGARSS47720.2021.9554934.

Stoate, C., Báldi, A., Beja, P., Boatman, N.D., Herzon, I., van Doorn, A., de Snoo, G.R., Rakosy, L. and Ramwell, C. (2009) Ecological impacts of early 21st century agricultural change in Europe – a review, *Journal of Environmental Management* 91, 22–46. https://doi.org/10.1016/j.jenvman.2009.07.005

Stumpf, A., Michéa, D. and Malet, J.-P. (2018) Improved co-registration of Sentinel-2 and Landsat-8 imagery for Earth surface motion measurements, *Remote Sensing* 10, 160. https://doi.org/10.3390/rs10020160

Ståhl, S., Allard, A., Esseen, P.-A., Glimskär, A., Ringvall, A., Svensson, J., Sundquist, S., Christensen, P., Gallegos Torell, Å., Högström, M., et al. (2011) National Inventory of Landscapes in Sweden (NILS) – scope, design, and experiences from establishing a multi-scale biodiversity monitoring system, *Environmental Monitoring and Assessment* 173(1–4), 579–595.

Swedish Environmental Protection Agency. (1987) *Metodbeskrivningar, Vegetation, BIN BIologiska Inventeringsnormer* [Descriptions of Methodologies, Vegetation, Biological Norms of Inventory]. Stockholm: Swedish Environmental Protection Agency.

Swedish Environmental Protection Agency. (2022) *Agenda för Landskapet, Implementeringsplan* [Agenda for the Landscape, Implementation Plan], Work Report. Swedish Environmental Protection Agency.

Tang, X., Dong, S., Luo, K., Guo, J., Li, L. and Sun, B. (2022) Noise removal and feature extraction in airborne radar sounding data of ice sheets, *Remote Sensing* 14, 399. https://doi.org/10.3390/rs14020399

UNESCO. (1973) *International Classification and Mapping of Vegetation*, UNESCO Ecology and Conservation, Series 6. Paris: UNESCO.

Vigo, J. (2005) *Les Comunitats Vegetals: Descripció i Classificació* [Plant Communities: Description and Classification]. Barcelona: Edicions Universitat Barcelona.

Vila-Viçosa, C., Arenas-Castro, S., Marcos, B., Honrado, J., García, C., Vázquez, F.M., Almeida, R. and Gonçalves, J. (2020) Combining satellite remote sensing and climate data in species distribution models to improve the conservation of Iberian white oaks (*Quercus* L.), *ISPRS International Journal of Geo-Information* 9(12), 735. https://doi.org/10.3390/ijgi9120735

Wastenson, L., Gustafsson, L. and Ahlén, I. (1996) *National Atlas of Sweden, Geography of Plants and Animals*. Stockholm: Norstedts.

10 Social data: what exists in reporting schemes for different land systems?

Claire Wood, Mats Sandewall, Stefan Sandström,
Göran Ståhl, Anna Allard, Andreas Eriksson,
Christian Isendahl, and Lisa Norton

Introduction

Social science is, in its broadest sense, the study of society and the manner in which people behave and influence the world around us. It tells us about the world beyond our immediate experience and can help explain how our own society works (Economic & Social Research Council, 2022). In this chapter, we consider social science data as a way of adding additional explanatory variables to trends and changes seen in environmental data.

Direct inclusion of social data in long-term environmental monitoring programmes is relatively rare, but in some cases, monitoring and socioeconomic data are combined in the phase of analyzing the data. However, the lack of social data is a weakness for many monitoring programmes, perpetuating the Cartesian separation between nature and culture. It limits the potential for robust analyses of socioecological system dynamics to explain observed variability and causation of change as well as to predict future change. Often the most successful studies linking social and environmental data tend to be at a sub-national scale or have a fairly narrow spatial, temporal, and/or thematic remit.

This chapter presents a series of case studies from a range of different countries, describing how social data have been incorporated into environmental studies. On the whole, these demonstrate the fairly narrow remit of this type of work, although some do offer suggestions for increasing the incorporation of social data into national-level monitoring, perhaps offering hope for the future.

Examples of the collection of social science data in relation to ecological monitoring from the UK are presented but illustrate the limited nature of this work. An example is given from Loweswater, in the English Lake District, where transdisciplinary science was undertaken at a small scale. This demonstrates an approach that could perhaps be expanded to national scales in some scenarios.

In Sweden, some attempts are being made to incorporate social data into monitoring, particularly in the realm of tourism, forestry scenario analysis, and the interaction between forestry and reindeer husbandry. The latter case demonstrates disparate social and ecological issues.

In Iceland, many environmental issues are caused by sheep grazing. Here, historic records are available, including written records, old aerial photos, and maps.

We also present a case study from Vietnam, an example of a low-income country where data regarding agricultural production and demography are reported annually.

Finally, the notion of *freelisting* is presented. Freelisting is an ethnographic tool that can be employed within interdisciplinary fields.

DOI: 10.4324/9781003179245-10

Case studies

The UK

Social data within national environmental monitoring schemes

As with most other countries, in the UK the collection of social data has traditionally been rare in many existing national long-term environmental monitoring programmes, such as the UK Countryside Survey (2022) or the National Forest Inventory (Forest Research 2022). Whilst some work has been undertaken to link social and ecological data in quantitative analyses, typically in short-term (1–5 years) to medium-term (5–12 years) research-driven projects aligned to these surveys, it is limited in spatial coverage and topical range. In stable, long-term (>10 years) national survey programmes, social data collection and integration with environmental parameters is not common practice among expert government agencies or stakeholder organizations. There are a number of examples where it may have been explored as an afterthought, or an "added-extra", as in Potter and Lobley (1996), who introduced a socioeconomic questionnaire into the UK Countryside Survey, following the 1990 edition of the survey, to shed light on farming practices. However, this type of work has not become a regular component of the survey, thus far.

Across England, from 2009 to 2019, the English conservation agency, Natural England, ran the Monitor of Engagement with the Natural Environment (MENE) survey (Gov.UK 2022c). It collected data about outdoor recreation, pro-environmental behaviours, and attitudes towards and engagement with the natural environment. It has now been re-named the People and Nature Survey, going forward. Whilst it is encouraging to see this type of data being collected in a national scheme, it does not necessarily offer much opportunity to integrate the data with other national monitoring schemes (such as the UK Countryside Survey, for example).

Agricultural surveys and agri-environment schemes

Often, social data that do exist are commonly collected in relation to the management of agricultural land, again carried out separately to long-term environmental monitoring schemes. There is an annual June Survey of Agriculture and Horticulture across the UK, carried out by the government, which collects detailed information, via a questionnaire, on arable and horticultural cropping activities, land usage, livestock populations, and labour force figures. The information includes long-term trends or detailed results for different types of farm, farm size, or geographical area (Gov.UK 2022b). In England, this is complemented by regular Farm Practices Surveys, looking at how English farming practices are affected by current agricultural and environmental issues, such as greenhouse gas mitigation and soil management. The content of the survey is agreed on each year in consultation with users to ensure the information collected remains relevant to current issues (Farm Practices Survey 2022a).

Another area where social data have been collected from land managers is in the area of agri-environment schemes (AESs). AESs offer farmers financial incentives to improve the conditions for semi-natural species dependent on less-intensive agriculture, such as insects and plants, and are implemented in several parts of the world with the goal of reversing biodiversity losses (McCracken et al. 2015). AESs are costly and have had variable success (Ansell et al. 2016). In an attempt to evaluate the role of the farmers' attitudes in relation to AES, McCracken et al. (2015) take a qualitative approach, proving a powerful link between biodiversity outcomes and farmer motivations. In addition to traditional counts of biodiversity (such as plant censuses and invertebrate richness) undertaken on a series of farms, interviews were completed with farmers to understand their experience and engagement with the AES. These qualitative data were given quantitative scores to enable analysis of the ecological data alongside the social data. Results revealed that farmer experience went a long way in determining habitat quality outcomes.

In the UK, the Department for the Environment and Rural Affairs (DEFRA) funds a range of other research relating to farmer attitudes and AESs (Mills et al. 2013), including research related to the introduction of a new post-Brexit scheme, the Environmental Land Management Scheme (ELMS). In Wales, the Welsh Government funded a large modelling and monitoring programme between 2013 and 2016 (the Glastir Monitoring and Evaluation Programme), which included a Farmer Practices Survey element; however, this was not closely linked to the field survey.

Environmental problems – water quality at Loweswater

Quite often, a need to collect social science information arises when a particular problem occurs, rather than being part of a longer-term monitoring programme. An example of this was at Loweswater in the English Lake District in 2004, where a pollution problem was identified in relation to runoff from farmland into the lake, causing toxic algal blooms indicating poor water quality. The scientific data collected were complemented by knowledge of local issues and tensions, current farming practices, and economic pressures collected from 13 farms in the Loweswater catchment (Waterton et al. 2006). This project was part of the Rural Economy and Land Use (RELU) programme, an interdisciplinary programme incorporating social science perspectives into research to enable better action to be taken to address environmental issues. The combination of an interdisciplinary, stakeholder-inclusive scientific approach and the positive stance towards understanding and managing the problem of pollution in Loweswater taken on by the farmers presented an opportunity to identify effective approaches to catchment management. Such approaches had already been pioneered elsewhere globally (for example, through the UNESCO Hydrology for the Environment, Life and Policy [HELP] programme; UNESCO 2022).

Sweden

The National Inventories of Landscapes

In recent years, the National Inventories of Landscapes in Sweden (NILS) has been trying to accommodate social aspects in its system of variables for monitoring (Allard 2017; Ståhl et al. 2011). The programme ran from 2003 to 2019 in the original form, after which the focus switched to monitoring habitats for the Natura 2000 reporting to the European Union (EU).

One of the initial thoughts was to provide data for researchers in the fields of recreation in nature and tourism, and variables were added to the survey, although the main focus of the survey was on ecology and the 16 national environmental quality objectives (Swedish Environmental Protection Agency [EPA] 2013). The programme was promoted at national conferences, such as the European Tourism Research Institute (ETOUR) at Mid-Sweden University. Another way to pursue the social aspect of the inventory was for co-applicants to provide additional funding in this field of research; for example, using remote sensing in the search for tracks from off-road driving (motorbikes or terrain vehicles) in mountainous terrain or researching older, inhabited settlements and their abandonment in the remote areas of the mountainous zone using stereo aerial photos from the 1950s.

An example of involvement of society on higher levels is the development of the National Land Cover Data, Nationella Marktäckedata (NMD; Swedish Environmental Protection Agency 2022), for version 2, with a long period of work involving the ad-ministrative boards of counties and municipalities, each looking out for the specialties of their own regions. A number of national authorities were also involved, to be part of the development into a map that would be of use for as many purposes as possible, such as statistics on green spaces in urban areas or planning where to build new housing areas. The Swedish Civil Contingencies Agency is another example, because they are responsible for fire mitigation and planning, a very real problem in a country relying on forestry as one of the main foundations of their economy. The collaborative work resulted in a new version with 48 classes instead of the previous 25 (see Figure 10.1). Lastly, the local involvement came in the shape of a competition, "Hack for Sweden", where one of the topics was combining the land cover map with social well-being in the form of recreation in the landscape. The winning solution was the creation of an app where you can choose, for example, the type of forest you want and find the nearest space to visit. The prize was funding to develop this app further (NewSeed IT Solutions 2002).

Figure 10.1 A section of the Swedish land cover mapping (under development), an open-source da-tabase from the area of Kristianstad, in southern Sweden. To be as useful as possible, the development for version 2 involved nearly 50 people from 25 different authorities and universities. The result of this work ended up in a version including 48 land cover classes instead of the previous 25.

Source: Swedish Environmental Protection Agency (2022).

Figure 10.2 The area (in red) used by the Swedish reindeer herding communities (RHCs) according to the Sami Parliament (2022).

Reindeer husbandry in Sweden

The topic of reindeer husbandry in Sweden encompasses a range of environmental, social, and cultural issues. The Sami are an indigenous people in Northern Europe in what is today Norway, Sweden, Finland, and Russia. In Sweden, Sami history, culture, traditions, legal rights, local economy, and well-being are closely connected to reindeer husbandry (Sami Parliament 2022). Sami people have an exclusive right to herd and graze their reindeer (*Rangifer tarandus*) within the reindeer husbandry area (RHA), which constitutes a large part of northern Sweden (Figure 10.2).

There are two major systems of reindeer husbandry. Mountain reindeer husbandry is dependent on summer pastures in the alpine region, whereas winter pastures are situated in the boreal region close to the Gulf of Bothnia. Forest reindeer husbandry relies on both summer and winter pastures in the boreal region in the eastern part of the RHA. In both reindeer husbandry systems, summer grazing includes a variety of plants, grasses, lichens, and fungi, whereas winter grazing mainly consists of lichens. Lichen availability is therefore considered a bottleneck resource for reindeer and thus reindeer husbandry (Sandström et al. 2016). In the 1970s, it became apparent that other land uses, mainly forestry, negatively impacted lichen pastures. Although there are monitoring data on lichens from the Swedish National Forest Inventory (NFI) dating back to the 1950s, there have been very few estimates on the status and trends of ground lichens in the RHA. However, in a recent paper, it was found that lichen–abundant forest had decreased (since the 1950s) by 70% (Sandström et al. 2016). The Swedish Environmental Quality Objective 14, "A Magnificent Mountain Landscape", states that a high reindeer grazing pressure is required to keep the landscape open, at the same time acknowledging that the objective is dependent on corresponding grazing opportunities in the forest region (Swedish EPA 2019). However, the government response so far has focused on facilitating consultations between RHCs and other land users, rather than to get involved in regulating or monitoring the lichen availability.

In Sweden, some social data are collected with respect to reindeer husbandry. These data are presented on the webpage of the Sami Parliament (2022) and include basic information regarding the number of reindeer, the composition of the herd, number of reindeer owners and their gender, produced reindeer meat in kilograms, and the value of reindeer meat in Swedish krona. There is an apparent ecological correlation between number of reindeer and available lichen winter pastures, but the authorities do not explore this correlation. According to the Reindeer Husbandry Act (Swedish Code of Statutes [Svensk Författningssamling] 1971) the county administrative boards are responsible for deciding the maximum number of reindeer for each RHC. This number should be decided based on the carrying capacity of the grazing lands but also consider other interests and land users. However, without reliable monitoring data, there is a risk that maximum permitted reindeer numbers will be reduced instead of addressing, for example, the loss of grazing lands due to competing land users (Horstkotte et al. 2022; Sarkki et al. 2022). Despite the decrease of lichen-rich forests, reindeer numbers have remained relatively constant over time. This has been attributed to different forms of adaptation (husbandry practices, mechanization, supplementary feeding, etc.). However, the RHCs are left with the responsibility to govern their pastures more or less independent of the government or government agencies. This has resulted in a situation where the RHCs have opted for a voluntary programme of mapping and monitoring their grazing lands on their own (see chapter 17).

Nevertheless, reindeer husbandry would undoubtedly benefit if the social and cultural aspects of reindeer husbandry were included in national environmental monitoring programmes. Such monitoring programmes could, for example, seek to answer the following questions (all of which are important to understanding future prospects for reindeer husbandry): how has lichen amount and distribution changed over time? What is the relationship between forest conditions, climate, and reindeer population sizes? What is the relationship between reindeer population sizes and the prospects of a continuous reindeer husbandry? What kind of habitats are important for reindeer husbandry (year-round), and are they monitored?

Monitoring as a basis for scenario analysis

Modelling scenarios can be an important tool for landscape analysis and can incorporate different types of data, including social data. Analysts may use monitoring data for several purposes. The state and trends of biodiversity indicators are an important basis for analyzing policy related to management or conservation of ecosystems. However, using monitoring data for scenario analysis also may add important insights when assessing policy options. This is particularly the case if we use monitoring data as part of broad analyses where socioeconomic impacts are considered as well.

Analysts may specify different scenarios and use monitoring and other data to predict the likely consequences in terms of ecosystem services under each scenario. Based on the results, decision makers may implement the policies that are assessed to lead to the scenario with the "best" future output of services. Compared to only assessing the state and trends, scenario analyses thus add important information for decision makers, because it is fundamental to know not only about the past and present but also about the likely future developments given different policies.

In this case example, we describe how the Swedish Forest Agency and the Swedish University of Agricultural Sciences conduct scenario analyses as an input to policy processes related to the management of Swedish forests. In these analyses, monitoring data from the Swedish National Forest Inventory (NFI; Fridman et al. 2014) are used in the Heureka system (Wikström et al. 2011) together with separately collected socioeconomic data.

Forest scenario analyses of the kind outlined above have been conducted for a long time in Sweden. In the most recent Forest Scenario Analysis (Skogliga Konsekvensanalyser, SKA 22; Swedish Forest Agency 2022), important study features involve the future outputs of ecosystem services from Swedish forests given different policy options for the forest management. At the heart of the analysis lies the trade-offs between future timber outputs, greenhouse gas fluxes, and habitat conditions. Because these issues are closely interlinked, it is important to study them within a unified scenario framework rather than making separate analyses addressing each of the services individually (The Swedish Forest Agency 2021, 2022).

The SKA22 project uses Swedish NFI monitoring data as an important input. These data are collected from several thousand field plots annually, distributed across Swedish forests to obtain a statistically correct baseline input regarding the state of the forests (Fridman et al. 2014). The measurements on each plot involve a large number of characteristics important for deriving indicators linked to, among other things, timber, carbon, and biodiversity. The Heureka system has a module specifically tailored for incorporating NFI data as a basis for regional and national forest scenario analyses. The system uses these data together with models for the development of forests to predict the future outcome of ecosystem services, given management options specified by the analyst. Each management option thus reflects a specific scenario for the future development of the forests.

To specify the management options and to provide socioeconomic data to Heureka that are not available from the NFI, the Swedish Forest Agency carries out a separate preparatory socioeconomic analysis. During this analysis, outlooks for relevant socio-economic parameters are made not only for Swedish conditions but also in a general international context because Swedish forestry is closely linked to international forestry developments.

The outlook for SKA22 identifies an increased demand for Swedish wood products; a need for adapting forest ecosystems to climate change, especially to avoid disturbances; and expectations to increase carbon sequestration and to halt biodiversity loss. There are also ongoing policy processes nationally, regionally (EU), and globally towards "more of everything". The scenario setup therefore ranges from business-as-usual (BAU) to more biodiversity, climate adaptation, increased growth, and a combination of these (The Swedish Forest Agency 2022).

The scenario analysis in Figure 10.3 shows larger trade-offs in the long-term perspective compared to the mid-term. It also shows larger effects for economy and biodiversity than for climate change mitigation. A strict focus on any single target (diversity, climate, or growth) would most likely generate larger societal conflicts compared to combining them, which, on the other hand, would not generate that much of a difference compared to the business-as-usual scenario.

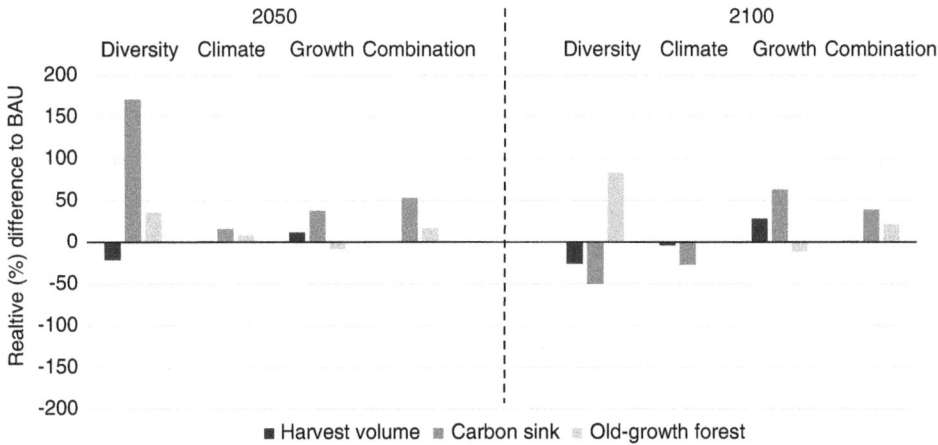

Figure 10.3 Scenario analyses in the project SKA 22, investigating possible outcomes in harvest volume, carbon sink, and area of old growth (virgin) forest in Sweden by the years 2050 and 2100, depending on different policy trajectories in Sweden, relative to business as usual (BAU). The scenarios show the outcome, should the policies demand focus on: diversity, climate, growth, or a combination of these.

Source: Swedish Forest Agency (2022).

Iceland

In some cases, historic records can give a valuable insight into the social elements of landscape change.

In Iceland, inventories of social data since first settlement (783 AD) relevant to environmental change over time have been made, searching down stratigraphic layers of soil, written records, old aerial photos, and maps. In the last millennium, together with shifts in demography and ownership of common grazing land for sheep, the use of common rangelands has resulted in severe degradation of areas of vegetated land over time (Gísladóttir 2001; Gísladóttir et al. 2010; Erlendsson et al. 2014; Sigurmundsson et al. 2014, 2021). In recent centuries, the remaining birch woodlands have also been much depleted, due to increases in numbers of sheep, in contrast to the forests of the island from the time of the settlements in the ninth century (Erlendsson et al. 2014). Figure 10.4 shows a highly degraded pasture, with only small remnants of vegetation, and Figure 10.5 shows a part of Thingvellir, the ancient parliament place, which is protected from grazing and is a representation of the original vegetation of the island. In Iceland, much effort is placed on soil research and the possibilities for the soil to retain surface water, because the very thin layer of soil can degrade quickly when the vegetation cover is degraded, and the underlying pumice stone will then not be able to hold surface water and the restoration of a viable soil layer will be very difficult. Because Iceland is situated on the Mid-Atlantic Ridge, there is also the added hazard of repeated volcanic eruptions (e.g. Arnalds 2015; Sigurmundsson et al. 2021). Vegetation cover and functioning ecosystems are the key to climate change mitigation but also provide carrying capacity (food and habitat) for grazing animals, so efforts to reverse the degradation of soils and vegetation by reducing stocking levels can lead to conflicts of interest and difficult choices for policymakers (Ágústsdóttir 2015; Halldórsson et al. 2017).

Figure 10.4 Grazing land at Tröllkonuhlaup in Iceland. Much of the area is heavily grazed and the soil is bare in many places.

Credit: Photo by Anna Allard.

Socioeconomic and production data – the case of Vietnam

In many low-income countries with centrally planned economies, governments assess and monitor agricultural production and other social, economic, and landscape-related parameters for purposes of governance; for example, as a basis for five-year plans and development strategies. One example is Vietnam, where strategies and policies firmly direct the resource management at central and local levels (Figure 10.6).

All over the country, data on status and change in, for example, agricultural production and demography are reported annually in a standardized format from local communes through districts and provinces to the central state level for the purpose of governance and monitoring of national plans and targets. Additionally, the area distribution of land use classes and some aggregated socioeconomic parameters are reported; for example, household income, poverty rate, and human health status. Such data are usually based on questionnaires. They can sometimes be quite detailed, but often they are not sufficiently consistent and unbiased for reflecting long-term changes and trends (Ohlsson et al. 2005, see more on this topic in chapter 3).

In a publication by the General Statistics Office of Vietnam (Text box 10.1, Figure 10.7), annual data collected from the agriculture, forestry, and fishery sectors during 1975–2000 in the 61 provinces of Vietnam were presented (a series of some older reference data for 1960–1975 was also included). The publication illustrates what type of standardized data

Figure 10.5 Land area protected from grazing at Thingvellir, Iceland, showing the arctic birch forest that once covered most of the land.

Credit: Photo by Anna Allard.

were collected within the government structure of Vietnam. When interpreting the data, one needs to be aware that production data are reported based on annual plans and targets issued by the government. They are therefore to some degree a reflection of government ambitions and not an objective assessment of the local situation (Ohlsson et al. 2005).

Freelisting: an ethnographic field method to monitor the diversity of biocultural knowledge

To understand and record place-based knowledge of local environments, in particular of environmental resources – wild, semi-domesticated, and domesticated – and their utilities and management, there are a series of ethnographic tools employed within interdisciplinary fields such as cultural anthropology, human ecology, and human geography. Typically used to study small-scale economies in the Global South where smallholder agriculture and horticulture dominate, these tools document what is commonly known as *traditional ecological knowledge* (TEK); related terms are *Indigenous knowledge, folk taxonomies,* and *ethnoecology* (e.g. Balée and Nolan 2019).

These qualitative fields of inquiry add to other approaches to monitor diversity; for instance, by supplementing species inventories with locally recognized varieties and by documenting the practices and management systems that contributed to their evolution.

Figure 10.6 Intensified agriculture combined with market economic policies since 1990 has released land
        and promoted plantation of forest and cash crops by households and enterprises in Vietnam.

Credit: Photo by Mats Sandewall.

They are based more on different sets of practical, functional, and morphological
characteristics than quantitative systematization. A major focus of inquiry lies in the
structure, characteristics, and variability of the adaptive, place-based knowledge systems
themselves, as well as how these relate to different cultural knowledge domains (such as
plants or *soils*). The term *cultural salience* of a domain is often used, which means the
degree to which members of a particular culture hold a domain to be of particular
importance (the rich taxonomic vocabulary to describe different kinds of snow and ice
among Inuit communities is a classic example), and although this is related to economic
significance, cultural importance should not be reduced to economic value (Austin et al.
2015; Balée and Nolan 2019).

    Among several tools of data collection, freelisting is a qualitative interviewing tech-
nique that has recently grown in popularity. It is a tool for rapidly exploring how groups
of people think about and define a particular domain and is well suited for engaging
communities and identifying shared priorities (Keddem et al. 2021). The domains are
defined by the data collector – for instance, *types of vegetation*, *edible root plants*, or *crop
varieties* – and collects the inside view (or *emic* view) of the people living and working in a
particular environment. The strength of this method is that it elicits unimagined,
spontaneous responses that can be collected, analyzed, and quantified, and results from
these analyses can be incorporated into mixed-methods studies in different populations

Text box 10.1: Forest reporting from Vietnam

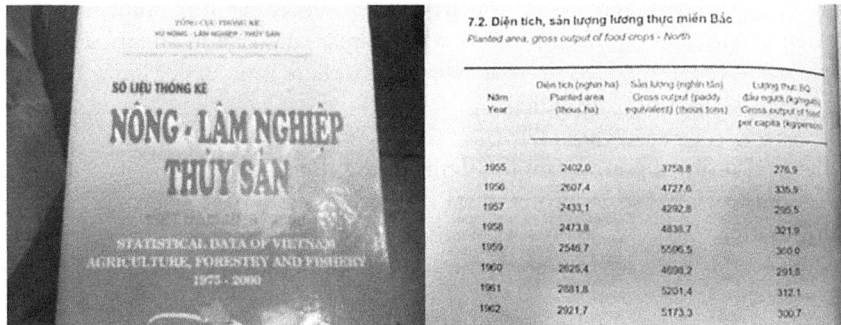

Figure 10.7 In Vietnam, data on status and change in, for example, agricultural production and demography were reported annually in a standardized format. Additionally, the area distribution of land use classes and some aggregated socioeconomic parameters were presented.

Various data (annual and for each province) presented in the numerous tables of the summary publication from Vietnam include:

Indicators

- Demography (year, province): population, households, and labour in agriculture, forestry, fishery
- Land use (by year and province): agricultural land (rice, other crops, total), forestry, fishery
- Vehicles and machines used in agriculture, forestry, fishery.

Agriculture

- Cultivated areas, yields, gross economic output, crops planted (households and industrial), food production, fruit trees, number of pigs, cattle.

Forestry

- Gross economic output (afforestation, exploitation), forestry area (status 1997), area of afforestation, quantity of logged wood

Fishery

- Gross output (breeding, capture, total), gross output of fish, shrimps, breeding water products

Farms and households

- Number of farms (annual crops, perennial crops, livestock, forestry, fishery, mixed activity
- Farm labour (owners' family, hired regular, temporary hired.

and settings to contribute valuable understanding of needs and priorities in the community (Bernard 1994; Balée and Nolan 2019; Meireles et al. 2021).

Freelisting is a simple method to carry out. It is an interview technique by which respondents who represent a given social group list all items in a given semantic domain. The social group and the semantic domain are defined according to monitoring aims; for instance, tree species among forest foragers who share a language in a particular region. The sample size is recommended to be at least 15, but there is no upper limit. Freelisting rests on several assumptions, including:

- People tend to list the most familiar items in the domain first.
- Items listed first tend to be prototypical and locally significant, and their names are usually shorter.
- Individuals who have more knowledge about a domain tend to list more items.

Freelists are analyzed statistically – for example, using the software ANTHROPAC 4.983/x (Borgatti 1992) – to assess the diversity of domains within, and among, social groups, with a particular focus on evaluating the salience, or cultural importance, of domains and items. For a comprehensive introduction to freelisting and an example of application, see Balée and Nolan (2019).

Key messages

- From the examples presented from a range of different countries outlined in this chapter, it is obvious that social factors are extremely relevant in almost all instances of land use and management.
- However, attempts to investigate and quantify these social factors as part of long-term environmental monitoring are not common, although social data and monitoring data are sometimes combined in the analysis phase linked to monitoring programmes.
- The importance of taking an integrated approach towards the management of the rural environment and incorporating local knowledge as part of that process has been recognized for some time. In rural policymaking, it has been recognized that there is a need to work with those who manage, use, or own land and other rural resources to try and understand stakeholder perceptions and practices around rural issues, in addition to traditional scientific data collection (Harrison and Burgess 2000; Stuiver et al. 2004; Toogood et al. 2004; Hooper 2005).
- Whilst some attempts to achieve this exist and have been described in this chapter, it is often a costly and complex undertaking (especially at scale), and there is still a long way to go before social and environmental data are fully integrated in ways that can help to address socioecological change (for example, see Austin et al. 2015).

Study questions

1   Why do you think it was important to enrol local knowledge for the pollution monitoring in the Loweswater catchment area?
2   How can scenario analyses contribute to the integration of social data in monitoring, and how do results benefit from it?
3   Why is it important to use the traditional ecological knowledge and cultural salience of the local population, and what does that bring to our knowledge when we want to understand possible future outcomes?

4   Considering what you know about national ecological monitoring programmes across Europe, what ideas might you put forward to incorporate additional social science into these existing schemes?

5   Given limited resources in low-income countries, can you think of additional ways of collecting social science data in relation to environmental monitoring in a cost-effective way?

Further reading

Balée, W. and Nolan, J.M. (2019) Freelisting as a tool for assessing cognitive realities of landscape transformation: a case study from Amazonia, in Isendahl, C. and Stump, D. (eds) *The Oxford Handbook of Historical Ecology and Applied Archaeology*. Oxford: Oxford University Press, pp. 366–387.

Horstkotte, T., Holand, Ø., Kumpula, J. and Moen, J. (eds) (2022) *Reindeer Husbandry and Global Environmental Change; Pastoralism in Fennoscandia*. Routledge.

Potter, C. and Lobley, M. (1996) The farm family life cycle, succession paths and environmental change in Britain's countryside. *Journal of Agricultural Economics* 47(1–4), 172–190. doi: https://doi.org/10.1111/j.1477-9552.1996.tb00683.x

References

Ágústsdóttir, A.M. (2015) Ecosystem approach for natural hazard mitigation of volcanic tephra in Iceland: building resilience and sustainability, *Natural Hazards* 78, 1669–1691. https://doi.org/10.1007/s11069-015-1795-6

Allard, A. (2017) NILS – a nationwide inventory program for monitoring the conditions and changes of the Swedish landscape, in Díaz-Delgado, R., Lucas, R. and Hurford, C. (eds) *The Roles of Remote Sensing in Nature Conservation*. Cham, Switzerland: Springer, pp. 79–90. https://doi.org/10.1007/978-3-319-64332-8_5

Ansell, D., Freudenberger, D., Munro, N. and Gibbons, P. (2016) The cost-effectiveness of agri-environment schemes for biodiversity conservation: a quantitative review, *Agriculture, Ecosystems & Environment* 225, 184–191.

Arnalds, O. (2015) *The Soils of Iceland*. Dordrecht, The Netherlands: Springer. https://doi.org/10.1007/978-94-017-9621-7

Austin, Z., McVittie, A., McCracken, D., Moxey, A., Moran, D. and White, P.C.L. (2015) Integrating quantitative and qualitative data in assessing the cost-effectiveness of biodiversity conservation programmes, *Biodiversity Conservation* 24, 1359–1375. https://doi.org/10.1007/s10531-015-0861-4

Balée, W. and Nolan, J.M. (2019) Freelisting as a tool for assessing cognitive realities of landscape transformation: a case study from Amazonia, in Isendahl, C. and Stump, D. (eds) *The Oxford Handbook of Historical Ecology and Applied Archaeology*. Oxford: Oxford University Press, pp. 366–387.

Bernard, H.R. (1994) *Research Methods in Anthropology: Qualitative and Quantitative Approaches*. Thousand Oaks, CA: Sage.

Borgatti, S.P. (1992) *ANTHROPAC 4.983/x*. Columbia, SC: Analytic Technologies.

Countryside Survey. (2022) Home page, https://countrysidesurvey.org.uk (Accessed May 13, 2022).

Economic & Social Research Council. (2022) What is social science?, https://esrc.ukri.org/about-us/what-is-social-science/ (Accessed February 14, 2022).

Erlendsson, E., Edwards, K.J. and Guðrún Gísladóttir, G. (2014) Landscape change, land use, and occupation patterns inferred from two palaeoenvironmental datasets from the Mosfell Valley, SW Iceland, in Davide, Z. and Byock, J. (eds) *Viking Archeology in Iceland: Mosfell Archeological Project Turnhout*. Turnhout, Belgium: Brepols Publishers, pp. 181–192. https://doi.org/10.1484/M.CURSOR-EB.1.102220

Forest Research. (2022) National Forest Inventory, https://www.forestresearch.gov.uk/tools-and-resources/national-forest-inventory/ (Accessed May 13, 2022).

Fridman, J., Holm, S., Nilsson, M., Nilsson, P., Ringvall, A.H. and Ståhl, G. (2014) Adapting national forest inventories to changing requirements – the case of the Swedish National Forest Inventory at the turn of the 20th century, *Silva Fennica* 48(3), 1–29. http://dx.doi.org/10.14214/sf.1095

Gísladóttir, G. (2001) Ecological disturbance and soil erosion on grazing land in southwest Iceland, in Conacher, A.J. (ed.) *Land Degradation*. Dordrecht, The Netherlands: Kluwer Academic, pp. 109–126.

Gísladóttir, G., Erlendsson, E., Lal, R. and Bigham, J. (2010) Erosional effects on terrestrial resources over the last millennium in Reykjanes, southwest Iceland, *Quaternary Research* 73(1), 20–32. https://doi.org/10.1016/j.yqres.2009.09.007

Gov.UK. (2022a) Farm Practices Survey, https://www.gov.uk/government/collections/farm-practices-survey (Accessed November 10, 2022).

Gov.UK. (2022b) June survey of agriculture and horticulture, https://www.gov.uk/agricultural-survey (Accessed November 10, 2022).

Gov.UK. (2022c) Monitor of Engagement with the Natural Environment (MENE). https://www.gov.uk/government/collections/monitor-of-engagement-with-the-natural-environment-survey-purpose-and-results (Accessed February 14, 2022).

Halldórsson, G., Ágústsdóttir, A.M., Aradóttir, Á.L., Arnalds, Ó., Hagen, G., Mortensen, L., Nilsson, C., Óskarsson, H., Pagneux, E., Pilli-Sihvola, K., et al. (2017) *Ecosystem Restoration for Mitigation of Natural Disasters*. Nordic Council of Ministers.

Harrison, C. and Burgess, J. (2000) Valuing nature in context: the contribution of common-good approaches, *Biodiversity & Conservation* 9(8), 1115–1130. https://doi.org/10.1023/A:1008930922198

Hooper, B. (2005) *Integrated River Basin Governance*. London: IWA Publishing.

Horstkotte, T., Kumpula, J., Sandström, P., Tømmervik, H., Kivinen, S., Skarin, A., Moen, J. and Sandström, S. (2022) Effects of other land users and the environment, in Horstkotte, T., Holand, Ø., Kumpula, J. and Moen, J. (eds) *Reindeer Husbandry and Global Environmental Change; Pastoralism in Fennoscandia*. Routledge, pp. 76–98. https://doi.org/10.4324/9781003118565

Keddem, S., Barg, F.K. and Frasso, R. (2021) Practical guidance for studies using freelisting interviews, *Preventing Chronic Disease* 18, 200355. http://dx.doi.org/10.5888/pcd18.200355

McCracken, M.E., Woodcock, B.A., Lobley, M., Pywell, R.F., Saratsi, E., Swetnam, R.D., Mortimer, S.R., Harris, S.J., Winter, M., Hinsley, S., et al. (2015). Social and ecological drivers of success in agri-environment schemes: the roles of farmers and environmental context, *Journal of Applied Ecology* 52(3), 696–705. https://doi.org/10.1111/1365-2664.12412

Meireles, M.P.A., de Albuquerque, U.P. and de Medeiros, P.M. (2021) What interferes with conducting free lists? A comparative ethnobotanical experiment, *Journal of Ethnobiology and Ethnomedicine* 17(1), 4. doi: 10.1186/s13002-021-00432-5

Mills, J., Gaskell, P., Reed, M., Short, C.J., Ingram, J., Boatman, N., Jones, N., Conyers, S., Carey, P., Winter, M., et al. (2013) *Farmer Attitudes and Evaluation of Outcomes to On-Farm Environmental Management*. Report to Defra, Gloucester.

NewSeed IT Solutions. (2002) Allemansappen, appen för Sveriges alla turer till nature [Allemansappen, the app for all Swedish treks into nature], https://allemansappen.se/home/ (Accessed November 10, 2022).

Ohlsson, B., Sandewall, M., Sandewall, R.K. and Phon, N.H. (2005) Government plans and farmers intentions – a study on forest land use planning in Vietnam, *Ambio* 34, 248–255.

Potter, C. and Lobley, M. (1996) The farm family life cycle, succession paths and environmental change in Britain's countryside, *Journal of Agricultural Economics* 47(1–4), 172–190. doi: https://doi.org/10.1111/j.1477-9552.1996.tb00683.x

Sami Parliament. (2022) Rennäringen i Sverige [Reindeer husbandry in Sweden], https://www.sametinget.se/rennaring_sverige (Accessed February 11, 2022).

Sandström, P., Cory, N., Svensson, J., Hedenäs, H., Jougda, L. and Brochert, N. (2016) On the decline of ground lichen forests in the Swedish boreal landscape – implications for reindeer husbandry and sustainable forest management, *Ambio* 45(4), 415–429. https://doi.org/10.1007/s13280-015-0759-0

Sarkki, S., Ivsett Johnsen, K., Löf, A., Pekkarinen, A.-J., Kumpula, J., Rasmus, S., Landauer, M. and Åhman, B. (2022) Governing maximum reindeer numbers in Fennoscandia, in Horstkotte, T., Holand, Ø., Kumpula, J. and Moen, J. (eds) *Reindeer Husbandry and Global Environmental Change; Pastoralism in Fennoscandia*. London: Routledge, pp. 173–187. https://doi.org/10.4324/9781003118565

Sigurmundsson, F., Gísladóttir, G. and Erlendsson, E. (2021) The roles of agriculture and climate in land degradation in southeast Iceland AD 1700–1900, *Geografiska Annaler: Series A, Physical Geography* 103(2), 132–150. https://doi.org/10.1080/04353676.2020.1813985

Sigurmundsson, F., Gísladóttir, G. and Óskarsson, H. (2014) Decline of birch woodland cover in Þjórsárdalur Iceland from 1587 to 1938, *Human Ecology* 42(4), 577–590. https://doi.org/10.1007/s1 0745-014-9670-8

Stuiver, M., Leeuwis, C. and van der Ploeg, J.D. (2004) The power of experience: farmers' knowledge and sustainable innovations in agriculture, in Wiskerke, J.D. and van der Ploeg, J.D. (eds) *Seeds of Transition*. Assen, The Netherlands: Royal Van Gorcum, pp. 93–118. https://library.wur.nl/ WebQuery/wurpubs/fulltext/338074 (Accessed February 14, 2022).

Ståhl, G., Allard, A., Esseen, P-A., Glimskär, A., Ringvall, A., Svensson, J., Sundquist, S., Christensen, P., Gallegos Torell, Å., Högström, M., Lagerqvist, K., Marklund, L., Nilsson, B., and Inghe, O. (2011) National Inventory of Landscapes in Sweden (NILS) – scope, design, and experiences from establishing a multiscale biodiversity monitoring system, *Environmental Monitoring and Assessment* 173, 579–595. https://doi.org/10.1007/s10661-010-1406-7

Swedish Code of Statutes (Svensk Författningssamling). (1971) *Reindeer Husbandry Act (Rennäringslag)*. *Svensk Författningssamling*. Stockholm: Allmänna Förlaget.

Swedish Environmental Protection Agency. (2013) Definitions of the 16 environmental quality objectives, https://www.regeringen.se/49bbb6/contentassets/179e266882ad451f9bbe28160c7e348f/ the-swedish-environmental-objectives-system-m2013.01 (Accessed February 14, 2022).

Swedish Environmental Protection Agency. (2019) *Storslagen Fjällmiljö - Underlag till den Fördjupade Utvärderingen av Miljömålen* [Magnificent Mountain Environment – Basis for the In-Depth Evaluation of the Quality Goal], Rapport 6872. Stockholm: Swedish Environmental Protection Agency.

Swedish Environmental Protection Agency. (2022) The National Land Cover Database, NMD, https:// www.naturvardsverket.se/en/services-and-permits/maps-and-map-services/national-land-cover-database?_t_hit.id=Boilerplate_Episerver_Features_EpiserverFind_Models_EpiserverFindDocument/ 8461_en&_t_q=land%20cover%20database (Accessed November 9, 2022).

The Swedish Forest Agency. (2021) *Skogliga Konsekvensanalyser 2022 – Bakgrund och Motiv till val av Scenarier* [Forest Scenario Analyses 2022 – Background and Motives to the Choices of Scenarios], Report 2021/6. https://www.skogsstyrelsen.se/globalassets/om-oss/rapporter/rapporter-202120 2020192018/rapport-2021-6-skogliga-konsekvensanalyser-2022--bakgrund-och-motiv-till-val-av-scenarier-.pdf (Accessed November 10, 2022).

The Swedish Forest Agency. (2022) Forest Scenario Analyses (Skogliga Konsekvensanalyser; SKA22), https://skogsstyrelsen.se/ska22 (Accessed November 10, 2022).

Toogood, M., Gilbert, K. and Rientjes, S. (2004) *Farmers and the Environment. Assessing the Factors That Affect Farmers' Willingness and Ability to Cooperate with Biodiversity Policies*. Biofact, European Centre for Nature Conservation, Wageningen, the Netherlands. https://www.researchgate.net/profile/Mark-Toogood-2/ publication/237115457_Farmers_and_the_Environment_Assessing_the_Factors_That_Affect_Farmers%27_ Willingness_and_Ability_to_Cooperate_with_Biodiversity_Policies/links/55e6c2ff08aea6823a7148d6/ Farmers-and-the-Environment-Assessing-the-Factors-That-Affect-Farmers-Willingness-and-Ability-to-Cooperate-with-Biodiversity-Policies.pdf (Accessed February 14, 2022).

UNESCO. (2022) Policy (HELP), https://en.unesco.org/themes/water-security/hydrology/ programmes/help (Accessed February 14, 2022).

Waterton, C., Norton, L. and Morris, J. (2006) Understanding Loweswater: interdisciplinary research in practice, *Journal of Agricultural Economics* 57(2), 277–293. https://doi.org/10.1111/j.1477-9552. 2006.00052.x

Wikström, P., Edenius, L., Elfving, B., Eriksson, L.O., Lämås, T., Sonesson, J., Öhman, K., Wallerman, J., Waller, C. and Klintebäck, F. (2011) The Heureka forestry decision support system: an overview, http:// mcfns.com/index.php/Journal/article/view/MCFNS.3-87 (Accessed February 14, 2022).

11 Understanding the social context of monitoring

E. Carina H. Keskitalo and Gun Lidestav

Introduction

Monitoring has, to a relatively limited extent, been concerned with social context, or the parameters that influence or even direct land use, both socially and economically as well as politically (Lindenmayer and Likens 2010). The focus of monitoring has instead often been the natural environment, based on historical assumptions on the environment as separate from society. Today, however, when the greatest impact on the environmental systems is inflicted by humans, it might not be possible to properly understand the monitored changes without also understanding the land use that affects them as well as the underlying drivers of land use change (see e.g. West 2016). As natural systems become increasingly impacted by humans, understanding social systems will play an increasing role in understanding land use change (e.g. Parr et al. 2003; West 2016).

This chapter aims to illustrate the importance of including social study techniques for the purpose understanding land use, allowing for an extension of the use and applicability of, as well as methods and data available for, monitoring. However, in addressing social issues, there are a number of factors that need to be taken into account. Among these is the need for a thorough understanding of issues relevant to social context and how these differ from a natural sciences context. Natural science may often consider social context in relation to direct participation on an individual level (for instance, in landscape studies, cf. Reed et al. 2016); however, social systems are so complex that it is almost impossible to conceive of them only on this level.[1] Instead, social systems comprise a number of actors from local to regional, national, and international levels – not least the legal drivers that are discussed in chapter 3. Further, social systems are interconnected. This means, for instance, that regulation on a higher level may delimit what can be done at lower levels – and consequently that the understanding of any one of these levels also requires an understanding of the other levels, including how any chosen lower-level case is impacted. Further, the monitoring activity itself is usually initiated and part of a system to inform policy and management (Parr et al. 2003), which is important to understand in order to contribute to it.

To correctly address social issues, it is essential to understand how individual stakeholders relate to each other as well as how they are integrated into different social systems (e.g., Keskitalo et al. 2016). Thus, this chapter will discuss the role of scale, the different actors in social systems, and the linkages between social actors or organizations and land use, while taking examples from different countries. In the chapter, the term *social* is generally applied in regard to social, political, economic, and cultural contexts, while also acknowledging that these contexts have been formed historically.

DOI: 10.4324/9781003179245-11

Understanding the spatial and temporal scale of social context

Social science generally acknowledges the role of systems beyond the individual in steering individual action, which means that individuals cannot be understood outside their context. Thus, the organizations or institutions they are part of or relate to, which over time may even have formed what different individuals value, should be considered. As a result of this, individuals may also have naturalized certain – organizationally or even nationally used – decision-making procedures or understandings. Seeing them as a given, an individual may not even think to mention them in direct participation. In addition, frequently used terms, such as *environment*, also have no clear, unambiguous meaning but rather reflect a particular or dominant discourse (Parr et al. 2003). Though this may sound abstract, it has extremely practical implications; for instance, an individual may not think to mention all factors that might play a role if they are asked what the influences on their land use are (e.g. Manicas 2006; Thornton et al. 2012). Instead, it may be necessary to tailor one's questions to understand what they usually do, supplementing this knowledge with studies of policy or legislation, or the like (for instance, literature reviews).

In social science, a number of different terms can be used to address the multiple social functions that individuals or organizations can serve. The wider focus takes in not only stakeholders who are seen as affected ("holding a stake") in an issue but also what actors may influence the issue. Who are the actors who can actually make decisions that influence the issue? The number and variety of groups that become relevant to, for instance, land use change thereby increase substantially from those included in the narrower focus on only those local stakeholders directly affected by an issue. Thus, there is much more to social context than merely acknowledging that a variety of actors may be stakeholders (cf. Keskitalo 2004).

Taken together, this means that applying direct participation methods among stakeholders is not sufficient for understanding social context – either on a sector level or elsewhere. What is more, the number and variety of groups that play a role in influencing land use change are seldom only local. More purely natural science studies and long-term ecological research (LTER) working at local scale may sometimes include only local actors, with a focus on, for instance, local co-management (cf. West 2016; Dick et al. 2018). However, seen from a decision-making perspective, very few decisions that affect larger-scale land uses are actually made at the local level. Though the local level, or any specific local case, may be one where problems manifest themselves, the solution may not necessarily be local (e.g. Keskitalo et al. 2016). Local co-management or local land use decision making is instead necessarily governed by, and will potentially also need to be analyzed on, higher levels. Most clearly, this can be illustrated in the types of legislative and policy drivers of land use also discussed in chapter 3.

Legislative and policy drivers of land use

Land use is typically governed through national (sometimes provisional) frameworks of legislation, regulation, policy, and practices reflecting a historically developed system of rights distribution. As historically formed in the context of different interests, it is often not "neutral": it may, for instance, favour one land use over another. Legislation as well as practices are also by their nature often difficult to change. Agreement and, indeed, political push from several types of actors on several scales, not least the national scale, may be needed to induce change. It is also often recognized that larger changes need to be seen as urgent and topical, as well as needing to be pushed by different influential actors and with available solutions, if they are to take place (e.g. Kingdon 1995).

Regulation and policy, though potentially easier to shift than legislation, are also often in line with the general logics and assumptions that permeate legislation. Practices, or "the ways that people do things", may sound simple to change but typically are not, because they are often historically developed and in line with other logics in the systems. They are often naturalized and taken for granted, which means that it is often difficult to conceive of changes to them (e.g. Thornton et al. 2012). Because of this, it is often also not possible for only some actors to change. Other actors and related systems or sub-systems that still act to preserve their original purpose or with the assumptions they have naturalized may not even accept or see changes as relevant and may even make the actors who are making changes seem incompetent or unknowing (e.g. Liebowitz and Margolis 1995).

Problems that manifest locally may also be created at higher levels because, for instance, international trade and market benefits may create the conflicts around land uses that manifest themselves locally (see Text box 11.1). In this regard, change, or solving a land use change problem, can seldom only be local but is rather embedded not only in legislative drivers such as European Union (EU) and national (and regional) legislation but also in policy and practice.

Text box 11.1:   The role of international trade in the local spread of invasive species

Not only legislative drivers but also how an international or national organization is set up may influence the way in which biodiversity protection can be managed. In particular, international trade and World Trade Organisation (WTO) rules mean that it is difficult to exempt trade in plants as long as it is not proven that the specific plant and location of origin (for instance, the soil that accompanies the plant) may result in invasive species spread. This means that even if different actors want to limit the trade in live plants, they still need to work within WTO rules (Pettersson and Keskitalo 2012; Pettersson et al. 2016).

This was a limiting factor when the Montesclaros Declaration, forwarded by scientists, suggested that trade in live plants be halted. What was not recognized here was that no national trade was permitted to change on its own but was made subject to regulation at other – higher – levels in the system (cf. Mackay et al. 2017).

As a result, managing invasive species spread may in some cases need to place more emphasis on monitoring and detection as well as persuading people not to buy plants in soil from foreign countries (but instead, for instance, grow from seed or not grow at all). In this way, managing invasive species spread may, in some cases, rely more on softer or more localized actions (e.g. Klapwijk et al. 2016; Keskitalo et al. 2018).

The role of international trade in the local spread of invasive species can be seen as an example of how a trade regime that originally formed primarily for trade purposes can impact – and potentially also limit – emerging areas that were not even part of the original concern in developing the trade regime. It may then also limit effective larger-scale change on plant trade.

As a result, a range of actors, owners, and interest groups, from large-scale multinational companies to small-scale owners and users, may potentially influence a solution. These actors and logics – what they organize for and how they define themselves – may also greatly differ between countries. For instance, comparing an industry sector in one country with the same sector in another country, even if the labels used are the same, may not be possible. Due to the different history of forest ownership, for example, a comparison of specific forest owner categories between countries may have to include clarification of the actual meaning in each case (see Text boxes 11.2 and 11.3).

Text box 11.2:   Forest logics: forest industry and forest owners in Sweden in comparison with other countries

Forestry is an industry that has developed largely within a national context, to the extent that forest use and the role of forestry in different countries will differ considerably. Influenced but not completely explained by differences in natural prerequisites, such as growing conditions, transport facilities by waterways, and frozen ground, the industrial structure has established a typical logic and logistic. Much has been written about how individual or family forest owners in Sweden, who possess half of the country's forest land, are far more integrated in a forestry framework than forest owners in other countries (Keskitalo 2017). One reason for this is that many of them, through membership in one of the major forest owner associations, are also part owners of wood processing industries and have access to a variety of management and advisory services (Lidestav and Arvidsson 2012). In addition, the large forestry corporations offer a range of forest services to both organized and non-organized forest owners to supply their industries. This interdependence between large-scale and small-scale forestry is further amplified through the current (but changing) forest policy and the authority (i.e. the Swedish Forest Agency) whose task is to both implement the policy and monitor its outcome.

Today, it is possible to purchase most services needed for forestry, to the extent that even inexperienced, sometimes called "new", forest owners are able to carry out forest management. Because data on forest owners and holdings are available through Sweden's well developed forest data services, the service-offering organizations can search out and approach the new forest owners (e.g. Andersson et al. 2020). Through this system, forest management in Sweden has naturalized an essentially even-aged forest regime, with largely planted forest and large-scale final harvest (clear-cutting). The harvesting operations are largely carried out by contractors and mechanized methods. In planting and pre-commercial thinning, the methods are manual or motor–manual, meaning that forest owners can still engage in this work, although these services can be also purchased (Lidestav et al. 2017). In Sweden, forest ownership is also recognized as a primary role, with land management based in forestry, whereas in other countries forest ownership may often be seen as secondary to agricultural management, with forest management assumed to be only on a small scale (Keskitalo et al. 2017).

Though the Swedish forest owner may thus be far from typical, there are in fact no "typical" forest owners because the category differs so much between countries. The context of forestry also varies: the role of forest owners' associations and the economic role of forestry, in a European context, are perhaps most pronounced in Sweden and Finland (Keskitalo et al. 2017).

However, taking this situation into account means that to understand much of forestry in Sweden, we need to understand it on both the local level and in the institutional or organizational context, as well as within the scope of the service provision that exists. For instance, it would be possible to cover all of the main organizational actors in Swedish forestry through qualitative interviews with some 15 actors (see Andersson and Keskitalo [2018] for an example). These actors include the large national and multi-national forest companies, forest industries and forest owners' associations, and forest management organizations. This means that qualitative interviews, for instance, can be a crucial instrument in understanding land use decisions (see chapter 14 on interview methodology).

Text box 11.3:   The meanings of *forest* and *ownership* are ambiguous

Forest and *ownership* may appear to be straightforward concepts; however, the understandings, definitions, and categorizations of these terms are often guided by the purpose for which they are used (Weiss and Nichiforel 2019). This means that the cross-country comparisons and overviews become complicated but also that longitudinal studies within a country can be challenging. In both cases it is important to be aware that concepts are not always applied consistently. In Sweden, for instance, forest is typically understood as "productive forest land"; that is, land that can produce an average of $1m^3$ of timber per hectare per year. This corresponds to 23 million hectares, or 82% of all forest land according to the Forestry Act definition, which then also includes 5 million hectares of "unproductive forest land". Because harvesting is not allowed on this land, it might just as well have been labelled "protected forest land"; however, along with "other wooded land", it is regarded as unproductive. With history in mind, the Swedish timber production–oriented definition has been useful, but it apparently also has some drawbacks in relation to international reporting on protected areas. Yet, international reporting on forest available for wood supply (FAWS) – that is, similar to the Swedish definition of productive forest land – is increasingly emphasized, and a definition of FAWS has been established for harmonized reporting (Alberdi et al. 2016).

This means that it is important to understand this variety, where also "[f]orest ownership means that different tenure arrangements are based on various combinations of property rights, which can be attributed, formally or informally, to the legitimate holder of the resource or to other resource users" (Weiss and Nichiforel 2019). The most basic categorization and assessment used regarding forest ownership draws a distinction between public and private. However, property rights theory distinguishes five legal ownership categories, also described as "resource regimes": state property, municipal or communal property, common property, private property, and open access property. Depending on the government, statistical bureau, or researcher in question, the resource regimes are interpreted and used differently. For instance, in Bulgaria, the Czech Republic, and Latvia, municipal forests are categorized as private, whereas in Estonia, Poland, and Romania they are classified as public. The significant differences in the understanding of private forests in Europe are illustrated in a recent comparative property rights analysis by Nichiforel et al. (2018). Furthermore, property can be simultaneously owned by several entities, because property ownership is organized through a "bundle of rights" giving the legitimate holder of the resource or other resource users access rights, withdrawal rights, management rights, exclusion rights, and/or alienation rights (Schlager and Ostrom 1992). This means that not only what forest is but what owners can do with it in different contexts will impact who can be asked about what – for instance, in a survey – but also what recommendations based on monitoring may be relevant.

In addition to this, social studies need to take into account the temporal as well as geographical scales. Temporal scale can be seen as being included particularly in disciplines such as history, anthropology, and archaeology. However, it is also well known in monitoring that the way in which specific land uses have developed is often largely dependent on historical factors. For these reasons, it may be very difficult to develop not only monitoring regimes across countries but also comparative studies – for instance, surveys (chapter 13) – across countries. Thus, for the purpose of land use–related research, it is crucial to understand the ways in which different systems have developed in different ways in different countries. Any attempt at direct comparison that ignores differing understandings or definitions of terms (such as *forest owner*) may in fact be like comparing apples to pears (e.g. Keskitalo 2017; see Text box 11.3).

Tailoring the means to the goals

To understand the social side of land use means to be aware of the properties of the systems as they have been described above and then to tailor the means to the ends. Though choices of scale or level in the social sciences, both geographically and temporally, are often made on a disciplinary basis, for monitoring purposes the choice of what levels to focus on – as well as what stakeholders to include – might need to be based on what groups most influence specific choices of land use. The forest management in a given country might then, for instance, also need to include the industry and national or even international levels (as shown in Text Box 11.2), whereas for local cases greater focus may be placed on local forest owners. However, the local case might also need to include a focus on the policy and legislative drivers and actors at higher levels.

To take into account this breadth, the study of social systems comprises both quantitative and qualitative approaches, which are both useful and equally valuable, and the choice (or combination) of which should depend on what one wants to study.

The following chapters in this book will provide examples from different types of social science studies. However, beyond the more specific methods described there, to develop an understanding of the systems, literature review will generally be important to know what has been written on the land use case and sector in focus. Policy documents by the different organizational actors can also be a useful source, although these are used more by some disciplines than others.

Quantitative approaches are regularly described in the literature as being more useful when one knows the exact parameters one is interested in, when they are clearly measurable, and when one knows how to obtain information about them. This would mean that quantitative approaches may be more useful when, for instance, the term or information sought is identified in the same way among those subject to investigation (including potential lay subjects) and when it is possible to measure clearly using more limited parameters. Thus, the following chapters will discuss, among other topics, the use of register data to understand property owners (chapter 12). Another common approach to gaining statistical information on groups is the use of surveys. These regularly aim at statistical sampling (but can also be used for smaller populations) and can generally be a way to understand priorities; for instance, in land use across a population (chapter 13).

Along the same lines, qualitative approaches are more usable when one is dealing with issues that may be identified differently by different participants in or subjects of research, who may then speak about them in different ways. A later chapter will discuss the use of interviews to obtain direct information from stakeholders in a way that, for instance, strategically selects groups or individuals who are the most relevant to a particular study (chapter 14). The focus here is not on generalization to a population but rather to theory; that is, on identifying the main factors that, for instance, influence land use decisions. For this reason, qualitative studies can be undertaken prior to quantitative studies, to ensure that all of the relevant factors one wants to cover in a larger survey are included and not missed (which might otherwise bias the results). However, qualitative studies can also be undertaken to gain a deeper understanding of specific experiences – and what is meant by specific factors, for instance – than quantitative studies allow for.

Text box 11.4 discusses the choice of approaches depending on what one wants to study with an application in the Swedish case, and Text box 11.5 contrasts the Swedish case with the Italian case in relation to how the forestry sector is organized.

Text box 11.4:   Designing a study of the forestry sector in Sweden: different approaches depending on aim

In the Swedish case (see Text box 11.2), the forest industry sector can be seen as comprising relatively few organized actors with great influence. Because these are highly organized, they can be understood by reference to legislation, regulation, and policy. They also produce their own policy papers that can be analyzed. The small number of actors means that selected functions can be interviewed in their entirety (full study of all relevant interviewees/functions in the country). Such a study has been undertaken by, for instance, Andersson and Keskitalo (2018).

By contrast, the small-scale private forest owners in Sweden are many and varied. Owning about half of all Swedish forest, the ways in which forestry is practised on their properties can potentially be studied or understood with reference to forest industry practices (as discussed in Text box 11.2). Their roles can also be understood with reference to legislation, regulation, and policies in which their roles are determined, as well as in relation to practices influencing land use through, for instance, questionnaires or interviews. Official data may be available in registers (see chapter 12), and questions addressed to the forest owner could be asked in generalized questionnaire studies. However, because the variation among forest owners is considerable, some may not even see themselves primarily as forest owners, meaning that some land use descriptors may not be familiar to them (see chapter 13).

If we are instead interested in the values forest owners hold, why they retain their properties even if they do not live on them, or the like, we may – for instance, by drawing from potential registers of forest owners – design interview studies for a maximal diversity in a given region (see chapter 14). An example of an interview study targeting individual forest owners in two areas of Sweden, with maximum variation in each, is Bergstén and Keskitalo (2018).

Text box 11.5:   Contrasting the Swedish and Italian forestry systems

If Sweden is very much a forestry-oriented system, Italy is perhaps less so. Forest holdings are regularly far smaller there than in the Swedish case and are perhaps much more often focused on values other than forestry per se, such as the food value of forests. The economic role of forestry in the country's gross domestic product is also comparatively far smaller, which has perhaps historically provided less of an impetus for developing the full forestry production system that is apparent in Sweden (see Text box 11.3; cf. Keskitalo et al. 2013, 2015).

As a result, developing an understanding of the forest-relevant occupations, forest use and values people may attach to forest, as well as the organizational context that governs this, might need to relate to far different actors in Italy than in Sweden. Social study in this context would need to be aware of these differences and not assume that solutions or policy recommendations in one context would be transferrable to the other – as is also illustrated in other chapters in regard to the integration of monitoring systems across countries.

Summing up

This chapter has been dedicated to showing why it is important to understand the social context with regard to, for instance, scale or variation between countries or land use systems in how they have developed. Stakeholders often cannot simply tell you about their system, because they may have naturalized it and will not know what to say. Social scientists, on the other hand, specialize in the description and analysis of these types of systems and in drawing out the type of information people within the systems may simply see as given. Direct stakeholder interaction without connected social analysis is thus no substitute for comprehensive social analysis using the tools that will be discussed in the following chapters in this volume. Designing social studies – or using existing social data – would thereby preferably take in knowledge of land use change drivers from the variety of social levels that influence the case. Designing this type of study or identifying the relevant social data would also need to be based on an understanding that one case may differ greatly from another.

Key messages

- It is important to understand how land use is governed and where decisions are taken to include all relevant levels for understanding land use. Local stakeholder studies should thus often be added to by including information from larger-scale studies and contexts.
- Cases will often vary widely due to their historical development. This both makes it difficult to design comparative monitoring programmes and means that social study or country comparisons need to be sensitive to how different terms are used in different countries (to avoid comparing apples to pears) and to how social context and land use drivers may differ. Being aware of these differences is key to developing comparative approaches that are sensitive to these variations.
- Making use of social studies and data may include reviewing what micro- or statistical data may be available, as well as what survey or interview data in specific areas may be available or working with social scientists in different disciplines to design studies.

Study questions

1   What are the types of factors one would need to take into account when designing a study of biodiversity in forests, to cover both spatial and temporal levels or scales from a social perspective?
2   Why is it important to not only speak with local stakeholders to understand what actions could be undertaken, for instance, to increase biodiversity?
3   Why it is important to be aware that basic concepts – for example, *forest* and *ownership* – are not always understood in the same way by different stakeholders or applied consistently in different surveys or monitoring programmes?

Note

1   This type of problem has been highlighted by Beck (2011) and Durant (2015), among others.

Further reading

Beck, S. (2011) Moving beyond the linear model of expertise? IPCC and the test of adaptation, *Regional Environmental Change* 11(2), 297–306.

This article problematizes some of the assumptions that have often been made on social knowledge in relation to decision making.

References

Alberdi, I., Michalak, R., Fischer, C., Gasparini, P., Brändli, U.-B., Tomter, S.M., Kuliesis, A., Snorrason, A., Redmond, J., Hernández, L., et al. (2016) Towards harmonized assessment of European forest availability for wood supply in Europe, *Forest Policy and Economics* 70, 20–29.

Andersson, E. and Keskitalo, E.C.H. (2018) Adaptation to climate change? Why business-as-usual remains the logical choice in Swedish forestry, *Global Environmental Change* 48(1), 76–85.

Andersson, E., Keskitalo, E.C.H. and Westin, K. (2020) Managing place and distance: restructuring sales and work relations to meet urbanisation-related challenges in Swedish forestry, *Forest Policy and Economics* 118, 1–7.

Beck, S. (2011) Moving beyond the linear model of expertise? IPCC and the test of adaptation, *Regional Environmental Change* 11(2), 297–306.

Bergstén, S. and Keskitalo, E.C.H. (2018) Feeling at home from a distance? How geographical distance and non-residency shape sense of place among private forest owners, *Society & Natural Resources* 32(3), 1–20.

Dick, J., Orenstein, D.E., Holzer, J.M., Wohner, C., Achard, A.L., Andrews, C., Avriel-Avni, N., Beja, P., Blond, N., Cabello, J., et al. (2018) What is socio-ecological research delivering? A literature survey across 25 international LTSER platforms, *Science of the Total Environment* 622, 1225–1240.

Durant, D. (2015) The undead linear model of expertise, in Heazle, M. and Kane, J. (eds) *Policy Legitimacy, Science and Political Authority: Knowledge and Action in Liberal Democracies*. London: Routledge, pp. 17–38.

Keskitalo, E.C.H. (2004) A framework for multi-level stakeholder studies in response to global change, *Local Environment* (Special Issue) 9(5), 425–435.

Keskitalo, E.C.H. (ed.) (2017) *Globalisation and Change in Forest Ownership and Forest Use: Natural Resource Management in Transition*. Basingstoke, UK: Palgrave Macmillan.

Keskitalo, E.C.H., Horstkotte, T., Kivinen, S., Forbes, B. and Käyhkö, J. (2016) "Generality of misfit"? The real-life difficulty of matching scales in an interconnected world, *Ambio* 45(6), 742–752.

Keskitalo, E.C.H., Legay, M., Marchetti, M., Nocentini, S. and Spathelf, P. (2015) The role of forestry in national climate change adaptation policy: cases from Sweden, Germany, France and Italy, *International Forestry Review* 17(1), 30–42.

Keskitalo, E.C.H., Lidestav, G., Karppinen, H. and Zivojinovic, I. (2017) Is there a new European forest owner? The institutional context, in Keskitalo, E.C.H. (ed) *Globalisation and Change in Forest Ownership and Forest Use: Natural Resource Management in Transition*. Basingstoke, UK: Palgrave Macmillan, pp. 17–56.

Keskitalo, E.C.H., Nocentini, S. and Bottalico, F. (2013) Adaptation to climate change in forest management: what role does national context and forest management tradition play?, in Lucas-Borja, M.E. (ed.) *Forest Management of Mediterranean Forest under the New Context of Climate Change*. New York: Nova Science Publishers, pp. 149–161.

Keskitalo, E.C.H., Strömberg, C., Pettersson, M., Boberg, J., Klapwijk, M., Oliva Palau, J. and Stenlid, J. (2018) Implementing plant health regulations with focus on invasive forest pests and pathogens: examples from Swedish forest nurseries, in Urquhart, J., Marzano, M. and Potter, C. (eds) *The Human Dimensions of Forest and Tree Health. Global Perspectives*. London: Palgrave-Macmillan, pp. 193–210.

Kingdon, J.W. (1995) *Agendas, Alternatives and Public Policies*. 2nd edn. New York: HarperCollins.

Klapwijk, M.J., Hopkins, A.J.M., Eriksson, L., Pettersson, M., Schroeder, M., Lindelöw, Å., Rönnberg, J., Keskitalo, E.C.H. and Kenis, M. (2016) Reducing the risk of invasive forest pests and pathogens: combining legislation, targeted management and public awareness, *Ambio* 45(2), 223–234.

Lidestav, G. and Arvidsson, A.-M. (2012) *Member, Owner, Customer, Supplier? – The Question of Perspective on Membership and Ownership in a Private Forest Owner Cooperative*. Rijeka, Croatia: INTECH.

Lidestav, G., Thellbro, C., Sandström, P., Lind, P., Holm, E., Olsson, O., Westin, K., Karppinen, H. and Ficko, A. (2017) Interactions between forest owners and their forests, in Keskitalo, E.C.H. (ed.) *Globalisation and Change in Forest Ownership and Forest Use. Natural Resource Management in Transition*. London: Palgrave Macmillan, pp. 57–96.

Liebowitz, S.J. and Margolis, S.E. (1995) Path dependence, lock-in, and history, *Journal of Law, Economics, & Organization* 11(1), 205–226.

Lindenmayer, D.B. and Likens, G.E. (2010) The science and application of ecological monitoring, *Biological Conservation* 143(6), 1317–1328.

Mackay, H., Keskitalo, E.C.H. and Pettersson, M. (2017) Getting invasive species on the political agenda: agenda-setting and policy formulation in the case of ash dieback in the UK, *Biological Invasions* 19(7), 1953–1970.

Manicas, P.T. (2006) *A Realist Philosophy of Social Science: Explanation and Understanding.* Cambridge: Cambridge University Press.

Nichiforel, L., Keary, K., Deuffic, P., Weiss, G. Jellesmark Thorsen, B., Winkel, G., Avdibegovićh, M., Dobšinská, Z., Feliciano, D., Gatto, P., et al. (2018) How private are Europe's private forests? A comparative property rights analysis, *Land Use Policy* 76, 535–552.

Parr, T.W., Sier, A.R.J., Battarbee, R.W., Mackay, A. and Burgess, J. (2003) Detecting environmental change: science and society – perspectives on long-term research and monitoring in the 21st century, *The Science of the Total Environment* 310(2003), 1–8.

Pettersson, M. and Keskitalo, E.C.H. (2012) Forest invasive species relating to climate change: the EU and Swedish regulatory framework, *Environmental Policy and Law* 42(1), 63–73.

Pettersson, M., Strömberg, C. and Keskitalo, E.C.H. (2016) Possibility to implement invasive species control in Swedish forests, *Ambio* 45(2), 214–222.

Reed, J., van Vianen, J., Deakin, E.L., Barlow, J. and Sunderland, T. (2016) Integrated landscape approaches to managing social and environmental issues in the tropics: learning from the past to guide the future, *Global Change Biology* 22, 2540–2554.

Schlager, E. and Ostrom, E. (1992) Property-rights regimes and natural resources: a conceptual analysis, *Land Economics* 68, 249–262.

Thornton, P.H., Ocasio, W. and Lounsbury, M. (2012) *The Institutional Logics Perspective. A New Approach to Culture, Structure and Process.* Oxford: Oxford University Press.

Weiss, G. and Nichiforel, L. (2019) Concepts and definitions of forest ownership, in UNECE and FAO (eds) *Who Owns Our forests? Forest Ownership in the ECE Region.* Geneva, Switzerland: Economic Commission for Europe, pp. 19–30.

West, S. (2016) *Meaning and Action in Sustainability Science: Interpretive Approaches for Social–Ecological Systems Research.* Doctoral Dissertation, Resilience Centre, Stockholm University.

12 Register data as a resource for analysis

Urban Lindgren and Einar Holm

Introduction

Most chapters of this book elaborate the use of observations of an elusive "nature" to enhance methods and practices in monitoring; for example, biodiversity status and development. As effectively demonstrated in the other chapters, it is a challenging task to explore and develop such a methodology. The target is often to measure the prevalence distribution of some biological species within and between certain areas. Results are often descriptively presented as a specific land cover map.

Beyond that, but just around the corner, more sociopolitical questions appear, such as who has the power to define a monitoring question and allocate resources for that monitoring to actually get off the ground? Who will use the monitoring results to support their own or others' interests? How much of the observed monitored outcomes in land cover is contingent on practices in land use by micro land users (like farmers or tourists)? Or is the outcome mainly the result of structural social forces like legislation, governmental agencies, or market force agents? Or is the outcome mainly a result of forces of "nature" itself? In that case, it is beyond human reach to influence the outcome.

A monitoring device is rarely set up and specified independent of current interests manifested by human agency. The target, the ambition, and the data, the methods, the budget, and personnel, are often mainly determined by organizations external to the core knowledge production unit, including different policy agencies.

A part of the human influence on outcome (e.g. drilling deep wells) as well as the environment's return impact on humans (e.g. desertification, food shortage, etc.) can, however, be studied by including human behaviour in the analysis. This requires data on individuals and an estimation of, for example, the relation between behaviour and distribution of environmental impact. This chapter provides a few examples of such studies. However, we mainly present some available data sources in the social, human end and leave it to the user to formulate and explore the huge amount of potential analysis emerging from combining social and monitoring data to find new answers.

The chapter draws on the example of Sweden, because the types of register data including register-based microdata and their organization differ a lot between countries (see Appendix 3 for a definition of these terms). We hope that this presentation illustrates the types of data and data uses that can be applied in other cases. Sweden is also an interesting case because population and land use registers go back a long time in history. The early population statistics (Tabellverket) covering the entire population started in 1749, and since then censuses have been carried out on a regular basis every five or ten years. Since 1990, this task has been administered by Statistics Sweden (SCB). The state had good reasons to keep detailed records

DOI: 10.4324/9781003179245-12

of the populace, not least for taxation of individuals, discharge to military services, and church records. In a pre-industrial agrarian society, agricultural production was one of the corner-stones of the economy, which for taxation purposes provided an incentive for the State to keep an active account of all agricultural properties (*hemman*, "homestead") and their pro-duction output. From the 16th century onwards, the list of agricultural properties (jordeböcker) was gradually developed to a complete property register. In the early 1900s this type of taxation – based on, for example, assessment unit of land (*mantal*) – was replaced by a property tax like the one abolished in 2007. The long and extensive history of collecting information on people, land, and many other domains has brought rich longitudinal records of great importance for the analysis of various issues and challenges in society. Examples made in this chapter will mainly be drawn from forest lands in Sweden, because this is the area most known to the authors, but there may exist corresponding types of data for other land uses.

The chapter is organized in three parts. In the first part (Aggregated data from Swedish sources), we describe some of the publicly available resources at Statistics Sweden and elsewhere containing data about land use and population. One of these resources is the online tool of Statistics Sweden designed for users making their own tables of aggregated statistics (*statistikdatabasen*), which, for example, may be a very helpful resource for students working on assignments and dissertations.

In the second part (Microdata), we broaden the perspectives by discussing some drawbacks of using aggregated data and how these issues may be ameliorated using mi-crodata. We also briefly address legal regulations that govern the use of microdata and procedures of data storage. Investigations based on such detailed data, whether from tables or individual records, only cover a fraction of what could be analysed using available data.

In the last part (Applications based on individual-level microdata), we present a mod-elling approach inspired by the Swedish geographer Torsten Hägerstrand and his devel-opment of time geography, which is a geographical and philosophical perspective on how people interact with each other and the environment. Spatial dynamic microsimulation models represent populations on an individual level and are developed to increase our understanding of demographic and socioeconomic patterns and processes. Microsimulation models require individual-level register data covering the entire population to produce policy-relevant results. Additionally, the models rely on a variety of specialised investigations from many different fields of research informing about human behaviour and geographical and economic contexts. Therefore, we provide several recent examples of research beneficial to the development of microsimulation models (see also Appendix 4). At the end of this part, an application of microsimulation models is presented.

Aggregated data from Swedish sources

The reason for collecting any data, whether on a micro or macro level, is the belief that it facilitates the response to questions related to certain topics and issues. Monitoring such a situation or chain of events simply requires appropriate observational data. Is this forest a coal sink or not? For instance, does forest ownership influence the propensity to move to the owned property or not? Such descriptive questions are sometimes justified by and embedded in a broader monitoring question connected to cause and effect and of consequences of certain activities and policy alternatives. For example, will increased cutting increase or decrease carbon dioxide emissions – in the short run or the long run? Will changed reg-ulations regarding individuals' opportunities to acquire forest properties affect migration patterns and investment incentives beneficial to the development of forest properties?

Some examples of Swedish and global open data sources

The main source of aggregate table data covering demographic and socioeconomic information for Sweden is the Statistical Database provided by Statistics Sweden (2022). Most public and many private agencies are obliged to report their activities and outcomes to SCB, and a large part of available aggregate data is made accessible by SCB's publications and web services. Other agencies such as the Swedish Tax Agency usually collect the information. SCB mostly uses data collected by other agencies for statistical purposes. This is quite different from many other countries where statistics agencies have a more direct responsibility for collecting data.

For rapid access, a large set of ready-made tables is available, but they are largely covered by the more flexible statistical database, which, due to its importance for studies targeting Sweden, is described in some detail below.

The main menu contains entry points like labour market, population, housing, democracy, energy, finances, trade, economy, health care, agriculture, forestry, fishing, culture, environment, national accounts, transport, education, research, and more. Under each main topic a more specific heading can be chosen. For the topic "population", one can select older census-based data or more recent register-based data. If register is selected, one of about 20 available variables can be selected; for example, "count of people". Then there is a choice of one out of five tables with different sets of variables as dimensions in the resulting frequency cross-table – number of people by region, civil status, age, and sex for the years 1968 to 2020. After selecting a table with certain dimensions, there is a final choice of different definitions of, and ranges for, the selected dimensions. For "region" there is choice of municipalities, counties, all Sweden, different kinds of local labour market regions, and NUTS (Nomenclature of Territorial Units for Statistics) regions for each or a selected subset of the regions. If all dimensions are selected, in this example the output might contain cell frequencies for each combination of municipality (290), year (52), age (100), sex (2), and civil status (4).[1] Thus, relatively detailed cross-tables with four to five dimensions (years included) for one of hundreds of selected variables can be chosen covering a substantial share of the demand for data in many studies.

However, it should be noticed that there is no free choice of surrounding dimensions. Many variables and tables allow for combining the target variable (e.g. educational level) with year, age, and municipality and not much more – and often less. Very few tables that can be selected from the statistical database enable output with combinations of more "rare" variables; for example, education level times profession times municipality times age or origin times migration destination municipality or commuting times age times sex, etc. If you are interested in an analysis of variables related to several interacting covariates be aware that the statistical database was not developed for supporting such analysis; rather, it was developed for simple description.

Another issue is the extent to which it is possible to retrieve the land use and environmental components of monitored and studied processes. For the statistical database the answer is simple. The finest level of spatial resolution offered in most tables is municipality and aggregates of municipality (e.g. county). This includes most aspects of environmental outcome and land use.

SCB has released two fixed neighbourhood-level divisions of the national space with much higher resolution compared to municipalities and parishes – DeSO (*demografiska statistikområden*, "demographic statistical areas") and RegSO (*regionala statistikområden*, "regional statistical areas"). The 5,984 DeSO neighbourhoods replace the older SAMS (small areas for market statistics) division and the DeSOs are entirely defined by SCB. The 3,363 RegSO neighbourhoods (aggregates of DeSO) are suitable for choropleth mapping, and each of them is given a label that is as close as possible to the current name of the neighbourhood. Because the divisions are fixed and corresponding shapefiles are available, mapping with borders, content, and label is straightforward. In the statistical database, a few indicators per DeSO or RegSO neighbourhoods are available: Swedish or foreign background times sex, type of household, citizenship, four birth regions, sex, and civil status. Each is available for the years 2015 to 2020. In addition, SCB offers – as open geodata – population number by square kilometres times five-year age groups times sex. Maps can be produced using GIS software (GeoPackage zip files).

Corresponding information about population, economy, and land use for other countries is available through web services maintained by the United Nations and the World Bank. Eurostat provides additional information for Europe. Tables like those available from SCB and its statistical database are obtainable for many other countries. However, there is usually no finer spatial resolution than entire countries (e.g. urban, rural, city size class). Moreover, for most countries and variables, no annual information is available, and most population data are based on census information. Therefore, data are available only for the census years, which sometimes differ between countries. However, there are fixed "census rounds". Most countries have at least decennial censuses (e.g. 2000, 2010, and 2020 or within one or two years).

Microdata

Microdata describe single units of observations such as persons, forest properties, firms, institutions, etc. Microdata could even be specific objects like individual trees. This type of data originates from different sources and data collection methods such as field observations (e.g. remote sensing, surveillance cameras, mobile phone movements, participant observations), interviews, survey questionnaires, and register data.

A micro database of a population is arranged as a sequence of rows (records, instances), one for each person. Each column contains the value of a certain attribute for each person; for example, "26" in the column denoting age for person 2.

Today the research community has good access to micro-level register data, but this is a rather recent development, partly due to a growing awareness among decision makers of the potential benefits of using this type of data for more refined analyses. More extensive production of annual register data is of a relatively recent date. Generally speaking, in the mid-1980s, Statistics Sweden started large-scale production of annual individual-level demographic and socioeconomic data. This was largely a result of political decisions to shift from census data – which was basically a questionnaire sent to the entire adult population every five years (the last census was carried out in 1990) – to register data. Censuses provide self-reported information, whereas register data originate from various data-collecting authorities (e.g. individual income data stem from income tax returns administered by the Swedish Tax Agency). The abolishment of the recurrent censuses was partly related to the introduction of register data, which includes more or less the same individual-level information. Moreover, self-reported information suffers from imperfections related to non-response and reliability issues.

Why should micro-level data be stored and used? In general, most questions are phrased – and an answer is expected – on a macro level. How old are the trees in this forest stand on average? How many forest cubic metres are there on this property? What proportion of the in-migrants to this municipality are local forest owners? This type of question dominates the background to the tables made accessible by Statistics Sweden and the Forestry Yearbook by the Swedish Forest Agency (Skogsstyrelsen). These publications are expected to provide some answers to questions asked by researchers, politicians, journalists, civil servants, and the general public.

Moreover, when table information is enough to provide answers to a study or a monitoring investigation, it is not certain whether the wanted table dimensions are available through tables possible to construct within a retrieval system like the statistical database. This might also be the case if the required dimension variable is actually present in the micro databases behind the retrieval system. Thus, from a practical point of view, this type of underlying micro database is much more useful as a complete, flexible, and efficient data container compared to a huge amount of aggregated tables. However, this constraint is partly unavoidable because a publicly available retrieval service should not give access to individual information under any circumstances.

Another reason to strive for access to "real" microdata for each individual in the population is that a table says very little about the reasons for change taking place in a situation it describes. In addition, policy is about how to manipulate change to obtain a different preferred outcome. A time series of cuttings per hectare in a municipality does not by itself give a clue to answering the question why cuttings increased. Including several municipalities is helpful, but impacts of factors such as forest age, age of owners, demand, and price are beyond what can be communicated by means of a simple descriptive table. Adding one possible covariate like the age of owners as a table dimension will likely just give a biased exaggerated estimate of the partial impact of owners' ages on cutting intensity, because age is correlated to omitted variables.

Contrary to popular belief, table relations between aggregated aspects of a population do not provide much help in identifying factors influencing change. Statements claiming causality between causes and effects require at least an analysis on the individual level as a necessary but not sufficient condition.

The use of microdata representing personal information of individuals is restricted by law (*EU General Data Protection Regulation* [GDPR] 2016). *Personal information* is defined as any information relating to an identifiable person. Concerning data collection for student assignments, dissertations, and other types of enquiries, GDPR includes requirements with regard to giving information to the research participants, asking for their consent, and informing them about their right to non-participation. Collected data should be carefully stored to avoid dispersal of personal information to the general public. Moreover, this is an important requirement for universities and researchers with access to register-based personal information produced by SCB or other agencies. These types of data repositories must be secured by servers in locked server rooms, and data users and technicians must be authorized for access to computer labs and other facilities according to specific routines. Research using register data including sensitive personal information (e.g. health, ethnicity, political opinions) needs to be approved by an ethics vetting board (Etikprövningsmyndigheten).

Applications based on individual-level microdata

A large fraction of studies based on register-based microdata use individual attributes connected only to a specific person; that is, year, sex, age, *x* and *y* coordinates for the place of residence, educational level, profession, earnings, number of hectares of forest owned, etc. Many studies and detailed monitoring indicators are constructed in this way by utilizing the possibility to freely select and combine such attributes in the analysis. This would not be possible with access only to aggregated table information from publicly available web services like the statistical database.

In addition, in some countries, Sweden included, there is also the possibility to obtain and use certain relational attributes; that is, attributes connecting individuals to each other or to other objects like a place of residence. Despite having large micro databases with an individual identifier, many countries do not store the identifier in several tables to enable combinations. This type of useful information is inherently connected to availability of individual register data for the entire population. So far, such information has not been explored to any large extent despite its potential to enable new traits of knowledge formation. A few suggestions are (1) a pointer (identifier) to oneself. This makes the database longitudinal and enables an entirely different set of statistical methods for exploring contrafactual effects compared to just having a series of disconnected annual cross sections. (2) Pointers to mother, father, family, extended family, schoolmates, workplace, dwelling, or owned forest property. A large set of behaviours is socially and biologically inherited from parents and early living environment. Such pointers give access to all of the mothers', fathers' (and other relevant individuals') own attributes (for example, mother's work income), and the analysis of a certain individual outcome (e.g. educational level) can be made contingent on the combination of one's own and parental attributes.

Microsimulation modelling

Microsimulation models[2] can be applied to further use this type of data and are useful for exploring "what if" questions like: What is the long-term impact on the distribution over individuals and regions of employment, income, wealth, health, forest ownership, carbon release and so on of changed taxes, ageing, migration, improved medical services, and nutrition habits? Microsimulation focuses on research questions dealing with the demographic and socioeconomic development of populations, firms, forests, and other entities. Who will be born, who will obtain an education and employment, and who will start a family over the years to come? What happens with income, wealth, and health? Who will die, when, and where? Who moves from and to specific places? Microsimulation models things like education, labour supply, income distribution, family formation, use of welfare benefits, spatial distribution of the population, settlement structure, residential segregation, etc. What are the impacts on all of these processes of, for example, policy changes in other parts of the system, such as changes in taxes, transfer payments, immigration policy, quality of medical services, attitudes to fertility, nutrition habits, and forest policy?

Such "what if" questions are studied extensively by means of microsimulation. In a paper published in the *International Journal of Microsimulation*, Li and O'Donoghue (2013) defined a dynamic microsimulation model simply as "a model that simulates the behaviour of micro-units over time". With the help of such models, "what if" questions are often answered by comparing outcome of experimental simulations with and without the proposed policy, behaviour, or structural change implemented in the model. Or, differently put, the simulation produces an alternative contrafactual development for a historical period to be compared with observed history for the same period.

A dynamic microsimulation moves an observed start population of individuals with their attributes forward in time – year by year, person by person, and event by event – while updating their existence and attributes by means of rules and estimated transition probabilities. A microsimulation model maintains and projects heterogeneity between individuals. Each person is a main character, interacting with other individuals and the environment. Their individual choices and actions are different and constrained by their own properties, preferences, and abilities and by resources and people in their vicinity. Estimated equations and postulated rules for all intertwined individual events are put to work together and the outcome for each person becomes contingent on the entire web of earlier life events and characteristics.

Microsimulation is about what could have happened or what might happen in the future to a large set of individuals, as is much other research and fictional accounts of individuals and societies of which they are part. Comparing the life biography of a fiction writer's main character with the series of events occurring to the agents in a microsimulation reveals distinguishing features. The talented fiction writer's detailed story, told about the main character, is unbeaten in its internal logic of reason, action, and agency and is often of a larger immediate interest to the reader compared to the story told by the relatively stripped-down sequence of life events produced by a simulation. Therefore, the bulk of biographies, fictional stories, and movies probably have had a much greater impact as decision support as have all microsimulation and other quantitative impact analysis studies together.

But the fiction writer's substantive and communicative advantage comes with a price. The people surrounding the main character cannot be as thoroughly described as the protagonist. They pop in when convenient for the storyline of the main character and then disappear – without background history or a future life. They do not need to have a coherent life history of their own. Therefore, the invented life path of the main character is fiction in two ways. The person's life trajectory relies on interaction, at certain points in time and space, with other persons who eventually could not be there in the right moment with the right set of attributes because of constraints and events in their own life paths. These interactions are easy to manipulate in fiction but harder in life outside the book covers.

In a microsimulation model, however, each person is a main character. Each and every one develops their own life in parallel and interacts occasionally with the others. Their existence is not constrained to what is required for other persons' actions. But within that equal frame, their individual choices and actions are different because they are constrained by their own properties, preferences, and abilities – inherited or attained – and by resources and people in their vicinity and further away.

So, in a certain place, at a certain point in time, a person about to mate might not find a suitable and agreeable potential partner (and he or she does not have the fiction writer's ability to invent one) and therefore must postpone mating and family formation. The same goes for employment. If there is no vacancy in the person's profession in the local labour market or if competition from other applicants is too fierce given the person's experience and ability, then such a job is not available for the person at that time and place and a series of secondary adaptations might become necessary: commute to a more distant workplace, move to a residence closer to that workplace, abandon work for a while, change your educational or professional focus instead, and so on. Similar constraints and differentiation apply to choice of education, place of residence, forest properties, etc.

A discrete choice regression equation might tell about the probability that a female gets a certain education. Another equation gives the probability that she moves if she gets that education, conditioned on additional characteristics like family status. A third equation (or a mating algorithm) might relate her education to the probability of family formation. When time goes on in the model (as in life), all those and other events occur, partly simultaneously and partly in a sequence, constraining some future choices and enabling others. The major advantage of microsimulation over a single regression equation applied on observables then is that the single equation does not answer the "what if" questions about a person some years ahead, because then many of the other drivers in that equation have also changed.

The ability to handle a myriad of "what ifs" is precisely the work performed by a dynamic microsimulation. Estimated equations and postulated rules for all intertwined individual events are put to work together and the later outcome becomes contingent on the entire web of earlier life events and characteristics for the person in focus and for those other persons and vicinities he or she was and became linked to. There are – at least for smaller models – workable alternatives for this kind of system–wide consistent computation of the development of each person's existence and attributes in the pure statistical toolbox, including approaches attempting to enable sequencing and relation-ships among entities and attributes, such as propensity score matching, panel data re-gression, and structural equation modelling. Such methods might also be helpful while constructing large microsimulation models and exploring their interpreted causality and bias created by endogeneity and selection errors.[3]

As discussed above, microsimulation models simulate individual behaviour over time. Commonly these models are individual-level representations of populations such as people living in a region or a country. The model keeps track of each individual, his or her demographic and socioeconomic attributes, and how these attributes change over time. Technically, the microsimulation model consists of, apart from the core module, a number of specific modules that consider different parts of life related to aspects such as family, housing, and working life. These modules are driven by estimated behavioural equations that have to be based on careful analyses of relationships between theoretically justified factors. Therefore, microsimulation models rely on specialized studies delving into many different relevant phenomena such as patterns of population distribution, geographical mobility, local labour market dynamism, etc.

To illustrate what can be done with this type of data, in the next section we provide an example of geographical research that can be used to develop modules in micro-simulation models as well as directly provide results. (Three further examples are also provided in Appendix 4.)

Do trees make people more rooted? Private forest owners' migration behaviours

Forestland is a tangible asset, likely both indicating and creating attachment to the forest site for the owners.[4] Forest ownership can both create and maintain a strong motive for developing the forest holding and its surroundings. Decisions made by non-industrial private forest (NIPF) owners can therefore be expected to influence population devel-opment in the local communities. Westin and Holm (2018) addressed forest owners' migration propensity and whether forest ownership influences migration to and from the municipality where the forest holding is located.

The study is based on the 8,592,367 individuals who were nationally registered in Sweden on both December 31, 2007, and December 31, 2012. The analyses are based on the ASTRID database, a longitudinal panel containing attributes for each individual, property, household, and workplace in Sweden for many years. Important for this study is that the entities are linked individually. Each owner of each forest property is contained in the dataset, which links individuals to properties each year. The term *forest owners* refers to NIPF owners, or family forest owners, which excludes private companies, public owners such as the state and municipalities, commons, trusts, and the Church of Sweden. In total, there were 309,441 forest owners in 2007 and 317,291 in 2012. In most descriptions and analyses in this study, forest owners are classified into three groups: *persistent forest owners*, who owned forest in both 2007 and 2012 (*n* = 274,713); *new forest owners*, who bought or inherited forest during the period 2007–2012 (*n* = 42,578); and *former forest owners*, who sold or in some way transferred their ownership during the period 2007–2012 (*n* = 34,728).

A migration event means a change in the municipality of residence for the individual. The migrant's place of origin (municipality) experiences out-migration while the destination faces in-migration. *In-migration* is the result seen from the receiving municipality and *out-migration* from the municipality of departure.

To pinpoint the partial influence of forest ownership and ownership change on migration, the description is supplemented with an analysis based on logit regression and propensity score matching. The logit regression here relates the probability of out-migration to a number of variables that previous migration studies have shown to affect migration in general. Answering the research question whether forest owners are more likely to migrate to the municipality where their forest is located compared to non-forest owners requires an analysis beyond simple descriptive statistics. Simply comparing the average proportion of movers selecting a certain destination if they did ("treatment") or did not (no treatment, control group) become new forest owners in that destination would likely overestimate the effect on the choice of destination. People without new forest but similar to the new forest owners might be over-represented among those selecting the same destination as the new forest owners. An approach to avoid such non-observable selection bias is propensity score matching (PSM; Rosenbaum and Rubin 1983; Dehejia and Wajba 2002). *Propensity score* is defined as the conditional probability of assignment to a particular treatment given a vector of covariates. The chosen treatment in this analysis is becoming a forest owner. The analysis is done using STATA's treatment-effects procedure, with sex, age, age square, income, and relative size of population within 50km of the destination as drivers against new owner as the target treatment variable.

The total population of Swedish forest owners in 2012 was 329,541, and they owned half of the country's productive forestland – 23.2 million hectares (Swedish Forest Agency 2014). The Swedish forest owners in 2012 – both local and absentee – were on average 58 years old. Female ownership increased from 25% in 1990 to 38% in 2012 (Lidestav et al. 2017). The share of female owners was higher among absentee owners (44%) compared to local owners (36%). Half of the absentee owners had a university degree, whereas 23% of local owners did.

Over the five-year period 2007–2012, 12.9% of the total population (out of all people living in Sweden in both years) migrated, which is approximately 2.6% per year. New forest owners migrated slightly more – and former owners somewhat less – than non-forest owners, and those giving up their ownership migrated slightly less. The least mobile group was persistent forest owners, whose migration rate was only a third of the rate of other groups. Though just over 3% of the Swedish population was persistent forest owners, these contributed only 1% of all moves between municipalities from 2007 to 2012. The low average mobility level for persistent owners remains in most age groups. It should be noted, though, that only a tiny share of the forest owner population is younger than 20. Acquiring or giving up one's forest property seems to be associated with increased mobility at most ages and even exceeds the level observed for the majority of those not owning forest at any time. To analyze the propensity to move to one's own forest, PSM has been used to control for unobserved selection processes; that is, that new forest owners are more similar to those migrating to their forest municipality compared to movers in general. The main average result is that the odds of selecting the forest municipality as the destination is twice as large for new forest owners compared to that for all other movers. Although a substantial part of forest owners select the municipality containing their forest holding when migrating, the influence of migrating forest owners on local population developments is minuscule because most movers are not forest owners.

Comparing non-forest owners to the group of local NIPF owners showed that the latter are less likely to move. Forest owners living in their forest municipalities seldom move out – about a third annually compared to others in the same age group. When moving, about half of absentee forest owners select their forest municipality as their destination and thus become local forest owners. So, forest owners are different from other movers. The new in-migrating forest owner's first thought is likely not to try to re-create an urban vicinity around the property but more likely the opposite: to escape the urban buzz and instead get close to the owned property as an activity place that is preferred over access to most urban amenities. Although private forest ownership significantly contributes to population development in small, remote rural municipalities, policies for local and rural development rarely acknowledge the potential that private forest owners represent for economic and population development in rural areas.

Moreover, implementing findings presented above in the migration and labour market equations of a simulation reveals indirect effects of hypothetically changed conditions for acquiring forest properties on education, employment, and regional development.

A microsimulation model application

The above presented study (Westin and Holm 2018) and the studies presented in Appendix 4 (Boschma et al. 2009; Dean et al. 2019; Eimermann et al. 2021) provide a gallery of potential inputs needed for constructing internal modules in a microsimulation model as well as some of the "what if" questions. This is the main practical reason for implementing such modules in a simulation. The following example application therefore demonstrates the experimental, "what if" aspect more extensively: How large are the socioeconomic impacts of infrastructure projects on economic growth?

Spatial, socioeconomic impact of infrastructure endowments – a case for agent-based spatial microsimulation

The presented application aims to further the methodology for assessments of the impact of infrastructure projects on local and national growth, employment, earnings, commuting, and migration by means of spatial microsimulation. Results from many aggregated studies on endogenous growth impacts of infrastructure investments vary considerably while generally showing a positive correlation between economic performance and the amount of infrastructure. In the light of results from over a hundred macro models, Lakshmanan (2011) claimed that "these macroeconomic models offer little clue to the mechanisms linking transport improvements and the broader economy". In this application, we suggest a model covering core parts of Lakshmanan's suggestions – the agent-based spatial microsimulation model "InfraSim". Individual agents representing each person in Sweden perform events like giving birth, ageing, dying, getting work, earning wages, and moving to new places of residence and work. The model contains a representation of actual links in the road network. The shortest path between each pair of municipalities is used as a distance measure. Improvement of infrastructure is implemented by artificially decreasing the road "distance" on selected links. Because distances are shorter on certain paths, numbers, compositions, and attributes of commuters and movers will change and, as a consequence, work and settlement patterns, income levels, etc., will change, too.

If an endogenous system-wide growth effect exists, it adds to intrinsic benefits and costs of investments and should be integrated into the decision support system of large investments in new infrastructure. In the model, growth emerges bottom-up from individual agency. The effect of labour market matching is shortcut into a "hedonistic" reduced form of person-based behaviour equations providing the result of, but not mimicking, market clearing. In addition to individual attributes, these equations are included as drivers of aggregate, agglomeration economies–related attributes of the environment, and they work as proxies for the impact of the demand side of the labour market (see also the section The importance of labour mobility for firm performance and regional development).

InfraSim dynamics and interaction

Exogenous changes in accessibility occur instantly, but agents' travel behaviours do not necessarily adapt rapidly. However, gradually some commuters adapt by moving closer to work. The induced slow change in settlement pattern creates new impacts of agglomeration economies on staying or moving and on destination choice and growth. The dynamics of changing workplace, commuting, moving, earning higher income, etc., develop at different speeds for a specific person and for different persons. Such choices and events interact with each other within and between individuals and municipalities.

The choice of destination is based on a production-constrained gravity interaction model using road distance and destination size, measured as the number of people working in the destination, as drivers. The target variable is the number of movers or commuters from each origin to each destination. In effect, the impact of agglomeration economies on mobility, destination attraction, employment, and income level is simplified and estimated as a result of employing an interaction model with distance to and density of employment as drivers.

Many of the shortest routes between the 290 municipalities of Sweden become even shorter by decreasing the distance on one and only one link in Middle Sweden (i.e. Eskilstuna–Västerås). The lines in Figure 12.1 connect origin and destination municipality for each such shortest route, taking advantage of and using the new shorter Eskilstuna–Västerås link. So, improved infrastructure often also shortens distances on routes within a wide surrounding area of the core improved link, thus directly multiplying the immediate local effect into a first round of improved accessibility influencing commuting and mobility in a large area.

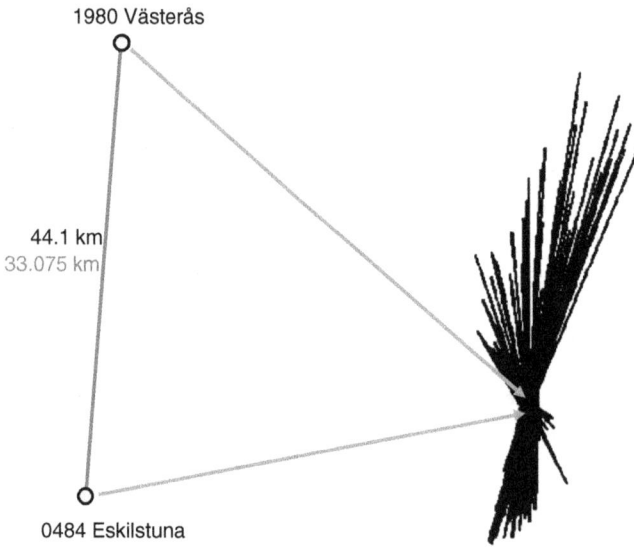

Figure 12.1 Lines connecting origin and destination municipalities for each pair of municipalities gaining a shorter route to the other by taking advantage of and using the new shorter Eskilstuna–Västerås link.

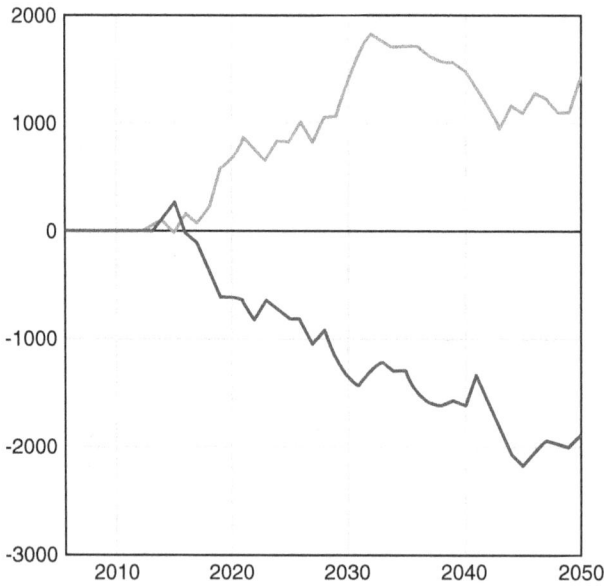

Figure 12.2 Effects on total population in Västerås and Eskilstuna of reducing travel time between the cities with 25%, 2015–2050.

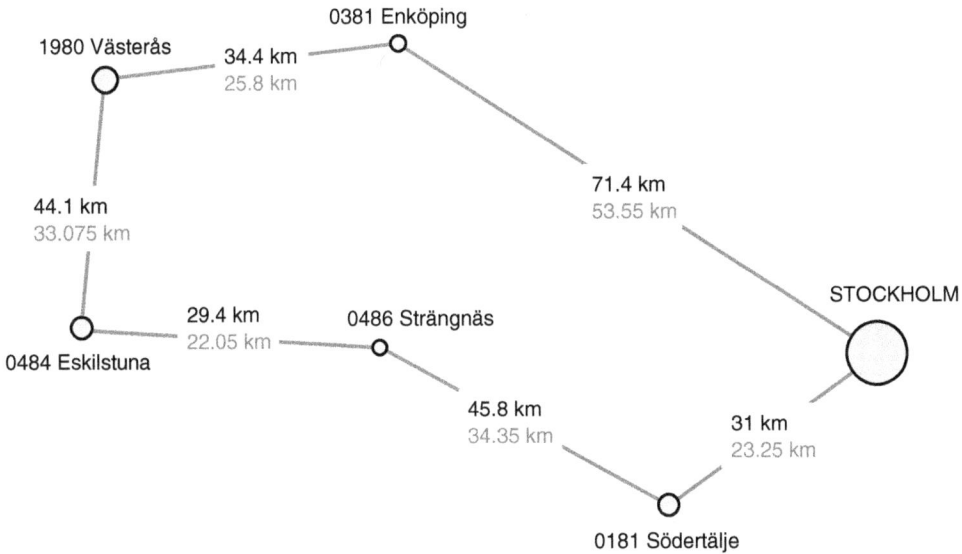

Figure 12.3 Experiment reducing travel time on each of the six links connecting seven cities in eastern Middle Sweden.

In an experiment, the distance/travel time for Västerås–Eskilstuna was reduced by 25% (see Figure 12.2). This outcome indicates the possibility of an unsymmetrical effect between places in different locations on the improved route. In this case, Västerås seems to be the winner regarding the net effect on population number. More people used the new road link to leave Eskilstuna as a place of residence and commute to work rather than the opposite. The impact on employment was similar.

In another experiment, potential effects of substantially improving infrastructure in Mälardalen (Middle Sweden) are explored (Figures 12.3 and 12.4). Most of the municipalities on the improved route around Mälaren gain population substantially over time. Exceptions are Enköping and Eskilstuna (again). Södertälje seems to have an advantageous location in the network for capitalizing on this kind of infrastructure improvements.

After some 40 years, the calculated net effect on earnings per employed contingent on this Middle Sweden infrastructure development reaches circa $100 annually ($300 in the surrounding county).

Figure 12.6 presents the development of commuting and employment in the whole of Sweden from the same experiment. The number of commuters increases twice as much as the number of employed does – after some 30 years the increase amounts to circa 8,000 commuters and 4,000 employees. It should be observed that the figures demonstrate a net effect of improved infrastructure on incomes, commuting, and employment when aggregated over the whole of Sweden. Behind this general growth effect hides places with negative outcome (see Figures 12.2 and 12.4). Whether or not one believes the modelled outcomes, they illustrate the conflict between growth and distribution. In this case there seem to be losers in the game, but the winners win more than the losers lose.

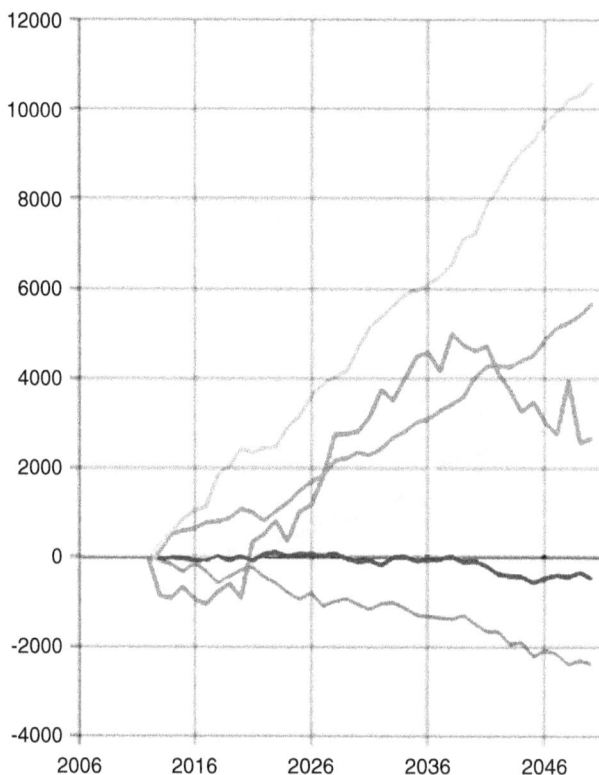

Figure 12.4 Net effects on total population in municipalities on the improved route if travel time between Stockholm–Linköping–Västerås–Eskilstuna–Strängnäs–Södertälje–Stockholm is reduced by 25% as indicated in Figure 12.3.

Can effects justify costs?

The income effect of the experiment adds up to an annual increase in gross domestic product of some 6 billion SEK (approximately $700 million). One metre of highway costs some 50,000 SEK to build. Based on that, the cost for 200km of improved road in the experiment would correspond to less than two years' added growth as a result of the amended accessibility. Presented initial cost estimates for new infrastructure often triple by the end of the construction process. Even so, this suggests a fair opportunity for the country – if somehow the beneficiaries can be convinced to pay.

Thus, results indicate a substantial local and national growth impact of large infrastructure investments in densely populated areas. The growth impact alone of large infrastructure investments in such regions would often suffice to pay for the investment within a few years. However, the spatial and social distribution of the effect is quite heterogeneous with winners as well as losers. Moreover, as usual, the simulation is a construction, largely creating an artificial, under-constrained fantasy world. One should not instantly trust its claims on the world outside the model.

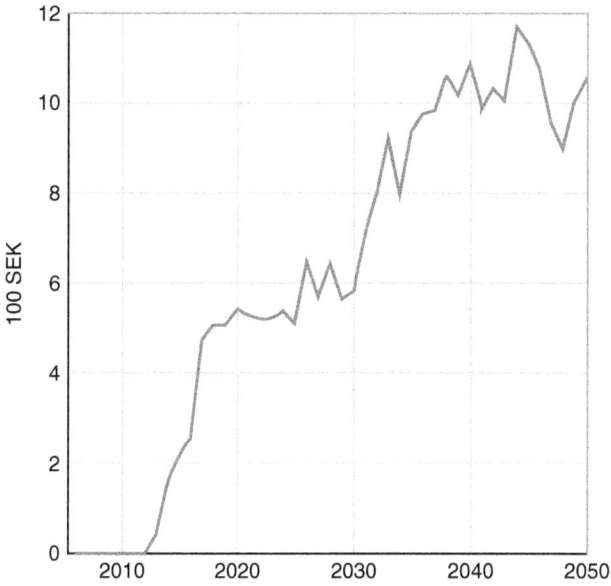

Figure 12.5 Net effect on income (employed) in the whole of Sweden if travel time between Stockholm–Enköping–Västerås–Eskilstuna–Strängnäs–Södertälje–Stockholm is reduced by 25% (100 SEK).

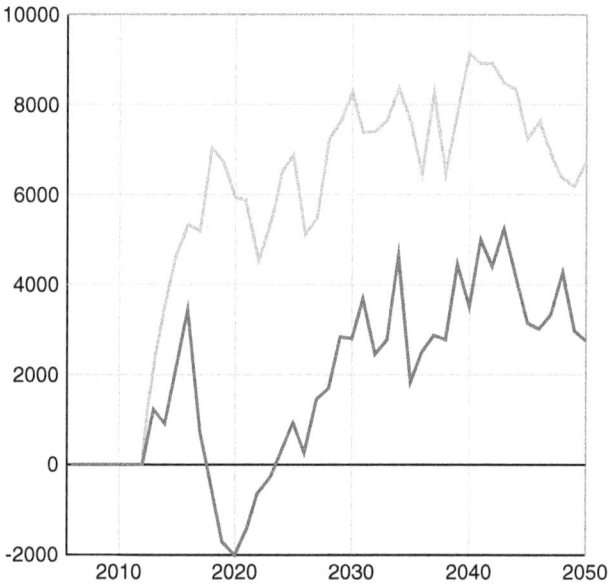

Figure 12.6 Net effects on commuting (green) and employment (blue) in the whole of Sweden if travel time between Stockholm–Enköping–Västerås–Eskilstuna–Strängnäs–Södertälje–Stockholm is reduced by 25%.

Conclusions

In this chapter, we provided an overview of how register data can be used as a resource for analysis. Register data can be used with different levels of detail. Aggregated register data are made publicly available by, for example, SCB and Eurostat, which offer pre-specified tables for anyone interested. Micro-level register data can be made available for research and other purposes, but the use of such data is restricted by regulations to protect personal information from the public eye.

We also presented a research agenda in geography and related disciplines that draws heavily on micro-level register data. Microsimulation modelling departing from time geography adopts an individual-level representation of people and artefacts, which implies a strong dependence on access to longitudinal individual-level register data covering the entire population. A significant reason for the need for such large-scale datasets is that human interactions across different domains of everyday life (e.g. family bonds, colleagues at work, neighbours at previous places of residence, etc.) cut through virtually all types of samples. For each and every individual these links are finely distributed across time, space, and many other dimensions. A microsimulation model can have many different purposes and applications. In previous studies, for example, we have developed microsimulation models for analyzing population dynamics (Holm et al. 1996) and long-term impacts of large investments on local population and labour market development (Lindgren 1999). Here we presented an application focusing on socioeconomic impacts of building infrastructure projects in the Mälardalen region. The results suggested that these investments generate large local and national impacts on economic growth.

To sum up, register data are a vital resource for monitoring and developing a wide array of functions of society. It has even been claimed that register data are a goldmine for research and education (e.g. Otterblad Olausson et al. 2004). The richness of combined individual-level register data from different domains (demographic, socio-economic, health-related, property ownership–related characteristics, etc.) provides an outstanding opportunity to deepen our understanding of the smallest facets of people and society. This type of empirical data enables new research questions to be posed and theoretical and methodological advancements to be made. In this euphoria it is easy to forget that the goldmine is conditioned on the trust people have given to the research community and other users of data. Linked individual-level data from registers reflecting many different life domains could potentially be harmful to the individual's personal integrity if misused. Using sensitive personal data for other purposes than approved and disseminating such individual-specific data to the public could bring about an erosion of trust, jeopardizing progress made for the benefit of individuals and society.

Key messages

- The collection of individual-level information on people and economic activities goes back several hundred years; the early population statistics is a good example thereof. Today, this type of data is of vital use for authorities monitoring demographic and economic processes in society, for universities and research communities, and for the general public.
- Simulating the dynamics of a large population requires information about how individuals' characteristics change over time.

- We provided examples of studies from different fields of research and whose analyses offer input to the estimation of behavioural equations of a microsimulation model. Broadly, the simulation model needs this type of investigation relating to nativity, mortality, family formation and dissolution, migration, education, labour markets, housing markets, etc.

Study questions

1 What sources of micro-level register data may exist in your cases?
2 How could you use micro-level register data in relation to monitoring?
3 What are the ethical considerations you would need to take into account in order to use micro-level register data?

Notes

1 There is a limit to the number of cell values possible to retrieve in a single run. Retrieving all information for a large table like the one exemplified here might require dividing the intended table in several smaller tables while retrieving data; that is, if the limit is 50,000 cells and, if one dimension in the intended table is sex and produces 60,000 cells with all dimensions represented, data for females and males (30,000 plus 30,000 cells) are retrieved separately and thereafter combined in the intended table.
2 This section draws partly on Holm (2017).
3 The origins of microsimulation modelling can be traced back to the works by Guy Orcutt (1957). Since then, a large number of dynamic microsimulation models that enhance the original Orcutt model have been put in operation all over the globe. Example models along these lines are DYNASIM, CORSIM, APPSIM, SfB3, MOSART, DYNCAN, LIAM, SESIM, LIFEMOD, MIDAS, DYNAMOD, SAGE, LifePaths, FAMSIM, SVERIGE, SMILE, and DYNAMITE. This list of acronyms can be extended into at least some 60 large, dynamic microsimulation models all over the globe but mainly developed in larger and richer countries (Li and O'Donoghue 2013). Historically, dynamic microsimulation applications were more often based on detailed empirical data compared to the more recently developed agent-based models. However, today the approaches converge. Long before agent-based models appeared, substantial parts of their conceptual and theoretical underpinnings were formulated by the Swedish geographer Torsten Hägerstrand in his "time geography" (Hägerstrand 1970). The Swedish HÖMSKE and SVERIGE models were developed as much based on his time geography as on dynamic microsimulation (Clarke and Holm 1987; Holm et al. 1996; Holm and Sanders 2007). Spatial microsimulation requires large start populations to provide fine-grained spatial resolution. Because of the reluctance to give analysts access to existing microlevel data due to privacy concerns, considerable research has focused on developing "reverse engineering" methods to recreate microdata from available table and sample data.
4 This section is based on Westin and Holm (2018).

Further reading

Westin, K. and Holm, E. (2018) Do trees make people more rooted? Private forest owners' migration behaviour. *Forest Policy and Economics* 94, 11–20.
This article illustrates an application of micro-level register data in relation to land use.

References

Boschma, R., Eriksson, R. and Lindgren, U. (2009) How does labour mobility affect the performance of plants? – the importance of relatedness and geographical proximity, *Journal of Economic Geography* 9(2), 169–190.

Clarke, M. and Holm, E. (1987) Microsimulation methods in spatial analysis and planning, *Geografiska Annaler B* 69(2), 145–164.

Dean, N., Dong, G., Piekut, A. and Pryce, G. (2019) Frontiers in residential segregation: understanding neighbourhood boundaries and their impacts, *Tijdschrift voor Economische en Sociale Geografie* 110(3), 271–288.

Dehejia, R. and Wahba, S. (2002) Propensity score-matching methods for non-experimental causal studies, *Review of Economics and Statistics* 84(1), 151–161.

Eimermann, M., Lindgren, U. and Lundmark, L. (2021) Nuancing holistic simplicity in Sweden: a statistical exploration of consumption, age and gender, *Sustainability* 13, 8340.

EU General Data Protection Regulation (GDPR): Regulation (EU) 2016/679 of the European Parliament and of the Council of 27 April 2016 on the protection of natural persons with regard to the processing of personal data and on the free movement of such data, and repealing Directive 95/46/EC (General Data Protection Regulation), OJ 2016 L 119/1.

Holm, E. (2017) Microsimulation, in *The International Encyclopedia of Geography: People, the Earth, Environment, and Technology*. Wiley.

Holm, E., Lindgren, U., Mäkilä, K. and Malmberg, G. (1996) Simulating an entire nation, in Clarke, G.P. (ed.) *Microsimulation for Urban and Regional Policy Analysis*, European Research in Regional Science, Vol. 6. London: Pion, pp. 164–186.

Holm, E. and Sanders, L. (2007) Spatial microsimulation models, in Sanders, L. (ed.) *Models in Spatial Analysis*. London: Wiley, 159–195.

Hägerstrand, T. (1970) What about people in regional science? *Papers of the Regional Science Association* 24, 6–21.

Lakshmanan, T.R. (2011) The broader economic consequences of transport infrastructure investments, *Journal of Transport Geography* 19(1), 1–12.

Li, J. and O'Donoghue, C. (2013) A survey of dynamic microsimulation models: uses, model structure and methodology, *International Journal of Microsimulation* 6(2), 3–55.

Lidestav, G., Thellbro, C., Sandström, P., Lind, T., Holm, E., Olsson, O., Westin, K., Karppinen, H. and Ficko, A. (2017) Interactions between forest owners and their forests, in Keskitalo, C. (ed.) *Globalisation and Change in Forest Ownership and Forest Use. Natural Resource Management in Transition*. London: Palgrave Macmillan, pp. 97–138.

Lindgren, U. (1999) Simulating the long-term labour market effects of an industrial investment – a microsimulation approach, *Erdkunde* 53(2), 150–162.

Orcutt, G. (1957) A new type of socio-economic system, *Review of Economics and Statistics* 58, 773–797.

Otterblad Olausson, P., Spetz, C.L. and Rosén, M. (2004) Stor användning av registerdata i svensk forskning – en nordisk konkurrensfördel. *Norsk Epidemiologi* 14(1), 125–128.

Rosenbaum, R. and Rubin, D.B. (1983) The central role of the propensity score in observational studies for causal effects, *Biometrika* 70, 41–55.

Statistics Sweden. (2022) Statistikdatabasen [Statistical Database], http://www.statistikdatabasen.scb.se/pxweb/en/ssd/ (Accessed November 17, 2022).

Swedish Forest Agency. (2014) *Swedish Statistical Yearbook of Forestry*, tables 2.3 and 2.4.

Westin, K. and Holm, E. (2018) Do trees make people more rooted? Private forest owners' migration behaviour, *Forest Policy and Economics* 94, 11–20.

13 Survey questionnaires: data collection for understanding management conditions

Kerstin Westin, Claire Wood, Urša Vilhar, and Marcus Hedblom

Understanding land management

Landscapes are constantly changing due to physical drivers such as geological processes and climate change as well as anthropogenic actions such as management and land use. These changes affect not only ecosystem services linked to providing, regulating, and supporting service but also cultural ecosystem services (UK National Ecosystem Assessment 2022) such as recreation, ecotourism, and spiritual experiences linked to the aesthetics of landscapes. Cultural services are rarely included in the present monitoring schemes. The existing long-term monitoring of landscapes is highly linked to natural sciences and mainly ecological indicators (Fry et al. 2009; Hansen and Loveland 2012). Yet, the way we perceive and experience the landscape is directly linked to health and well-being and willingness to participate in outdoor recreation. For example, a greater extent of specific habitats increases happiness for many people, and spending time in high-quality natural habitats increases well-being (Sonntag-Öström et al. 2015). Thus, to develop strategies for landscape management and land use policies that account for public perception, it is important to understand the consequences of land use changes (Schirpke et al. 2018). A major reason for the lack of detailed monitoring programmes linking social and natural science data over time is the high costs of gathering such data (Kienast et al. 2015), as well as a lack of researchers willing to cross interdisciplinary boundaries. Whilst funding is also a problem, Schirpke et al. (2021) noted that it is challenging to model aesthetic landscape values over time due to complex interactions between human observers and the landscape. Norton et al. (2012) also highlighted the inherent difficulty of combining different types of data, a lack of appropriate data, and a lack of scientists to broach disciplinary boundaries.

In designing a questionnaire aimed at understanding the consequences of land management strategies, there are many aspects to take into account and many different types of information that could be collected from the people involved in managing or using the land. Land varies widely in use, ownership, geography, and environmental quality. All of these are factors in determining the management objectives of a site and therefore the type of information that might be collected in surveys, which both extract information regarding current management practices and gather requirements from users of the land who might benefit from the way the land is managed.

Land under intensive use is likely to be complex in terms of management. For example, intensively farmed land can have a huge range of factors to investigate and take into account, such as livestock intensity, fertilizer and other inputs, farm staffing, water usage, crop yields, and forage types. Agri-environment schemes on farmland will have a

DOI: 10.4324/9781003179245-13

direct bearing on how land is managed. Forestry is also a type of land use that needs to be monitored quite carefully in terms of management. Forests may be managed for timber and financial gain (production oriented) or to promote biodiversity and provide ecosystem services (nature oriented), but they also may be used for recreation by the public and managed for conservation objectives. It is also possible that they are neglected when owners have neither the time nor inclination to proactively manage them (see Keskitalo 2017).

Certain types of land may face multiple pressures of use; for example, coastal zones are popular for recreational purposes but also have high conservation value in terms of habitats and biodiversity. Land nearer to urban areas may also face greater pressures for recreational use; for example, the Peak District National Park in Great Britain. Depending on the landowner, and more often their tenants, some land is managed only for financial gain, whereas other land may have a wider range of management objectives, such as increasing amenity or conservation value. Forests can be managed intensively by large-scale private owners as well as commercial owners promoting production or management can be more nature oriented, emphasizing biodiversity and preservation – often by small-scale private forest owners (Forest Research 2022).

It is clear that there are a wide range of issues connected to land management, nearly all with a social dimension, such as upland vegetation burning and grazing, invasive species, rewilding, development, and pollution. In short, there are many drivers for investigating land management, and in designing a questionnaire aimed at people associated with that land, researchers must be clear on what aspect of management they want to investigate and why.

Why do we need surveys?

Obtaining information from land managers, landowners, and land users, as well as from policymakers, is important on several levels. Management information adds an additional explanatory variable in understanding data on environmental measurements associated with the land, such as vegetation surveys, soil analyses, water samples, and habitat surveys. This can help identify current and past trends in environmental change. Understanding environmental change, preferably with additional management information, can also help to direct policy for future improvements. A good example of this from Great Britain is the post–World War II loss of hedgerows identified by the Countryside Survey (Barr et al. 1991), which led to new policies of hedgerow creation (The Hedgerows Regulations 1997). Other policies for improving the quality of land might include pollutant controls or schemes to reduce invasive species or habitat loss.

Further, a survey may seek to identify the effect of implemented changes such as agri-environment schemes, as in Wales (Emmett and GMEP Team 2017). It may also help to identify conflicts regarding land use, such as tourism and recreation versus conservation or productivity versus biodiversity in forest management (Nordlund and Westin 2011).

What is a questionnaire survey?

Each land manager will have different objectives in managing his or her land, which will therefore determine the type of information collected in a land management survey. Surveys provide a tool to understand the intentions of land managers and users. In other chapters in this book, field inventories (chapter 5), citizen science (chapter 6), and interviews (chapter 14) have been described. This chapter focuses on questionnaire

(a)

PLOT DESCRIPTION AND HABITATS

1 Site No	2 Plot No.	3 Recorder	4 Date
5 Slope ° or %	6 Aspect · Mag.		

A TREES - MANAGEMENT
7 Cop. Stool	8 Singled cop.	9 Rec. cut. cop.	10 Stump hard new
11 Stump hard old	12 Stump con new	13 Stump con old	14

B TREES - REGENERATION
15 Alder	16 Ash	17 Aspen	18 Beech
19 Birch	20 Hawthorn	21 Hazel	22 Holly
23 Hornbeam	24 Lime	25 Oak	26 Rowan
27 Rhododendron	28 Sweet chestnut	29 Sycamore	30 Wych elm
31 Other hrwd.	32 Scots pine	33 Yew	34 Other con.

C TREES - DEAD (- HABITATS)
35 Fallen brkn	36 Fallen uprtd.	37 Leg.v.rotten	38 Fall. bnh >10cm
39 Hollow tree	40 Rot hole	41 Stump<10cm	42 Stump >10cm

D TREES - EPIPHYTES AND LIANES
43 Bryo.base	44 Bryo.trunk	45 Bryo.branch	46 Lichen trunk
47 Lichen branch	48 Fern	49 Ivy	50 Macrofungi

E HABITATS - ROCK
51 Stone.<5cm	52 Rocks 5-50cm	53 Boulders >50cm	54 Scree
55 Rock outcp.>5m	56 Cliff >5m	57 Rock ledges	58 Bryo.covd.rock
59 Gully	60 Rock piles	61 Exp.grav/sand	62 Exp.min.soil

F HABITATS - AQUATIC
63 Sml.pool <1m²	64 Pond 1-20 m²	65 Pon.lake>20 m²	66 Strm.riv.slow
67 Strm.riv. fast	68 Aquatic veg	69 Spring	70 Marsh/bog
71 Ditch drain dry	72 Ditch drain wet	73	74

G HABITATS - OPEN
75 Gld.5-12m	76 Gld.>12m	77 Rky.knoll<12m	78 Rky.knoll>12m
79 Path <5m	80 Ride >5m	81 Track non prop	82 Track metalled

H HABITATS - HUMAN
83 Wall dry	84 Wall mortared	85 Wall ruined	86 Embankment
87 Soil excav.	88 Quarry/mine	89 Rubbish dom.	90 Rubbish other

I HABITATS - VEGETATION
91 Blkthorn thkt.	92 Hawthorn thkt.	93 Rhodo.thkt.	94 Bramble clump
95 Nettle clump	96 Rose clump	97 W.herb clump	98 Umbel.clump
99 Bracken dense	100 Moss bank	101 Fern bank	102 Grass bank
103 Leaf drift	104 Herb veg.>1m	105 Macfungi.soil	106 Macfungi.wood

J ANIMALS (mainly signs)
107 Sheep	108 Cattle	109 Horse-pony	110 Pig
111 Red deer	112 Other deer	113 Rabbit	114 Badger
115 Fox	116 Mole	117 Squirrel	118 Anthill
119 Copse bones	120 Spent ctrdgs.	121	122

COMMENTS

(b)

ArcGIS Survey123 — Plot Information

Page 2: Plots

- Plot descriptions
  - Trees - management

Trees - management

Cop. Stool	Singled cop.	Rec. cut. cop.
Stump hard.new	Stump hard.old	Stump con.new
Stump con.old	Other	

- Trees - regeneration
- Trees - dead
- Habitats - rock
- Habitats - aquatic
- Habitats - open
- Habitats - human
- Animals

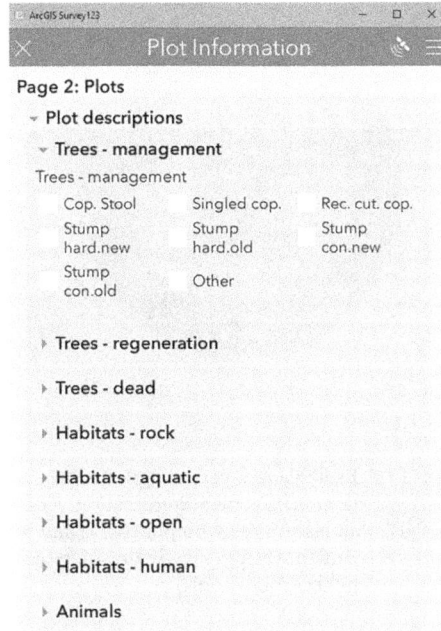

Figure 13.1 Example of a very simple survey capturing woodland management information, from a long-term monitoring programme in Great Britain, the "Bunce" Woodland Survey (Wood et al. 2015). (a) Information was collected on paper form in 1971 but (b) has now progressed to electronic capture.

surveys, which often aim to find patterns of behaviour and attitudes by obtaining a larger set of observations/responses. There are a range of ways in which this information may be collected. For example, simple observations of management may be collected at the same time as other field data as part of long-term monitoring, as in the Bunce Woodland Survey (Wood et al. 2015). This might be collected on paper or via digital mobile applications (see Figures 13.1 and 13.2).

If there is an intention of incorporating the questionnaire survey into a long-term monitoring programme, it may be wise to keep the survey simple to ensure repeatability. This is especially true if the survey is part of a national programme, rather than a small-scale site or regional survey. On the whole, long-term environmental monitoring incorporating social and management information is extremely rare.

How do we define and find the population for a survey?

All of the individuals or entities that share the characteristics that are defined by the study we want to carry out constitute the *population* (see also Appendix 1). The first task is therefore to define the population. This can be fairly uncomplicated in some cases. For example, when investigating how people would vote in an election, the population consists of all eligible voters. However, defining the population is usually not that

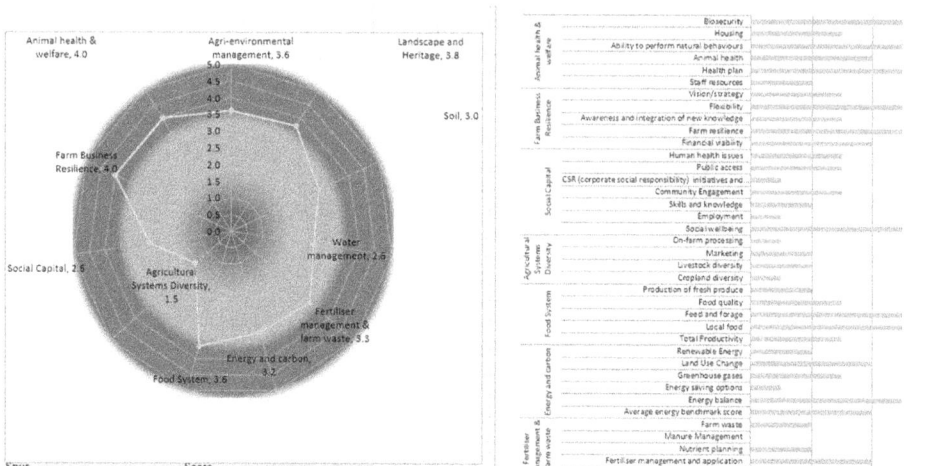

Figure 13.2 Example of a very complex survey designed in Microsoft Excel to capture many aspects of farm management information. Known as a "public goods tool", it incorporates analysis as an instant visual output (right) and was designed by the Organic Research Centre in Great Britain (https://www.organicresearchcentre.com/our-research/research-project-library/public-goods-tool/).

straightforward; for example, when surveying landowners, private owners, companies that own land, and public owners such as cities or states. But what about commons? Is it every single member or the board who makes the decisions on management? What if a holding is co-owned by two or three people – are all part of the population or should only one owner per holding be included in the population? If co-owners are included, sampling might lead to two or more people owning a holding being included, which would then lead to different levels of representation within the population compared to single-owner holdings. When surveying, for example, farmers, we also must define what and who a farmer is. Is it someone who has a business registered for farming or is it also someone who owns farmland and maybe keeps some livestock to provide food for the family? When land is leased to someone who manages it, is the landowner or the tenant part of the population? The answer to questions like these partly lies within the aim of the survey but needs to be considered before starting the sampling.

Surveying a whole population is seldom possible unless our population is limited to, for example, all landowners in a village. *Sampling* is the process of selecting participants who are to represent the population (see chapter 4 and Appendix 1). This can be done as a random sampling, where each member we pick has the same probability of being chosen but there is no limitation as to being representative with respect to, for example, size of holding, owner's education level, etc. If we want to make sure that we have representation of different specific groups that are important for our survey, we use stratified sampling to ensure that we have a proportional or at least a minimum number of participants in every subgroup that is important for the investigation.

A list of all members in the population constitutes the *sampling frame*. How do we find the members of a population? Are there cadastres or registers available? When it comes to landownership, one can assume that all holdings are linked to an owner, but how the registers are organized differs. They can be national as, for example, in Sweden and Finland or on a regional or federal level as, for example, in Germany. It is not always possible to have information about the sample members; for example, people's addresses are not always knows, the registers or cadastres do immediately record change of ownership, or property identifications are not linked to an exact geographical location. All personal data handling has to follow the *EU General Data Protection Regulation* (GDPR 2016), which in some cases limits access to personal information (see the section "How do we ask questions linked to monitoring?").

Once we have defined the population, established the sampling frame, and decided on the sampling strategy (unless we aim to survey a total population), the respondents have to be approached. The design of the survey is influenced by the method of interaction with the respondents. A complex questionnaire survey – for example, an activity diary on time used for different management activities during a specific period – often needs some instant feedback from the researcher. The diary can have inconsistencies that can be checked while talking to the respondent. In these cases, a physical face-to-face meeting is most effective. It is advantageous to send out the questionnaire in advance, so that the respondents can review the questions. An advantage with face-to-face interviews is the interaction between the respondent and researcher. However, face-to-face questionnaire surveys are time-consuming and thus expensive. Setting up meetings takes time, and because the time needed for meetings is difficult to estimate, the researcher needs good margins between meetings. The respondents can be geographically spread, meaning long travel times between meetings, which adds to time and reduces the number of respondents who can be surveyed. Not all respondents are comfortable with having a stranger coming to their home or workplace, which can result in respondents dropping out. Conversely, face-to-face meetings with respondents can be perceived as threatening for the researcher and often require two persons for the visit. Finally, the data have to be entered into a file, which is time-consuming.

Telephone surveys are more time efficient than face-to-face meeting, and inconsistencies and misunderstandings can be sorted out. It is possible to have follow-up questions, as in face-to-face meetings. Telephone surveys are usually preceded by sending the questionnaire in paper format to the respondent, which enables the respondent to review the questions in advance and refer to the questions on paper during the interview. Telephone surveys have higher response rates and lower item non-response rates (partial non-response) than other survey modes (Lesser et al. 2012). However, most sampling frames and registers lack up-to-date cell phone numbers (which is the most common telephone device these days), meaning that the sampling frame is not always suited for the way we want to reach the respondents and we end up with a bias. As with face-to-face meetings, data have to be entered into a data file afterwards.

Mail-out surveys consist of paper questionnaires distributed by mail. Physical addresses are usually available in registers, which means that most respondents can be reached (unless they have no known address or have not updated their address). The mail-out surveys need to be printed on paper, copied, and sent out, which is costly. The respondents are asked to return the completed survey in a pre-paid envelope, which adds to postage costs. Paper questionnaires soon run into many pages, and the task of filling them out can be perceived as too time-consuming. After two or three weeks, a reminder is sent to respondents who have not answered the survey, which adds to costs and time. Once a completed questionnaire is received, it has to be registered manually to a data file. Missing answers have to be treated as just missing; similarly, multiple answers to a question when only one answer is valid are regarded as missing. A combination of mail-out and digital questionnaires is increasingly more often applied. The respondent receives a paper survey by mail but can choose to respond digitally, which cuts costs for postage and data entry. In contrast to face-to face and telephone interviews, follow-up questions are not possible, and misunderstandings cannot be sorted out. However, an evaluation by Lesser et al. (2012) showed that the response rate was lower for a combination of digital and mail methods compared to mail only.

Digital surveys are becoming more common, and one advantage is lower costs and less time than face-to-face or telephone interviews and mail-out surveys. The sample can be large without adding extra costs except for the postage for a postcard, and when the respondents answer the survey their answers are automatically registered in a data file. In that respect, they are self-administered (Belisario et al. 2015). Registers often lack e-mail addresses, but this problem can be overcome by sending a postcard to the respondent with a unique code and asking the respondent to log in and access the survey. When designing the survey, attention has to be paid to readability; that is, whether the questions and their alternatives are suitable for a computer screen and a smartphone. A large number of respondents can be reached, but response rates are generally lower compared to mail-out surveys (D.M. Shannon and Bradshaw 2002; Zahl-Thanem et al. 2021). For example, Belisario et al. (2015) found that response rates were between 10% and 20% lower for digital surveys compared to other delivery modes. Although many people have the knowledge and experience to work with digital platforms, some groups are excluded from internet and digital applications due to economic restrictions, age, etc. This leads to the issue of representativity (see the section How do we ask questions linked to monitoring?).

We strive for as high a response rate (number of responses divided by total sample number) as possible (see section How do we ask questions linked to monitoring? for more on validity). At the same time, budgets are generally not unlimited, so the researcher has to balance a number of requirements or characteristics (see Table 13.1).

How do we ask questions linked to monitoring?

Using surveys, we strive for answers to our questions, and the questions must both mirror what we want to know (*validity* – to measure what we want to measure) and be understood by the respondents. Some questions are contextual and assume that the respondent understands and has an experience of the context. When asking, for example, how important it is that a forest be characterized by biodiversity, a straight question is often difficult to formulate – what is a *forest*? Do we mean a forested area of a certain size? What does *biodiversity* mean? The question needs to be divided into more concrete questions that together – via an index, for example – can describe biodiversity (C.E. Shannon and Weaver 1949).

Table 13.1 Characteristics of different survey data collection methods

	Time	Cost	No. of respondents	Response rate	Data entry
Face-to-face	High	High	Low	High	Manual
Telephone	High	Medium	Medium	High	Manual
Mail-out survey	Medium	High	Medium	Medium	Manual
Digital/web	Low	Low	High	Low	Automatic

Even the seemingly simple question can be open for misinterpretation. A survey on how many hours per week people spent on computer games showed that some elderly respondents were very active. It turned out that some respondents had misunderstood the question regarding how old they were and answered with their year of birth; that is, some younger respondents who were born in, for example, 1983, answered "83" instead of their actual age at the time of the survey. In this case, this could be controlled by matching with the registers.

Open answers are possible but should be limited. They can be perceived as time-consuming for the respondent, the answers tend to be very short, and each respondent formulates their answers differently and may use words that have different meaning for different people. For example, in a migration survey, in open answers to a question about reason for moving (domestic), some respondents answered that they wanted to move closer to family, and others stated that their parents were getting older. Do these reasons express the same rationale, or do they indicate different aspects of life? In general, questions with fixed alternatives are easier to answer for the respondent, but at the same time the alternatives can be too few, too broad, or understood differently by different respondents. In the end, to design policies etc. we need to ask questions to understand what people think and why they behave in a certain way. Pilot surveys, preferably combined with interviews, are important to test how the questions are understood. Pilot surveys also provide an idea of how time-consuming the survey is for the respondent and can lead to limiting the number of questions.

There is a balance between the number of questions and time a respondent spends answering the questions. The more questions there are, the harder it is to get sufficient numbers of responses. But what is a sufficient number? In general, we need an acceptable response rate; that is, the number of people who responded divided by the number of people in the sample. The higher the response rate, the more likely it is that the respondents are a good representation of the population. Over time, response rates have dropped. Reasons for decreasing rates include survey fatigue, targeting the wrong population, and poorly designed surveys that "make no sense" to the potential respondents. Of course, response rate varies with survey mode and is, in general, highest with face-to-face interviews and lowest for digital/online surveys. *Reliability* – that is, the extent to which the survey or test will give the same result if repeated. Repeatability is likely to increase where the response rate increases provided that the sample is large enough to capture the population. However, when surveying people, we must be aware that their responses are dynamic, because their socioeconomic situation as well as attitudes might have changed, and the context may be different. Another important issue is *validity*; that is, how well we manage to measure what we intend to measure.

The sample size is dependent on the size of the population, the variation in responses, the analysis you want to perform, and the confidence you want to place on your results. The more respondents, the better, but, as stated in chapter 4, a small sample is enough if it is well chosen. Still, there are some rules of thumb: if we have a sample of 30 or more observations, the sampling distribution of the mean can be assumed to be normal (Mordkoff 2016), and a sample of at least 50 respondents is desirable to have statistical significance (Trost 2001). Because we often need to know how different groups of people perceive, for example, land management and how they act, we need questions that distinguish different characteristics among the respondents. Background questions on sociodemographic data, employment, resources, etc., are useful in this respect. To avoid too many questions, we need to decide what sub-groups are of interest for the study. However, the more sub-groups there are, the more respondents we need. If we are interested in differences between, for example, men and women, young and old, or urban and rural residents, questions on gender, age, and place of residence are important background questions. If we want to combine the sub-groups to understand, for example, how young women in rural areas perceive a forest's qualities, we need a large number of respondents to analyze the data quantitatively, and a rule of thumb is 50 respondents in each sub-group to assume statistical significance. In this example, we have eight sub-groups, we expect a response rate of 30%, and we need 50 respondents in each group, so we need to invite 1328 respondents (8 × (50/30 × 100)). However, it is not certain that the response rate will be the same in all sub-groups, so to ensure that we will receive at least 50 respondents in each sub-group, it is a good idea to invite a larger number.

The GDPR went into effect May 25, 2018, and states that participants in a survey must give their consent for participating, unless the survey is conducted anonymously and there are no personal data that can identify the respondent. Examples of personal data are e-mail and physical addresses, phone number, registration plate number, or information that enables you to trace a respondent by combining bits of data. GDPR is valid for most surveys. GDPR is applicable if, for example, you have a population frame from which you draw a sample for which you have information on the individuals.

Long-term survey of landscape management – linking survey data to physical monitoring data

As discussed, social factors are rarely included in long-term ecological monitoring schemes. From a social science perspective, *visual beauty*, linked to scenic beauty, aesthetic quality, and visual preferences, is determined by both subjective responses and to some extent objective criteria (Dronova 2017). Subjective views vary depending on one's age, profession, background, cultural heritage, environmental expertise, and other social dimensions (Kaplan 1995; de Val et al. 2006; Dramstad et al. 2006; Gunnarsson et al. 2017).

One could ask whether it is possible to have objective criteria of a landscape. Fry et al. (2009) suggested that there are a number of broad common "evolutionary" landscape properties that seem to be preferred irrespective of culture and personal preferences, meaning that to some extent they are "objective". Here we will provide you with some examples from England (Norton et al. 2012), Sweden (Hedblom et al. 2019), and Switzerland (Schirpke et al. 2021).

Norton et al. (2012) combined interviews with national monitoring data in the British UKCEH Countryside Survey. Their aim was to demonstrate a novel approach for providing measures of cultural services at a national scale in England, creating a map indicating high cultural values. They conducted a telephone survey complemented with a survey of 150 members of the public in 16 focus groups and extended creative sessions, as well as 16 in-depth interviews (see Natural England 2009). A focus group interview involves a small number of demographically similar public participants who have other common experiences. Their reactions to specific evaluator-posed questions are studied. In Norton et al. (2012), participants were from a mix of socioeconomic, gender, and age groups, including people living and working in or using the areas concerned. Participants were asked to identify types of landscape, landscape features, and emotions that they linked to each of eight cultural services (history, place, inspiration, calm, leisure/activities, spiritual, learning, escape). In total 20 landscape features were selected; for example, waters, coast, mountains, grassland, woodland, hedges, etc. The data from interviews (qualitative data) and data from the monitoring program (spatial areas) were selected in isolation, meaning that, for example, spatial scales were not referred to in interviews. Thus, there was partly a mismatch in study scales and one of the major challenges of the project of how to integrate qualitative data with spatial biophysical data. Linking the datasets was done by experts who subjectively transferred people's perceptions of cultural services (e.g. low-high inspiration of waters). The final product was a map revealing recreational potentials in the whole of England.

Hedblom et al. (2019) used a similar approach as Norton et al. (2012), although they narrowed down the landscape to only include mountain areas. One of the aims was to suggest indicators based on the National Inventories of Landscapes in Sweden (NILS) physical monitoring data and link that to perceived landscape properties. The NILS data are hierarchical and include very detailed data from lichen species ($0.28m^2$) in the field to remote sensing habitat classifications ($1km^2$). The idea was to use a more detailed physical monitoring data than Norton et al. (2012) because they suggested much finer-grained information of the local landscape. The survey questions in Hedblom et al. (2019) were highly linked to the existing monitoring data. Questions about perception were linked to photos taken in the field in the same places where the vegetation was recorded. Thus, the respondents' answers could be directly linked to specific existing landscapes. The 39 respondents in this case were all active in mountain areas (either as company representatives or working for a national agency) participating a conference on the theme linked to "Swedish mountain areas". Interestingly, one of the main findings was echoed from Norton et al. (2012); that is, an open landscape with views was ranked highly. Yet, the main finding was that it was possible to link specific detailed physical data (qualitative, such a birch cover) to perception and appreciation. This method, however, can be complex in long-term studies. For example, a rather low tree cover in the alpine region that is perceived as positive today may be perceived as negative in 100 years when the tree cover has doubled (ongoing trend due to climate change; Pearson et al. 2013), and we would not be able to ascertain whether this negative perception was the result of the tree cover per se or the result of the changes in people's attitudes. Thus, it is important to evaluate not only previous and present questionnaires but also previous and present visual landscapes using photos or physical data. This was done in an innovative way by Schirpke and colleagues in 2021.

The aim for Schirpke et al. (2021) was to analyze changes in aesthetic landscape values for periods between 1950 and 2010 across the European Alps. They did this by combining three former surveys where respondents answered a 5- to 10-minute survey and rated a number ($N = 187$) of 360° photos (in total $N = 2209$ respondents). A mean preference score for each landscape photograph was calculated and linked to 1 of 19 landscape types; for example, urban areas, vineyards, pastures, etc. The landscape types were derived from maps from 1950, 1980, and 2010. They then randomly selected 30,000 viewpoints and evaluated what was potentially seen at different distances from this point (0–60m, 60m–1.5km, 1.5–60km, and 10–50km). The end product was three maps showing how the aesthetic values changed over 60 years (1950–2010) of different management in the European Alps.

In summary, it is possible to combine physical monitoring data with rather short social surveys (short surveys are preferred because people are busy today and longer surveys have lower response rates; *short* means approximately 5–15 minutes) to reveal people's perceptions of rather large areas (England, Swedish forest landscapes and mountains, and the

European Alps). The main obstacles are (1) interpreting respondents' subjective percep-tions of physical monitoring data such as cover of trees, cover of shrubs, coniferous versus deciduous forest, etc. and (2) following people's perceptions over time. There are no long-term monitoring programs that link people's perceptions with data collection. Although Schirpke et al. (2021) showed that it is possible to have a present perception of landscape and also look back at historical perceptions, but we cannot know exactly how people perceived the landscape in past times (something both Hedblom et al. (2019) and Schirpke et al. (2021) emphasized). A way forward is thus to have repeated surveys in representative parts of physical surveys, including respondents from different demographic groups. Using photos in combination with questions seems to be the best way to interpret perceptions of physical monitoring data. The more detailed the photos are, the easier it is to link to specific physical data; for example, Schirpke et al. (2016, 2021) used 360° photos but suggested that future studies use so-called tangential visibility analyses where viable pixels are measured according to their size and distance from the observer's point of view.

How do we carry out international surveys, and how do we ask questions?

Sometimes we are interested in land use or attitudes towards land management in several countries. The goal of such international surveys might be to observe the different perceptions or preferences of citizens in different countries over time or to propose European Union (EU)-wide policies based on citizens' attitudes about a particular issue. Such international surveys require very good agreement on the terms used, because their meanings may differ across disciplines, as well as across countries, cultures, and languages. In addition, a very good translation of the questions and often an adaptation to country-specific characteristics is very important. In international surveys, it is advisable to work with national experts who can help with country-specific questions to avoid mis-understandings. Usually, they also take care of the sampling at the national level and evaluation of the responses.

An example is an international survey on private forest owners' willingness to adhere to different management activities in the process of transition to a wood-based bioec-onomy in five EU countries: Austria, Finland, Germany, Slovenia, and Sweden (Juutinen et al. 2022; see Table 13.2). The survey was conducted as part of the project ValoFor: Small Forests – Big Players: Valorising small-scale forestry for a bio-based economy, funded by ERA-NET ForestValue and the European Union's Horizon 2020 research and innovation programme (Grant Agreement N° 773324). The survey, with a common structure and questions for the five participating countries, was originally developed in Sweden and then translated from Swedish into English and other lan-guages. In each country, the translated survey was pre-tested to ensure clarity of ques-tions and overall structure and adapted by forestry experts from the participating countries with alternatives for the local language.

The survey included questions on forest owner characteristics (e.g., sex, age, educa-tion, years of ownership), the forest holding (e.g., size, management plan, certification), motives for ownership, and perceived utilities (forest values) of the holding. One section included questions on to what degree different management activities were applied and whether changes in these were expected. Some management activities that were assessed aimed to promote a production-oriented management, some activities were purely nature oriented, and some were a mix of both. Not all forest management activities were

Table 13.2 International survey on private forest owners' willingness to adhere to different management activities in the process of transition to a wood-based bioeconomy – the case of five EU countries: Austria, Finland, Germany, Slovenia, and Sweden

Population	*Small-scale private forest owners in Austria, Finland, Germany, Slovenia, and Sweden*				
Type of questions	Direct questions: multiple choice, yes/no; Likert scale; One selection; Open questions				
Collected data	2524 responses				
	Austria	Finland	Germany	Slovenia	Sweden
Sampling design	A web-based questionnaire sent to a market research company's address list. The survey was kept open until the required sample size of a minimum 300 respondents was reached.	A nationwide questionnaire sent by mail to a random sample of 3000 Finnish family forest owners with at least 2 hectares of forestry land	An online questionnaire, using the tool LimeSurvey. The survey was kept open until the required sample size of a minimum 300 respondents was reached.	Printed invitation to the online questionnaire and printed questionnaire sent to a random sample of 2000 forest owners; additional personal interviews until the required sample size of a minimum 300 respondents was reached	Printed questionnaire sent to random sample of 2000 forest owners
Survey distribution	A private market research company	A private market company	The von Thunen institute	Slovenian Forestry Institute	The University of Agriculture in Umeå
Response rate	Not recorded	31.4%	Not recorded	Online questionnaire: 13.3%	31.7%
Number of responses	300	942	307	323	652
% Male respondents	85.9	76.2	89.0	70.6	76.5
Owner's age, mean (years)	50.1	64.1	48.6	53.8	65.1
Years of ownership, mean	21.6	27.1	17.1	17.9	27.9
Size of forest property, mean (hectares)	29.1	55.5	111.6	13.0	92.0

Source: Juutinen et al. (2022).

applicable in all five countries. For example, protection against browsing is not applicable in Finland and Sweden but is important in Austria, Germany, and Slovenia and was therefore part of the country-specific analyses.

Due to different sample sizes and data collection methods, the respondents represent themselves and are not assumed to be representative of all forest owners in the five countries. However, their attitudes and perceptions of forest ownership and forest management provide important insight into the current and possible future behaviours of private forest owners and the possibilities to direct European forests in the direction presented by the New EU Forest Strategy. It also provides an insightful snapshot of the current situation in small private forests.

Conclusion

In conclusion, understanding land management information is an important aspect of understanding drivers of environmental change. In long-term ecological monitoring, linkages between social and natural science are rare due to the costs involved and a lack of willingness amongst researchers to cross interdisciplinary boundaries. In terms of the practicalities of designing land management surveys, there are many aspects to consider. Questionnaire surveys aim to gather information about people – land managers, land-owners, land users, policymakers, and politicians – and how they perceive, for example, land management. Their attitudes and behaviours can only be measured by asking questions. Designing a questionnaire requires the researcher to have a clear idea about what is to be examined and who to survey and how. Important steps are therefore to identify a population, a sampling frame, and a method for how to reach the respondents.

Key messages

- We need questionnaire surveys to acquire knowledge of people's preferences, attitudes, behaviours, and reasoning.
- To draw conclusions from surveys, we need correct sampling procedures and well-designed questionnaires.
- We need to acknowledge ethical concerns and follow the GDPR whenever respondents are identifiable.

Study questions

1 How would you define a population of visitors to a forest or a Natura 2000 area?
2 Where could you find the sampling frame for a study on that population?
3 What group differences would you examine, and what would that mean for sampling?

Suggested reading

Krosnick, J.A. and Presser, S. (2018) Question and questionnaire design, in Vannette, D. and Krosnick, J.A. (eds) *The Palgrave Handbook of Survey Research*. London: Palgrave Macmillan, pp. 439–455. doi.org/10.1007/978-3-319-54395-6_53
Krosnick and Presser (2018) highlights how to construct a questionnaire and what type of questions (open versus closed) to include, scales, bias, etc.

References

Barr, C., Howard, D., Bunce, B., Gillespie, M. and Hallam, C. (1991) *Changes in Hedgerows in Britain between 1984 and 1990*. Contract Report to the Department of The Environment. Grange-over-Sands, UK: Institute of Terrestrial Ecology.

Belisario, J.S.M., Jamsek, J., Huckvale, K., O'Donoghue, J, Morrison, C.P. and Car, J. (2015) Comparison of self-administered survey questionnaire responses collected using mobile apps versus other methods, *Cochrane Database of Systematic Reviews* 7, MR000042. https://doi.org/10.1002/14 651858.MR000042.pub2

de Val, G., Atauri, J.A. and de Lucio, J.V. (2006) Relationship between landscape visual attributes and spatial pattern indices: A test study in Mediterranean-climate landscapes, *Landscape and Urban Planning* 77(4), 393–407. doi: 10.1016/jlandurbplan.2005.05.003

Dramstad, W., Tveit, M.S., Fjellstad, M. and Fry, G. (2006) Relationships between visual landscape preferences and map-based indicators of landscape structure, *Landscape and Urban Planning* 78(4), 465–474. doi: 10.1016/j.landurbplan.2005.12.006

Dronova, I. (2017) Environmental heterogeneity as a bridge between ecosystem service and visual quality objectives in management, planning and design, *Landscape and Urban Planning* 163, 90–106. https://doi.org/10.1016/j.landurbplan.2017.03.005

Emmett, B.E. and GMEP Team. (2017) *Glastir Monitoring & Evaluation Programme. Final Report to Welsh Government – Executive Summary*. Contract Reference: C147/2010/11. Bangor, Wales: NERC/Centre for Ecology & Hydrology.

EU General Data Protection Regulation (GDPR): Regulation (EU) 2016/679 of the European Parliament and of the Council of 27 April 2016 on the protection of natural persons with regard to the processing of personal data and on the free movement of such data, and repealing Directive 95/46/EC (General Data Protection Regulation), OJ 2016 L 119/1.

Forest Research. (2022) Behaviour of private land owners and managers, https://www.forestresearch.gov.uk/research/behaviour-of-private-land-owners-and-managers/ (Accessed May 13, 2022).

Fry, G., Tveit, M.S., Ode, A. and Velarde, M.D. (2009) The ecology of visual landscapes: exploring the conceptual common ground of visual and ecological landscape indicators, *Ecological Indicators* 9(5), 933–947. https://doi.org/10.1016/j.ecolind.2008.11.008

Gunnarsson, B., Knez, I., Hedblom, M. and Sand, Å.O. (2017) Effects of biodiversity and environment-related attitude on perception of urban green space, *Urban Ecosystems* 20(1), 37–49. doi: 10.1007/s11252-016-0581-x

Hansen, M.C. and Loveland, T.R. (2012) A review of large area monitoring of land cover change using Landsat data, *Remote Sensing of Environment* 122, 66–74. https://doi.org/10.1016/j.rse.2011.08.024

Hedblom, M., Gunnarsson, B., Iravani, B., Knez, I., Schaefer, M., Thorsson, P. and Lundström, J.N. (2019) Reduction of physiological stress by urban green space in a multisensory virtual experiment, *Scientific Reports* 9, 10113. https://doi.org/10.1038/s41598-019-46099-7

The Hedgerows Regulations. (1997) London: The Stationery Office.

Juutinen, A., Haeler, E., Jandl, R., Kuhlmey, K., Kurttila, M., Mäkipää, R., Pohjanmies, T., Rosenkranz, L., Skudnik, M., Triplat, M., et al. (2022) Common preferences of European small-scale forest owners towards contract-based management, *Forest Policy and Economics* 144, 102839. doi.org/10.1016/forpol.0200.102839

Kaplan, S. (1995) The restorative benefits of nature: toward an integrative framework, *Journal of Environmental Psychology* 15(3), 169–182. https://doi.org/10.1016/0272-4944(95)90001-2

Keskitalo, E.C.H. (ed.) (2017) *Globalisation and Change in Forest Ownership and Forest Use: Natural Resource Management in Transition*. London: Palgrave Macmillan.

Kienast, F., Frick, J., van Strien, M.J. and Hunziker, M. (2015) The Swiss landscape monitoring program – a comprehensive indicator set to measure landscape change, *Ecological Modelling* 295, 136–150. https://doi.org/10.1016/j.ecolmodel.2014.08.008

Lesser, V., Newton, L. and Young, D. (2012) Comparing item nonresponse across different delivery modes in general population surveys, *Survey Practice* 5(2). https://doi.org/10.29115/SP-2012-0009

Mordkoff, J.T. (2016) The assumption(s) of normality, http://www2.psychology.uiowa.edu/faculty/mordkoff/GradStats/part%201/I.07%20normal.pdf (Accessed November 18, 2022).

Natural England. (2009) Experiencing landscapes; capturing the cultural services and experiential qualities of landscape, http://publications.naturalengland.org.uk/publication/48001

Nordlund, A. and Westin, K. (2011) Forest values and forest management attitudes among private forest owners in Sweden, *Forests* 2, 30–50. https://doi.org/10.3390/f2010030

Norton, L.R., Inwood, H., Crowe, A. and Baker, A. (2012) Trialling a method to quantify the 'cultural services' of the English landscape using Countryside Survey data, *Land Use Policy* 29(2), 449–455. https://doi.org/10.1016/j.landusepol.2011.09.002

Pearson, R.G., Phillips, S.J., Loranty, M.M., Beck, P.S.A., Damoulas, T., Knight, S.J. and Goetz, S.J. (2013) Shifts in arctic vegetation and associated feedbacks under climate change. *Nature Climate Change* 3(7), 673–677.

Schirpke, U., Meisch, C., Marsoner, T. and Tappeiner, U. (2018) Revealing spatial and temporal patterns of outdoor recreation in the European Alps and their surroundings, *Ecosystem Services* 31, 336–350.

Schirpke, U., Timmermann, F., Tappeiner, U. and Tasser, E. (2016) Cultural ecosystem services of mountain regions: modelling the aesthetic value, *Ecological Indicators* 69, 78–90. https://doi.org/10.1016/j.ecoser.2017.11.017.

Schirpke, U., Zoderer, B.M., Tappeiner, U. and Tasser, E. (2021) Effects of past landscape changes on aesthetic landscape values in the European Alps, *Landscape and Urban Planning* 212, 104109. https://doi.org/10.1016/j.landurbplan.2021.104109.

Shannon, C.E. and Weaver, W. (1949) *The Mathematical Theory of Communication*. Urbana: Universiy of Illinois Press.

Shannon, D.M. and Bradshaw, C.C. (2002) A comparison of response rate, response time, and costs of mail and electronic surveys, *The Journal of Experimental Education* 70(2), 179–192. doi: 10.1080/00220970209599505

Sonntag-Öström, E., Stenlund, T., Nordin, M., Lundell, Y., Ahlgren, C., Fjellman-Wiklund, A., Slunga Järvholn, E. and Dolling, A. (2015) "Nature's effect on my mind" – patients' qualitative experiences of a forest-based rehabilitation programme, *Urban Forestry & Urban Greening* 14(3), 607–614. https://doi.org/10.1016/j.ufug.2015.06.002

Trost, J. (2001) *Enkätboken* [The Questionnaire Book]. Lund, Sweden: Studentlitteratur.

UK National Ecosystem Assessment. (2022) Ecosystem services, http://uknea.unep-wcmc.org/EcosystemAssessmentConcepts/EcosystemServices/tabid/103/Default.aspx (Accessed May 13, 2022).

Wood, C.M., Smart, S.M. and Bunce, R.G.H. (2015) Woodland Survey of Great Britain 1971–2001. *Earth System Science Data* 7(2), 203–214. https://doi.org/10.5194/essd-7-203-2015

Zahl-Thanem, A., Burton, R.J.F. and Vik, J. (2021) Should we use email for farm surveys? A comparative study of email and postal survey response rate and non-response bias, *Journal of Rural Studies* 87, 352–360. https://doi.org/10.1016/j.jrurstud.2021.09.029

14 Interviews with landowners and managers – what can they provide?

E. Carina H. Keskitalo and Elias Andersson

Introduction

In this chapter, interviews are introduced as a qualitative research method applied within the field of monitoring. The benefits, limitations, implementation, and application will be highlighted, and some vital reflexive questions and decisions will be raised. In this sense, the chapter provides a start and a description of a "bare bones" general methodology.

Similar to other research processes, qualitative research using interviews is influenced by multiple choices that will have consequences for the results. Therefore, these choices should be acknowledged, motivated, and discussed to constitute an integrated part of the process. This transparency and awareness will enable a better, solid knowledge production, and the close and integrated nature of interviews makes these aspects especially significant.

Because an interview can be conducted in many different ways, more rigour is required on the part of the researcher in clarifying how the study is designed, motivated, conducted, and analyzed. It is the researcher who constitutes the instrument or tool for information gathering and analysis, meaning that he or she needs to be "well calibrated" in order to measure correctly and not introduce bias.

The chapter will cover most of the steps in the process of conducting interviews but cannot include all of the factors that are relevant to interviews or all of the great variety of more specific methods and analysis. In planning and implementing an interview study, we urge the reader to read up on the vast introductive methodological literature on interviews and qualitative research. To this end, the reference list includes a number of sources specifically targeted towards environmental studies and conservation research and recommends, as general sources, for instance, Kvale (1996; or Holme et al. (1997) for Scandinavian language speakers) and, for a mixed methods approach, Creswell (2009).

Interviews – when and why?

Different methods and methodologies are designed and optimized to see, explore, and analyze specific aspects of reality. This is often based in different theoretical perspectives on everyday life and the environment and how these can and should be studied to reach a better informed understanding.

The general difference between quantitative and qualitative methods, where interviews fall under the latter, is that quantitative methods focus on few variables in multiple observations, whereas qualitative methods are focused on the opposite; that is, multiple variables in few observations.

DOI: 10.4324/9781003179245-14

Much quantitative study is thereby *deductive*; that is, it proceeds from an assumed hypothesis on how things work and draws only those conclusions that follow from the facts and premises of the hypothesis. However, if the hypothesis is wrong or does not include all relevant factors, in the worst case this may lead to a misconception of a situation or incorrect policy recommendations. As a result, it is crucial to use *inductive* research, based on identifying what factors are considered to be relevant to the situation itself, based on open-ended research, to give assurance that the factors reviewed actually are the important ones. A deductive research design is still possible but provides a less flexible and reflexive process, similar to a survey, and more limited data that need to be handled and analyzed accordingly. Text box 14.1 exemplifies how a deductive study could be revised to be an inductive study and shows the difference between the types.

Undertaking interviews may thus be most relevant when a research topic or question requires in-depth, detailed information that is difficult to measure or quantify; that is, on perceptions and interpretations of the same aspect and how these are related. This type of data might be particularly important when issues can be defined differently by different actors, individuals, or groups, which means that, for instance, a questionnaire will be difficult to use. From a more exploratory perspective, interviews might also be suitable when the knowledge on details within an area is limited or insufficient for specifying specific and relevant factors. This means that interviews are fruitful in a large number of cases and situations but also that they need to be tailored to the specific situation.

For each study in which interviews are to be undertaken, it is important to first identify the research questions (or title, aims, and objectives to be highlighted, as illustrated in Text box 14.1). In relation to this, it is then also important to identify the interview type: are you after individual understanding (person interviews) or group understanding (focus group interviews)? Based on this, you need to devise initial interview questions, select whom you will interview, and potentially submit your study for ethical review, after which you can conduct pilot interviews to refine your research design and finally undertake the study, followed by analysis and write-up. Considerations in the different stages are outlined in the remainder of the chapter.

Relevant questions for interviews

Based on your research questions, relevant theories, and existing literature, the next step is to start developing questions that help you produce relevant knowledge and understanding. The purpose of an interview is to gain an understanding of phenomena and their connections, as they are conceived by the research subjects/individuals you interview. The aim of the interview is to capture "the actor's point of view" to understand their actions, behaviours, thoughts, and motivations. Shulamit Reinharz (1992) stressed that "interviewing offers researchers access to people's ideas, thoughts, and memories in their own words rather than in the words of the researcher". In this way, interviews are about not just "talking with people" to extract specific information but rather gaining their knowledge, experiences, and insights on topics and understanding these in relation to the interviewee's perspective, situation, etc. Conducting interviews constitutes a fine balance between interacting and influencing, which requires an awareness of one's own actions and behaviour within the interview setting. This direct relation is also a prerequisite for a trustful and confident interview and exchange. For example, in an interview situation, the person being interviewed can tend to try to say what they think the interviewer wants to hear or is looking for.

Text box 14.1:   Examples of the title, aims, and objectives using deductive and inductive research strategies

Deductive study:

> *Title:* The effects of income from tourism on support for conservation: a case study from [the study site]
> *Aim:* To test whether income from tourism at [the study site] has an effect on support for conservation.
> *Null hypothesis:* There is no difference in support for conservation between those who receive income from tourism and those who do not.

Objectives:

1   To identify and define the different ways in which local people in [the study site] support conservation.
2   To develop an index of "support for conservation" based on the above.
3   To identify different sources of income for local people related to tourism.
4   To collect data on levels of support and tourism-related income from a representative sample of local people.
5   To analyze the patterns of variation in income and support within the study sample and test for statistically significant relationships between them.
6   To make recommendations for tourism management at [the study site] based on the results.

Inductive study:

> *Title:* Factors affecting support for conservation in [study site].
> *Aim:* To explore what factors affect people's levels of support for conservation.

Objectives:

1   To explore local people's perspectives on "conservation" – what they understand by the term, their experiences of it in practice, what value they place upon it, and what they regard as "support".
2   To document different forms of "support for conservation".
3   To explore people's explanations of why they do or do not give support.
4   To trace back the events and motivations that led people to give support (or not) by focusing in on particular situations that called for support.
5   To analyze the different factors that influence support (or lack of it) from the above.
6   To make recommendations based on the results.

In relation to developing an interview question guide, it is also important to consider the specific situation and positions of the persons you are interviewing with regard to what types of questions are both relevant and possible in order for the interviewees to provide insight into the phenomenon examined the study. This means that you need to break your research questions down into questions that directly relate to the experiences and knowledge of the interviewees. If, for example, you are interested in climate change impacts and interview a non-expert on the topic, you could ask questions like, "Will it affect you if spring comes earlier? If autumn comes later? If summers are warmer and wetter? It there are more extreme weather events such as storms and fires?", and so on. You thus do not assume that there will be an impact (in which case, you may bias the interviewee to say that there will be one) but rather ask an open question about whether there will be an impact. What is more, you do not assume that the interviewee will have the same understanding of *climate change* that you have but instead specify the question so that a lay interviewee will be able to understand it based on their immediate experience (see, for instance, Keskitalo (2008) for a study of different types of land or resource users that focused on these types of questions and Axelsson Linkowski et al. (2017) for an example of a conservation study integrating interview data). Understanding the interviewee is also important to better understand the information generated. It is therefore vital to include background questions about the person interviewed to reflect on the source of the information, not mainly to assess the source but to understand the person in relation to the information produced. For example, how long has a person been in contact with the phenomenon, from what perspective, and what practices shape their experience and understanding in different ways?

Once you have started thinking about these types of questions, which also implies thinking about whom you want to interview and in what way, it is time to start thinking about the most relevant type of interview and how to select whom to interview, before you then go back to developing very clear questions for these persons.

Types of interviews

Perhaps the most common tool in the qualitative researcher's toolbox is the personal interview. *Personal* implies that the interviewer, or interviewers, meet with a single interviewee. These types of interviews are often done, for instance, when you want perspectives from specific organizational positions or from persons who work in specific occupations or have specific experiences. You are then after their unique organizational, occupational, or experience-based viewpoint and understanding of the phenomenon. For instance, you may interview private forest owners who are of different ages and genders, have different sized forest, and have acquired their forest in different ways. In their individual interviews they may describe the reasons why they are forest owners and how they use their forest in a different way to others. This will give you an understanding of the breadth of the phenomena of reasons for forest ownership, and, if many of them share the same reasons for their forest ownership and use, you may also be able to identify crucial factors that can be studied in yet other cases to examine how general they are or to what groups and in what circumstances they may apply (see Bergstén and Keskitalo (2018) for a study to this purpose).

On the other hand, if you instead are interested in how a group (for instance, members of an organisation or a board or a smaller community) perceive or argue about specific topics, focus group interviews may be relevant. The focus group interview is frequently used in market research and product development to explore the perceptions and drivers of different consumer segments. In environmental monitoring, a similar focus can be placed on, for example, segments of landowners, stakeholders, or land users to better understand things like socioecological knowledge, peri-urban forest use, and decision making on a landscape level. Thus, these types of interviews are commonly conducted when you are after group perspectives, dynamics, and interactions that highlight the dual interest in both the specific topics and the relations within the group (e.g. different stakeholders). Therefore, similar to the importance of understanding the interviewee in a personal interview, it is important to understand the group in a focus group interview to also better understand the knowledge that it produces. Although the focus group interview is widely used in socioecological systems and conservation research, it is vital to be aware that a group perspective is formed by the group dynamics and specific norms in the situation and does not necessarily provide a deeper personal understanding of why someone becomes a forest owner (which may sometimes relate to relatively personal family or inheritance issues or reasons that might be less socially common). Focus group settings are thus not necessarily suited to more sensitive issues or to issues in which political, status, or power discrepancies or other relations between group members may limit their discussion of certain issues and steer your results (for a discussion on focus groups in a conservation setting, see O.Nyumba et al. 2018).

Finally, you may even come to the conclusion that what you are after is so sensitive or difficult to speak about that you cannot get at it by asking people questions. For instance, you may have started to suspect that there is a dynamic between large landowners or different interest groups, on a board, or in a local community, water council, or the like, that results in what land uses are supported and how. Or you may be interested in investigating how land is used and what qualities are appreciated (following a person or group out gathering, fishing, or the like). For such purposes, qualitative scholars often use *observation*; that is, participation in a setting where the dynamics one is interested in are playing out. This type of research is subject to multiple difficulties, not least gaining access, although this may be easier in some situations than others. The researcher also has to be very "well-tuned" in order to observe and catalogue only the observations that are relevant and needs to have defined observation protocols beforehand. For instance, is it the length of intervention and topic of intervention that board members make that you want to note? Or is it the way participants argue and what value words they use? Or is it their status or rhetoric? In the field, what uses do you record, and how? How dependent are these on the seasons or other factors that impact what they can show you right now? Participant observation is also subject to decisions about whether to participate fully – for instance, in a local observation situation in the field – or observe fully, without taking part. For these reasons, looking further into participative observation would need to draw on extensive method literature in the area (see Crandall et al. (2018) for a brief discussion of observation and method choice in relation to the conservation case).

Many of these methods may include data that you could potentially find in other forms, such as protocols from meetings or the like, and for this purpose text analysis is often a relevant supplement to the methods noted above. Text analysis can focus on, for instance, media, protocols, policy, or even legal studies and is subject to a number of detailed methodologies in each case (see, for instance, Kuckartz (2014) for an overview focused on general coding methods similar to those that will be described later in this chapter or Toivonen et al. (2019) for examples related to conservation).

Sampling – who is relevant, how, and why?

It is not the number of interviews you undertake that is the important thing but rather that they cover the phenomena you are studying. The previous section illustrated that whom you select to speak with or observe is of crucial importance both for what results you will get and for the relevance of your research. Thus, it is essential to also define the selection purpose, sampling or selection of interviewees, and, to some extent, interviewee characteristics and how many interviews/to what extent interviews are to be undertaken.

Contrary to quantitative research, which aims at generalizing to a given population (such as a country), qualitative research aims at generalizing to theory. This means that the important features of qualitative research are not exactly how many participants emphasize a factor but rather what factors are emphasized and, only as a secondary concern (to be shown, for instance, in additional studies or followed up in quantitative studies), how important they are. Thus, you aim to get an understanding of the full range of factors that influence a phenomenon; for instance, why someone is a forest owner. Assuming that the interviewees cover the relevant range of characteristics (age, gender, size of property owned, forest property use, etc.), once new subjects no longer add information on additional factors that might influence forest ownership, you are said to have reached theoretical saturation. This means you have gained a full understanding of what factors influence forest ownership in this case. You may also have a good understanding of what characteristics among interviewees influence what factors are more important to them. You can then generalize this to theory (established literature) on forest ownership and what factors influence it. The factors you have found may, however, differ in other cases – perhaps forest ownership is built on different parameters in Germany than in Sweden? – but in that case you have added to the literature as an important contribution that needs to be considered in other studies and that can be used to build hypotheses that support quantitative studies on a solid basis of understanding the full potential range of factors that may impact forest ownership.

Although this case of forest ownership may not be what you are after, the dynamics would be the same for research on, for example, what factors people value in landscapes or what factors influence farmers' decision making in a given case.

Qualitative studies, and specifically interviews, thus do not generally rest on random sampling as quantitative, generalizable studies do but on strategic sampling or selection for specific characteristics; for instance, to include those who have a say in a certain area or those who are in a group you want to research. For the forest owner study, you might thus design a maximum variation study (maximal variation in characteristics of forest owners) in which the number of interviews is enough when you reach theoretical saturation. Or you may not reach theoretical saturation but may at least be able to identify which factors everyone mentioned, which most mentioned, and which some mentioned, clarifying that there may be additional features you have missed. Therefore, it would be important to know what characteristics among forest owners you included and what type of forest owners suggested factors that were unusual in context. Perhaps the next study needs to highlight some specific types of forest owners to add to this.

The issue of interviewee characteristics is thus crucial. For a study of farmers in a certain area, perhaps you want to get all of the farmers in the area (a full case study with its focus on this characteristic), or maximum variation (such as in the "Swedish forest owner case" discussed in chapter 11, in which you cannot interview all of the several hundred thousand forest owners). Within a specific setting or context, key informant interviews with central and knowledgeable persons might be an option to a full case study. Identifying the key informants is crucial here and requires some insight into the relations and dynamics of the context. For a study of specific organizations, similar practices might be relevant based on the person organizationally responsible for the specific area. The sample is then motivated by the organizational position of the interviewee (e.g. head of advisory, forest management, etc.) and can then be selected across a number of organizations (strategic selection). For some studies there may be membership or statistical registers available; for instance, to support a maximum variation selection of interviewees. However, if you are after persons who have been involved in a network or active on an issue for which there is no membership listing or any type of official basis for selection, perhaps snowball sampling is the most accurate method for selection. In snowball sampling, you start with some – perhaps central – member you have identified in the network and ask this person (or persons) for the contact information for others who have been active. You might then also ask these persons when you interview them to identify others who have been active. Snowball sampling – so named because it emulates the mass a rolling snowball generates – is thus not designed to be inaccurate or random in any way but to constitute the most precise means of selection possible in some circumstances (see e.g. Rust et al. (2017) for examples from the conservation research area).

To recap, then: what should be evident from this discussion is that interviewees are generally selected from those who may have this highly specific information you are after. Qualitative interviews are thereby generally subject to strategic selection; unlike quantitative studies, they are not intended to be generalizable to a population. Instead, the generalization you want to make in a qualitative study is to *theory*: do you have enough and as varied a selection of persons to truly cover the important factors you are interested in? If your focus is on how different sizes or types of farm owners reason about environmental protection in an area or take measures to protect biodiversity, you may want to include both large- and small-scale owners with different types of farms. In that case, if very different owners highlight the same types of factors, you can theorize that these factors may be important. As someone in monitoring who may be more interested in practical application than in social science theory, by strategically selecting for variation, you may gain an understanding of the great many different factors that impact, for instance, management as it influences biodiversity (see, for instance, Rust et al. 2017). These types of interviewees can sometimes be found through official sources and, if these do not exist, "snowball sampling". (In some studies, such as the one described in Text box 11.2, one may also select interviewees on an organizational basis; for instance, those working in a specific capacity at each larger forest industry actor.)

Designing the interview guide and undertaking the interview

You are now – having broadly defined types of interviewing and interviewee selection – ready to construct your interview guide.

Because you do not know which interviewees have the information you seek on in-depth factors, and the impact of additional factors is also unknown, interviews are generally designed so that you introduce as little bias as possible in the role of interviewer. You are after the specific knowledge of your interviewees, not applying your concepts to them (other than in a later interpretation stage, in a transparent and clear way in your write-up).

Some of the most crucial issues in interviewing involve ensuring not only that no bias is introduced but also that research ethics are guaranteed, by making sure that you have informed consent at the outset of an interview. This includes informing the interviewee about the interview's purpose, the way the data will be used, how they will be referred to (anonymously, by type of organization, or by organization name, for instance) and that they can at any point retract their data. In some cases, interviewees are asked to sign a consent form, and different countries also have their own requirements. In the European Union (EU), General Data Protection Regulation (GDPR) regulation influences interview requirements, and many countries have transparency legislation that may impact how much you can promise your interviewee with regard to anonymity. Make sure you are well read up on what is relevant in your case and on what ethics boards (for example, at your university or institute) need to review before approving your study, once it has been designed and can be assessed. Whether formal ethics board approval is needed will depend on your proposal and country or region of operation.

In designing your interview guide, the questions directly following an introduction and clarification of informed consent are typically relatively broad, to avoid biasing your interviewee and to ensure that they identify what is important to them in relation to the phenomenon you are studying. These introductory interview questions can be of the type "Can you tell me about your work as a … ?", "Can you describe what a typical day is like for you?", or the like. These allow the interviewee to answer the questions based on their own situation, rather than you imposing your terms on them. The open questions should, of course, be adjusted to suit your study: if you want information on what natural features or other features of their agricultural land are important to them, you might ask about how they use the land they have and what (features) they feel are important for them to have. However, a qualitative interviewer might not necessarily ask directly about one specific type of land that they themselves then define in a way that differs from how the interviewee does. Different studies may have different needs, however, and for this reason there are different types of interviews, ranging from more structured (with more specific questions) to more open or unstructured (where you are after the interviewee's description and analysis in their own terms and thus keep the questions as open as possible). These are generally defined and developed in an interview guide, sometimes with follow-up questions that may or may not be used depending on what guide you are using (and to ensure that the interviewer does not use terms that introduce bias; see Text box 14.2).

Given this generally rather broad way in which questions are posed, with a focus on the interviewee's situation and definition of features, it may be difficult – and might introduce bias – for the interviewer to simply note what issues are brought up that seem important to the interviewee. Instead, qualitative interviews are generally recorded, with the interviewee's permission and noting that they may withdraw their data at any point. Here, EU GDPR regulation as well as open access developments have led to an increasing formalization of how research ethics and data management are set up. In much of social science, interviews have been anonymized, and descriptions and quotations provided in the results have often been limited to protect the interviewee (i.e. not describing sensitive issues in a way that can have direct negative impacts on the interviewee). Qualitative interviews are also often transcribed; that is, the full interview is written out verbatim to ensure that none of the interviewee's information is missed. This may result in a transcription of perhaps 20 full pages from a one-hour interview – a large amount of detailed data for a full interview study.

Text box 14.2:   Examples of questions in qualitative interviews

Questions asked in qualitative interviews are highly variable. Kvale (1996) has suggested nine different types of questions. Most interviews will contain virtually all of them, although interviews that rely on lists of topics are likely to follow a somewhat looser format.

Kvale's nine types of questions and our examples of questions:

1 Introducing questions: "Please tell me about your interest in bird watching."; "Have you ever seen a moose?"; "Why did you go to the National Park?"
2 Follow-up questions: getting the interviewee to elaborate his or her answer, such as "Could you say some more about that?"; "What do you mean by that?"; or even "Yeeees?"
3 Probing questions: following up on what has been said through direct questioning.
4 Specifying questions: "What did you do then?"; "How did he react to what you said?"
5 Direct questions: "Do you have an opinion on the lethal control of large carnivores in Norway?"; "Are you happy with the way you and your husband were treated while visiting the park's interpretation centre?" Such questions are perhaps best left until towards the end of the interview to avoid influencing the direction of the interview too much.
6 Indirect questions: "What do most people around here think of the ways park rangers treat local people living in the park?", perhaps followed up by "Is that the way you feel too?" to get the individual's own view of the situation.
7 Structuring questions: "I would now like to move on to a different topic".
8 Silence: allow pauses to signal that you want to give the interviewee the opportunity to reflect and amplify an answer.
9 Interpreting questions: "Do you mean that your opinion has changed because of the recent conservational actions?"; "Is it fair to say that what you are suggesting is that you don't mind having wolves in the area where you live, but when they are causing economic damage you should be compensated?"

Source: Text reproduced and partly revised from Torkar et al. (2011). Reproduced with permission from the authors.

However, at the time of writing, open access data management is placing unprecedented requirements on qualitative interviewing. The requirement by the EU to provide open access to data may mean that conducting interviews that may include information that is sensitive for the interviewee but does not fall under formal confidentiality requirements may be difficult. For instance, in the Swedish system it is becoming increasingly difficult to guarantee interviewees' anonymity (Keskitalo 2022). This may result in problems arising during the interview, because it cannot be foreseen when sensitive issues may arise. If anonymity of research data is now more limited and data are more openly shared, an interviewee who speaks openly about their land use and mention problems with neighbouring fields or practices could have problems if neighbours become aware of these comments. Even a study that did not foresee problems with sensitivity might end up with data management, anonymization, and related ethical issues. For these reasons, it is important to clarify how you will be recording data and review what requirements are placed on data access and sharing before the interview takes place. If you will not be recording data but only making notes, it may be crucial to have a note-taking protocol prepared, similar to what you may apply during participant observation. Make sure that you are actually writing down all of the factors the interviewee notes as important to the phenomena you are studying, and be prepared for how much these can vary (for instance, by undertaking pilot or test interviews before the study begins with people with characteristics similar to those included in the study). It is important, however, to be aware of the large amount of data that interviews generate and that notes will never be able to capture everything, which results in a reduction in underlying data quality for your analysis. Given that listening and note-taking while following an interview guide might be challenging, a full transcript is vital to reduce bias and ensure that crucial aspects are not missed and will be fully available for the analysis.

Once you have all of this information in place, with all of the requisite permissions, and have a fully designed study, it is time to contact your interviewees. If you have a maximum variation design, you should have alternative interviewees available in case some refuse. An introductory email can be sent with information on the study and what it will be used for, as well as details on setting up a time and place for the interview and how long the interview might take. Some may not have time or may otherwise not want to participate. If you cannot find replacement interviewees with your requisite features, you will need to state the resultant limitations in your final write-up.

Finding patterns – analyzing data and producing results

Finally, you have all of the data you have been able to gather. It has taken months, perhaps, and you (or your research assistants or your team at large) have spent weeks transcribing (a one-hour interview can take about a day to transcribe manually). Although voice recognition software can speed this up, you may still need to check the transcription against the recording. So, either you have your transcribed data – perhaps verbatim, so that you can check whether you have rephrased questions that might bias the interviewee – or you have your notes. What do you do now?

Given that the aim of much qualitative interviewing is to allow the interviewees themselves to provide information in their own words, reflecting, for instance, on what factors they see as important, coding and analysis constitute a crucial part of the work. Interviewees may describe a large array of factors and often describe some factors that are not important to you as a researcher or to your work with monitoring. For this reason, those who work with qualitative interviewing often develop a coding guide, typically with a first version developed at the same time as the interview guide. The coding guide might list the types of broad factors and themes you are interested in (for instance, ecosystem services and forest planning and management). Based on these main themes, you can then carry out your initial round of coding of the material (i.e. all of the interviews combined). You will then have a large amount of data under each broader code, which will require you to create sub-codes to sort the material under each code. The sub-codes can either be derived from your theoretical framework or be based on communalities in the material (e.g. perceptions of/willingness to pay for ecosystem services or tools/technologies for forest planning and management). It is also important to reflect on whether there is any important information that falls outside the different levels of coding and whether it is relevant to include these. The structured coding also helps you to acknowledge factors or topics that you did not expect and thus did not notice during the interview. As a qualitative interviewer, it is through the analytical work that you take precautions against introducing bias into the data your interviewee has provided. To support this work, there are a large number of software applications that can be used. This is especially relevant in dealing with large amounts of data. For smaller amounts of data, the same work can be done on paper with notes and highlighters in different colours. Once you have gone through all of your interviews in this way, you will have large lists of quotations under each code and sub-code. At this stage, each sub-code should be so clearly developed that you can then, in the description of the results, describe what it constitutes and how common this consideration was among the interviewees in a way that highlights your analysis, and basis for analysis, to the reader in a transparent way (see e.g. Holzer et al. (2019) for an example of coding in relation to a transdisciplinary study). See Text box 14.3.

Given the general selection mechanisms applied in a qualitative study, in which what is important is, for instance, the variety within a focused group or a strategic selection, it is not relevant to use percentages for analysis. Often, given the large data basis provided by even a few qualitative interviews, interview studies seldom include very large numbers of interviews, because the strategic selection can be designed to cover a manageable number of interview subjects (for instance, a certain area and different types of property owners in the area or strategic selection of those with similar assignments in the same type of large company or agency). This means that percentages are not relevant to analysis: what is important instead is what all or most of the (varying) interviewees mention. If some issues are only mentioned by some groups of interviewees, this may also be important, because this could indicate an area of interest particular to these groups, which might be subject to additional studies (for instance, are small-scale owners in a specific area more open to undertaking specific types of environmental protection measures?).

This type of analysis constitutes only a "bare bones" outline of a general method for coding. Coding and analysis can also be adapted to a number of more specific purposes, such as discourse analysis (see, for instance, Crandall et al. (2018) for a discussion).

Text box 14.3:   Describing the results

A typical description of results in a qualitative study includes the themes and sub-themes you have found and provides clear indications as to whether the factors described were found among the interviewees at large or among a specific group or are atypical. The factors that you can say something about with the most confidence – for instance, in a study based on maximum variation in a category of interviewees – would be those that all or most of the interviewees mentioned.

You will often provide quotations to illustrate your interpretation of the interviews and what was said. For instance, you might note that "a typical comment among the participants was … " and then provide a quotation that exemplifies the factor and how it was discussed.

For a monitoring study, you may want to provide some shorter analysis like this in an appendix to illustrate, for instance, to decision makers that your study is well founded and describe the factors that were relevant in the text.

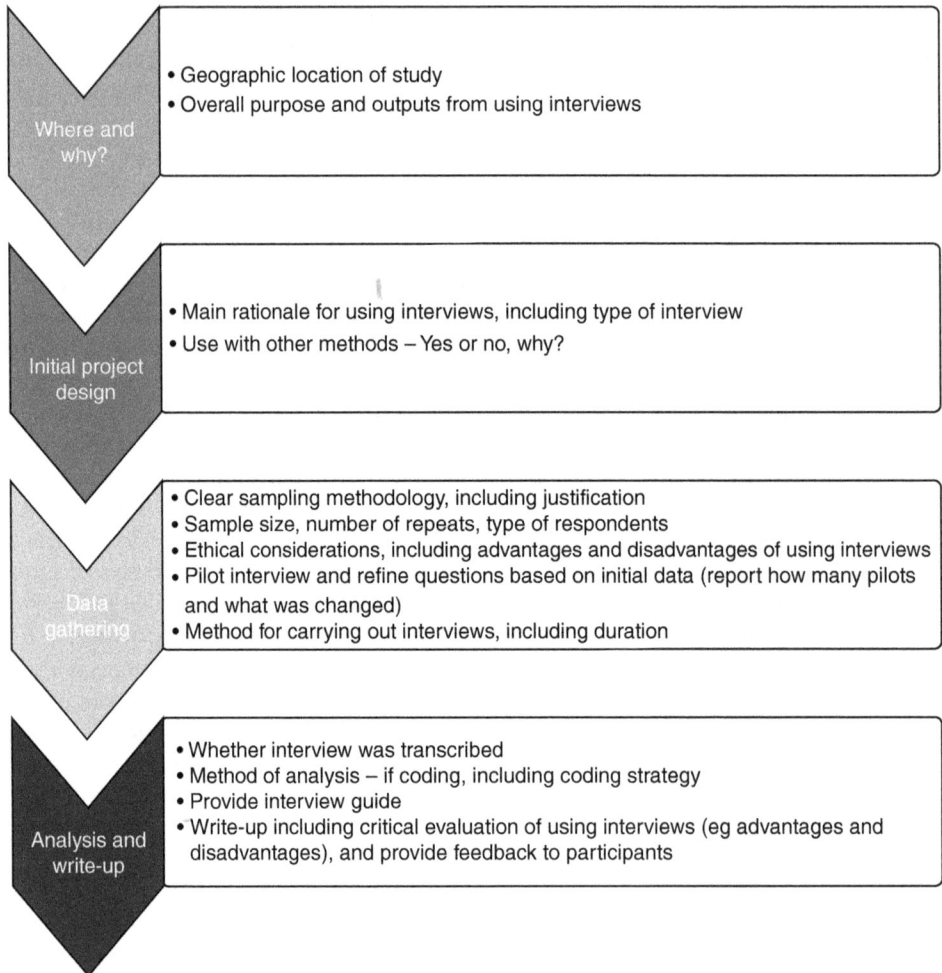

Figure 14.1 Key data to provide when reporting on interviews.

Source: Figure and caption reproduced from Young et al. (2018).

Given all of these potential choices in study design, it is important to clarify all of the types of factors that have been discussed in this chapter in your write-up. The research design should be oriented to your aim and purpose, and the interview type and interviewee selection as well as the design of interview guide and chosen coding should follow from this (see Figure 14.1 for an outline of general considerations and data to provide when reporting on interviews).

Summing up

As this chapter has shown, qualitative interviewing is about reflexivity – for instance, considering how to avoid biasing a study and how to keep the questions open – and awareness of what choices are made to support a research design that addresses the purpose of the study. The write-up of a qualitative study demands no less rigour than that of a quantitative study and may even require more rigour and transparency on the part of the researcher as the instrument. Qualitative studies can significantly support an understanding of land use factors (see e.g. Bennett et al. (2017) for a discussion), but the researcher needs to either develop a thorough study or contact groups that regularly fulfil such roles and are well versed in the method. All choices in the process have implications for the results, and the study needs to be designed and undertaken with attention to all of these factors to provide the deep and broad understanding that is the strength of qualitative studies.

Key messages

- Like quantitative methodology, qualitative methodology requires a high degree of rigour and exactness; for instance, in determining what type of interviews to undertake; what to ask, to whom, and why; and how to code and analyze the material.
- Interviews are sometimes the most exact way to obtain social information, and interviewee selection targets these key sources through different means of strategic selection.
- Avoiding bias in interviews or analysis is crucial, and the design of the interview guide, pilot interviews, and formal coding are important steps.

Study questions

1   When may interviewing be a particularly important methodology to use?
2   Why and when is strategic selection important for selecting interviewees?
3   When is snowball sampling the most accurate way to identify relevant interviewees?
4   What is interview coding and why is it important?

Further reading

Kvale, S. (1996) *Interviews: An Introduction to Qualitative Research Interviewing*. Thousand Oaks, CA: Sage.

This book is sometimes used as one of the standard references for interview methodology.

References

Axelsson Linkowski, W., Kvarnström, M., Westin, A., Moen, J. and Östlund, L. (2017) Wolf and bear depredation on livestock in northern Sweden 1827–2014: combining history, ecology and interviews, *Land* 6(3), 63.

Bennett, N.J., Roth, R., Klain, S.C., Chan, K.M., Clark, D.A., Cullman, G., et al. (2017) Mainstreaming the social sciences in conservation, *Conservation Biology* 31(1), 56–66.

Bergstén, S. and Keskitalo, E.C.H. (2018) Feeling at home from a distance? How geographical distance and non-residency shape sense of place among private forest owners, *Society & Natural Resources* 32(3), 1–20.

Crandall, S.G., Ohayon, J.L., de Wit, L.A., Hammond, J.E., Melanson, K.L., Moritsch, M.M., Davenport, R., Ruiz, D., Keitt, B., Holmes, N.D., et al. (2018) Best practices: social research methods to inform biological conservation, *Australasian Journal of Environmental Management* 25(1), 6–23.

Creswell, J.W. (2009) *Research Design: Qualitative, Quantitative, and Mixed Methods Approaches*. Thousand Oaks, CA: Sage.

Holme, I.M., Solvang, B.K. and Nilsson, B. (1997) *Forskningsmetodik: Om Kvalitativa och Kvantitativa Metoder*. Studentlitteratur.

Holzer, J.M., Adamescu, C.M., Cazacu, C., Díaz-Delgado, R., Dick, J., Méndez, P.F., Santamaría, L. and Orenstein, D.E. (2019) Evaluating transdisciplinary science to open research-implementation spaces in European social–ecological systems. *Biological Conservation* 238, 108228.

Keskitalo, E.C.H. (2008) *Climate Change and Globalization in the Arctic: An Integrated Approach to Vulnerability Assessment*. London: Earthscan.

Keskitalo, E.C.H. (2022) Open access and sensitive social sciences data in different legislative contexts: the case of strategic selection "elite" interviewing in Sweden, *International Journal of Qualitative Methods* 21, 1–10.

Kuckartz, U. (2014) *Qualitative Text Analysis: A Guide to Methods, Practice and Using Software*. Thousand Oaks, CA: Sage.

Kvale, S. (1996) *Interviews: An Introduction to Qualitative Research Interviewing*. Thousand Oaks, CA: Sage.

Newing, H., Eagle, C.M., Puri, R.K. and Watson, C.W. (2011) *Conducting Research in Conservation*. London and New York: Routledge.

O. Nyumba, T., Wilson, K., Derrick, C.J., and Mukherjee, N. (2018) The use of focus group discussion methodology: insights from two decades of application in conservation, *Methods in Ecology and Evolution* 9(1), 20–32.

Reinharz, S. (1992) *Feminist Methods in Social Research*. New York: Oxford University.

Rust, N.A., Abrams, A., Challender, D.W., Chapron, G., Ghoddousi, A., Glikman, J.A., Gowan, C.H., Hughes, C., Rastogi, A., Said, A., et al. (2017) Quantity does not always mean quality: the importance of qualitative social science in conservation research, *Society & Natural Resources* 30(10), 1304–1310.

Toivonen, T., Heikinheimo, V., Fink, C., Hausmann, A., Hiippala, T., Järv, O., Tenkanen, H. and Di Minin, E. (2019) Social media data for conservation science: a methodological overview, *Biological Conservation* 233, 298–315.

Torkar, G., Zimmermann, B. and Willebrand, T. (2011) Qualitative interviews in human dimensions studies about nature conservation, *Varstvo Narave* 25(2011), 39–52.

Young, J.C., Rose, D.C., Mumby, H.S., Benitez-Capistros, F., Derrick, C.J., Finch, T., Garcia, C., Home, C., Marwaha, E., Morgans, C., et al. (2018) A methodological guide to using and reporting on interviews in conservation science research, *Methods in Ecology and Evolution* 9(1), 10–19.

15 Designing and adapting biodiversity monitoring schemes

Alan Brown, Henrik Hedenås, Einar Holm, Torgny Lind, Anna E. Richards, Suzanne M. Prober, and Becky Schmidt

Introduction

This chapter brings together the ideas and recommendations from previous chapters, to look at best practice, innovation, and adaptation in biodiversity monitoring. Is our monitoring project effective, sustainable, and able to answer the questions being asked now? How do we keep it relevant to the needs of science, society, and policy while maintaining continuity of measurements and interpretation with the past so we can be confident about the accuracy and reliability of results in the future?

The chapter starts by briefly reviewing some different types of monitoring, stressing that the focus of the book is on the overlap between "question-driven" and "mandated" monitoring, and shows how monitoring can be seen as having different cycles of design, data collection, analysis and reporting, innovation, and adaptation – even though most long-term practical schemes are messy and these cycles are often overlapping and incomplete, because innovation in data collection and other external pressures can quickly change what was originally planned.

This nested sequence of activities is used to structure the rest of the chapter, looking first at three components of good design that will influence adaptive monitoring: developing (and revising) a conceptual model of ecosystem change, using predictive models to extend and focus the objectives of monitoring, and managing different sources of uncertainty in monitoring design.

The chapter goes on to look at some of the essential elements of continuity and innovation in monitoring projects, showing how the design and adaptation of projects can be seen as a series of stages from stand-alone designs that take advantage of new datasets to the incorporation of the results into wider schemes of hybrid analysis and collaboration. An important point is how innovation both makes existing projects better able to answer existing questions and enlarges the scope for adapting to new questions, while maintaining the essential elements of continuity.

A key type of conceptual model is the "state-and-transition model", involving habitats, land cover or land use, or population states. These can be set in the wider context of drivers of change and pressures from society that affect biodiversity. An example is given for Australian forests and woodlands (Text box 15.1). It is also possible to model future societal changes and preferences that influence both what we monitor and how the results are interpreted. An example of this microsimulation is shown for forest management in Sweden (Text box 15.2).

Every monitoring project has to adapt in its own way, so rather than trying to set out what would have to be unfocused general principles, the chapter illustrates how

DOI: 10.4324/9781003179245-15

continuity with past monitoring is balanced against adaptation to meet future require-ments with examples taken from other chapters and, in this chapter, from the National Inventories of Landscapes in Sweden (NILS) land cover–land use project and the Swedish National Forest Inventory (Text boxes 4.1, 15.3 and 15.4).

Leaving technologies aside, as chapter 17 illustrates for reindeer husbandry, the most important component of monitoring is people: the people who need the information, who help frame the questions and use the results, and the experts who come up with a survey-sampling design, carry out the fieldwork, and interpret the data.

Expertise in ecology, forestry, and land management is critical for drawing up a conceptual model that isolates and focuses the questions (hypotheses) being addressed in the statistical design, and statisticians and field staff should work together with the people who use the information, to decide which observations are both practical and necessary and how to allocate effort and time across the range of chosen measures. As Lindenmayer and Likens (2018) stressed, continuity of staffing, shared understanding, and partnerships between a range of experts are all marks of a successful project.

Different types of monitoring and some related activities

There are a number of useful ways of categorizing and understanding monitoring projects and schemes. The European Environment Agency definition of monitoring introduced in chapter 2 has two parts, referring both to how well a plan, programme, or measure complies with environmental policy and to the way in which information can be used to make sure projects meet their objectives. Monitoring is seen as part of a regulatory mechanism, giving feedback that helps steer policies and practical manage-ment towards achieving their goals.

Monitoring can also be found in many different settings, including projects whose primary purpose is long-term monitoring (such as the UK Countryside Survey in-troduced in chapter 2 or the NILS project mentioned in chapters 4 and 5), monitoring schemes combined with experimental approaches and modelling in research projects looking at ecosystem functions, monitoring to find a source of pollution or detect illegal waste disposal, monitoring to look at the environmental impacts of development and industry, and monitoring to say whether or not the maintenance or restoration of a habitat or species population has been achieved (see chapters 16 and 17).

Looking more widely at the motivation, sponsorship, and operation of monitoring schemes, Lindemayer and Likens (2018) suggested three types:

* *Curiosity-driven or passive monitoring* typically has no statistical design and is not linked to answering specific questions or triggering any particular management intervention.
* *Mandated monitoring* is carried out in response to the requirements of government legislation or directives, usually with some general specification on what habitats, species, or environmental measurements to include but often leaving the details of sampling design and methods to governments or regional agencies.
* *Question-driven monitoring* has a rigorous statistical design that is able to answer pre-determined questions or hypotheses, including effect sizes, statistical significance, and statistical power.

Lindenmayer and Likens (2018) used these categories to point out some of the weak-nesses of some government-mandated monitoring projects devoted to data collection often at a large geographical scale and covering very many variables but without testing predictions or revealing mechanisms of change.

The main focus of this book has been on the second and third types, and especially where they overlap: mandated monitoring but also question-driven monitoring, including those larger mandated monitoring projects that answer the needs of both policy and long-term habitat management. With some care taken in the design, sufficient funding, and continuity of staffing, we argue that mandated monitoring can and should be question driven.

However motivated, monitoring looks at the state – and changes in state – of some variables of interest. For biodiversity monitoring, the EEA definition's requirement to test compliance with policies or objectives can be made more specific: "using an intermittent (regular or irregular) series of observations over time, to show the extent of compliance with a standard or the degree of deviation from an expected norm" (paraphrased from Hellawell 1991).

In practical habitat management, this standard might be defined from the start as a set of directly measurable objectives (chapter 16), and here monitoring can be as little as a single survey followed by a comparison (of the inferred population state) with the standard. More generally, biodiversity monitoring uses a series of repeated surveys to detect changes and trends in condition; that is, comparing estimates of population parameters from successive surveys. Both versions are clearly question driven and can be expressed as testable hypotheses that a standard has been achieved or a specified minimum degree of change has been detected and – crucially – knowing that the sample survey design is very likely to detect this change, if it exists.

The power of question-driven monitoring is the way in which the hypotheses to be tested are incorporated in the sample-survey design and field methods from the start, contrasting with the type of data collection where these are only asked retrospectively and with less certainty that data collection will allow them to be answered. Hellawell (1991) used the term *surveillance*[1] for an extended programme of surveys systematically undertaken to provide a series of observations in time, where no specific questions (in the form of hypotheses about the direction and extent of change) are being asked even at the start.

In a well-designed (and well-funded) scheme, some of the opportunities for both general surveillance and question-driven monitoring can be taken up. An optimal design will be able to accommodate both of these; for example, using key replicated contrasts to address specific initial questions using a core set of measures but accommodating additional measures and new technology over time that enables a broader suite of questions to be addressed. Criteria to facilitate the latter will include a strong underpinning design (e.g. control and managed plots); appropriate (potentially nested) plot or transect sizes for measuring different types of organisms (acknowledging that you may not be able to accommodate all); orientation, shape, and permanent marking to best align with remote sensing (e.g. large square plots with <1m accuracy of corner locations); collection of good metadata; and, potentially (but not essential), collection and storage of initial baseline samples such as soils that can be measured later (see Further reading section for some practical texts on methods).

As well as new questions being raised, relevant new datasets can become available that have the potential to improve the accuracy, precision, detail, and scope of existing monitoring; that is, auxiliary data that could not be incorporated into the original sample-survey design. Chapter 4 made a general distinction between *design-based* monitoring and a more recent *model-based* approach (originally developed out of geostatistical methods used by soil scientists) that takes advantage of the types of auxiliary variables now becoming available from satellite and airborne remote sensing. Even though, as stressed in chapter 4, the most effective use of any information is to incorporate it into the design from the start, model-based or *model-assisted* methods (along with some of the hybrid methods described in chapter 9) can be used to improve the spatial detail, supplement gaps in data collection, and make more precise estimates by incorporating new auxiliary variables into an existing design, working with data already collected (Ståhl et al. 2016; note the difference between hybrid methods and hybrid estimation).

New technologies such as remote sensing and environmental DNA (Cristescu and Hebert 2018) can be used directly to produce thematic maps and spatial indexes or as complementary methods to in situ survey, monitoring, and field sampling (Lausch et al. 2016; Anderson 2018; Ruppert et al. 2019).

It is also important to include other forms of evidence, especially from surveys (chapter 13) and interviews with landowners and other interested parties (chapter 14) to make sure that monitoring is truly cross-sectoral and inter-disciplinary and that projects service the needs of both policy and the communities that policymakers represent. To monitor effectively, we need to understand systems of land use (chapter 11) and the wider social and economic context that frames both the questions asked and how the results will be used. This type of co-production is exemplified in chapter 17.

There are also types of investigation related to monitoring, including schemes of evidence collection that are similar to experimental designs: the before–after–control–impact (BACI) approach mentioned in chapter 2 and modifications of BACI designed to cope with missing prior trend data or reference sites that converge on the more typical statistical survey sampling methods. In these BACI designs, which are used to assess the effects of environmental impacts and recovery, the "standard or expected norm" is set more dynamically by the control sites, which are assumed to track natural variation year on year. More generally, combining monitoring with research methods and experiments that look at ecosystem functions and cause and effect can both inform predictions and models of future change and identify new key measures as candidates for monitoring.

The important role of curiosity-driven monitoring

Even though curiosity-driven monitoring is to some extent relegated by Lindenmayer and Likens (2018) to a passive role, starting with curiosity and remaining curious through the life of any monitoring scheme is one of the most important human ingredients of all schemes. Curiosity and attention to detail, especially in field inventory, both maintains our interest and personal satisfaction in a project and allows us to modify and adapt the project to answer new questions prompted by everything we observe, helping us to view the narrow focus of recording specified in the monitoring protocol in its wider context (see, for example, National Biodiversity Network 2022). Curiosity-driven projects, especially on managed sites, can also be much more effective in making observations of rare and important species, especially in those cases where some searching and local knowledge can enable the sort of complete census that would be highly unlikely in a probability sample. Notice how this focus is similar to the qualitative survey methods discussed in chapter 14.

Cycles of decision and action

The early chapters in the book introduced monitoring as having repeated cycles of survey, data collection, and analysis, within longer term cycles of redesign and adaptation to new questions and policies. These can be seen as cycles of first making decisions and then carrying out actions (Figure 15.1a). In chapters 2, 4, and 5, we saw the steps in sample-survey design developing at the same time as specifying practical in situ field methods and other types of data collection. As envisaged, this design process leads to routine data collection and reporting, as well as training, minor adjustment, and patching up of methods – what builders call "snagging" – to make monitoring work. Though getting a project off the ground might have to be done in several stages, notably where pilot studies are needed to test methods and estimate variability and sample size, once up and running, this could be seen as business as usual.

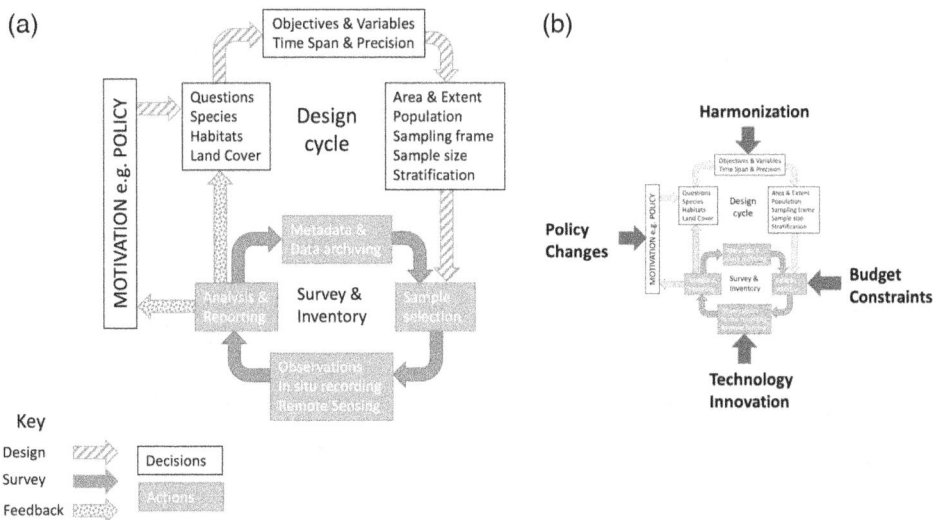

Figure 15.1 Adaptive monitoring cycles of data collection, reporting, and design over different time-scales (compare Figure 4.1 in chapter 4 and figure 2.1 in Gitzen et al. 2012). Monitoring can be seen as (a) a cycle of decisions, some of which are repeated as schemes are revised, and a cycle of actions such as survey and inventory, including in situ observations, analysis, and data archiving. Feedback from repeated surveys is returned to the agencies responsible for policy – or other stakeholders and clients – via a feedback loop from analysis and reporting. Feedback can also prompt revised decisions about the importance of different species, habitats, and land covers; the objectives, variables, time span, and precision needed; and the sampling frame and sample size. In practice, there are subsidiary feedback loops that have been left out; for example, training and quality control around field data collection. There will also be (b) external pressures that are outside the immediate scope of the monitoring design but can nevertheless influence how both the design and survey adapt over time. These include policy changes, the need for harmonization of variables and classifications, budgetary constraints affecting the sample selection, and the way in which new technologies update the type of observations and data collection. Other external factors, not shown in the diagram, can influence the form of analysis, notably innovations in software such as machine learning methods.

However, real long-term projects suggest that this picture is far too neat, and a steady state of business as usual is rarely achieved for more than a few years. Instead, pressures to change the objectives and thematic classes, cut funding, and adopt innovative methods of survey and inventory can often change the design and in situ protocols even before the first survey has been repeated (Figure 15.1b). This seems to have been especially true of the last two decades with the realization of climate change and the growth of new technologies. Making neat like-for-like comparisons of data may not always be possible.

Earlier chapters have introduced a number of long-term monitoring projects, notably Britain's Countryside Survey (in chapter 2), the National Inventories of Landscapes in Sweden (in chapters 5, 8, and 9). There are also a number of products from notable long-term forest inventories mentioned in several chapters, such as Sentinel-1 and -2 plus lidar (laser scanning) used together with data from the inventory to create the latest forest map in chapter 7 and as one of several data sources for validation/mapping of land cover in chapters 8 and 9. Of course, we can expect that, at the time of conception, these were designed using the best of (then) current knowledge, within the constraints of funding to answer a set of what were (then) the most important questions. These are the initial cycles from design, through initial data collection and snagging to routine reporting.

Over a period of decades, there will have been changes in the focus of policy – for example, from just forest products to include biodiversity in forestry; in analytical methods, including the more widespread use of power analysis; and in the availability of new types of data and processing described in chapters 6 to 9. As chapter 8 hints, innovation will continue as new technologies such as eDNA become routine, just as we have seen satellite remote sensing becoming routine for biodiversity monitoring over the past few years.

Even without changing the questions (hypotheses) being addressed, monitoring schemes innovate new and better methods of doing the same thing but with more precision, accuracy (eliminating bias), efficiency, and information content. For example, by adding routine photography to recording plots, using Global Positioning System (GPS) for positioning (supplemented by local markers such as buried metal plates) and relocating plots, and mobile phones to record observations instead of paper and pencil. As chapters 6, 7, and 8 show, innovation can include using citizen science volunteers to make bulk observations, followed by using remote sensing and modelling to better target and weight the results to remove bias, and using new machine learning and deep learning methods to interpret the results. If well designed, citizen science observations can also be used to validate satellite data analysis. Hybrid methods (chapter 9) can be used to combine the results of different types of observation, and model-based and model-assisted methods of inference can be used to combine in situ observations and auxiliary data using statistics and informatics (Kwong et al. 2022 is a nice example). Notice how innovation in biodiversity monitoring is constantly pulled along by new technologies, many of which were developed in other fields such as medicine, space exploration, and defence, and by new methods developed in disciplines such as statistics and the social sciences. An important difference here is between innovation in the methods of measurement (mapping habitat extent from satellites rather than air photographs) while keeping the measures themselves (extent, habitat classes) consistent.

Other factors, including changes in policy, shift the focus of monitoring (and might add to the budget or take some of it away) and can be seen to push monitoring projects to adopt new approaches (Lindenmayer and Likens 2009; Burton et al. 2014). This might mean answering new questions, answering existing questions with greater accuracy and precision or to a shorter timetable, or trying to preserve the full scope of a scheme despite a reducing budget. Ideally, adaptation[2] will make the most appropriate use of innovative methods, and adaptation itself might be very positive if it can free up resources to fund urgently needed biodiversity monitoring elsewhere. It is essential for the basic survey methodology to have some flexibility, so that the data can be re-interpreted in the future. On the smaller scale of a site or habitat patch, monitoring will need to be flexible enough so that it can answer new questions that come up in future site management. The examples of land monitoring and forest inventory in Text boxes 15.3 and 15.4 show very long-established inventories being adapted to provide information and analysis matching the new priorities of society.

We should also consider the bigger picture and cycles of collaboration and harmonization of data collection with other projects in the same area or with similar projects in other countries, contributing towards local and global-scale models that attempt to predict the likelihood of future trends in biodiversity, land use, climate, and the environment. Collaboration might mean, for example, collecting variables and using standards that are better able to support data integration, including common semantic classes for land cover and land use. This possibility has been encouraged by strategies and directives across countries, such as the Food and Agriculture Organization's (FAO) forest monitoring programme and European Union (EU) Biodiversity Strategy 2030 discussed in chapter 3, and by continental-scale initiatives to create common frames of reference for harmonized data collection, such as the EU INSPIRE directive and the CORINE land cover classification. Lucas et al. (2022) proposed a globally relevant taxonomy of habitat and land cover change classes which show how harmonized analysis and reporting can be extended even further.

In addition to these synoptic long-term monitoring projects adapting to meet new requirements, the existence of stable long-term projects can allow layers of more specialized, complementary projects to be set up – with parallels to the way in which commercial satellites can complement the major remote sensing projects. These monitoring schemes can either be seen as filling in the gaps left in biodiversity monitoring or, better, monitoring some things very well precisely because they do not need to monitor everything. For example, in the UK, the Countryside Survey is complemented by habitat monitoring set up by the nature conservation agencies to focus on areas of scarcer habitats and high diversity within managed, protected sites. Another component of this survey, Land Cover Map, is complemented by mapping from both in situ observations and remote sensing in large parts of England and Scotland and the whole of Wales, using more detailed habitat classes. Other UK agencies produce maps of forestry (the National Forest Inventory), agricultural land use, and even individual trees. Long-standing surveys include breeding birds, butterflies, and other groups of insects; sites dedicated to tracking long-term environmental change such as the Environmental Change Network (2022) – part of the International Long-Term Ecological Research network (2022); and, of course, a large number of environmental and meteorological measurements. More generally, field observations are brought together by the National Biodiversity Network of Britain (2022).

This ecosystem of monitoring projects includes those sponsored by central government alongside monitoring carried out by charities and lobby groups, including highly knowledgeable enthusiasts and volunteers, so the motivation and involvement of the individuals varies as much as the observations they carry out.

Modelling what we need to monitor

An important part of planning monitoring is making decisions that narrow down the target species or habitats, variables, and objectives. Conceptual models can be used to convert a very general idea of what may be happening to biodiversity, habitats, and populations to something that is measurable and capable of being monitored (see Lindenmayer and Likens 2018). This is the first step in deciding what we need to monitor, before we can specify the sample-survey design, statistical estimators, and type of in situ or other measurement. A different type of model attempts to predict future changes in the state of land cover, land use, and habitats by modelling trends in social preferences for different land uses.

Though other important models are also relevant – for example, modelling climate change or forest productivity – the two examples of conceptual and predictive models already illustrate how both ecological studies of biodiversity and social studies need to consider not only the current state of biodiversity and land use but the possible range of future states in order to decide what should be monitored.

These different aspects can be combined in a DIPSIR (driver, impact, pressure, state, impact, and response) model, which looks at the dynamics of changes in the state of biodiversity as a process of cause and effect. Though there is no space to go into detail here, a useful review of DIPSIR looking at the socioecological context is Gari et al. (2015).

Conceptual models of habitat states and transitions

State-and-transition models (Bestelmeyer et al. 2010; Brandon et al. 2021) are a form of conceptual model that can be used to test our understanding of a habitat (or other land cover class) before making important decisions that will influence how it is managed and monitored. They illustrate how any given habitat is likely to respond to the various environmental pressures and management actions most likely to affect it (Rumpff et al. 2011). For practical management and monitoring purposes, state-and-transition models are best developed and applied at the individual site level, or general models can be adapted to a particular site. At this level, we can have a good understanding of how known pressures and management activities will impact the habitat. As an example, Figure 15.2 (see Text box 15.1) shows a summary of the current understanding of dynamics of floodplain eucalypt forests and woodlands in southern Australia.

While acknowledging that habitats change gradually across ecotones and over time, the state-transition simplifies the picture into components with starting and ending states, shown as boxes, connected by arrows indicating the transitions and the pressures or types of management that cause them or reverse them. These two distinct states are first described ecologically and then translated into quantitative descriptions that can be monitored; for example, listing indicator species or thresholds for cover values that might define the successful restoration of habitat condition as the desired end state or other triggers of management action (see de Bie et al. 2018). Of course, the models must be applied with caution, because they describe transitions between past or known states, and future changes may be very different. The point is not so much that the model captures the complete range of possible dynamics but that it allows enough simplification for decisions to be made about which species or measurements to monitor.

This is especially helpful when we want to isolate that part of the system that is currently of most interest. In analytical terms, the two states and the transition are treated as being *conditionally independent* of the rest of the model, making the assumption that the starting state summarizes everything we need to know about the potential for any transitions that might follow. This sort of simplification and focus is essential in making a monitoring scheme fit for purpose, though a model needs to be selected with care (Sato and Lindenmayer 2021). In particular, the model is typically based on past and current dynamics, and the future dynamics may be different, affecting different states on different timescales, as they are influenced by climate change and other factors outside the control of site managers.

The models (such as the Gunbower–Koondrook–Perricoota [GKP] case study in Text box 15.1) usefully combine measures of ecosystem condition and extent and can be used for ecosystem accounting, and the incorporation of endogenous (i.e. outside the model) ecosystem dynamics as a context avoids drawing false conclusions from the measures being monitored.

Text box 15.1:   State-and-transition model of inland floodplain eucalypt forests and woodlands at the Gunbower–Koondrook–Perricoota Forest Icon Site, Australia

Gunbower–Koondrook–Perricoota (GKP) forest icon site is a 56,020ha area consisting of national parks and state forests (managed for native timber harvest) located on the River Murray in southern Australia. It is a nesting and breeding site for internationally protected migratory waterbirds and contains the second largest extent of river red gum (*Eucalyptus camaldulensis*) forests in Australia (Hale and Butcher 2011; Harrington and Hale 2011).

GKP sits within the Murray-Darling basin, Australia's largest river system with significant cultural, environmental, and economic value. Over the past 150 years, there has been extensive extraction of water and modification to endogenous flow regimes in the basin (that were previously regulated by climate and management by First Nations Australians) to provide water for agriculture, towns, and industries. As a result of these modifications, the connectivity of rivers to floodplains and to groundwater has been diminished, adversely impacting the health, abundance, and range of water-dependent species and ecosystems in the basin (Murray-Darling Basin Authority [MDBA] 2019).

In 2012, the Murray-Darling Basin Plan (the Basin Plan; MDBA 2012b) was introduced with the aim of returning the basin to a healthy working system by improving its environment, while balancing social and economic needs in a sustainable way. Within the Basin Plan there is a strategy for delivery of environmental water to GKP to meet high-level environmental objectives including maintaining and restoring healthy wetlands and river red gum communities, providing conditions suitable for successful waterbird breeding events, and maintaining healthy native fish populations in wetlands (MDBA 2012a; Hale and SKM [Sinclair Knight Merz] 2011).

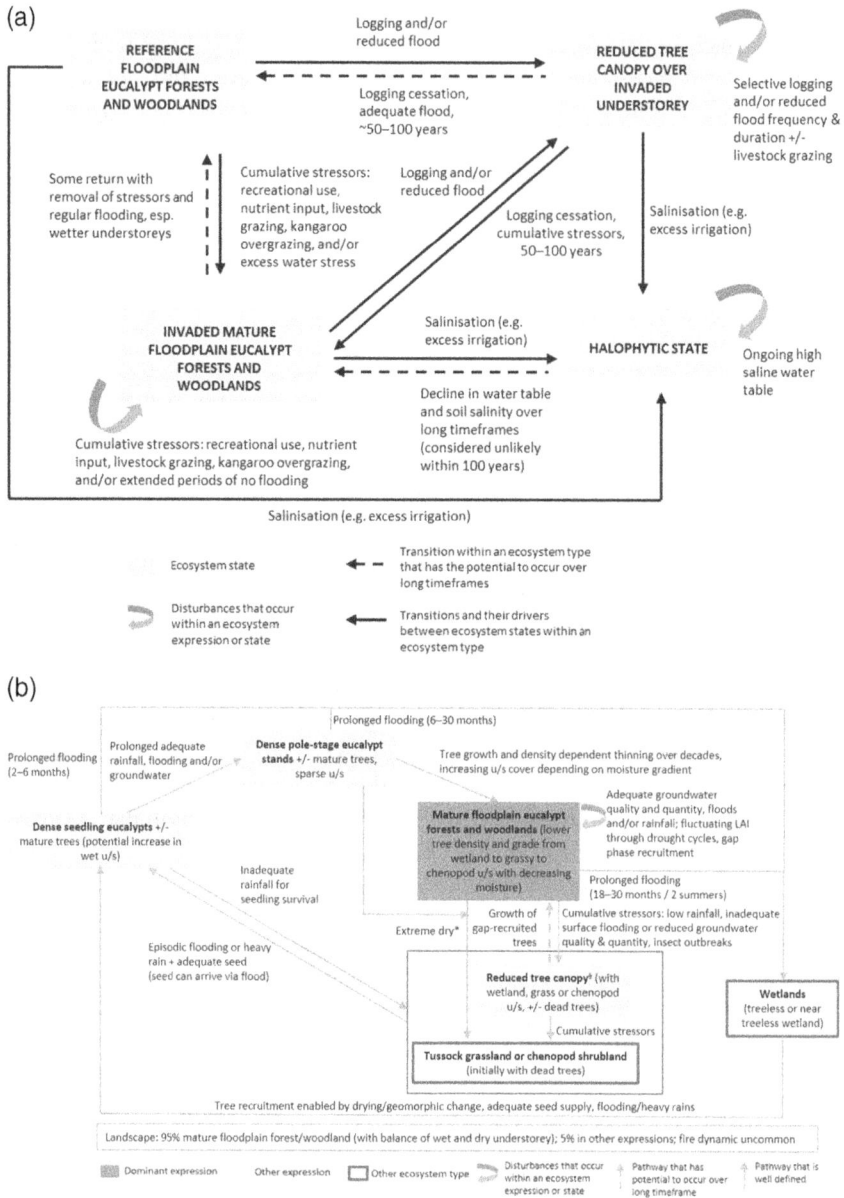

Figure 15.2 (a) State-and-transition model for the inland floodplain eucalypt forests and woodlands ecosystem type at GKP. An example of the detail of within-state dynamics, not shown in the top panel, is depicted in (b) a conceptual model of the reference state for inland floodplain eucalypt forests and woodlands at GKP. This figure shows dynamic shifts between expressions that capture variability in the reference state across space and time. The box labelled "Landscape" describes the proportion of the landscape in each expression for a landscape that has ecological integrity. For further details, see Richards et al. (2021). *Historical record suggests it has occurred twice in the last 500 years. §Reduced canopy of large mature trees. LAI = leaf area index; u/s = understorey.

The types and drivers of change observed in ecosystems at GKP in recent times (between 2010 and 2015) were conceptualized in a set of dynamic state-and-transition models (Richards et al. 2021). Figure 15.2a is an example of a state-and-transition model for the inland floodplain eucalypt forests and woodlands ecosystem type, which cover more than 80% of GKP. The models in Figure 15.2 were developed through expert elicitation using methods and templates from the Australian Ecosystem Models (AusEcoModels) Framework (Richards et al. 2020). The AusEcoModels Framework systematically synthesizes best available scientific knowledge about the dynamic characteristics and drivers of Australian ecosystems and the degree to which they display ecological integrity (Kay 1991; Kandziora et al. 2013). This knowledge is captured in a set of dynamic conceptual "archetype" models of ecosystem types that are used as templates for the development of state-and-transition models. Here, ecosystem types are subdivided into ecosystem states (including reference and modified states). Ecosystem states encapsulate a relatively stable set of ecosystem expressions linked by pathways of disturbance and recovery. Ecosystem expressions record transient variability of an ecosystem at any point in space or time, and, together, characteristics of ecosystem expressions within a state capture all possible combinations of abiotic and biotic ecosystem characteristics of an ecosystem state. Figure 15.2b is an example of a dynamic ecosystem reference state (a state that has the highest level of ecological integrity) showing six ecosystem expressions, two of which demonstrate shifts to other ecosystem types that may result from certain flood or drought disturbance regimes.

The models depicted in Figure 15.2 may be used as a conceptual underpinning for the development of ecosystem accounts (United Nations 2021) through quantification of characteristics of ecosystem expressions to enable spatial mapping of the extent of ecosystem states over time. This, in turn, provides an indication of ecosystem condition by linking each ecosystem state to an ecosystem condition score denoting departure from the reference state. Ecosystem states can also be described in terms of their capacity to supply specific ecosystem services, thus providing a coherent link between ecosystem extent, condition, and the supply of ecosystem services.

Model-based[3] monitoring of interacting social and forestry dynamics

The purpose of policy-driven monitoring is often to (1) describe the changing state of the environment, (2) assess threats, (3) provide a basis for follow-up of chosen measures and investments, and (4) provide a basis for analysis of the national and international environmental impacts such as from different emission sources. There is a great need for comprehensive map information about, for example, Annex I habitats, ecosystem services, or red-listed species as a planning basis at the landscape level.

Using predictive models, environmental monitoring field data can be matched with different types of comprehensive data to generate maps. The advantage is that it is quick to produce these maps and it will also be possible to present future forecasts or scenarios in map form.

Earlier in the chapter, different definitions of monitoring human and environmental processes were described and labelled as passive, mandated, question driven, design based, model based, etc. This section gives an example of a model framework that is potentially useful for giving decision support for influencing some long-term human–environment interactions, some of which will need to be monitored. This example targets forests and how their development and characteristics might be influenced by (private) forest owners' management preferences and practices (discussed in Eggers et al. 2014) and thus illustrates how ecological and social data can be combined, using, amongst others, micro simulation, as discussed in chapter 12. Such a projection reveals the long-term consequences of owners' chosen management alternatives on future forest attributes like volume, quality, and location of forest capital; volume of harvesting for different forest products; and assessment of environmental impact (carbon sink, biodiversity) of predicted forest use. The effort was divided in two tracks: (1) constructing the ForestPop model, simulating the development of population, forest owners, and their relation to owned forestland and (2) modifying the existing Heureka forest development model to accept time-varying input from ForestPop.

The chosen methodological frame is time-driven micro-simulation (Clarke and Holm 1987; Holm 2017; see also chapter 12) as originally suggested by Orcutt (1957) and further developed by Caldwell and Morrison (2000). JinJing and O'Donoghue (2013) defined a dynamic micro-simulation model simply as "a model that simulates the behaviour of micro-units over time". Chapter 12 gives a more detailed description of how micro-simulation is used to model large-scale individual populations based on register and "big" data. In this chapter, the model description starts directly with the ForestPop management application. The core output required as input by a Heureka-type model is the current owner's choice of management strategy for the next 5-year period within 100 years' time horizon for each privately owned forest property in Sweden.

One basic property of the approach is using and updating an individual representation of each instance in the studied population – often individuals or families or firms. Superficially, this might be regarded as a waste of storage space. In practice, and if one wants to maintain more than some seven attributes per instance (i.e. age, sex, earnings, education, profession, workplace, forest property, family), a row for each instance is actually a more condensed storage compared to a corresponding multi-dimensional table (sparse matrix) with seven dimensions. Moreover, most studies, findings, and theories in social science are about individuals. Assigning that directly to model persons minimizes bias in model representation of states, events, and causal assessments.

Despite not being as extensively used in natural science applications (climate and weather models can be regarded as types of micro-simulation models with localized air cubes as units), the same kind of argument would still apply. In this combined exemplified application, the instance unit selected is each privately owned forest property. One could easily envision other applications where it would be appropriate to extend resolution down to single individuals and single trees.

Managing risk and uncertainty in data collection and inference

Biodiversity monitoring is fundamentally concerned with recognizing and managing the state of the environment, identifying the pressures leading to change, and minimizing the risks of losses going undetected, not just as abstract statistical expressions but as actual risks of changes that we do not want happening in the real world (Pe'er et al. 2014).

Text box 15.2:   ForestPop

The main entity to simulate in this version of ForestPop is each property of forestland, not each individual in the population. By this simplification, individuals only appear as a temporarily created asset of the forest property in the shape of a current private owner with some characteristics influencing future management alternatives for the forest property. The simulation assigns management to each one of the 270,000 instances of forest properties, each one pointed at by one of the current forest owners (330,000). All of those forest properties are contained individually in the model and some of their attributes are changed during simulation. This simple version of ForestPop is entirely implemented in Excel with values in one row for each individual property attribute over each time period.

Five different management strategies labelled as follows summarize forest owner responses to detailed survey questions: Strategy 1: *Passive*, Strategy 2: *Conservation*, Strategy 3: *Intensive*, Strategy 4: *Productivity*, and Strategy 5: *Save*. The main issue for the simulation is to assign one of the five management alternatives to each property in each period by relating the probability of a certain choice to a set of owner and property characteristics.

The ForestPop management simulation model

Figure 15.3 shows a sketch of the simulation model. Most events are formulated and estimated as discrete choice equations; that is, a binary or multinomial logit.

Because all properties are simulated simultaneously, any municipality can be selected for output. Figure 15.4 shows the development of management alternatives for Sölvesborg municipality.

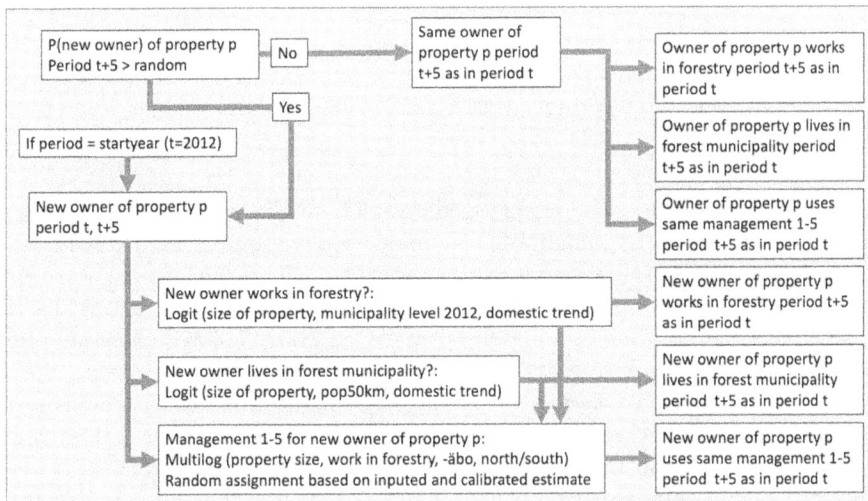

Figure 15.3 Structure of the ForestPop management simulation model.

The development of Management Alternatives over time

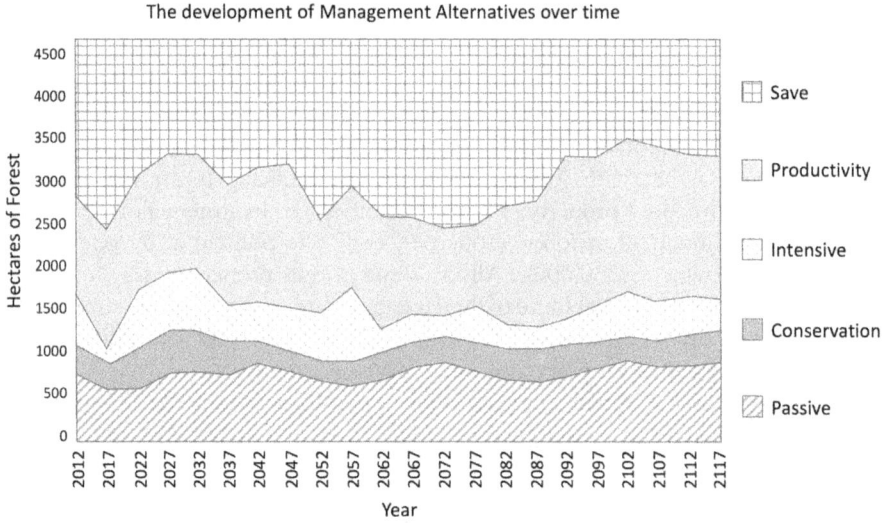

Figure 15.4 The development of management alternatives for Sölvesborg municipality over the next 100 years using the ForestPop simulation model.

As a result, with the given estimates and assumptions for the simulation, not much of a systematic large-scale long-term change in the distribution of management strategies for private forest properties seems to happen. On a smaller scale, for single municipalities, larger changes might happen, but more definite conclusions require running several replications of the model with different random seeds to discern to what extent the result is an impact of random noise. Whether or not the differences in management strategy development produced by the simulation for single municipalities reflect random noise or a systematic change on local level, it is of interest to discover whether that also will produce substantially different outcomes in forest characteristics via a Heureka-type forest model (Heureka 2022).

Some risks fall out of scope of being managed: most obviously, where a project is needed but unfunded or projects that might be well funded but we suspect only exist to put off actions that governments already know are necessary. Monitoring fails if it does not answer the target questions (although it may still address others) or answers the right questions when it is too late to respond or when the results are ignored or misinterpreted by decision makers.

More typically, we are concerned with managing risks captured within the scope of a project using scientific and especially statistical methods (Gitzen et al. 2012; Daniel et al. 2017). The statistical basis of survey sampling designs (chapter 4; Stauffer et al. 2021) includes probability sampling and unbiased estimators where, for the same design, the result is on average accurate but becomes more precise in proportion to the square root of the sample size. To halve the uncertainty of the estimated population statistics, we need to square the number of sample units or adopt a more complex design that, for example, introduces stratification. The UK Countryside Survey (chapter 2) is a good example of gradually increasing the sample size (from 256 in 1978 to 591 in 2007) to increase the statistical power and ability to report separately on trends in part of the UK.

In this statistical approach to risk, the challenge is how to design a survey so that the questions such as change detection can be answered most efficiently, including how we use auxiliary data to improve the spatial detail and precision of estimates without increasing the amount of field observation so much. NILS (see Text box 8.1) used remote sensing data to select the most informative set of plots for field inventory, in a form of "balanced sampling" stratification (based on the Normalized Difference Vegetation Index [NDVI]) as well as experimenting with drone imagery for automatic cover estimates and quality control. Other sorts of auxiliary information can be invaluable; for example, using geomorphology and cultural landscapes to target searches for scarce habitat, illustrated in the case of Swedish landscape monitoring (chapter 5), and modeling using environmental predictors to target drone flights for monitoring *Geum radiatum* (chapter 8).

Throughout the examples in the book, we see some quite complex designs that take advantage of stratification at different scales to target the field inventory.

Any statistical design depends on the reliability of the data, which in turn depends on the repeatability of field observations between years, habitats, seasons, and observers, discussed in chapter 5 (for a review, see Morrison 2016). Though some of this can be standardized by adding metadata (to allow for seasonal differences) and training, observer variation is different to statistical error in the sense that we cannot design unbiased estimators to remove it, and increasing the number of observations may simply increase the bias. To control this type of uncertainty, we might need better plot design, better choice of "detectable" species to record, multiple observers, to record time-to-detection, or to use image analysis to measure plot photographs. Good training and quality control are essential, and most best practice in field inventory is not written in scientific reports, and certainly not in academic papers. Here we need both experience of our own and the wisdom to consult more experienced field ecologists.

In a similar way, citizen science observations (chapter 6) can include both observer bias and poorly balanced distribution of samples (Kosmala et al. 2016) and observation protocols need to be designed and quality controlled in a similar way to any field survey. An excellent, readable example of what needs to be considered in a design for a vascular plant monitoring scheme is Walker et al. (2010) describing the UK National Plant Monitoring Scheme (2022).

Values taken from satellite imagery have a component of uncertainty, some of which remains in the processing chain (chapter 7). In the case of optical imagery, the processed values depend on modelling the atmosphere through which the downwelling light and reflected return passes and on the calibration of the sensor. Calibration and some random noise ("speckle") can affect radar images, and lidar depends on both calibration and the stability of the platform. In all three cases, we have increasingly sophisticated software to take this into account, and the ability to stack and compare tens or hundreds of images covering the same area gives more and more ability to smooth out random errors using statistical methods. Using standard software, algorithms (e.g. for vegetation indexes), and pre-processing in the form of analysis-ready data (ARD) helps reduce uncertainty in corrections and calibrations and makes the results more comparable between projects and between regions. The new services providing ARD and shared processing environments are discussed in chapter 7.

Continuity and innovation in monitoring schemes

In looking at the design and adaptation of biodiversity monitoring, it is helpful to suggest a series of steps or stages in which a scheme starts off with a design that takes into account everything known at the time, selecting the most up-to-date methods and existing data sources, and gradually adapts and becomes part of a wider ecosystems of projects. Notice that these only appear as stages of development, because over the past two or three decades, both long-term monitoring projects and new technologies have matured alongside one another, giving the impression of novel uses of data and analysis evolving over time. We should expect future designs to already make allowance for what are now innovations in new types of data, analysis, and modelling.

Stage #1: the original design and conception

As illustrated in Figure 15.1, a monitoring scheme is a set of decisions about what is being monitored, where, and how often, taking into account the best current knowledge to define the questions being asked, population, sampling frame, variables and measures, and the details of data collection. Where available, remote sensing imagery and map data might be used for stratification and model-building, but the key design question is that of chapters 4 and 5: the survey sampling design (and analysis) and the details of field inventory. Both in situ methods and related data collection that has to be tasked (carried out specifically for this project) are key to the design because they are typically not – at this stage – shared with any other project and because their cost is proportional to the scale of data collection. Notice that remote sensing using drones comes into this category of cost, but the use of freely available satellite imagery does not.

Though "cost" might be thought of as funding, this is also the use of a limited pool of experts, the potential for damage to species and habitats, the cost of delaying decisions (perhaps we already know enough to take action), and the opportunity cost of funding and carrying out this project rather than another one.

Stage #2: adaptation to new questions and new methods

After the initial design and data collection, the monitoring schemes will be modified and adapted, to answer either new policy-driven questions or the existing questions more efficiently and/or with greater precision. Some of the innovative methods mentioned in earlier chapters will be used as better ways of making existing measurements; for example, improved field instruments or the substitution of lidar for photogrammetry. Some of the variables and classes in the project will be dropped and others added, and still others will have their sampling frames and sample sizes adjusted. New datasets might be included to improve targeting or stratification but, in general, the original objectives of monitoring remain the same.

Here, continuity can be seen in several aspects of monitoring: continuing to answer the same questions in the same way but also maintaining the standards, quality control, and expertise to do this. Notably, the breadth of expertise within a project has to be greater than the minimum knowledge required to carry out the monitoring protocol. It has to include the contextual understanding to say whether methods and the sampling scheme are still competent to give unbiased answers. Expert field staff must not only be able to maintain their skills but be able to pass these on to new generations in a way that guarantees an unbroken set of observations.

Good practice and good design principles are equally important in the original monitoring scheme and in the way it is adapted, or innovation is introduced, making sure to avoid compromising the logic and continuity of the original design. Some of these practical issues are discussed in chapter 5, stressing the need for comparability of measures, metadata, good training, and the maintenance of both field skills and expert field teams.

It is useful to distinguish between the *measures*, which are the target variables for monitoring, and the *measurements* used to make the in situ observations. For example, vegetation cover (measure) could be recorded subjectively by field staff, by taking and analyzing photographs or using a needlepoint frame (a wooden frame holding needles that are pushed down to intercept plant parts). Of course, some in situ observations for data collection will always need to be measured in the same way to give comparability, even though some aspects may change, such as how we now use smartphones and tablets to record, rather than paper and pencil. In other cases, new methods of measurement can be substituted but keeping the measure the same.

Here there is also a more general principle of overlap. At meteorological stations, replacement instruments need an overlap period when both old and new models are run together. The ability to detect small trends over long periods can be compromised by changes in the performance of instruments (for a good example of the effort needed to recover comparable statistics for just one measure, see Brocard et al. 2013). For example, quicker responses in instruments or electronics, smaller detection limits for chemical analysis, or an improvement in the signal-to-noise ratio can result in a false signal of change.

The collection of soil and water samples has to be planned in the same way as any field activity and depends just as much on funding and expertise.

It is not sufficient to have continuity in observation methods and sample collection; this needs to be supported by continuity in metadata and quality control. In fact, it is very common for government agencies and site managers to have good archival records of observations from previous decades but no reliable record of where these observations were made as the markers and reference points are lost. Though GPS (post "selective availability") has solved this to some extent, the spatial references can still be too uncertain for accurate co-location with satellite data, and GPS coordinates can vary with the reference frame. Care must be taken if no ground markers are used not to trample or disturb the site trying to relocate it. Using permanent ground markers such as metal pegs that can be found again with a metal detector reduces the risk.

Essential metadata can include recent weather conditions (which might affect remote sensing) and who is doing the recording. Has the observer ever seen this species or habitat before? More generally, is the growth season, or loss of snow cover, delayed or early, droughted or wet? It might be better to make some measurements at a comparable time in the seasonal development, rather than at precisely the same time each year. Habitat cover changes with season, and flowering can make the difference between a plant being easy to count and hard to find. The key point here is that real continuity means making comparable observations, using ecological expertise, not just observing in exactly the same way on each repeat visit. Similarly, expert knowledge might suggest field recording designs with partial replacement of sample units to maintain a representative sample.

Text box 15.3: Adapting NILS (see also Text box 4.1)

The National Inventories of Landscapes in Sweden, NILS, was initiated in 2003 as the first nationally comprehensive inventory of Swedish landscapes (Ståhl et al. 2011). Before the inventory was started, extensive development work was carried out, which included an information analysis (Esseen et al. 2004) in which a large number of researchers and authorities had the opportunity to make suggestions on content and methods. Further, strength of change estimates were conducted as a basis for discussions on the appropriate dimensioning and design of the inventory (Ringvall et al. 2004). The result was a comprehensive inventory that focused on gathering information on how land use affects natural and cultural landscape values in a landscape context. NILS was later evaluated on behalf of the Swedish Environmental Protection Agency in 2012 and was considered to be of great importance in environmental monitoring in Sweden as the only program that follows the everyday landscape. However, the main criticism of NILS has been that the everyday landscape in Sweden largely consists of "trivial" forest and the Swedish National Forest Inventory (NFI) already collects sufficient data on these forests. Thus, data from NILS are considered partly redundant. At the same time, there is an increasing need for information on the status and change of habitat types in the Species and Habitats Directive Annex I. NILS could not answer these new questions with its relatively sparse sample size. Discussions about how NILS could be improved and made more efficient to follow more specific and rare habitats were intensified and resulted in a new sample design, developed in 2020 (Adler et al. 2020).

The new sample design is a general framework for national inventories in Sweden where it is possible to include supplementary inventories based on new needs, at both national and regional levels, within the same design. To be able to follow both unusual and common habitat types, it is possible to choose different densities of samples depending on how common the focal habitat type is. Because the sample design is the same, the same estimation procedure and estimation algorithms can be used at both national and regional levels as well as at different sample densities.

NILS is today an umbrella for four complementary national inventories, based on the same design, focusing on following specific and rare habitats that are not covered in any other national inventory:

- The NILS deciduous forest inventory, initiated in 2020, is a national inventory that supplement the Swedish National Forest Inventory with regard to deciduous forests. However, there is a particular need to increase the information about certain types of Annex I deciduous forests such as broadleaf forests, alluvial forests, and old "trivial" deciduous forests.
- The NILS grassland inventory, initiated in 2020, is a nationally comprehensive and general inventory of grasslands with a focus on habitat types with high natural values. Together with the alpine inventory and the seashore inventory, the inventories together cover all types of grasslands in Sweden.
- The NILS alpine inventory, initiated in 2021, covers both common and rare habitats in the alpine area of Sweden.
- The seashore inventory is a "sister" inventory within the sample frame run by THUF (Terrester Habitatuppföljning; Terrestrial Habitat Monitoring).

Stage #3: incorporating types of data not in the original design

New sources of data, notably remote sensing, have developed and matured over the past decade. The crucial difference is that the costs of many of these new types of data are not scaled or proportional to the area being monitored. Seen from the perspective of the user (if we ignore the costs to government), the coverage of hundreds or thousands of square kilometres every few days costs no more than the coverage of just a few kilometres once a year. Similarly, software resources and cloud-based data storage can be free, or at least very inexpensive, and costs can be shared between multiple users online.

Rather than making decisions that have to trade off potentially costly alternatives for in situ measurements, any amount of data and analysis is close to no-cost and analysis can be done at almost any geographical scale, or for any time of year, with the ability to co-register and stack imagery and in situ measurements. Instead of just being able to make observations only in the present and future, archival imagery and scanned maps may give some access to the condition of habitats and landscapes from the past. All of this enlarges the scope of an existing project to make more precise and spatially detailed inferences and to answer new questions retrospectively. The limits and costs are now only time and analytical expertise, provided the analysis does not try to answer questions that are unsupported by the limited in situ (or "ground truth") observations. We also have to be careful to follow good practice in research and be careful not to both define and test hypotheses retrospectively using the same set of observations.

New technologies, and especially remote sensing instruments, improve all the time, often breaking continuity with older instruments and acquisitions. This has been both necessary and inevitable because conventional satellite instruments may be years behind the technology when launched, because they take so long to build, whereas their newly designed replacements have to be specified to last as long as possible and may be on smaller platforms that can be designed, built, and launched much more quickly.

However, comparability of measurements is still important for monitoring, and government-sponsored satellite programmes are designed to maintain elements of continuity alongside innovation to have a time series of well calibrated, comparable observations. The National Aeronautics and Space Administration's (NASA) Landsat series of satellites with the Thematic Mapper (TM, ETM) instrument have been designed to maintain continuity in wavebands, spatial resolution, and instrument calibrations. Even though there has been an improvement in the radiometric resolution of the latest sensor from 8 to 12 (recorded as 16) bits, meaning much better recording of differences across the range of illumination, the spatial resolution of the pixels remains at 30m for most bands.

In practice, there is no need for Landsat-9 to have as good a resolution as the European Space Agency's (ESA) optical instrument on Sentinel-2 (10m for several of the bands) precisely because the Copernicus series already exists alongside Landsat. In turn, the constellations of Sentinel satellites have more or less fixed their design to provide the continuity of precise radiometry, and swarms of hundreds of very small commercial satellites such as PlanetScope are newly able to provide a complementary very high spatial resolution service.

Continuity, comparability, and innovation are also well balanced in the increasing number of services providing standardized ARD from these satellites, with the sharing of both data and analysis enabled by free, open-source imagery and analytical software.

Where new requirements replace or reduce existing field observations, innovation in new technologies such as satellite remote sensing to acquire wall-to-wall imagery can make supplementary observations of at least some variables – for example, detecting losses and gains in habitat or tree cover. For the often cloudy landscapes of Northern and Western Europe, an especially beneficial innovation is the increasing coverage of high spatial resolution synthetic aperture radar (SAR) imagery, which can penetrate cloud cover and show seasonal variation in ground surface properties, soil moisture and vegetation structure, depending on wavelength.

Notice the distinction between satellite remote sensing, which now has routinely scheduled coverage of almost everywhere and builds up an archive day by day, and instruments on airborne platforms, including drones, which have until recently been much more limited in the range of sensors but can now carry small multi-spectral, hyperspectral, lidar, and even radar instruments. Optical remote sensing using aircraft can still be very expensive and hard to carry out, given the difficulties of access and waiting for light winds and cloud-free conditions around midday. Drones are generally cheaper and can fly underneath clouds but have a limited range and endurance, so, again, these are not necessarily low-cost.

Future high-altitude long-endurance (HALE) drones may become more like satellite platforms, able to build an archive of continuous observations over wide areas. What are now spatially limited environmental measurements might in the future be collected everywhere, all the time, by future networks of micro-sensors. The recent swarms of micro-satellites are now giving continuous recording of high spatial resolution optical imagery but, unlike drones, these have to look through the full thickness of the atmosphere, where some spatial detail is lost to scattering and aerosols.

Remote sensing technologies are also becoming highly effective closer to the ground, including automatic camera traps, cameras recording seasonal phenology (phenocams), and automatic sound recorders. The large datasets these produce can be analyzed automatically, giving continuous, simultaneous recording at any number of points. Cameras will also record metadata, such as light levels and air temperature, which can be used to help make the results more comparable across seasons and between sites.

Other types of data allow questions to be asked retrospectively, looking back in time at past changes, most obviously, soil samples stored for future analysis, natural deposits of sediments, and glacial ice that can be cored to extract pollen, diatoms, macrofossils, and fossil gases. Dendrochronology (studying tree growth rings) can be used both to date ancient buildings and reveal past climates. Even postcards from 150 years ago can be photographic contact prints showing high-interest landscapes in great detail, provided there is a relevant set of contemporary field observations to help interpret the data.

We can only know what has already happened – and only imperfectly – but this understanding can influence what might happen next in at least two ways: as a general description of what is now lost or degraded but might be restored and as a record of the pattern and rate of change in both the state of biodiversity and environmental pressures that bring about these responses. Old, legacy observations driven by curiosity and attention to detail can be invaluable as descriptions of more or less undisturbed habitats and landscapes. Digitized historic maps and air photographs can be invaluable as a data source.

Future patterns and rate of change can be modelled (Conroy et al. 2011; Honrado et al. 2016). Our ability to do this depends on how relevant and statistically sound the observations are in the first place and how well estimates from field studies can be improved and supplemented by well-informed assumptions, auxiliary datasets from new technologies such as remote sensing and environmental DNA (including the outputs of other models), citizen science, and new forms of analysis, data mining, and deep learning.

The ForestPop example showed how it is possible to model the preferences of landowners and land managers (Text box 15.2) and use future states of forest habitats to predict how different scenarios might influence the future state of biodiversity, given the background trends in environmental pressures and climate change. This may raise new questions for monitoring, including shifting the focus of data collection onto variables and places where social data suggest change is most likely to take place.

Text box 15.4: Adapting NFI

Balancing continuity with adaptation: the Swedish NFI as an example

In the beginning of the 20th century, the Swedish government and forest industries were concerned that wood as a resource would not cover the increased needs mainly for the forest industry. Estimation of annual growth and possible fellings showed higher rates of felling than growth. The estimations of growth varied a lot because of uncertainties in forest data available with no systematic forest inventory monitoring the forest throughout the country (von Segebaden 1998). With this as a background, the National Forest Inventory (NFI) was established in 1923 with a focus on inventory of the wood as a resource for the industry (Fridman et al. 2014).

Gradually, the scope of the NFI widened from only wood resources to describing multifunctional forestry with data on wood, bioenergy, biodiversity, carbon, and recreation. One main challenge has been to maintain and expand the long time series of traditional data while at the same time making modifications to accommodate new kinds of information. To meet these challenges, there have been changes in inventory design, definitions of variables, additional measurements, how to report data, and the methods for measurement.

The inventory design has been modified several times. The sampled area was reduced when changing from belt inventory (1923–1952) to rectangular tracts with sample plots comprising approximately one day's work from 1953 onwards. In 1953, the inventory also changed from a county-by-county inventory to annually covering the entire country, providing possibilities to report up-to-date information as statistics on fellings each year. In 1983, permanent plots remeasured every fifth year were introduced, increasing the accuracy of change estimation but also making the data more applicable to research and improving estimates of current state (Ranneby et al. 1987). Since 2003, inventory has also covered protected areas enabling comparisons of managed and unmanaged forests. Since 2018, auxiliary data have been used to improve the location of temporary inventory plots (Grafström et al. 2017).

The number of variables assessed in the field has increased over the years. Many of the newer variables are related to biodiversity, forest damage, and social values. Examples are habitat classification, damage by moose grazing, occurrence of fen species on wetland, amount of deadwood, and occurrence and amount of berries. Another development is more modelled variables estimated by inventory data in the NFI database. Examples are biomass content for tree fractions, volume estimations, and growth.

The first report of the forest resource was published in 1932 (Statens Offentliga Utredningar 1932). It reported forest area on land use classes, growing stock distributed on tree species and dimensions, area of vegetation types, and growth. Since the 1950s, data based on NFI have been reported annually by the Forest Agency. Since 1981, the NFI has published its own annual report, called SKOGSDATA. The demand for up-to-date reporting and availability has increased, and today much of the data is available for own analysis and also possible to download (Swedish University of Agricultural Sciences 2022).

Today, there are many users of NFI data, such as authorities, companies, forest organizations, non-governmental organizations (NGOs, researchers, and international organizations. The NFI has always provided forest data for forest scenario analyses. From the start, the focus was on growth and potential felling from an industrial wood consumption perspective, transitioning to scenario analyses with broader scope from the 1980s onwards. Until the 1970s, NFI data were used in manual calculations and thereafter as input to computerized decision support systems. With technical development such as the use of GPS for plot location and fast development of remote sensing sensors, the NFI sample plot data have become a very important data source of reference data for remote sensing and the production of wall-to-wall data (e.g. Tomppo et al. 2008; Nilsson et al. 2017). The NFI data are also the main source for Sweden's greenhouse gas reporting for the land use, land use change, and forestry (LULUCF) reporting (Swedish Environmental Protection Agency 2019).

As described, the scope of society's demand for up-to-date data answering many forest-related questions has widened considerably. Meeting this need without destroying the continuity or compromising the quality of the variables already included have been a challenge. The user must also understand that NFI estimates are statistical estimates that typically can only be used as averages and totals for fairly large areas (Gregoire and Valentine 2008).

Stage #4: part of an ecosystem of monitoring schemes

Over a period of decades, biodiversity monitoring projects become part of a wider "ecosystem" of monitoring schemes, where, for example, a government-mandated land cover land use scheme is complemented by bespoke biodiversity monitoring schemes – for example, for birds, butterflies, and vascular plants (discussed in relation to the UK Countryside Survey in chapter 2; see also chapter 16) – and similar land cover monitoring schemes other regions and countries can contribute data to make comparisons, combine results, and attempt to map the state of biodiversity across whole regions, up to the global scale.

One approach to this is data harmonization and the adoption of similar classifications and standards. Though field measurements and observations can be in common – at least within the limits of species having the same preferences across different biogeographical regions – habit classes are often very different and require conversions sometimes known as "cross-walks" (see chapter 9 for examples). For habitat and thematic land use land cover (LULC) classes, there is a difference between optimizing comparability in data collection, where we prioritize continuity over time and using locally adapted classification schemes, and optimizing comparability across a regional scale, where we prefer broader, universal thematic classes. One of the lessons from the Danish Small Biotopes project (Text box 4.2, chapter 16) is the problems created by varying the set of thematic classes, making it more difficult to analyze change over time. In the same way, there are also dangers in designing monitoring to maximize opportunities for collaboration if the balance of data collection shifts too far away from answering specific questions about the species and habitats of direct interest (see the discussion on data integration in Lindenmayer and Likens 2018, chapter 5).

Fortunately, many of these conflicts can be reconciled using machine learning (chapter 8) and informatics (the science of data storage and analysis), and we can use both local vegetation types with universally applicable land cover and habitat classes such as the EODHaM General Habitat Classes (Lucas et al. 2015).

As explained in chapter 9, we can also combine the results of different types of monitoring. The uncertainty in thematic classes can also be modelled (Tsutsumida and Comber 2015; Stritih et al. 2019) and classes combined. This type of processing must consider different sources of error and uncertainty (Comber et al. 2004), notably situations where a comparison between two classification schemes ("semantics" or "ontologies") leads to ambiguity; for example, a single class in one scheme corresponds with two or more classes in another scheme.

Expert systems such as eCognition can incorporate ecological knowledge into a rule base working with objects (chapter 9) at multiple scales, geo-referencing, stacking, and combining information from different sources. Data can be used to create standard products such as vegetation indexes (NDVI; chapter 7) and variables expressing seasonal phenology, soil moisture, and structure. These indexes can be used to help disambiguate (select between) possible thematic classes by adding new information, and the results can be analyzed using functions other than probability in a field of methods known as informatics (for an introduction, see chapter 18 in Liu and Mason 2016). This wider range of analytical functions will be an essential part of combining data and hybrid methods in the future.

Looking beyond conventional analysis, the possibility of combining data into a virtual global model such as Destination Earth (see chapter 8) suggests new ways of using the results of monitoring that leave behind the questions addressed in the original design.

Innovation and continuity in the design of future monitoring schemes

Because of the parallel development of biodiversity monitoring schemes and new, innovative technologies, it can look as though all of the existing schemes start off mainly focusing on in situ measurements and incorporate remote sensing imagery and other new data sources only in later stages. This is not always true in detail; for example, the UK Countryside Survey (chapter 2) had a remote sensing component in the land cover map from the early 1990s – though at the time imagery was expensive and some had to be specially tasked. However, the general point is that what was previously innovation will become part of the design in the future monitoring schemes, because the most efficient designs use all of the available information to specify stratification and modelling in survey sampling. We should expect new monitoring schemes to incorporate current technologies from the start, just as NILS has adapted to use satellite and drone imagery as part of sample selection for field inventory. Over time, innovation becomes part of normal good practice, and results from existing monitoring schemes can be taken into account in the design of future schemes.

New technologies are increasingly enabling one another, particularly in the way that software innovations facilitate the handling of what would otherwise be overwhelmingly large datasets. Expert systems, machine learning, and deep learning (discussed in chapter 8) are used to identify species from camera traps (Willi et al. 2019; Whytock et al. 2021) and for habitat mapping (Lidberg et al. 2020). In the same way, data collected for one purpose can show the potential for other uses; for example, the way in which lidar used to update the digital elevation model (DEM) of Sweden was adopted for the forest inventory (see Text box 7.4).

However, new forms of analysis can be severely limited by the availability of in situ observations for training and testing the accuracy of the classifier. It is an open question whether or not recent trends in "big data", data mining, and machine learning (reviewed in Jeansoulin 2016) discussed in chapter 8 can replace more targeted monitoring. The basic approach (shown as stage #1, above) is to design a system focusing on the collection of in situ data and field observations as the starting place for any biodiversity monitoring scheme. Not everything can be seen or measured from space!

We need continuity – for comparability with past data – along with those innovations that allow us to keep up with the pace of change in the future. We need a full "ecosystem" of general, synoptic land cover and land use monitoring together with more specialized schemes looking at rarer species and habitats, and we need to involve scientists and experts of different sorts with policymakers and the people who live in, and depend on, a biodiverse world.

Key messages

- The principles of designing an entirely new project are the same as those we use to adapt and innovate those we already have.
- All successful long-term land cover and habitat monitoring projects have family resemblances because they follow the same principles of good practice and because the design of earlier projects influences the later ones. Some of the best reviews of existing projects can be found in the proposals for new projects.
- At least in Europe, survey data, archival aerial photography, satellite imagery, historical maps, and community knowledge are bound to exist and will influence the design of any new biodiversity monitoring scheme. In Wales, for example, some of the most useful air photography for looking at landscape change was taken by the German Luftwaffe between 1939 and 1942.
- Now that new technologies and datasets are widely available, sometimes at little or no cost, existing projects adapt to the same new types of observation and analysis in ways that more closely resemble one another.
- Not all of this "best practice" can be learned from the academic literature. For example, field inventory staff are best trained on more general habitat survey work, so they become familiar with how to see and classify the landscape before learning rarer species and closer observational skills.
- Breaks in continuity can come from changes in field staff, the loss of funding for analysis and storage of samples, mistakenly throwing out records, or the failure to maintain markers of field plots and migrate datasets to new software versions. Biodiversity monitoring asks us to keep up with new technologies and ideas while maintaining our archives and knowledge about past observations – sometimes we need the skills of ecologists, and sometimes librarians.

- Examples throughout the book show how adapting to new technologies can maintain continuity of recording, add metadata, and improve the reliability and scope of inferences from long-established monitoring projects. New technologies help open up the possibility of reliable, repeatable citizen science observations. We can use hybrid methods to combine and harmonize thematic classes without diluting their information content. Results can be linked and layered using wall-to-wall observations from remote sensing, with future technologies such as small sensor arrays, deep learning for extracting and classifying objects on images, and eDNA likely to revolutionize (and democratize) in situ observations.
- Whether adapting existing monitoring projects or starting new ones, our job is to combine old and new in the best way possible.

Study questions

This short chapter has only limited space to describe a very large subject, and what is missing is a whole range of detailed examples of how long-term projects were originally designed and have adapted over time. However, the best way of following these is in their websites and publications – but bearing in mind that some of the key points of detail may not be published.

1   Look at any of the examples in the book online and try to decide (a) what they were originally designed to monitor and (b) whether subsequent modifications to the design, field methods, sampling frame, or sample size came about as improvements to a working scheme, as a result of changes in objectives, or because the original design did not work well enough. Look in particular for references to statistical power calculations and observer variation.

2   For the big biodiversity monitoring schemes in Europe, the United States, and Australia/New Zealand, see if you can get hold of the field manuals and compare them. How do the instructions vary, and is this because of differences in the typical habitats or species being monitored? What arrangements do they have for quality control?

3   Read some of the more accessible books and websites on artificial intelligence (AI), machine learning and deep learning (for example, Domingos 2015, which is highly readable). Do you understand how the accuracy of the results can be assessed and, where they are inaccurate, how the algorithm can be adjusted to improve the result? Look for published examples of habitat mapping using machine learning (eg random forest algorithms) and using expert systems (probably software called eCognition). Which do you think works better if you have experts who are familiar with mapping habitats from air photos?

4   Look for papers on mapping habitat, land cover, and land cover change. Is there a universal set of classes that can work across more than one country? Across the world? If not, how should we compare changes across several countries with similar directives; for example, across Europe?

5   Taxonomic skills are hard to acquire, and expertise is scarce. Based on what the literature suggests, do you think this expertise will be completely replaced by DNA analysis? How would a field inventory look in the future using these new technologies? Will we still need to be able to make field identifications, and why?

Notes

1 Not to be confused with the use of the term "surveillance" used in Article 11 of the EU Habitats Directive (European Commission 1992), which refers to monitoring.
2 As in adaptive monitoring.
3 Notice this is a different use of *model-based* to the statistical survey sampling term.

Further reading

For further reading complementary to the approach of this book, there are some excellent well-established textbooks that have systematic descriptions of biodiversity monitoring methods, including:

Hill, D., Fasham, M., Tucker, G., Shewry, M. and Shaw, P. (2005) *Handbook of Biodiversity Methods: Survey, Evaluation and Monitoring*. Cambridge: Cambridge University Press. 588 pp.
Sutherland, J.W. (ed.) (2006) *Ecological Census Techniques*. 2nd edn. Cambridge: Cambridge University Press. 450 pp.
Along with Gitzen et al. (2012) in the main references, Henderson's book is the latest edition of the classic text by one of the pioneers of ecological research, T.R.E. Southwood:
Henderson, P.A. (2021) *Southwood's Ecological Methods*. 5th edn. Oxford: Oxford University Press. 528 pp.

There are also some recent texts on landscape ecology; for example:

Francis, R.A., Millington, J.D.A., Perry, G.L.W. and Minor, E.S. (eds) (2022) *The Routledge Handbook of Landscape Ecology*. 1st edn. London & New York: Routledge, Taylor & Francis Group. 502 pp.

In addition to books, there are countless useful websites covering science topics; see, for example:

NASA Share the Science. https://science.nasa.gov
GIS Geography. https://gisgeography.com

References

Adler, S., Christensen, P., Gardfjell, H., Grafström, A., Hagner, Å., Hedenås, H., Ranlund, Å. (2020). *Ny Design för Riktade Naturtypsinventeringar inom NILS och THUF* [A New Design for Targeted Nature Inventories in NILS and THUF], Working Report 513, Swedish University of Agricultural Sciences, Department of Forest Resource Management, Umeå. https://pub.epsilon.slu.se/17091/
Anderson, C.B. (2018) Biodiversity monitoring, Earth observations and the ecology of scale, *Ecology Letters* 21, 1572–1585. doi: 10.1111/ele.13106
Bestelmeyer, B., Moseley, K., Shaver, P., Sánchez, H., Briske, D. and Fernández-Giménez, M. (2010) Practical guidance for developing state-and-transition models, *Rangelands* 32, 23–30.
Brandon, B., Fernández-Giménez, M., Densambuu, B. and Bruegger, R. (2021) State-and-transition modelling, in Biggs, R., de Vos, A., Preiser, R., Clements, H., Maciejewski, K. and Schlüter, M. (eds) *The Routledge Handbook of Research Methods for Social–Ecological Systems*. Routledge, pp. 371–382. https://doi.org/10.4324/9781003021339-32
Brocard, E., Jeannet, P., Begert, M., Levrat, G., Philipona, R., Romanens, G. and Scherrer, S.C. (2013) Upper air temperature trends above Switzerland 1959–2011, *Journal of Geophysical Research: Atmospheres* 118, 4303–4317. doi: 10.1002/jgrd.50438
Burton, A.C., Huggard, D., Bayne, E., Schieck, J., Sólymos, P., Muhly, T., Farr, D. and Boutin, S. (2014) A framework for adaptive monitoring of the cumulative effects of human footprint on biodiversity, *Environmental Monitoring and Assessment* 186(6), 3605–3617. doi: 10.1007/s10661-014-3643-7
Caldwell, S. and Morrison, R. (2000) Validation of longitudinal microsimulation models: experience with CORSIM and DYNACAN, in Mitton, L., Sutherland, H. and Weeks, M.J. (eds) *Microsimulation in the New Millennium*. Cambridge: Cambridge University Press, pp. 200–225.

Clarke, M. and Holm, E. (1987) Micro-simulation methods in spatial analysis and planning, *Geografiska Annaler. Series B, Human Geography* 69(2), 145–164.

Comber, A., Fisher, P. and Wadsworth, R. (2004) Integrating land-cover data with different ontologies: identifying change from inconsistency, *International Journal of Geographical Information Science* 18(7), 691–708. doi: 10.1080/13658810410001705316

Conroy, M.J., Runge, M.C., Nichols, J.D., Stodola, K.W. and Cooper, R.J. (2011) Conservation in the face of climate change: the roles of alternative models, monitoring, and adaptation in confronting and reducing uncertainty, *Biological Conservation* 144 (4), 1204–1213. doi: 10.1016/j.biocon.2010.10.019

Cristescu, M.E. and Hebert, P.D.N. (2018) Uses and misuses of environmental DNA in biodiversity science and conservation, *Annual Review of Ecology, Evolution, and Systematics* 49(1), 209–230. https://doi.org/10.1146/annurev-ecolsys-110617-062306

Daniel, C.J., Ter-Mikaelian, M.T., Wotton, B.M., Rayfield, B. and Fortin, M.-J. (2017) Incorporating uncertainty into forest management planning: timber harvest, wildfire and climate change in the boreal forest, *Forest Ecology and Management* 400, 542–554. http://doi.org/10.1016/j.foreco.2017.06.039

de Bie, K., Addison, P.F.E. and Cook, C.N. (2018) Integrating decision triggers into conservation management practice, *Journal of Applied Ecology* 55, 494–502. http://doi.org/10.1111/1365-2664.13042

Domingos, P. (2015) *The Master Algorithm: How the Quest for the Ultimate Learning Machine Will Remake Our World*. Allen Lane – Penguin Books.

Eggers, J., Lämås, T., Lind, T. and Öhman, K. (2014) Factors influencing the choice of management strategy among small-scale private forest owners in Sweden, *Forests* 5, 1696–1716.

Environmental Change Network. https://ecn.ac.uk/(Accessed 2022).

Esseen, P.-A., Glimskär, A., Moen, J., Söderström, B. and Weibull, A. (2004) *Analys av Informationsbehov för Nationell Inventering av Landskapet i Sverige (NILS)* [Analysis of the need for information from National Inventory of Landscapes in Sweden], Work Reportt 132, Umeå University. https://www.slu.se/centrumbildningar-och-projekt/nils/publikationer/rapporter-fran-nils/

European Commission. (1992) *The Habitats Directive*, Council Directive 92/43/EEC of 21 May 1992, https://ec.europa.eu/environment/nature/legislation/habitatsdirective/index_en.htm (Accessed November 11, 2022).

Fridman, J., Holm, S., Nilsson, M., Nilsson, P., Ringvall, A. and Ståhl, G. (2014) Adapting national forest inventories to changing requirements – the case of the Swedish National Forest Inventory at the turn of the 20th century, *Silva Fennica* 48, 1–29. doi: 10.14214/sf.1095

Gari, S.R., Newton, A. and Icely, J.D. (2015) A review of the application and evolution of the DPSIR framework with an emphasis on coastal social–ecological systems, *Ocean & Coastal Management* 103, 63–77. http://dx.doi.org/10.1016/j.ocecoaman.2014.11.013

Gitzen, R.A., Millspaugh, J.J., Cooper, A.B. and Licht, D.S. (eds) (2012) *Design and Analysis of Long-Term Ecological Monitoring Studies*. Cambridge University Press. https://doi.org/10.1017/CBO9781139022422

Grafström, A., Zhao, X., Nylander, M. and Petersson, H. (2017) A new sampling strategy for forest inventories applied to the temporary clusters of the Swedish National Forest Inventory, *Canadian Journal of Forest Research* 47, 1161–1167. doi: 10.1139/cjfr-2017-0095

Gregoire, T. and Valentine, H. (2008) *Sampling Strategies for Natural Resources and the Environment*. Chapman and Hall.

Hale, J. and Butcher, R. (2011) *Ecological Character Description for the Gunbower Forest Ramsar Site*. Report to the Department of Sustainability, Environment, Water, Population and Communities (DSEWPaC), Canberra, Australia.

Hale, J. and SKM (Sinclair Knight Merz). (2011) *Environmental Water Delivery: Koondrook–Perricoota Forest*. Prepared for Commonwealth Environmental Water, Department of Sustainability, Environment, Water, Population and Communities, Canberra, Australia.

Harrington, B. and Hale, J. (2011) *Ecological Character Description for the NSW Central Murray Forests Ramsar Site*. Report to the Department of Sustainability, Environment, Water, Population and Communities (DSEWPaC), Canberra, Australia.

Hellawell, J.M. (1991) Development of a rationale for monitoring, in Goldsmith, B. (ed.) *Monitoring for Conservation and Ecology*. London: Chapman and Hall, pp. 1–14.

Heureka. (2022) Forest sustainability analysis, https://www.slu.se/en/departments/forest-resource-management/program-project/forest-sustainability-analysis/ (Accessed 2022).

Holm, E. (2017) Microsimulation, in *The International Encyclopedia of Geography: People, the Earth, Environment, and Technology*. Wiley.

Holm, E. and Sanders, L. (2007) Spatial microsimulation models, in Sanders, L. (ed.) *Models in Spatial Analysis*. Geographical Information Systems Series. London: ISTE Group, pp. 159–195.

Honrado, J.P., Pereira, H.M. and Guisan, A. (2016) Fostering integration between biodiversity monitoring and modelling, *Journal of Applied Ecology* 53(5), 1299–1304. https://doi.org/10.1111/1365-2664.12777

International Long-Term Ecological Research network. (2022) Home page, https://www.ilter.network/ (Accessed 2022).

Jeansoulin, R. (2016) Review of forty years of technological changes in geomatics towards the big data paradigm, *International Journal of Geo-Information* 5, 155. doi: 10.3390/ijgi5090155

Jinjing, L. and O'Donoghue, C. (2013) A survey of dynamic microsimulation models: uses, model structure and methodology, *International Journal of Microsimulation* 6(2), 3–55.

Kandziora, M., Burkhard, B. and Müller, F. (2013) Interactions of ecosystem properties, ecosystem integrity and ecosystem service indicators – a theoretical matrix exercise, *Ecological Indicators* 28, 54–78.

Kay, J.J. (1991) A nonequilibrium thermodynamic framework for discussing ecosystem integrity, *Environmental Management* 15(4), 483–495.

Kosmala, M., Wiggins, A., Swanson, A. and Simmons, B. (2016) Assessing data quality in citizen science, *Frontiers in Ecology and the Environment* 14(10), 551–560. https://doi.org/10.1002/fee.1436

Kwong, I.H.Y., Wong, F.K.K., Fung, T., Liu, E.K.Y., Lee, R.H. and Ng, T.P.T. (2022) A multi-stage approach combining very high-resolution satellite image, GIS database and post-classification modification rules for habitat mapping in Hong Kong, *Remote Sensing* 14, 67. https://doi.org/10.3390/rs14010067

Lausch, A., Bannehr, L., Beckmann, M., Boehm, C., Feilhauer, H., Hacker, J.M., Heurich, M., Jung, A., Klenke, R., Neumann, C., et al. (2016) Linking Earth observation and taxonomic, structural and functional biodiversity: local to ecosystem perspectives, *Ecological Indicators* 70, 317–339.

Lidberg, W., Nilsson, M. and Ågren, A. (2020) Using machine learning to generate high-resolution wet area maps for planning forest management: a study in a boreal forest landscape, *Ambio* 49, 475–486. https://doi.org/10.1007/s13280-019-01196-9

Lindenmayer, D.B. and Likens, G.E. (2009) Adaptive monitoring: a new paradigm for long-term research and monitoring, *Trends in Ecology and Evolution* 24(9), 482–486.

Lindenmayer, D.B. and Likens, G.E. (2018) *Effective Ecological Monitoring*. CSIRO Publishing.

Liu, J.G. and Mason, P. (2016) *Image Processing and GIS for Remote Sensing: Techniques and Applications*. John Wiley & Sons.

Lucas, R.M., Blonda, P., Bunting, P., Jones, G., Inglada, J., Arias, M., Kosmidou, V., Petrou, Z.I., Manakos, I., Adamo, M., et al. (2015) The Earth Observation Data for Habitat Monitoring (EODHaM) system, *International Journal of Applied Earth Observation and Geoinformation* 37, 17–28. https://doi.org/10.1016/j.jag.2014.10.011

Lucas, R.M., German, S., Metternicht, G., Schmidt, R.K., Owers, C., Prober, S.M., Richards, A., Tetreault Campbell, S., Williams, K.J., Mueller, N., et al. (2022) A globally relevant change taxonomy and evidence-based change framework for land monitoring, *Global Change Biology* 28(21), 6293–6317.

Morrison, L.W. (2016) Observer error in vegetation survey: a review, *Journal of Plant Ecology* 9(4), 367–379. doi: 10.1093/jpe/rtv077

Murray-Darling Basin Authority. (2012a) *Gunbower Forest Environmental Water Management Plan*. Canberra, Australia: MDBA.

Murray-Darling Basin Authority. (2012b) *The Murray-Darling Basin Plan*. Canberra, Australia: MDBA.

Murray-Darling Basin Authority. (2019) *Basin-wide Environmental Watering Strategy*. Canberra, Australia: MDBA.

National Biodiversity Network. (2022) The National Biodiversity Network, https://nbn.org.uk/the-national-biodiversity-network/archive-information/nbn-gateway/ (Accessed 2022).

Nilsson, M., Nordkvist, K., Jonzén, J., Lindgren, N., Axensten, P., Wallerman, J., Egberth, M., Larsson, S., Nilsson, L., Eriksson, J. and Olsson, H. (2017) A nationwide forest attribute map of Sweden predicted using airborne laser scanning data and field data from the National Forest Inventory, *Remote Sensing of Environment* 194, 447–454.

Orcutt, G.H. (1957) A new type of socio-economic system, *The Review of Economics and Statistics* 39, 116–123.

Pe'er, G., Mihoub, J.-B., Dislich, C. and Matsinos, Y.G. (2014) Towards a different attitude to uncertainty, *Nature Conservation* 8, 95–114. https://doi.org/10.3897/natureconservation.8.8388

Ranneby, B., Cruse, T., Hägglund, B., Jonasson, H. and Swärd, J. (1987) Designing a new national forest survey for Sweden, *Studia Forestalia Suecica* 177, 1–29. https://pub.epsilon.slu.se/4634/

Richards, A.E., Dickson, F., Williams, K., Cook, G.D., Roxburgh, S., Murphy, H., Doherty, M., Warnick, A., Metcalfe, D. and Prober, S. (2020) *The Australian Ecosystem Models Framework Project: A Conceptual Framework.* Clayton, Australia: CSIRO. https://doi.org/10.25919/f61q-1386

Richards, A.E., Prober, S.M., Schmidt, R.K., Sengupta, A., McInerney, P. and Tetreault-Campbell, S. (2021) *Ecosystem Classification and Conceptual Models for the Gunbower-Koondrook-Perricoota Forest Icon Site.* A Technical Report from the Land and Ecosystem Accounts Project. Clayton, Australia: CSIRO. https://doi.org/10.25919%2F7zf8-7073

Ringvall, A., Ståhl, G., Löfgren, P. and Fridman, J. (2004) *Skattningar och Precisionsberäkning i NILS – Underlag för Diskussion om Lämplig Dimensionering* [Estimates and the Calculation of Precision in NILS – The Basis for Sample Size Determination], Arbetsrapport 128 2004. Sveriges Landtbruksuniversitet, Institutionen för Skoglig Resurshushållning och Geomatik, Uppsala.

Rumpff, L., Duncan, D.H., Vesk, P.A., Keith, D.A. and Wintle, B.A. (2011) State-and-transition modelling for adaptive management of native woodlands, Biological Conservation 144, 1224–1236. https://doi.org/10.1016/j.biocon.2010.10.026

Ruppert, K.M., Kline, R.J. and Rahman, M.S. (2019) Past, present, and future perspectives of environmental DNA (eDNA) metabarcoding: a systematic review in methods, monitoring, and applications of global eDNA, *Global Ecology and Conservation* 17, e00547. https://doi.org/10.1016/j.gecco.2019.e00547

Sato, C.F. and Lindenmayer, D.B. (2021) The use of state-and-transition models in assessing management success, *Conservation Science and Practice* 3, e519. https://doi.org/10.1111/csp2.519

Statens Offentliga Utredningar. (1932) *Uppskattning av Sveriges Skogstillgångar Verkställd åren 1923–1929: Redogörelse Avgiven av Riksskogstaxeringsnämnden* [Estimation of Sweden's Forest Assets in the Years 1923–1929: Report Issued by the National Forest Inventory Board]. P.A. Norstedt & Söner.

Stauffer, N.G., Duniway, M.C., Karl, J.W. and Nauman, T. (2022) Sampling design workflows and tools to support adaptive monitoring and management, *Rangelands* 44(1), 8–16. https://doi.org/10.1016/j.rala.2021.08.005

Stritih, A., Bebi, P. and Grêt-Regamey, A. (2019) Quantifying uncertainties in Earth observation–based ecosystem services assessments, *Environmental Modelling and Software* 111, 300–310. https://doi.org/10.1016/j.envsoft.2018.09.005

Ståhl, S., Allard, A., Esseen, P.-A., Glimskär, A., Ringvall, A., Svensson, J., Sundquist, S., Christensen, P., Gallegos Torell, Å., Högström, M., et al. (2011) National Inventory of Landscapes in Sweden (NILS) – scope, design, and experiences from establishing a multi-scale biodiversity monitoring system, Environmental Monitoring and Assessment 173(1–4), 579–595.

Ståhl, G., Saarela, S., Schnell, S., Holm, S., Breidenbach, J., Healey, S.P., Patterson, P.L., Magnussen, S., Næsset, E., McRoberts, R.E., et al. (2016) Use of models in large-area forest surveys: comparing model-assisted, model-based and hybrid estimation, *Forest Ecosystems* 3, 5. doi: 10.1186/s40663-016-0064-9

Swedish Environmental Protection Agency. (2019) *Greenhouse Gas Emission Inventories 1990–2018. National Inventory Report Sweden 2020.* Submitted under the United Nations Framework Convention on Climate Change and the Kyoto Protocol.

Swedish University of Agricultural Sciences. (2022) Forest statistics, https://www.slu.se/en/Collaborative-Centres-and-Projects/the-swedish-national-forest-inventory/foreststatistics/forest-statistics/ (Accessed February 22, 2022).

Tomppo, E., Olsson, H., Ståhl, G., Nilsson, M., Hagner, O. and Katila, M. (2008) Combining national forest inventory field plots and remote sensing data for forest databases, *Remote Sensing of Environment* 112(5), 1982–1999.

Tsutsumida, N. and Comber, A.J. (2015) Measures of spatio-temporal accuracy for time series land cover data, *International Journal of Applied Earth Observation and Geoinformation* 41(2015), 46–55. http://dx.doi.org/10.1016/j.jag.2015.04.018

UK National Plant Monitoring Scheme. (2022) What is the National Plant Monitoring Scheme, https://www.npms.org.uk (Accessed 2022).

United Nations. (2021) System of environmental-economic accounting—ecosystem accounting (SEEA EA), https://seea.un.org/ecosystem-accounting (Accessed February 2022).

von Segebaden, G. (ed.) (1998) *Rikstaxen 75 år. Utvecklingen 1923–1998.* Institutionen för Skoglig Resurshushållning och Geomatik, Rapport 8. Uppsala, Sweden: Sveriges Landtbruksuniversitet (SLU), Institutionen för Skoglig Resurshushållning och Geomatik.

Walker, K., Dines, T., Hutchinson, N. and Freeman, S. (2010) *Designing a New Plant Surveillance Scheme for the UK.* JNCC Report No. 440. Peterborough, UK: Joint Nature Conservation Committee.

Whytock, R.C., Świeżewski, J., Zwerts, J.A., Bara-Słupski, T., Koumba Pambo, A.F., Rogala, M., Bahaa-el-din, L., Boekee, K., Brittain, S., Cardoso, A.W., et al. (2021) Robust ecological analysis of camera trap data labelled by a machine learning model, *Methods in Ecology and Evolution* 12, 1080–1092. https://doi.org/10.1111/2041-210X.13576

Willi, M., Pitman, R.T., Cardoso, A.W., Locke, C., Swanson, A., Boyer, A., Veldthuis, M. and Fortson, L. (2019) Identifying animal species in camera trap images using deep learning and citizen science, *Methods in Ecology and Evolution* 10(1), 80–91. https://doi. org/10.1111/2041-210X.13099

16 Monitoring small biotopes and habitats with a history of cultural management

Clive Hurford and Gregor Levin

Small biotopes in the Danish rural landscape

With agricultural land use comprising around two-thirds of the Danish terrestrial landscape, habitats are often characterized by small, isolated biotopes in a matrix of intensive agriculture. For 32 case areas of $2 \times 2km^2$, the Danish small biotope monitoring program includes a wall-to-wall mapping of land use and land cover with specific focus on registration of small, uncultivated biotopes (Agger and Brandt 1988; Brandt and Levin 2006; Fredshavn et al. 2015; for a detailed description of the monitoring program, see Text box 4.2). For 13 areas on Funen, Zealand, and the southern islands, registrations were carried out in 1981, 1986, 1991, 2007, and 2013.

Figure 16.1 summarizes the change in area proportion of major land use and areal biotope types. Due to the low quality of registrations, line biotopes are not included here. Between 1981 and 2013, land in agricultural rotation decreased from 89.3% to 82.2%. Built areas increased from 6.6% to 10.9% with the largest increase between 1996 and 2003. Forested biotopes (uncultivated land with ≥50% woody vegetation) increased from 2.1% to 3.5%, and open biotopes (uncultivated land with <50% woody vegetation) increased from 1.5% to 2.5%. Lakes and ponds with an open water surface increased from 0.5% to 0.9%. In summary, the development was characterized by a growing urbanization and increase in uncultivated biotopes at the expense of agricultural land.

Figure 16.2 illustrates that the development of open biotopes was characterized by an increase in periodically wet biotopes (bogs and wet meadow), particularly after 1996, and a decrease in dry biotopes (dry grassland and heather). This development can be partly ascribed to an increasing focus on wetland restoration as a measure to decrease nutrient leaching to the freshwater environment and to increase carbon sequestration.

Figure 16.3 illustrates the development of different types of forested biotopes. Over the whole period, the area of coniferous forest and swamp forest remained fairly stable. The considerable increase of forested biotopes between 1996 and 2007 was mainly due to a growth of the proportion mixed forest. After 2007, the proportion of broadleaf forest increased substantially. The overall increase in forested biotopes can be linked to a variety of driving forces such as subsidies for afforestation since the early 1990s; a growing interest in hunting and, consequently, planting of small woodlots to attract game; and vegetation succession due to termination of grazing and mowing of uncultivated biotopes. The general increase of broadleaf and mixed forest reflects general a trend towards planting of native broadleaf species, which are of higher value to biodiversity and more resistant to climatic stress and strong winds.

DOI: 10.4324/9781003179245-16

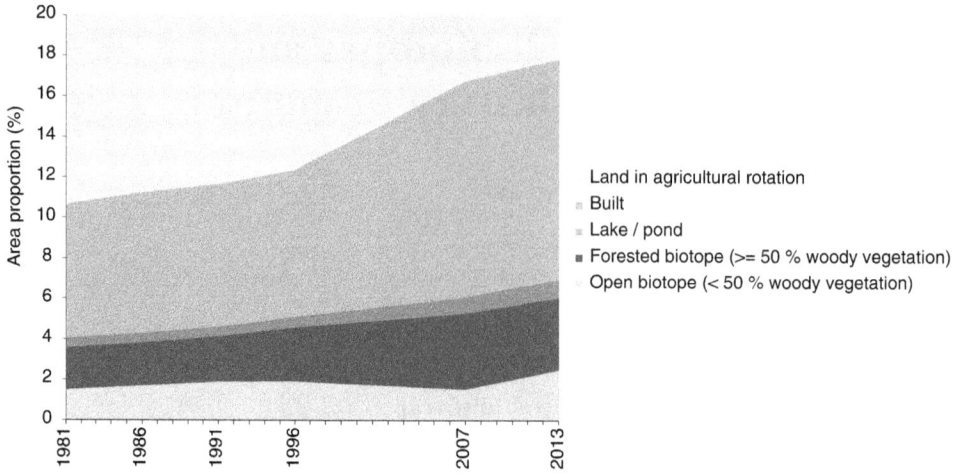

Figure 16.1 Development of major land use and areal biotope types for 13 areas from 1981 to 2017.

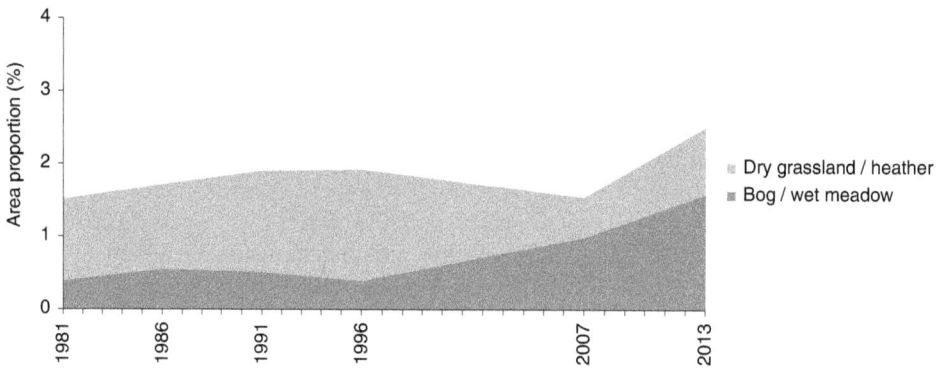

Figure 16.2 Development of dry and periodically wet open biotopes for 13 areas from 1981 to 2017.

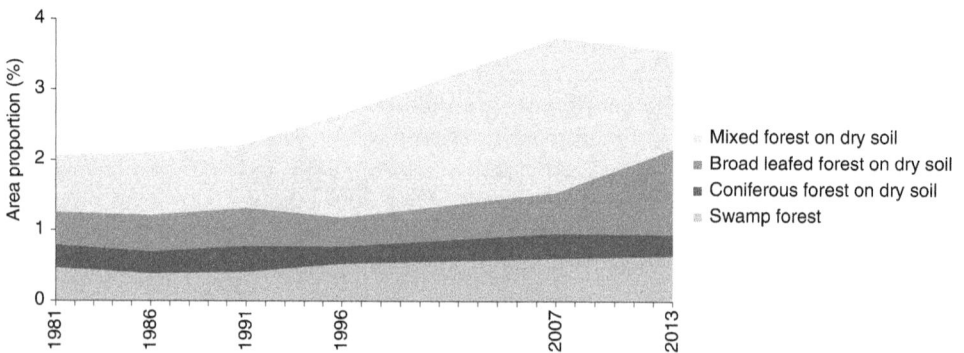

Figure 16.3 Development of dry forested biotopes for 13 areas from 1981 to 2017.

Monitoring habitats with a history of cultural management

Most land in Europe has a history of cultural management and remains under some form of management: arable habitats alone account for circa 30% of the European land area, with this likely to increase as food demand increases. We then have land under forestry operations, heathland management, and agricultural practices, such as hay meadows and most other grassland habitats. Despite this, examples of these habitats supporting high levels of biodiversity are so few that they are already well known to local naturalists, especially in Western Europe and increasingly so in Central and Eastern Europe.

Adopting a random or grid sampling programme on a national scale is inappropriate for detecting and monitoring these sparsely distributed, high biodiversity examples of cultural habitats. We need to adopt a far more selective approach, ideally by collating a register of the most important sites for biodiversity and applying the appropriate maintenance or restoration management regimes at the earliest opportunity.

Habitats with a history of cultural management differ from habitats considered to be wilderness or "natural", because we have well-informed expectations of the biodiversity that the available management options can deliver. By considering the extent and severity of the damage on a site, the proposed restoration method(s), and the proximity of the species that we would like to recolonize the habitats, we can predict, with a reasonable degree of confidence, the likelihood of maintaining or restoring the biodiversity of a habitat. Success is more likely if the biodiversity that we are aiming to restore is still present on the site or on land adjacent to it. For this reason, we should prioritize the relatively few examples of habitats that are already in a favourable state for management security. We do not need to carry out restoration management on these sites, but we do need to secure the management into the future.

A model for habitat management and monitoring

The model in Figure 16.4 was adapted from a model originally developed as a tool for Natura 2000 reporting (Rowell 1993), which recommended setting upper and lower limits to define when a habitat is in a favourable state (condition) and the need for a restoration target if the habitat or species is found to be outside the upper or lower limits.

It is also a simple but logical model for monitoring conservation management that illustrates the point at which the management must switch from a maintenance phase, which is applied when the habitat is in a favourable state, to a restoration phase when the state is no longer considered favourable (Figures 16.5 and 16.6). The restoration target is invariably higher than the lower limit for favourable status because the point at which we consider a habitat to be threatened cannot, logically, be the same point at which we consider it to be restored.

The need for prioritization

We do not have the time, financial resources, or expertise to monitor everything everywhere; therefore, transparent prioritization is essential for any successful management project. For example, we must prioritize (a) which sites to monitor, (b) which habitat(s) and/or species to manage and monitor on those sites, and (c) which indicator species to monitor. We know from sampling trial data that attempting to record all species invokes unacceptable levels of observer variation.

It is not unusual for protected sites to comprise several habitats, so if you only have the resources to undertake one management action on a site, which habitat should be prioritized? We would recommend that the preferred management actions, and what

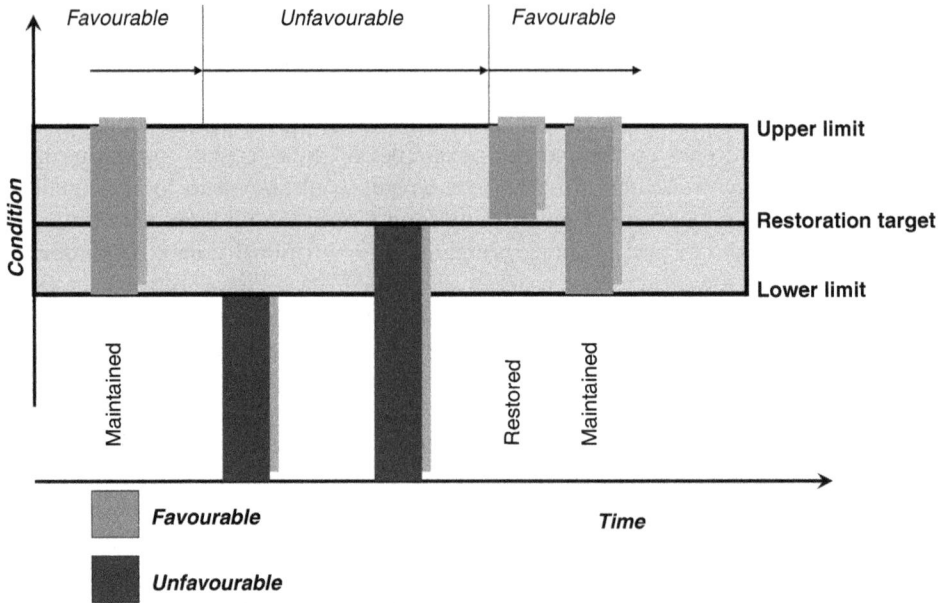

Figure 16.4 This model promotes good practice in biodiversity management and monitoring. Before it can be applied, the land managers must define how to recognize when the habitat is in a favourable state and the point at which it will need restoration, though this could well be after the first monitoring event.

Source: Adapted from Brown (2000) and Rowell (2006).

they would be expected to achieve, be listed in priority order in the site management plan. A dispassionate way to prioritize the habitats on a site (see Text box 16.3) could take into account the following:

1   The percentage of the national resource that each habitat represents.
2   The percentage of the national resource that each key species represents.
3   The number of key species dependent on each habitat.
4   A focus on habitats that are known to respond positively to management.

We must also consider whether managing the priority habitat on the site would have a negative impact on adjacent habitats. Often, the management that benefits the priority habitat will also have a positive impact on adjacent habitats, but this is not always the case.

Developing management objectives for cultural habitats

Any responsible management project will have clearly expressed aims underpinned by a sound rationale. The key questions for any management project are as follows:

1   What do you want your management to achieve?
2   Where do you want to achieve it?
3   How will you know when you have achieved it?

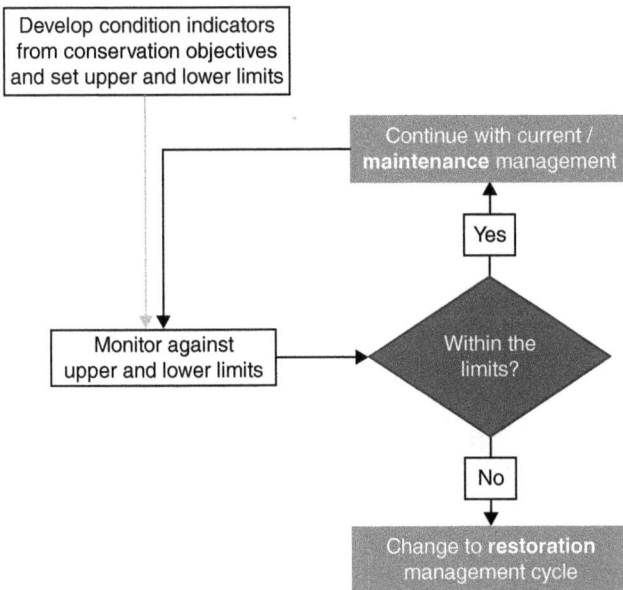

Figure 16.5 A flowchart showing the sequence of management actions and monitoring events applied when a habitat is in the desired (favourable) state. This equates to the process applied when the habitat state is between the upper and lower limits in the model shown in Figure 16.4. Note that the first action is to set the upper and lower limits and to develop sample targets for monitoring.

Source: Adapted from Hurford (2006a).

The answers to these questions are typically informed by our prior knowledge of the following:

1   The historic management of the habitats on the site.
2   How the priority habitat responds to management.
3   The environmental factors most likely to impact the habitat on that site.
4   The historic composition of the biodiversity on the site.
5   The likelihood of maintaining or restoring the historic biodiversity on the site.

After collating the existing information, we can consider what we want to achieve and where. Generally, habitat management focuses on two broad attributes: the extent of the habitat(s) and the state or condition of the habitat(s). The potential for increasing the extent of a habitat is often limited by parameters such as suitable substrate, appropriate topography, natural barriers, and non-natural barriers. In contrast, the state of a habitat is often defined by the key species (positive and/or negative) that we would expect to be present or absent if the habitat was in a favourable state. The case study in this chapter includes an example of a management objective.

The monitoring project

To monitor a management project accurately and precisely, we must phrase the management aims in measurable terms: this minimizes the risk of introducing unacceptable

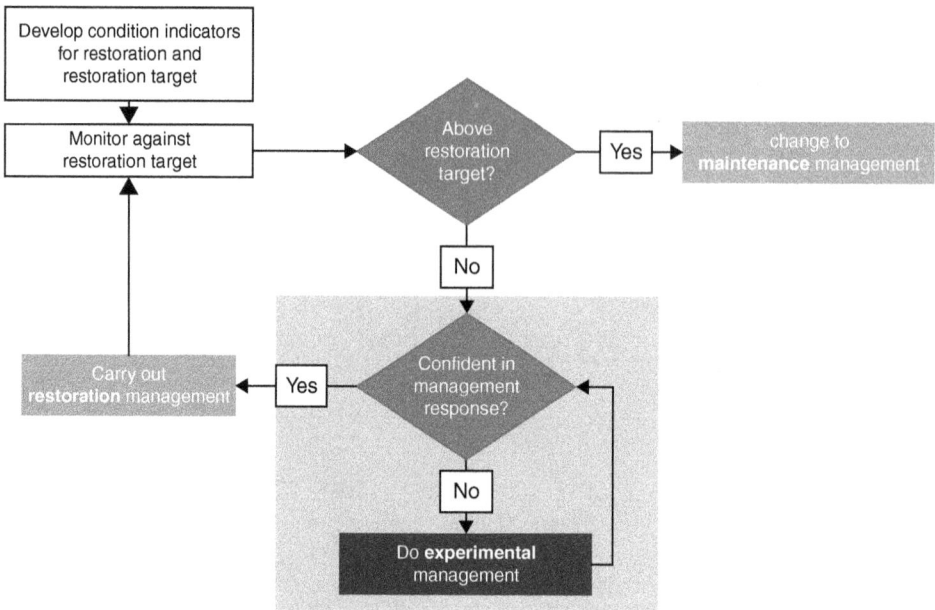

Figure 16.6 A flowchart showing the sequence of management actions and monitoring events applied when a habitat is in an unfavourable state. This equates to the process applied when the habitat state is outside the upper or lower limits in the model shown in Figure 16.4. Here, the first action is to set the restoration target and develop the sample target for monitoring the restoration of the habitat.

Source: Adapted from Hurford (2006a).

levels of observer variation to the monitoring project (see more in chapter 5). This information is provided in a table stating the sample targets or condition indicators. The development of the sample targets is the critical phase of a monitoring project, because it will state what to monitor and where to monitor in clear and unambiguous terms. Examples of measurable, unambiguous sample targets are provided in Text boxes 16.1 and 16.2 and in Table 16.4 in the case study in the The humid dune slack surveillance project section.

Deciding what to monitor

When deciding what to monitor, we recommend recording only a small subset of species with high detection rates. This species assemblage is selected to tell us what we need to know as reliably and efficiently as possible; that is, whether the management has been successful or not. These species will be expected to respond either positively or negatively depending on the success of the management. Ideally, the species assemblage will comprise both plant and animal species (see Text box 16.1). These indicator species assemblages, which are ecological shorthand for the desired habitat state, are a key component of the sample targets used to inform a monitoring project.

The examples in the presented case study all focus on assemblages of species that we expect to co-occur within a small area (typically either a 50cm radius or 1m radius) when the habitat is in the desired state. This approach has two distinct advantages over setting targets for individual species frequencies. Firstly, the recorders carrying out the monitoring soon learn what the desired state of habitat looks like, which means that they soon reach the point where a quick walkabout over the site will tell them whether sampling is needed to detect the status of the habitat. This allows the biodiversity managers to allocate their monitoring resources to best effect. Secondly, and more importantly, we can show the person responsible for managing the land (who will most likely not be an ecologist) the habitat state that the management aims to deliver.

Where to collect evidence for the monitoring project

If we are familiar with the site and the habitat that we are monitoring, then we do not need to collect evidence across the whole of the site – we can use logical inference to derive the state of the habitat outside the sampling areas and, therefore, whether the management has achieved its aim.

A straightforward example would be where the main threat to a meadow is eutrophication caused by runoff from the field above. In this situation, the place that we are most likely to find evidence of this is along the edge of field adjacent to the source of the runoff: if this part of the meadow is unaffected, it is unlikely that other parts will be. Similarly, if we are concerned about scrub encroachment, or perhaps encroachment by *Pteridium aquilinum*, then we are most like to detect this in habitat patches adjacent to the existing stands of scrub or *Pteridium*.

In less obvious situations, assuming that all of the priority habitat on the site is under the same management, we could, for example, decide to collect evidence from the areas of the site where the habitat is most likely to meet the criteria in the sample targets. If the habitat at these locations fails to meet the criteria in the sample targets, we can logically assume that the habitat outside the sampling area(s) would also fail. Conversely, we can collect evidence from the areas of habitat least likely to meet the criteria in the sample targets, and if the habitats at these locations do meet the criteria, then we can logically assume that the habitat outside the sampling areas would also meet them. If, however, the management situation is more complex and involves different levels of management in different management compartments, any assumptions of state can only safely apply to the compartment(s) we have sampled. Ideally, in this situation, we would locate at least one sampling plot in each compartment under different management. If the purpose of the monitoring project is to inform the habitat management on the site, then this would be the case regardless of the sampling model applied.

This approach allows us to avoid intensive (and expensive) data collection exercises if we already know what the result of sampling would be. In this case, our time is better spent collecting data on sites where the outcome is less certain.

Data collection design

Although there are many options for collecting monitoring evidence (see chapter 2), in habitat management projects we should always remember that we are not carrying out research: we are primarily looking for the most efficient and reliable means of feeding back into the management process.

Text box 16.1:   A sample target for the Ranunculion river habitat in a Welsh river

	The Ranunculion habitat of the Western Cleddau will be in a favourable state when:	
Habitat extent	**Lower limit**	In each of Sections 1–5 (see map in management plan), during periods of low flow and good water clarity in the month of July: The major cover-forming aquatic plants cover >150m² of river channel
Habitat quality	**Lower limit**	1 Four or more aquatic mesotrophic indicator species are present in each of Sections 1–5. 2 On average, five or more clean water benthic invertebrate families should be present in Sections 1–5, with no less than three families present in any one section. 3 *Gammarus* spp. are present in all sections and *Asellus* spp. are rare or absent in all sections. 4 Fresh signs of *Lutra lutra* activity are present in each section. 5 Salmonids and *Cottus gobio* are present in each section. 6 Either *Calopteryx virgo* or *Calopteryx splendens* (or both) is present in each section, and both species should be recorded in at least one section.

Site-specific definitions

Major cover-forming aquatic plants	Batrachian *Ranunculus* spp., *Myriophyllum alterniflorum*, *Callitriche brutia*, and *Fontinalis* spp.
Aquatic mesotrophic indicator plant species	Batrachian *Ranunculus* spp., *Myriophyllum alterniflorum*, *Callitriche brutia*, *Fontinalis squamosa*, *Chiloscyphus polyanthos*, *Lemanea fluviatilis*
Clean water benthic invertebrate families	Leuctridae, Perlodidae, Chloroperlidae, Ephemerellidae, Heptageniidae, Odontoceridae, Goeridae, Brachycentridae, Sericostomatidae
Fresh signs of *L. lutra* activity	Tracks in silt or mud in the river channel, spraints still oily
Salmonids	*Salmo trutta* or *Salmo salar*
River channel	A gently sloping bed of substrate submerged under water
Rare	Less than five individuals per completed kick sample

The sample targets in this table were set on the basis that the sampling would take place under optimum conditions in the month of July. In effect, the targets state that the Ranunculion habitat of the Western Cleddau will be in a favourable state if:

1   There is sufficient channel cover of Ranunculion macrophyte species.
2   The macrophyte species present suggest that the trophic status of the river is stable.
3   Enough families of clean water benthic invertebrates co-occur along the length of the river.
4   The fauna that we expect to be associated with the Ranunculion is present along the length of the river.

In this case, the data are collected from five 100m stretches of river distributed at regular intervals from the source to the mouth of the river. This monitoring approach will reveal whether the species that we would expect to be present are distributed along the length of the river or whether the biodiversity of the river declines in the lower sections of the river because of point source pollution incidents (perhaps originating from sewage treatment plants) or water abstractions further upstream.

Five of these six mesotrophic indicator plant species have been shown to have high detection rates in observer variation sampling trials (see Figure 5.11). The remaining targets were informed by the results of baseline sampling in 2006. The site-specific definitions are essential to minimize the risk of observer variation.

Source: Adapted from Hurford and Guest (2010).

Often the sampling approach can be determined by how the question is phrased. For example, the sample target tables for monitoring the Ranunculion river habitat and arable habitat (Text boxes 16.1 and 16.2, respectively) specify how and where the monitoring data should be collected.

Text box 16.2: The sample target table for the arable weed flora on a farm in Wales

Sample targets		To maintain the arable weed flora at Newton Farm in a favourable state where:
Extent and distribution	**Lower limit**	In one margin of at least three different fields planted with cereal or root crops:
		>50% of the sample points have two or more of *Kickxia elatine*, *Spergula arvensis*, and *Stachys arvensis* within a 50cm radius.
		And when
		>30% of the sample points have *Kickxia elatine*, *Spergula arvensis*, and *Stachys arvensis* present within a 50cm radius.
		And when
		Chrysanthemum segetum, *Fallopia convolvulus*, *Lamium amplexicaule*, *Lamium hybridum*, and *Misopates orontium* are present in at least one of the cereal or root crop field margins.

Site-specific habitat definitions
Field margin	Vegetation within 4m of any field boundary
Root crop	Fields planted with swedes or turnips
Cereal crop	Fields planted with spring barley, wheat, or oats

Rationale underpinning the sample targets

The eight species in the sample target table were chosen for one or more of the following reasons:

- Sensitive to herbicides and/or nitrogen applications.
- Listed among the 100 most rapidly declining species in the UK.
- Locally scarce.

The aim of the sample target is to detect whether there has been any change in soil pH because of nitrogen and/or phosphate applications. All three species selected for sampling in the field margins are sensitive to changes in soil pH. The sample target also requires the already rare species to persist but does not specify where, because this will depend on which fields are cultivated in the year of the monitoring event. The site-specific habitat definitions are essential in helping to minimize the risk of observer variation influencing the monitoring result.

Source: Adapted from Hurford (2006b).

When we place sampling plots in managed stands of priority habitat, we often opt to record point data on a systematic grid design (Figure 16.7). This allows us to record the number of points so that the habitat meets the criteria set out in the sampling target. To be of any practical use for feeding back into management decisions, the data collection points should be at an appropriate spacing. Active management will almost certainly introduce heterogeneity into even the most homogeneous stand of habitat (noting that habitats in poor condition are typically less diverse), so if the management aims to create more homogeneity, points need to be closer together to detect this. In our experience, the optimum distance between data points will be either 5m or 10m, the former in habitats with a fine-grained mosaic of states.

A "management monitoring" case study – Kenfig Special Area for Conservation

This case study demonstrates how to work through the process of setting conservation objectives on a complex site. The detail, however, is site specific and should not be applied to other dune systems without working through a similar process.

Background information on Kenfig Special Area for Conservation

The dune system at Kenfig extends over an area of circa 602ha on the south Wales coast and has attracted the attention of naturalists since the early 1900s. Consequently, there is a relatively well-documented natural history archive of botanical, entomological, and ornithological records.

The primary concern at Kenfig is the Habitats Directive Annex 1 habitat 2190 (European Commission 1992) humid dune slack habitat, which supports two dependent internationally rare species: the fen orchid *Liparis loeselii* and the petalwort *Petalophyllum ralfsii*. The species-rich dune grassland is also of international importance. More than 550 species of higher plant have been recorded at Kenfig, including at least 16 species of orchid. Several locally scarce insects also occur, including the hairy dragonfly *Brachytron pratense* and the shrill carder bee *Bombus sylvarum*.

During World War II, military training created open ground for pioneering plant species. However, aerial photographs show that, since military activities ceased at Kenfig (in the mid-1940s), the hind dunes have stabilized and open sandy habitats declined to <3% by 2003.

Figure 16.7 Sampling on a systematic grid, within a square or rectangular outline as appropriate, will reveal any directional changes of habitat state. Each star in the diagram represents a data collection point.

Management issues at Kenfig

The scope for the dune system to expand is restricted by a golf course to the south, a haul road to the west, a river to the north, and a road with adjacent villages to the east. Within these boundaries, the major threat to the biodiversity at Kenfig is geomorphological stability, resulting in species loss through habitat succession. Several factors have contributed to this situation:

* Offshore sand dredging and intertidal sand extractions.
* A decline in the frequency of summer storms, which has reduced scouring by windblown sand.

- A steep decline in rabbit numbers after outbreaks of myxomatosis: Kenfig was formerly managed as a warren and the rabbits helped to maintain the open sandy habitats by grazing, burrowing, and scraping.
- Reduced levels of stock grazing: graziers are increasingly reluctant to put animals on common land that is not fenced.
- Increasing visitor pressure (>150,000 visitors annually), which effectively restricts livestock to the northern parts of the site (well away from the main footpaths used by dog walkers).
- A policy of non-intervention management from 1978 to 1992.

Existing knowledge

The most relevant sources of existing knowledge were a recent PhD on the fen orchid (*Liparis loeselii* var. *ovata*) population at Kenfig and a National Vegetation Classification (NVC; Rodwell 2000) map of Kenfig that outlined the extent and distribution of the NVC vegetation communities in 1988. The site manager was concerned, however, that significant successional changes had occurred since then.

The survey process

While considering the survey options at Kenfig, we realized that, even if we carried out a new NVC survey, it would not provide us with the information needed to inform the management objectives. This is because the critical phases of dune slack development at Kenfig occur as seral stages within the NVC communities and the highest priority species at Kenfig are strongly associated with the early successional phases of humid dune slack development. Furthermore, had we set targets for the extent of the NVC communities at Kenfig, further problems would await us at the monitoring stage because we could not see a practical way of measuring it, because the priority SD14 NVC vegetation occurs as part of a tight mosaic with several other NVC communities.

Generally, the main problem with setting conservation objectives at plant community level is that conservation management is typically applied at the broader habitat level.

Developing appropriate habitat survey methods for Kenfig

Against the known background of dune stabilization and advanced habitat succession, we decided that we needed a survey to map the distribution of clearly recognizable seral stages of dune development at Kenfig. We started this process by asking the site manager to show us examples of dune grassland and dune slack vegetation that he considered to be of high conservation interest. Without exception, these examples were the species rich, successionally young seral stages of dune development. We then asked the site manager to show us examples of vegetation that he considered to be of low conservation interest: these tended to be the more mature, species-poor seral stages. Finally, we asked him to show us examples of vegetation that would cause him concern: these were the open, successionally young stands that were starting to close up.

Initially, we mapped the approximate extent and distribution of dune slack habitat from aerial photographs and then drafted a definition for each distinct successional stage of dune grassland and dune slack on the site. In all, we defined seven stages of dune slack (Table 16.1). We then visited the site on three separate days to test the definitions and

Table 16.1 The vegetation key for the dune slack seral stages at Kenfig NNR

Vegetation type	Definitions of dune slack states at Kenfig
Embryo slack	Some 25%–50% open ground present in the immediate area of an active blowout, *Salix repens* occurring in distinct clonal patches, with *Carex arenaria* an obvious associate and with either *Sagina nodosa* or *Juncus articulatus* present within 2 m of any point in the stand
Successionally young slack	A mosaic of patchy bare soil with thalloid liverworts and low, closed vegetation, with patches of moss cover, mostly of *Campylium stellatum* or *Calliergon cuspidatum*, but bryophytes not forming a dense mat. *Salix repens* can be abundant but not canopy forming and grasses should be generally scarce. At least two of *Carex viridula, Juncus articulatus, Anagallis tenella, Samolus valerandi*, and *Eleocharis quinqueflora* should be present within 2 m of open soil, and *Liparis loeselii* may also be present with other orchids; for example, *Epipactis palustris, Dactylorhiza praetermissa, D. incarnata*
Orchid-rich slack	Little or no bare soil evident, though orchids – that is, *Dactylorhiza praetermissa, D. incarnata, Epipactis palustris*, and *Gymnadenia conopsea* – are patchily common. *Salix repens* can be canopy forming and *Calliergon cuspidatum* can form dense mats in places. At least two of *Holcus lanatus, Poa subcaerulea, Pyrola rotundifolia*, and *Galium palustre* should be present within 2 m of any point; *Phragmites australis, Calamagrostis epigejos*, and *Molinia caerulea* can be evident, though none of these will dominant
Species-poor wet slack	Either, as above, but species poor with few orchids, or *Salix repens* co-dominant with *Carex nigra*, typically with a dense cover of *Hydrocotyle vulgaris* under the *Salix repens*
Dry slack	A drier, species-poor slack type, where *Salix repens* forms a shrubby canopy with *Holcus lanatus* and *Festuca rubra* notable among the associates, prone to invasion by *Betula* and taller *Salix* shrubs; for example, *S. cinerea* or *S. caprea*. Slacks where grasses such as *Festuca rubra* and *Elymus repens* are locally co-dominant with the *Salix repens* should be placed in this category
Brackish slack	Stands of *Juncus maritimus* present
Single-species stands	*Calamagrostis epigejos* or *Phragmites australis* forming dense stands or *Molinia caerulea* tussocks dominant

modify them as necessary before we were confident that wherever we were on the site, the dune slack and dune grassland vegetation conformed to one of the defined states.

We then visited the site with two MSc students from Swansea University who were going to carry out the survey as part of their thesis. This visit was spent training them to identify the species named in the definitions and providing them with specimens for reference. The student mapping the seral stages of dune slack development needed to recognize 30 species.

The students worked together on the site, using the definitions that we provided (Table 16.1) to map the approximate extent and distribution of each seral phase onto photocopies of aerial photographs. The survey took five weeks to complete. The field data were then transferred from the aerial photos into the geographic information systems (GIS) programme ArcView to produce the map shown in Figure 16.8. This map, which shows the balance between the successionally young and more mature dune vegetation at Kenfig, played a key role in the objective setting process. It is clear from the map that both the mature and orchid-rich seral stages of the humid dune slack habitat

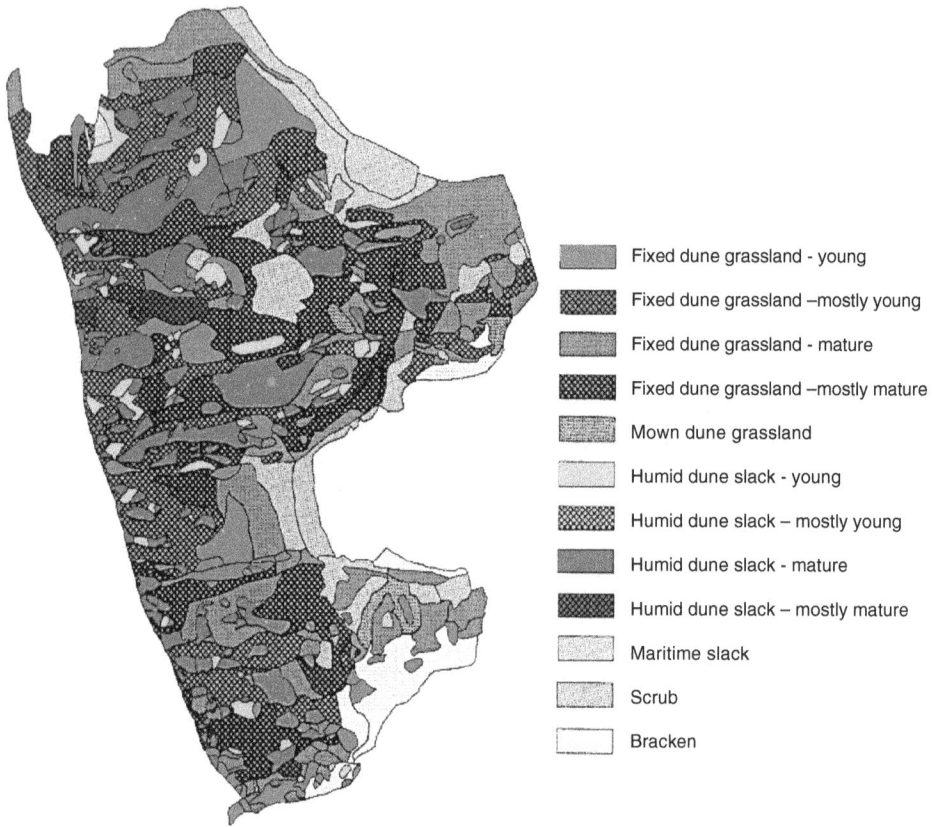

Figure 16.8 This habitat map shows the distribution of dune slack and dune grassland seral stages at Kenfig National Nature Reserve (NNR). This shows that successionally young dune slack vegetation was only locally distributed and in small patches. This habitat state is essential for supporting the internationally rare fen orchid (*Liparis loeselii*) and petalwort (*Petalophyllum ralfsii*) populations at Kenfig. Note that we did not consider this map to be either precise or repeatable, but we did trust it to provide a reasonable representation of the distribution and extent of the habitat states on the site.

were plentiful but that successionally young dune slack vegetation was much scarcer, with embryo slack vegetation restricted to just two small areas (Figure 16.9).

At the end of the survey phase and ahead of the objective setting phase, we developed a site-specific habitat change diagram for the dunes at Kenfig (Figure 16.10) to check our understanding of the habitat transition sequence there. In the absence of active management on the site, all change was predicted to be in one direction, from successionally young species–rich vegetation to mature species–poor vegetation.

Prioritization

The results of the habitat condition survey and an exercise to dispassionately identify the management and monitoring priorities at Kenfig (shown in Text box 16.3) both identified the humid dune slack habitat as the priority.

Figure 16.9 Embryo slack vegetation in the vicinity of the last naturally formed and active dune blowout at Kenfig NNR.

Credit: Photo by Clive Hurford.

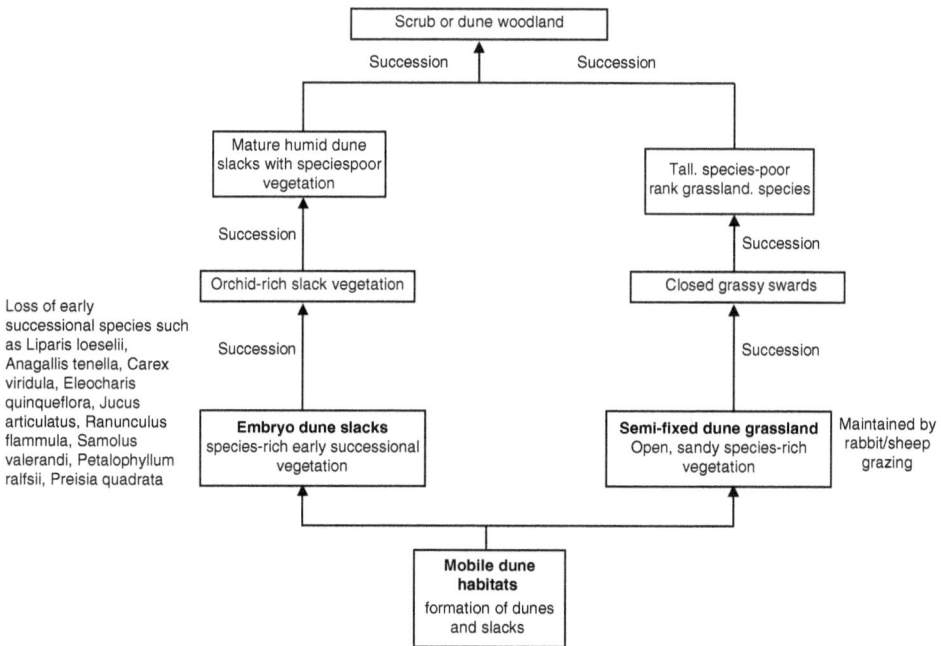

Figure 16.10 A diagram illustrating successional change for the dune habitats under a non-intervention management policy at Kenfig NNR. Damp dune habitats near the water table follow the route of succession on the left; the dry sandy habitats follow the route on the right.

Text box 16.3:   Habitat priority assessment

This table shows an example of a dispassionate scoring system for identifying biodiversity monitoring priorities. SSSI stands for Site of Special Scientific Interest, areas that are protected for their contents of scientific values (flora, fauna, or physiological features) under the Wildlife and Countryside Act (Legislation.gov.uk 1981).

Habitat priority assessment table

Site: Kenfig NNR

Habitat: Humid dune slacks

Habitat designation	Value	Site score	Dependent species designation	Value	Site score
International priority habitat and special UK responsibility	10	0	International priority species and special UK responsibility	10	0
International priority habitat	9	0	International priority species	9	0
Annex I habitat and special UK responsibility	8	0	Annex II species and special UK responsibility	8	16
Annex I habitat	6	6	Annex II species	6	0
SSSI habitat	3	0	SSSI species	3	0
Area of habitat			**Population size**		
50% of national resource	10	0	>50% of national resource	10	10
26%–50% of national resource	8	8	26%–50% of national resource	8	0
11%–25% of national resource	6	0	11%–25% of national resource	6	0
6%–10% of national resource	4	0	6%–10% of national resource	4	0
1%–5% of national resource	3	0	1%–5% of national resource	3	0
<1% of national resource	1	0	<1% of national resource	1	1
Habitat not under threat	0	0	Species not under threat	0	0
Habitat total		**14**	**Dependent species total**		**27**

Overall score = 41

This dispassionate scoring system not only identifies the management priority within each site but also allows us to prioritize resource allocation between sites; for example, a site where the highest overall score for a habitat was 38 would be prioritized for resources ahead of a site where the highest overall score was 25.

Note that the scoring system focuses on three key variables:

1   Level of designation.
2   Presence of dependent species.
3   Percentage of the national resource present on the site.

The humid dune slack habitat was the highest scoring/priority habitat at Kenfig because the site supports more than 25% of the Welsh resource and because it has two Annex II species that are dependent on the habitat. Finally, note that if the habitat or species is not under any threat or able to be managed – for example, cliff ledge species – they do not score because there is no scope for a management response.

Surveillance

Because we already knew that the successionally young seral stage of humid dune slack development was the management priority at Kenfig, we concentrated our surveillance effort on the successionally young dune slack vegetation. This vegetation type can be very species rich, however, and attempting to record all species would have been a time-consuming exercise. Therefore, we recorded the assemblage of carefully selected attributes described below.

The humid dune slack surveillance project

This detailed surveillance allows us to keep track of habitat loss in the more important areas of the site and will provide us with information to support the case for destabilisation should the need arise. Our surveillance recording involved the following:

- Placing ten L-shaped 5 × 5 m linear plots in the more extensive areas of successionally young dune slack vegetation.
- Locating five 50 cm × 50 cm quadrats (each divided into 16 cells) at set distances along each plot.
- Recording the presence in each cell of the following positive indicators of successionally young dune slack: *Anagallis tenella, Carex viridula* ssp. *viridula, Eleocharis quinqueflora, Juncus articulatus, Ranunculus flammula, Samolus valerandi, Preissia quadrata,* and thalloid liverworts (Table 16.2).
- Estimating by eye whether more than 10% bare soil and less than 25% bryophyte cover was present in each cell, because these are also positive attributes of embryo slacks.
- Using quadrat photographs to record cover-forming species and minimize observer variation.

Marking the corners of each plot with buried wooden markers and aluminum tags and then locating the markers using a Differential Global Positioning System (DGPS) accurate to circa 3 cm.

We chose the indicator species for the sample target by analyzing relevé data collected from successionally young and orchid-rich slack vegetation: the species in the indicator

Table 16.2 Quadrat data from a successionally young dune slack surveillance plot at Kenfig

Species	Cell number																%
	1	2	3	4	5	6	7	8	9	10	11	12	13	14	15	16	
Carex viridula	1	1	1	0	0	0	0	1	1	0	0	0	1	1	0	0	44
Juncus articulatus	0	0	1	1	1	1	1	0	1	0	1	1	0	1	0	0	56
Eleocharis quinqueflora	0	0	0	0	0	0	0	0	0	0	0	0	0	0	0	0	0
Ranunculus flammula	1	1	1	1	0	1	0	1	1	1	1	1	1	1	1	1	87
Anagallis tenella	0	0	0	0	0	0	0	0	0	0	0	0	0	0	0	0	0
Samolus valerandi	0	0	0	0	0	0	0	0	0	0	0	0	0	0	0	0	0
Preissia quadrata	0	1	1	0	0	1	1	1	0	0	1	0	0	0	1	1	44
Thalloid liverworts	1	1	1	1	0	1	1	1	0	0	1	1	1	1	1	1	81
Bryophytes >25%	0	0	0	0	0	0	0	0	0	0	0	0	0	0	0	0	0
Bare soil >10%	1	1	1	1	0	1	1	1	0	1	1	1	1	0	1	1	81

Note. 1 = present, 0 = absent.

assemblage were present in the successionally young relevés but absent from the orchid-rich relevés. This surveillance sampling helped us to develop the site-specific definition of successionally young dune slack vegetation in the management objective by:

- Increasing our confidence in the species assemblage that we selected for monitoring successionally young dune slack vegetation.
- Confirming how many of the selected species we should expect to find within our area of search (which in this case was a 50cm radius).

The humid dune slack management aims

In 1994, we initiated a *Liparis* recovery project at Kenfig (Hurford 1997). The early signs from this experimental management project were encouraging and suggested that maintaining and restoring both the mature and orchid-rich slack seral stages of dune slack vegetation was not going to be a problem if there was a sufficient area of successionally young dune slack habitat present; for examples of the habitat at Kenfig, see Figures 16.11 and 16.12. This meant that, from this point forward, the management could prioritize the scarce embryo and successionally young dune slack vegetation at Kenfig. We knew from other dune systems that sheep and rabbit grazing could maintain successionally young vegetation, but there was no evidence that we could restore embryo slack vegetation if we lost it. Therefore, our best chance of restoration would involve

Figure 16.11 The humid dune slack habitat at Kenfig in the 2012–2013 winter. In wet winters, the dune slacks at Kenfig can flood to a depth of 2m or more.

Credit: Photo by Clive Hurford.

Figure 16.12 Liparis loeselii flowering in mown humid dune slack habitat at Kenfig. Mowing played an important role at Kenfig by allowing the *Liparis* population to persist, albeit in sub-optimal habitat, until the mechanical excavations were sanctioned to increase the area of successionally young dune slack habitat.

Credit: Photo by Clive Hurford.

destabilization of the dunes in areas where the water table was conducive to dune slack formation. In the past, however, buildings had been buried by drifting sand at Kenfig, and major dune destabilization work was unlikely to be approved by the trustees in the short term.

The management objective for the humid dune slack habitats at Kenfig in Table 16.3 was informed by the habitat survey described above.

The rationale underpinning the management objective for the dune slacks at Kenfig

The current geomorphological stability at Kenfig means that until the system reverts to a geomorphologically dynamic state, the site must be managed as a stable system. The conservation objective considers the following:

- The extent of the successionally young vegetation will continue to diminish in the absence of appropriate management.
- While the current geomorphological state persists, new embryo dune slack formation is unlikely.
- Increases in the extent of orchid-rich and the more mature stages of dune slack succession can be achieved through management practices and successional processes at Kenfig.
- Open access to people and dogs leads to grazing stock congregating in the north of the site, limiting the options for effective management elsewhere.

Table 16.3 The management objective for humid dune slack vegetation at Kenfig

Management objective	To maintain the dune slack habitat at Kenfig in a favourable state where:	
Extent	Lower limit	Extent mapped in 1997 (Figure 16.11)
Quality	Lower limit	In Section 1
		>30% of the dune slack habitat in Area Y and >45% of the dune slack habitat in Area Z is either embryo or successionally young slack vegetation and >70% of the dune slack vegetation outside of Areas Y and Z is either successionally young or orchid-rich

- Only two sizeable areas of dune slack with both embryo and successionally young vegetation remained at Kenfig, Areas Y and Z. The corners of these areas were recorded with as high-accuracy DGPS, plotted in a GIS, and transferred onto an aerial image of the site in the management plan

The humid dune slack monitoring project

The first monitoring event at Kenfig was carried out in 1998, shortly after the appointment of a new site manager. We delayed the objective setting and monitoring processes until he was familiar with the site.

The sampling approach

The process of defining the vegetation key for the habitat survey (Table 16.1), when combined with the survey results, provided the site manager with the information needed to:

- Form an opinion on the current state of the dune habitats.
- Define how to recognize when the habitats are in a favourable state.
- Set informed management objectives, stating not only how much of the key habitats are required but also where they are required (i.e., their extent and distribution).

By the end of the survey phase, three stages of dune slack succession were recognized to be of high biodiversity interest at Kenfig: embryo, successionally young, and orchid rich. All embryo slack vegetation was found within the defined Areas Y and Z. The successionally young and orchid-rich dune slacks were more widely distributed (though mostly in the northern half of the site). Because the northern section of Kenfig was under one grazing regime, the dune slack sampling could focus initially on Areas Y and Z, allowing inferences to be made about the state of the remaining areas based on the monitoring results). This is a *selective* approach to monitoring.

The sample targets

The limits in this sample target reflect those set out in the management objective. The sample target in Table 16.4 differs from the conservation objective by referring to the

Table 16.4 The sample target for dune slack vegetation at Kenfig

Sample target for when the habitat state is favourable	*The* dune slack *habitat at Kenfig will be in a favourable state when:*	
Quality	Lower limit	>30% of the sampling points in Area Y and >45% of the sampling points in Area Z meet the criteria below for embryo or successionally young slack vegetation and
		>70% of the dune slack habitat outside of Areas Y and Z comprises vegetation where >50% of the sampling points meet the criteria for either successionally young or orchid-rich dune slack
Site-specific habitat definitions		
Dune slack vegetation	Moist vegetation on level ground between sloping dunes, typically with *Salix repens* present	
Successionally young dune slack vegetation	Bare soil and thalloid liverworts present and at least four of the following present: *Carex viridula* ssp. *viridula, Juncus articulatus, Anagallis tenella, Samolus valerandi, Eleocharis quinqueflora, Ranunculus flammula,* or *Liparis loeselii* within any 50 cm radius and	
	none of the following present: *Phragmites australis, Molinia caerulea, Calamagrostis epigejos* within any 1 m radius	
Embryo slack vegetation	25%–50% open ground with *Salix repens* forming clonal patches and	
	at least two of *Carex arenaria, Sagina nodosa,* or *Juncus articulatus* present within any 1m radius	
Orchid-rich dune slack vegetation	At least two of the following present: *Epipactis palustris, Dactylorhiza incarnata, Gymnadenia conopsea,* or *Pyrola rotundifolia* in any 50 cm radius and	
	none of the following present: *Phragmites australis, Molinia caerulea, Calamagrostis epigejos* within any 1 m radius	
Sampling approach	Selective	
Sampling method	Systematic grid recording (non-relocatable grid mapping)	

percentage of sampling points (as opposed to the percentage of habitat), by identifying the sampling areas, by clearly defining how to recognize the habitat states mentioned, and by specifying the sampling approach and sampling method.

The sampling method

We used a systematic grid design to monitor the extent and state of the dune slack habitats in Areas Y and Z, locating the boundary of each sampling area with DGPS. This method would effectively map the distribution of dune slack states in each area. The sampling procedure involved pacing 10m between recording points on a regularly spaced grid and recording whether the vegetation within a 50cm radius at each point met the criteria set out in the sample target.

On completing the monitoring exercise, the number of points meeting the criteria for embryo and successionally young dune slack vegetation was converted to a percentage of the total number of sampling points taken from slack vegetation. The dune slack monitoring in Areas Y and Z could be repeated in less than three days.

Table 16.5 The results of the humid dune slack monitoring in Areas Y and Z at Kenfig NNR

Humid dune slack monitoring results	Area Y	Area Z	Kenfig total
Number of sampling points	89	53	142
Embryo dune slack points	3	0	3
Successionally young slack points	2	4	6
Percentage of recording points meeting sample target criteria	6	8	
Lower limit	30%	45%	
Status	Unfavourable	Unfavourable	

Monitoring results and analysis

The results in Table 16.5 show that the dune slack habitat at Kenfig was in an un-favourable state, with the extent of embryo and successionally young dune slack vege-tation in Areas Y and Z in the areas most likely to meet the lower limit in the sample target falling well short. We did not monitor the dune slacks outside of Areas Y and Z because the overall state of the slack vegetation had already been determined by the sampling results in Areas Y and Z.

The restoration target for humid dune slack vegetation at Kenfig

Because monitoring found that the dune slack habitat state was unfavourable, the site manager needed to consider the target for restoration (Table 16.6). This restoration target differs from the management objective by abandoning Areas Y and Z and setting a more specific and more demanding target for successionally-young and orchid-rich dune slack vegetation across the site. This restoration target ensures that neither habitat is considered restored until the condition is well above the respective lower limit stated in the conservation objective.

The rationale underpinning the restoration target

By the time the monitoring project was completed and the restoration target setting process had begun, we estimated that there was <3% bare sand left at Kenfig, that embryo dune slack vegetation was restricted to a single location, and that successionally young dune slack vegetation had disappeared. With the benefit of hindsight, the original management objective should have placed more emphasis on the embryo and succes-sionally young states, because these are under immediate threat. Furthermore, because *Liparis loeselii* and *Petalophyllum ralfsii* are dependent on the embryo and successionally young dune slack habitats, the restoration target also considers the restoration targets for these Annex II species. This is good practice: all maintenance and restoration targets must consider the needs of dependent species.

The management response

The monitoring result in 1999 prompted a series of management actions. The immediate response was to expand the close-mowing programme in the south of the site and to

Table 16.6 The restoration target for the dune slack vegetation at Kenfig

Restoration target		To restore the dune slack habitat at Kenfig to favourable status where
Extent	Target	Extent mapped in 1997
Quality	Target	>40% of the dune slacks comprise habitat where >70% of sampling points meet the criteria for successionally young dune slack
		and
		>80% of the dune slacks comprise habitat where >70% of sampling points meet the criteria for either successionally young or orchid-rich slack vegetation
Habitat definitions		
Dune slack vegetation		Moist vegetation on level ground between sloping dunes, typically with *Salix repens* present
Successionally young dune slack vegetation		Bare soil and thalloid liverworts present and at least four of the following present: *Carex viridula ssp. viridula, Juncus articulatus, Anagallis tenella, Samolus valerandi, Eleocharis quinqueflora, Ranunculus flammula,* or *Liparis loeselii* within any 50 cm radius
		and
		none of the following present: *Phragmites australis, Molinia caerulea, Calamagrostis epigejos* within any 1 m radius
Embryo slack vegetation		25%–50% open ground with *Salix repens* forming clonal patches
		and
		at least two of *Carex arenaria, Sagina nodosa,* or *Juncus articulatus* present within any 1 m radius
Orchid-rich dune slack vegetation		At least two of the following present: *Epipactis palustris, Dactylorhiza incarnata, Gymnadenia conopsea,* or *Pyrola rotundifolia* in any 50 cm radius
		and
		none of the following present: *Phragmites australis, Molinia caerulea, Calamagrostis epigejos* within any 1m radius

increase the sheep grazing levels in the north: these were doubled from 100 to 200 sheep. In the background, there were also long-term negotiations with the trustees over the installation of a fence that would enable the introduction of cattle grazing at Kenfig. While these negotiations were underway, the close-mowing programme allowed the *Liparis loeselii* population to persist in the south of the site, but the increased level of sheep grazing had the opposite effect on the population in the north, with only a handful of *Liparis* plants surviving until 2008, when a stock-proof fence was installed (Carrington et al. 2010).

However, this fence around the northern section of the site, combined with the decision to allow mechanical habitat restoration, made a huge difference to the potential for restoring the dune slack habitats on the site. As soon as the fence line was completed, the first dune slack scrapes were excavated (Figure 16.13) and most of the sheep were removed and replaced by smaller numbers of cattle (Figure 16.14). Thereafter, the restoration management regime settled into a regular pattern until the onset of the COVID-19 pandemic in 2019, with close-mowing continuing outside the fenced area in the south of the site, cattle grazing continuing in the north of the site, and one or two restoration slacks being excavated annually at suitable locations across the site.

Despite the ongoing restoration work, the *Liparis loeselii* population at Kenfig continued to decline until 2013, when only 44 were recorded. In 2012, however,

Figure 16.13 The dune slack excavations took place in winter, often while the slacks were under water. The stand of *Molinia* in front of the excavator shows the habitat structure before the excavation work started.

Credit: Photo by Clive Hurford.

Figure 16.14 Cattle were introduced to the site as soon as the fencing work was completed. These animals used the first two scraped dune slacks as watering holes, but this ceased to be an issue as more scrapes were excavated.

Credit: Photo by Clive Hurford.

Figure 16.15 The first *Liparis loeselii* plants to colonize the excavated restoration slacks at Kenfig were seen in 2012. These plants were probably already persisting in the slack as pseudobulbs but were brought closer to the surface by the management work.

Credit: Photo by Clive Hurford.

three years after the start of the excavation programme, the first *Liparis* plants had been seen in a restoration slack (Figure 16.15). These plants had probably persisted as pseudobulbs buried under the soil but had been brought closer to the surface by the management work.

At the time of writing, in 2022, more than 5ha of successionally young dune slack has been created through mechanical excavations, cattle grazing continues in the fenced-off area, and there have been two attempts at destabilizing the foredunes, though without significant success. With regards to the *Liparis loeselii* population, in 2019, the *Liparis* population was estimated at 4250 plants, the majority in the areas of successionally young dune slack habitat created by excavations, with the remaining plants in the mown slacks. Without these management interventions, driven by the management and monitoring process described here, *Liparis loeselii* var. *ovata* would be on the brink of extinction in the UK.

Summary

This chapter outlines approaches for detecting and monitoring small biotopes in cultural landscapes. These habitats, their management options, and likely outcomes are well known. This allows monitoring projects to adopt a selective approach and focus on what the management aims to deliver and where. There are real consequences if monitoring delivers the wrong result. Site managers could change management unnecessarily, damaging habitats and species, or persist with an inappropriate and damaging management programme that does not allow restoration. Furthermore, incorrect reporting could create additional political pressures or lead to complacency and loss of funding. Therefore, we must do everything we can to minimize the potential to deliver the wrong monitoring result. This includes incorporating existing knowledge of what is already known about the management of the habitat, the historical management of the site and the potential for biodiversity gains. Critically, the sample design and data collection methods must address issues relating to observer variation.

Key messages

- Small biotopes and habitats with a history of cultural management need special consideration for monitoring both their changing extent and distribution and the biodiversity interest.
- Managing and monitoring these habitats is not a research exercise – the management options and their likely outcomes are already well known. This allows monitoring to focus on whether the management aims have been met.
- In these situations, monitoring is an integral component of the site management process and feeds back into it.
- Because there are real consequences to getting monitoring wrong, data collection should avoid the use of subjective methods and focus on applying objective approaches combined with photographic support.

Study questions

1   What are the key stages in the development of a management project?
2   Why is prioritization important, and how would you approach prioritizing sites, habitats, and species for management and monitoring in your region?
3   How would you test the reliability of your preferred data collection methods?

Further reading

The following books were published to provide guidance on the development of monitoring projects in cultural habitats. Both include case studies covering a wide range of habitats and species.

Hurford, C. and Schneider, M. (eds) (2006) *Monitoring Nature Conservation in Cultural Habitats: A Practical Guide and Case Studies*. Dordrecht, The Netherlands: Springer.

Hurford, C., Schneider, M. and Cowx, I. (eds) (2010) *Conservation Monitoring in Freshwater Habitats: Practical Guide and Case Studies*. Dordrecht, The Netherlands: Springer.

References

Agger, P. and Brandt, J. (1988) Dynamics of small biotopes in Danish agricultural landscapes, *Landscape Ecology* 1(4), 227–240.

Brandt, J. and Levin, G. (2006) *Indikatorer for Landskabsændringer: Analyser af Komplekse Landskabsændringer på Baggrund af RUCs Småbiotopundersøgelser* [Indicators for Landscape Changes: Analyses of Complex Landscape Changes from Investigations of Small Biotopes at Roskilde University]. Roskilde Universitet.

Brown, A. (2000) *Habitat Monitoring for Conservation Monitoring and Reporting. 3: Technical Guide*. Life Nature Project No. LIFE95 NAT/UK/000821. Integrating Monitoring with Management Planning: A Demonstration of Good Practice in Wales. Bangor: Countryside Council for Wales.

Carrington, D., Hurford, C., Jones, A. and Pankhurst, T. (2010) The fen orchid – a species on the brink, *British Wildlife* 22(1), 1–8.

European Commission. (1992) *The Habitats Directive*, Council Directive 92/43/EEC of 21 May 1992, https://ec.europa.eu/environment/nature/legislation/habitatsdirective/index_en.htm (Accessed November 11, 2022).

Fredshavn, J.F., Levin, G. and Nygaard, B. (2015) *Småbiotoper 2007 og 2013* [Small Biotopes 2007 and 2013], Scientific Report from DCE – National Centre for Environment end Energy, No 143. The National Monitoring Program for Terrestrial Nature Types (NOVANA) Aarhus, Aarhus University. https://dce2.au.dk/pub/SR143.pdf

Hurford, C. (1997) *Year 1 Report on the Fen Orchid* Liparis loeselii *Species Recovery Programme at Kenfig NNR, Glamorgan*. Unpublished report. Bangor: Countryside Council for Wales.

Hurford, C. (2006a) Developing a habitat monitoring project, in Hurford, C. and Schneider, M. (eds) *Monitoring Nature Conservation in Cultural Habitats: A Practical Guide and Case Studies*. Dordrecht, The Netherlands: Springer, pp. 57–59.

Hurford, C. (2006b) Monitoring arable weeds at Newton Farm, in Hurford, C. and Schneider, M. (eds) *Monitoring Nature Conservation in Cultural Habitats: A Practical Guide and Case Studies*. Dordrecht, The Netherlands: Springer, pp. 169–184.

Hurford, C. and Guest, D. (2010) Monitoring the Ranunculion habitat of the Western Cleddau: a case study, in Hurford, C., Schneider, M. and Cowx, I. (eds) *Conservation Monitoring in Freshwater Habitats: Practical Guide and Case Studies*. Dordrecht, The Netherlands: Springer, pp. 125–136.

Legislation.gov.uk. (1981) *Wildlife and Countryside Act 1981*, chapter 69. https://www.legislation.gov.uk/ukpga/1981/69 (Accessed November 21, 2022).

Rodwell, J.S. (ed.) (2000) *British Plant Communities: Vol. 5. Maritime Communities and Vegetation of Open Habitats*. Cambridge: Cambridge University Press.

Rowell, T.A. (1993) *Common Standards for Monitoring SSSIs*. Peterborough, UK: Joint Nature Conservation Committee.

Rowell, T.A. (2006) The relationship between management and monitoring, in Hurford, C. and Schneider, M. (eds) *Monitoring Nature Conservation in Cultural Habitats: A Practical Guide and Case Studies*. Dordrecht, The Netherlands: Springer, pp. 13–22.

17 Case study: reindeer husbandry plans – "Is this even monitoring?"

Per Sandström, Stefan Sandström, Ulrika Roos, and Erik Cronvall

Introduction

As demonstrated in earlier chapters, there are widely varying types and applications of monitoring. This chapter provides an insight into the somewhat non-conventional monitoring programme termed Renbruksplan (RBP; in English, reindeer husbandry planning). Even though RBP is founded on many of the common monitoring methodologies described in earlier chapters, RBP did not start out as a monitoring programme and is often not recognized as such. RBP also lies outside the common national reporting systems. It has also sometimes been seen as an example of community-based monitoring, where the Sami reindeer herders, as Indigenous people and as the significant "knowledge holders" (Wilson et al. 2018), are carrying out the data collection and driving the demand towards its outcome (see chapter 2).

This chapter describes the history and development of this programme, how methods were developed and used, and some of the results, impacts, and contributions to land use decisions, management, and policy.

We start this chapter by providing a short background about Sami reindeer husbandry as a land use system, to give context to the RBP effort. To understand the situation for reindeer husbandry today, some background of the historical perspective and how the interpretation of legal rights for the Sami people has changed over time is needed (Mörkenstam 1999; Brännlund and Axelsson 2011). Sami life and culture are in many ways centred on reindeer and the reindeer husbandry system (Text box 17.1) and thus is the focus of this programme. Reindeer husbandry has existed in various forms for millennia (Aronsson 1991) and throughout the 19th century was encouraged and privileged by the government of Sweden as the best-suited land use form in northern Sweden (Cramér and Ryd 2012). Reindeer husbandry as a land use system originated from the taming of a few reindeer for milking, as draft animals, and as decoys to attract wild reindeer (Björklund 2013), from which it gradually moved towards the land use system we see today. The formation of a formally organized reindeer husbandry system became necessary to increase production capacity in response to an increased government tax burden (Cramér and Ryd 2012).

Today, reindeer husbandry represents a land use system that ultimately depends on the implementation and consequent footprint of other land uses such as infrastructure development, mining, and energy exploitation or specific forestry activities that affect lichen conditions, the key winter food source for reindeer (P. Sandström 2015; P. Sandström et al. 2016). In strict monetary terms, all of these competing land uses are more significant than reindeer husbandry. As all these other land use systems expand and intensify, the hope and

DOI: 10.4324/9781003179245-17

Text box 17.1:   Sami reindeer husbandry

The reindeer husbandry system constitutes an extensive, complex, and unique land use form carried out by the Indigenous Sami people across Sapmi, an area covering northern Sweden, Norway, Finland, and parts of the Kola Peninsula in Russia. There are also more than 20 other indigenous groups across the Russian north and Mongolia who depend on reindeer husbandry (Oskal et al. 2009). With seasonal grazing lands connected via annual long-range migration routes, reindeer constitute the last remaining large ungulate migrations in the Northern Hemisphere (Vors and Boyce 2009). The reindeer husbandry system is also the only enduring grazing system of semi-domesticated or domesticated animals that still use their native range and seasonal movement patterns to access and use grazing resources in the same or similar way as its native ancestor species. Maintaining such a land use system today offers both challenges and opportunities for managers as well as for policymakers.

Reindeer husbandry in Sweden is allowed on 55% (constituting a 22.6 million ha area) of the land base, including 49% of the productive forest lands, more than 40% of the standing forest volume, and 99% of the alpine biomes (P. Sandström et al. 2016). Because no part of the reindeer husbandry area is set aside exclusively for reindeer husbandry, it is always carried out in conjunction with other land uses. The reindeer husbandry area is divided into 51 reindeer herding communities (in Swedish, *sameby*). A reindeer herding community constitutes a large geographic area usually stretching from the mountains in the west to the Bay of Bothnia in the east. A reindeer herding community is also an administrative and financial organizing association for the Sami reindeer herders' companies. Many reindeer herding communities are further divided into *siidas* (winter groups), each containing one or several reindeer herding companies.

Currently there are about 240,000 reindeer in Sweden, a number that has remained relatively constant over the last decades, distributed among 4598 reindeer owners, working full time or part time as herders (Sametinget 2021c).

---

prospect for a balanced and functioning coexistence alongside reindeer husbandry are continually and increasingly challenged. Of particular concern, because it overlaps entirely with reindeer husbandry, are the intensification and expansion of forestry into areas not previously impacted by modern forestry practices (Text box 17.2). Initiatives and discussions to mitigate the conflict between reindeer husbandry and forestry have been going on for a long time. The need to find a balance between the impacts of forestry and the needs of reindeer husbandry was addressed as far back as 1954 at a meeting in Jokkmokk (Skuncke 1955). At this meeting, reindeer herders expressed concerns about future negative impacts from forestry on grazing resources and the need to address issues with a landscape perspective. Since then, the need for a knowledge-based dialogue, supported by maps and field inventories in various forms as tools towards solutions, has been put forward (e.g. Heikka 1981). Reindeer herders have long expressed their needs for such tools to support planning and negotiations in relation to the forest industry and more recently in relation to other ongoing land uses; hence the development of the RBP programme.

Text box 17.2:   Impacts of forestry on reindeer husbandry

Forestry and reindeer husbandry each represent two extensive and overlapping land use forms. On these lands, forestry is the prioritized and dominant land use system because of its economic importance. As the demand for forest products has increased and modern forestry practices have intensified and spread into new areas, the negative impacts on reindeer husbandry have been amplified. Forestry affects reindeer husbandry in several ways. First and most important, modern forestry has profound impacts on mat-forming (*Cladina* spp. or *Cladonia* spp.) and arboreal lichens (e.g. *Bryoria fuscescens* and *Bryoria fremontii*), which are pivotal key winter grazing resources (P. Sandström et al. 2016). Much of the conflict and dialogue between reindeer husbandry and forestry revolves around management of lichen forests. Other negative impacts include the densification of forests, which, in addition to having negative effects on ground lichen growth, also inhibits reindeer, as well as herders' ability to move through the forest landscape. Forest regeneration after harvest usually begins with soil scarification, which removes ground vegetation cover, with consequent effects on ground lichen (Roturier and Bergsten 2006; Roturier 2009). The impacts of timber harvests on canopy cover also affect the conditions of the snow cover. Even though forestry can cause such negative impacts, it is no doubt technically possible to adapt forest management practices to the needs of reindeer husbandry while maintaining profitable forestry operations (Roturier 2009; Korosuo et al. 2014; Lundström 2016). However, many of the identified solutions such as increased levels of cleaning and thinning and gentler soil scarification methods have not yet been broadly implemented. Lack of implementation of such solutions is partly connected to the forest sector's poor understanding of reindeer use of grazing areas and the movement patterns of the reindeer. Solutions are also held up because of drawn-out consultation procedures (P. Sandström et al. 2003; C. Sandström and Widmark 2007; Roos et al. 2022). After 70 years of modern forestry practices, forestry is still seen as a threat to future reindeer husbandry. However, at the same time, if correctly carried out, forestry could be the saviour, because thinning and cleaning of dense young forests is essential for the ground lichen resource. Hence, there is a need for programmes such as the RBP to contribute data and knowledge to highlight key issues so that they can be better addressed.

The story behind *Renbruksplan*

The birth and naming of the Renbruksplan process stems from a meeting in a reindeer herder's kitchen in a Malå reindeer herding community (for definition, see Text box 17.1) in 1998. A discussion between reindeer herders and Swedish Forest Agency (SFA) personnel around how reindeer husbandry could match the forestry sector's well-established strategic forest planning tool, *skogsbruksplan*, led participants towards a matching reindeer herder's tool, *Renbruksplan*. Until then, the forestry sector put forward data and maps during consultation meetings, whereas reindeer herders could only present their point of view in words.

The first official RBP working meeting, organized in 2000, included representatives from the two reindeer herding communities of Malå and Vilhelmina Norra, personnel from SFA, the Swedish Board of Agriculture, the County Administrative Board of Västerbotten, and the Swedish University of Agricultural Sciences (SLU). During the first meetings of the original RBP group, the overarching goals of the process were defined to provide and improve the basis for (a) operational reindeer herding and (b) consultations with other land uses. The initial focus was on forestry as the "other" land user.

Because we knew that mapping would be central to our approach, the tools brought to the first meeting included printouts of recent satellite images, plastic overlays, and coloured pens. In general, at this first meeting, it was clear that the reindeer herders were the experts on reindeer husbandry and everybody else there knew very little about this. However, each one of the participants brought some new knowledge to the meeting. A mutual learning process, later termed *co-production of knowledge* (P. Sandström 2015), started immediately. Initially, the group had no specific working strategy or a clear final product in mind. Instead, the approach and methods were developed in a stepwise fashion where we iteratively invented, evaluated, and re-invented our strategies (Poudyal et al. 2015). The reindeer herders' mappings of seasonal grazing areas for each of the eight seasons of the well-established cycle of the reindeers' year were a central part of the process (Manker and Pehrson 1953; P. Sandström et al. 2003). For each local portion of each of the pilot reindeer herding communities, we used collared pens to delineate the most important grazing lands with the initial focus on the forested winter grazing areas. We produced plastic overlays with coloured mappings of important grazing areas and delivered these to the County Administrative Board of Västerbotten to be digitized. However, interpreting mappings and digitizing the drawings was not a simple task. This working practice was soon rejected. We realized that instead it was necessary for the Sami reindeer herders to digitize and describe their grazing lands themselves. Consequently, we developed the first version of a custom-made geographic information system (GIS), named RenGIS (in English, ReindeerGIS). RenGIS v1.0 was developed in the standard commercial ArcView 3.0 GIS environment. As the number of participating reindeer herding communities increased from the original 2, then to 6, later to 10, and finally to today's 50 reindeer herding communities, the need for an even more specific and custom-made tool became clear. Also, in response to the increasing number of users, license costs for a commercial GIS platform became too high, prompting us to convert the RenGIS system to the programming environment of TatukGIS, resulting in the current freeware RenGIS v2.0 currently in use (TatukGIS 2014). RenGIS is now freely available to all users (Sametinget 2022).

What is a Renbruksplan?

As participating reindeer herding communities' specific needs became clearer, separate sections of the RBP work process crystallized. All sections were not part of an original plan of the programme but emerged from iterative development and testing together with participating reindeer herding communities. Currently, we divide the monitoring programme into the following four sections:

- Reindeer herders' mappings of important grazing lands.
- Reindeer herders' field inventories.
- The collection and use of Global Positioning System (GPS) data from reindeer.
- The compilation of all other land use forms and available land cover data.

All information from each of these sections is created, analyzed, and visualized in RenGIS.

The work in each of these sections was preceded by the development and refinement of a number of common definitions and terminologies. Together with participating reindeer herders, we developed a common general division of the reindeers' grazing year. We also developed common terminologies for reindeer grazing lands. We divided the grazing lands into three main types: (a) general grazing areas representing the total seasonal grazing lands; (b) core areas, or important grazing land within general grazing areas; and (c) key areas, situated within core areas and of greatest importance to reindeer. Additionally, we divided and defined common vegetation classes as reindeer grazing types (P. Sandström 2015). Training involving reindeer herders and organized by personnel from SLU preceded all data collection, with a specific focus on coordinating work and maintaining common ground between all different reindeer herding communities and their data collection teams. During the 22 years of operation, 400 Sami reindeer herders from all reindeer herding communities have participated, contributed, and been involved in trainings in RenGIS, satellite image interpretation, field inventory techniques, forestry, drone use, and GPS. These training sessions have played central and necessary roles for each operational component of the process. Each session was focused less on teachings by experts but instead more on peers teaching peers (co-production of knowledge). The session organizers seldom delivered and instructed completely tested components. Instead, trainings consisted of continually testing and developing prototypes and learning from each other's experiences. Together we evaluated and improved the processes iteratively over time (Poudyal et al. 2015). Such mutual learning and mutual development have guided the project's efforts concerning improvements of most sections and components. In fact, most of the ideas regarding improvements of existing modules and the invention of new modules originate from these learning opportunities. The presence of Sami reindeer herders, researchers, as well as the programmer of RenGIS during all of the trainings ensured the tight connection between identified needs and production of new modules into RenGIS.

In the following paragraphs, we describe each of the four sections of the RBP.

Reindeer herders' mappings of important grazing lands

The first, and maybe the most central, section of the RBP monitoring programme is the delineation and description of important seasonal grazing lands into grazing types (key, core, and general grazing area). The Sami reindeer herder with the most knowledge of each specific, local area carries out the mapping by digitizing in RenGIS.

All delineation was done through on-screen digitizing, using the most up-to-date satellite images as background (P. Sandström et al. 2003; P. Sandström 2015). The specific delineation was then discussed and readjusted in consultations with other local experts from each *siida* (winter group). Through this process, each local area was mapped for each of the eight reindeer grazing seasons but with specific initial focus on the forested winter grazing areas. Mappings for each local area were subsequently merged to cover the entire grazing lands for each season. Finally, the digitized material for each season was merged to cover the entire reindeer herding community (Figure 17.1).

The delineation of important grazing lands has grown from a modest pilot project in two reindeer herding communities to cover the entire lands of 50 reindeer herding communities, representing an area of 225,000 km^2 and spanning more than half of

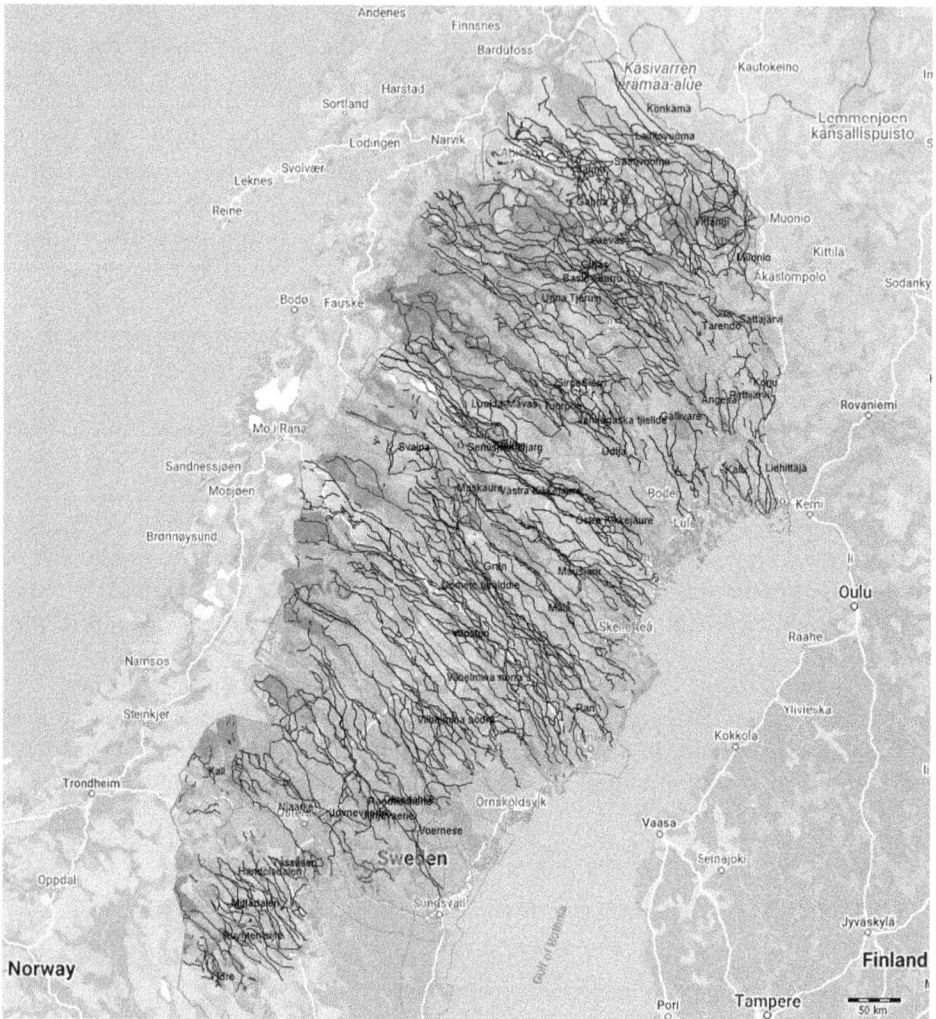

Figure 17.1 The combined map describing the reindeer herders' mappings of important seasonal grazing lands for 50 out of the 51 reindeer herding communities in Sweden for all grazing seasons. Key areas are shown in red, core areas in blue, general grazing areas as see through and reindeer migration routes are shown as green lines. Data and mapping compiled and visualized from RenGIS.

Sweden's land area (P. Sandström 2015). Today nearly 10,000 key areas have been digitized and described, with the large majority situated in the boreal forest.

Reindeer herders' field inventories

For most monitoring programmes, field inventorying constitutes the core of the programme. However, with the RBP programme, the actual field inventory work is just one section of the entire programme. As mapping of important seasonal grazing lands was carried out, the need to support and strengthen these mappings with field inventories became apparent.

Figure 17.2 View from within a key grazing area, where the reindeer grazing type is classified as lichen-rich pine forest. Each delineated key grazing area contains information from field inventories and a description by the reindeer herder. This area is described as "rich in trees with pendulous lichen in rolling terrain and with large variations in canopy cover providing good grazing even during periods with difficult snow conditions".

Credit: Photo by Per Sandström.

Partly developed out of the protocol used in the Swedish National Forest Inventory (Fridman et al. 2014) with respect to variables and inventory methods, we developed a specific vegetation classification system for reindeer grazing types (Sametinget 2017; Figure 17.2). We further adjusted this protocol and definitions of grazing types according to recommendations from reindeer herders.

Currently we have developed and implemented four different field inventory protocols. These include an inventory system for (a) forest lands, (b) wetlands, (c) mountain areas, and most recently (d) ground lichen. We have also developed specific inventory manuals (Sametinget 2017) and inventory instruction films to support the work (Sametinget 2021a). All field inventories were carried out by local reindeer herders from each respective area and reindeer herding community. Field inventory trainings organized by SLU preceded all inventory work.

Our first developed inventory protocol covering forest lands was launched in 2001. The objectives were to visit, estimate, and measure field inventory plots and document information to safeguard future re-inventories and, in some cases, by placing lichen cages as described in Figure 17.3. However, in addition, one objective was for the herders to gain additional overall and general knowledge about each visited delineated grazing area and to add such information for each mapped area (Figure 17.2). This was especially important because most of the inventories were carried out on winter grazing lands, previously mostly only visited by the herders when the ground was snow

(a)   (b)

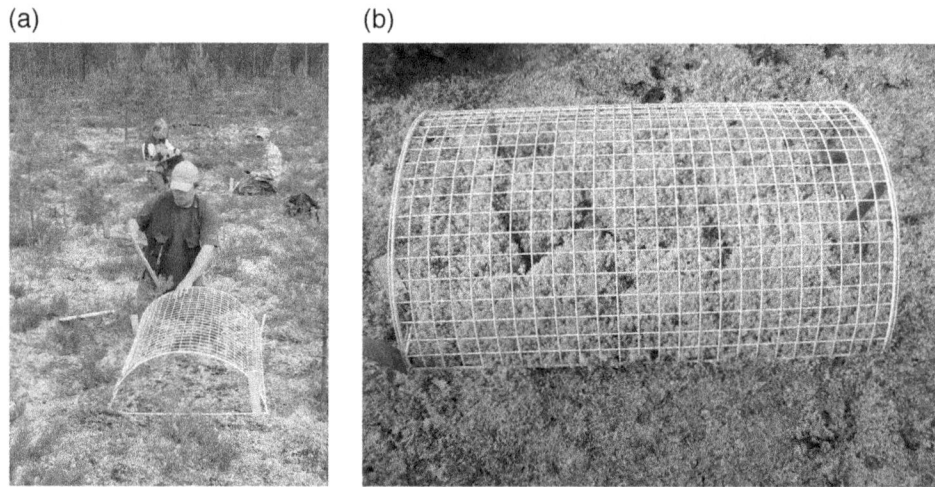

Figure 17.3 As part of the RBP grazing land inventory system initiated in 2001, we placed a series of
lichen cages in the field to be able to follow lichen growth over time: (a) an overview of
the field plot and (b) a 1.5m × 1m lichen cage.

Credit: Photos by Per Sandström.

covered. Another objective of the field visits was to improve herders' satellite image
interpretation skills further, because the field visits served as additional ground
truthing. Finally, the field visits gave the reindeer herders' specific information for final
adjustments of the boundaries of the delineated important seasonal grazing areas.
Consequently, the effort to map important grazing lands became closely linked to the
effort of the field inventory. General information on each delineated grazing area
gained during field visits was subsequently added into RenGIS. Such field-based and
specific information for each grazing area can be especially important to support and
strengthen the dialogue and land use negotiations with the forest industry as well as
other land users.

Based on requests from reindeer herding communities and other land users, SLU
designed and implemented a ground lichen inventory programme as part of the overall
RBP process during 2019. The goal with this inventory was specifically targeted towards
producing high-accuracy ground lichen maps for each reindeer herding community.
Both inventory data and produced lichen maps will provide important base information
for future re-inventories, mapping efforts as well as land use consultations. This is
especially important and sought after as a response to alarming reports of a 71% decline in
the lichen-rich forests grazing type during the last 65 years (P. Sandström et al. 2016).
Ground lichen, the key winter grazing resource for reindeer, has continually declined
since the introduction of modern forestry methods during the 1950s and, consequently,
inventory and mapping of this resource are much needed. However, because the ground
lichen–rich forests have become so rare (according to P. Sandström et al. (2016), only
3.8% of the reindeer husbandry area remains), the National Forest Inventory, which is
based on a random sample of field plots, provides too few lichen-rich plots.
Consequently, we needed additional field data to produce high-accuracy maps even for

lichen-rich areas. This need paved way for the RBP ground lichen inventory and map production system, here described through the following five steps:

1   SLU produces a preliminary lichen map based on NFI data, satellite images, and lidar data (Swedish Forest Agency 2022b).
2   SLU organizes field training in lichen inventories for reindeer herders.
3   Based on the preliminary lichen map and balanced sampling methods, SLU delivers coordinates to reindeer herding communities for additional field plots to improve the preliminary lichen maps.
4   Reindeer herders carry out field inventories based on the SLU lichen inventory protocol and deliver field data to SLU.
5   SLU carries out the final classification and produces the final lichen map for each participating reindeer herding community.

The ground lichen mapping and field inventory effort now constitute important components of the RBP programme. The ground lichen, field inventory and mapping can be repeated in the future to map and monitor changes of this important key resource. Earlier efforts to map ground lichen using satellite image data in combination with field data from the Swedish National Forest Inventory (Fridman et al. 2014) showed promising results in terms of general classification accuracy (P. Sandström et al. 2003; Gilichinsky et al. 2011). However, new methods developed through the RBP programme provide higher classifications accuracies than earlier efforts.

To date, we have visited, measured, and photographed more than 10,000 field plots. Most of the field plots are located in identified and digitized pre-winter and winter key grazing areas. Currently, field data collection has moved to being carried out via a cell phone app (Sametinget 2021b). This allows data from field inventories to be entered directly into RenGIS in the field. Over time, the field inventories of forest lands initiated in 2001 have expanded into new areas and new reindeer herding communities. During 2020, three reindeer herding communities initiated re-inventories of their older field plots. Such information will provide critical baseline knowledge about changes in lichen cover in relation to both forestry activities that may have occurred since the first inventory as well as changes because of successional changes in the forests.

Real-time GPS data

The use of real-time GPS collars on reindeer has become commonplace in most reindeer herding communities since it was first introduced in a project between SLU and the Vilhelmina Norra reindeer herding community in 2006. General technical solutions for real-time GPS already existed and were used, for example, in moose research (Dettki et al. 2004). However, the introduction of real-time GPS into reindeer husbandry became the first example of a fully participating data user – the reindeer herder – taking full operational advantage of real-time GPS data.

The GPS collar consists of a GPS unit that records the position of the animal according to a schedule defined by the user. The collars also include a communication unit that transfers the positions via the cellular network. Positions are then transferred to a web server, where the positions can be viewed on the reindeer herder's cell phone or computer. Hourly updates from GPS-equipped reindeer have provided important support and allowed herders to reduce their use of snowmobiles, all-terrain vehicles (ATVs), and helicopters; for example, during reindeer gatherings for calf markings, for slaughter, and during seasonal migrations. Adoption of the GPS technology has also reduced stress levels for reindeer and improved working conditions for the reindeer herders (Andersson and Keskitalo 2017). Nowadays, a herder's workday usually begins with checking the reindeer's movements since the day before on their cell phone or computer.

As important as the operational advantages, reindeer equipped with real-time GPS collars provide vital information for monitoring changes in reindeer habitat use over time, as data from GPS-equipped reindeer provide a continually feed of their positions into the system. Long-term data in the form of GPS positions, descriptions of reindeer habitat use, and reindeer movements have created an important data bank, used in numerous research projects, impact assessments, and environmental court cases (Skarin et al. 2015, 2018, 2021, 2022; Cambou et al. 2022). Because new GPS data are collected from many reindeer every hour, every day, the record of reindeer use has provided, and will continue to provide, valuable monitoring data. Furthermore, GPS positions from collared reindeer have helped herders refine and improve their delineation of important grazing lands. Currently we also use GPS data describing reindeer use to verify the quality of our lichen maps. Here older positions can show high reindeer use of areas now not in use by reindeer as the lichen have disappeared.

As of 2022, about 35 reindeer herding communities use and manage their real-time GPS-collars and, over time, several thousand reindeer have been equipped with GPS collars.

Mapping of all other land uses

In 2001, the RBP team identified the need to include data about other land uses in RenGIS. In 2010, SLU received funds from the Sami Parliament to carry out the first compilation of other land uses and incorporate this into RenGIS. This compilation, updated every year, now represents the only custom-made and easy-to-use land use database of its kind. Compiled data originate from different state agencies for their respective geographic areas and responsibilities, including the Swedish Forest Agency, the Swedish Mining Inspectorate, the Swedish Energy Agency, the Swedish Meteorological and Hydrological Institute, the Swedish Board of Agriculture, the county administrative boards, the Swedish Land Survey, the Swedish National Heritage Board, and the Sami Parliament. Most of the data are publicly available but not compiled in one place and into one database. Having all of these data compiled in RenGIS provides a common basis for the analysis of land use issues and activities as both individual and cumulative impacts. These data have played a major role in numerous land use dialogues (Herrmann et al. 2014; Skarin et al. 2015, 2018, 2021, 2022). Being able to illustrate, visualize, and explain land use changes over time provides data for yet another component of the RBP programme.

Outcome and impacts of the Renbruksplan programme

Usually monitoring programmes constitute a field data collection component, whereas analysis and use of data is left to others. In the work with RBP, a significant aim of the programme was also to provide tools and opportunities for the data collectors – in our case, the reindeer herders – to also analyze, visualize, and communicate the compiled data and to illustrate changes over time. In this way, the RBP is also a tool for learning and communication about the complex reindeer husbandry system within the reindeer husbandry collective as well as among other land users. This makes the RBP programme unique because the data collector, the data owner, and the data user, to some extent, is the same person.

The role of the tool – RenGIS in a monitoring programme

Visualizing and explaining the extensive and complex land use system of reindeer husbandry, with its long-range seasonal movements and large shifts in seasonal grazing lands, is best done using GIS-aided tools. To support communication about the content of the RBP, we custom-developed RenGIS as our tool for data collection, analysis, visualization, and dialogue.

RenGIS has become the tool used to analyze and illustrate landscape changes and shifts in habitat use over time and relate these to changes in vegetation composition and changes caused by other land users. It is important to understand that the RBP is not just a series of datasets that by itself support dialogue and planning and decision making. The RBP, through its communication tool RenGIS, only works to its full extent together with its presenter, the reindeer herder.

One important overall outcome of RBP and RenGIS is that they provide a platform for learning about reindeer husbandry. Numerous specifically developed modules in RenGIS, such as "play GPS-equipped reindeer movements", have been used in numerous public and professional meetings to communicate reindeer habitat use in relation to vegetation as well as other land uses (Skarin et al. 2021). Hence, there are several examples of how a co-produced RenGIS is more effective than tools developed outside the programme.

RenGIS is publicly downloadable from Sametinget (2022) with some common data available for all users. There are also some more specific data only available for the reindeer herding communities. In addition, each reindeer herding community owns and manage their own data, such as their digitized important grazing lands, field inventory data, and data from their GPS-equipped reindeer. Hence, this form of raw data from the RBP programme is not directly available for all. Instead, knowledge from such data is the common outcome of the programme.

Specific outcome and impacts

Data and knowledge from RBP communicated via RenGIS are well-established contributors to land use dialogues, negotiations, and consultations in Sweden. Data from RBP play a major part in governmental and non-governmental reports and strategies. Examples include the Swedish environmental objectives "A Magnificent Mountain Landscape" (Swedish Environmental Protection Agency [SEPA] 2014) and "Sustainable Forests" (SEPA 2013), the government commission "Follow-Up of the Attention to Reindeer Husbandry" (Swedish Forest Agency 2011), "Dialogue and Collaboration between Forestry and Reindeer Husbandry" (Swedish Forest Agency 2013), the Sami Parliament environmental programme *Eallinbiras* (Sametinget 2009), the Swedish Forest Stewardship Council (FSC 2010, 2020), the Swedish Sami Organization Forest Policy (SSR 2019), and "Indicators for the National Forest Programme of Sweden" (Swedish Forest Agency 2022a).

Data and knowledge from RBP have played central roles in the analysis of impacts from wind power developments (Skarin et al. 2015, 2018, 2021, 2022), forestry (Korosuo et al. 2014; Lundström 2016), climate impacts (Löf et al. 2012), mining (Herrmann et al. 2014), infrastructure development (P. Sandström et al. 2020), and cumulative impact assessments (Arctic Monitoring and Assessment Program 2017; Kløcker Larsen et al. 2020), as well as in several court cases (see e.g. Cambou et al. 2022; Skarin 2022). Furthermore, data from RBP play a central role in a number of ongoing research projects.

Conclusion

Initiated in 2000, the RBP programme can now be considered a long-term monitoring programme. Starting as a modest pilot project in two reindeer herding communities, with some initial funding from the Swedish Board of Agriculture, the programme has grown significantly to inventory and map more than half of Sweden's land area. Since 2005, the programme has received funding through earmarked money in the Swedish national budget, first via the Swedish Forest Agency and lately via the Sami Parliament. The work has been carried out by more than 400 data collectors and data users and contributes knowledge to planning and decision making from local to national scales. Furthermore, data and knowledge from the programme can contribute data for monitoring landscape changes, impact assessments, research projects, policymaking and strategic planning.

For Sami reindeer husbandry, the RBP programme plays an important role in explaining, maintaining, and incorporating the complex and geographically extensive reindeer husbandry system into the context of other land use systems. Thus, the RBP programme can be seen as a challenging real-life test case for advanced, sustainable landscape management. The challenges span geographic scales that range from single grazing patches to half of the land area in Sweden and can provide a stronger basis for sustainable landscapes as well as for the continuation of an ancient indigenous land use system.

Many of the environmental monitoring programmes described in this book provide key component data within large national data depositories. Here the RBP programme differs significantly, because some of the data remain with the data producer and thereby data owner. However, the RBP programme still provides important information that contributes towards better understanding of environmental change, as well as support to planning, decision making, and policy development. Though not recognized at the initiation of the RBP programme, one focus of the work has become to provide a broader interdisciplinary understanding among all land users. We recognize the reindeer husbandry system as a suitable indicator system for sustainable landscapes and sustainable land use. Hence, the RBP monitoring programme can provide knowledge to address and resolve important and complicated land use issues.

Key messages

- RBP can serve as an example of both a non-conventional and a community-based monitoring programme.
- A programme initially introduced as community-based mappings supported through field measurements has developed into providing the basis of the more conventional monitoring system to document changes at local and landscape levels.

- Specific components such as GPS tracking of reindeer have provided advanced possibilities for habitat monitoring.
- The RBP programme is unique because it was initiated and carried out by community members and the compiled information is owned and used by community members. There are both challenges and advantages with this arrangement.

Study questions

1 Can reindeer husbandry plans be seen as an example of monitoring? Why or why not?
2 Can you see any problems with the reindeer husbandry plans as a monitoring programme in relation to:

- Monitoring as a support for policymakers?
- Fields of conflicting interests?

3 How are the reindeer husbandry plans related to social/human understanding of monitoring?

Further reading

Sandström, P. (2015) *A Toolbox for Co-production of Knowledge and Improved Land Use Dialogues – The Perspective of Reindeer Husbandry*. Doctoral Thesis, Acta Universitatis Agriculturae Suecicae–Silvestra.

Sandström, P., Granqvist Pahlén, T., Edenius, L., Tømmervik, H., Hagner, O., Hemberg, L., Olsson, H., Baer, K., Stenlund, T., Brandt, L.-G., et al. (2003) Conflict resolution by participatory management: remote sensing and GIS as tools for communicating land use needs for reindeer herding in northern Sweden, *Ambio* 32(8), 557–567. https://doi.org/10.1579/0044-7447-32.8.557

Provide thorough descriptions of the RBP programme 12 years apart.

https://www.sametinget.se/renbruksplaner presents all components of the RBP programme including working manuals and instructive films (Accessed November 22, 2022).

References

Andersson, E. and Keskitalo, E.C.H. (2017) Technology use in Swedish reindeer husbandry through a social lens, *Polar Geography* 40(1), 19–34. https://doi.org/10.1080/1088937X.2016.1261195

Arctic Monitoring and Assessment Program. (2017) *Indigenous peoples' perspectives. Adaptation Actions for a Changing Arctic: Perspectives from the Barents Area*. AMAP.

Aronsson, K.A. (1991) *Forest Reindeer Herding AD 1–1800. An Archaeological and Palaeoecological Study in Northern Sweden*. Doctoral Thesis, Department of Archaeology, University of Umeå.

Björklund, I. (2013) Domestication, reindeer husbandry and the development of Sami pastoralism, *Acta Borealia* 30(2), 174–189. https://doi.org/10.1080/08003831.2013.847676

Brännlund, I. and Axelsson, P. (2011) Reindeer management during the colonization of Sami lands: a long-term perspective of vulnerability and adaptation strategies, *Global Environmental Change* 21(3), 1095–1105. https://doi.org/10.1016/j.gloenvcha.2011.03.005

Cambou, D., Sandström, P., Skarin, A. and Borg, E. (2022) Reindeer husbandry vs. wind energy. An interdisciplinary analysis of the Pauträsk and Norrbäck court decisions in Sweden, in Tennberg, M., Broderstad, E.G. and Hernes, H.-K. (eds) *Indigenous Peoples, Natural Resources and Governance: Agencies and Interactions*. Routledge, pp. 39–58.

Cramér, T. and Ryd, L. (2012) *Tusen år i Lappmarken: Juridik, Skatter, Handel och Storpolitik.* [Thousand Year in the Lappmark: Law, Tax, Trade and Politics.] Skellefteå, Sweden: Ord & Visor.

Dettki, H., Ericsson, G. and Edenius, L. (2004) Real-time moose tracking: an internet based mapping application using GPS/GSM-collars in Sweden, *Alces* 40, 13–21.

Forest Stewardship Council. (2010) *Svensk Skogsbruksstandard Enligt FSC med SLIMF-Indikatorer* [Swedish FSC Standard for Forest Certification including SLIMF Indicators]. Uppsala: Svenska FSC.

Forest Stewardship Council. (2020) *FSC-Standard för Skogsbruk i Sverige* [The FSC National Forest Stewardship Standard of Sweden]. https://se.fsc.org/se-sv/regler/skogsbruksstandard (Accessed 2022).

Fridman, J., Holm, S., Nilsson, M., Nilsson, P., Ringvall, A.H. and Ståhl, G. (2014) Adapting national forest inventories to changing requirements – the case of the Swedish National Forest Inventory at the turn of the 20th century, *Silva Fennica* 48(3), 1–29. http://dx.doi.org/10.14214/sf.1095

Gilichinsky, M., Sandström, P., Reese, H., Kivinen, S., Moen, J. and Nilsson, M. (2011) Mapping ground lichens using forest inventory and optical satellite data, *International Journal of Remote Sensing* 32(2), 455–472. https://doi.org/10.1080/01431160903474962

Heikka, G. (1981) *Jokkmokksmodellen för Samråd Rekommenderas av Gruppen* [The Jokkmokk Model for Consultation]. Vol. 11. Samefolket.

Herrmann, T.M., Sandström, P., Granqvist, K., D'Astous, N., Vannar, J., Asselin, H., Saganash, N., Mameamskum, J., Guanish, G., Loon, J.-B., et al. (2014) Effects of mining on reindeer/caribou populations and indigenous livelihoods: community-based monitoring by Sami reindeer herders in Sweden and First Nations in Canada, *The Polar Journal* 4(1), 28–51. https://doi.org/10.1080/2154 896X.2014.913917

Kløcker Larsen, R., Skarin, A., Stinnerbom, M., Vannar, J., Alam, M., Kuhmunen, M., Lawrence, R., Nygård, J., Raitio, K., Sandström, P., et al. (2020) *Omtvistade Landskap – Navigering Mellan Konkurrerande Markanvändning och Kumulativa Effekter* [Contested Landscapes - Navigating between Competing Land Uses and Cumulative Impacts]. Rapport 6908, Naturvårdsverket, Stockholm.

Korosuo, A., Sandström, P., Öhman, K. and Eriksson, L.O. (2014) Impacts of different forest management scenarios on forestry and reindeer husbandry, *Scandinavian Journal of Forest Research* 29, 234–251. https://doi.org/10.1080/02827581.2013.865782

Lundström, A. (2016) *Formulering och Utvärdering av Renskötselanpassad Skogsskötsel med Integrerad Geografisk Information från Beteslandsindelning* [Formulation and Evaluation of Reindeer Herding Adapted Forest Management with Integrated Information from Grazing Land Division]. Umeå: Sveriges Lantbruksuniversitet.

Löf, A., Sandström, P., Stinnerbom, M., Baer, K. and Sandström, C. (2012) *Renskötsel och Klimatförändring – Risker, Sårbarhet och Anpassningsmöjligheter i Vilhelmina Norra Sameby* [Reindeer Husbandry and Climate Change – Risks, Vulnerability and Adaptability in Vilhelmina North Reindeer Herding Community]. Forskningsrapport 2012:4, Statsvetenskapliga Institutionens Skriftserie, Umeå Universitet.

Manker, E. and Pehrson, R.N. (1953) *The Nomadism of the Swedish Mountain Lapps: The Siidas and Their Migratory Routes in 1945*. Stockholm: Gebers.

Mörkenstam, U. (1999) *Om "Lapparnes Privilegier": Föreställningar om Samiskhet i Svensk Samepolitik 1883–1997* [On "Lappish Privileges": Notions of Sami Identity in Swedish Sami Politics 1883–1997]. Stockholm University.

Oskal, A., Turi, J.M., Mathiesen, S.D. and Burgess, P. (2009). Reindeer herders voice: reindeer herding, traditional knowledge and adaptation to climate change and loss of grazing lands: International Centre for Reindeer Husbandry, in Oskal, A., Turi, J.M., Mathiesen, S.D. and Burgess, P. (eds) *EALÁT*. Kautokeaino: Sami Univ. College.

Poudyal, M., Lidestav, G., Sandström, S. and Sandström, P. (2015) Supporting community governance in boreal forests by introducing participatory GIS through action research, *International Journal of Action Research* 11(3), 236–264.

Roos, U., Lidestav G., Sandström S. and Sandström P. (2022). Samråd: an institutional arrangement in the context of forestry and reindeer husbandry in northern Sweden, *International Forestry Review* 24(3), 441–457.

Roturier, S. (2009) *Managing Reindeer Lichen during Forest Regeneration Procedures*. Doctoral Thesis, Acta Universitatis Agriculturae Suecicae–Silvestra.

Roturier, S. and Bergsten, U. (2006) Influence of soil scarification on reindeer foraging and damage to planted Pinus sylvestris seedlings, *Scandinavian Journal of Forest Research* 21(3), 209–220. https://doi.org/10.1080/02827580600759441

Sametinget. (2009) *Sametingets Livsmiljöprogram Eallinbiras* [The Sami Parlaiment Habitat Programme Eallinbiras]. http://www.sametinget.se/7366 (Accessed November 22, 2022).

Sametinget. (2017) *Manual för Fältinventering* [Manual for Field Inventory]. https://www.sametinget.se/113475 (Accessed November 22, 2022).

Sametinget. (2021a) *Instruktionsfilmer* [Instructional Movies]. https://www.sametinget.se/RenGIS_film (Accessed November 22, 2022).

Sametinget. (2021b) *Manual för Fältdatainsamling*, [Manual for Field Data Collection]. https://www.sametinget.se/159699 (Accessed November 22, 2022).

Sametinget. (2021c) *Statistik Rennäring*, [Reindeer Husbandry Statistics]. https://www.sametinget.se/renstatistik (Accessed November 22, 2022).

Sametinget. (2022) *RenGIS 2.0*, https://www.sametinget.se/renGIS (Accessed November 22, 2022).

Sandström, C. and Widmark, C. (2007) Stakeholders' perceptions of consultations as tools for co-management – a case study of the forestry and reindeer herding sectors in northern Sweden, *Forest Policy and Economics* 10(1–2), 25–35. https://doi.org/10.1016/j.forpol.2007.02.001

Sandström, P. (2015) *A Toolbox for Co-production of Knowledge and Improved Land Use Dialogues – The Perspective of Reindeer Husbandry*. Doctoral Thesis, Acta Universitatis Agriculturae Suecicae–Silvestra.

Sandström, P., Cory, N., Svensson, J., Hedenås, H, Jougda, L. and Brochert, N. (2016) On the decline of ground lichen forests in the Swedish boreal landscape – implications for reindeer husbandry and sustainable forest management, *Ambio* 45(4), 415–429. https://doi.org/10.1007/s13280-015-0759-0

Sandström, P., Granqvist Pahlén, T., Edenius, L., Tømmervik, H., Hagner, O., Hemberg, L., Olsson, H., Baer, K., Stenlund, T., Brandt, L.-G., et al. (2003) Conflict resolution by participatory management: remote sensing and GIS as tools for communicating land use needs for reindeer herding in northern Sweden, *Ambio* 32(8), 557–567. https://doi.org/10.1579/0044-7447-32.8.557

Sandström, P., Myntti, E.-L., Sandström, S., Jonsson, N., Lidestav, G. and Jonsson, T. (2020) Who knew digitizing and dialogue could change the course of reindeer husbandry rights? We know, now: building bridges between knowledge systems and over highways, in McDonagh, J. and Tuulentie, S. (eds) *Sharing Knowledge for Land Use Management – Decision-Making and Expertise in Europe's Northern Periphery*. Cheltenham, UK: Edward Elgar, pp. 130–143.

Skarin, A., Nellemann, C., Rönnegård, L., Sandström, P. and Lundqvist, H. (2015) Wind farm construction impacts reindeer migration and movement corridors, *Landscape Ecology* 30, 1527–1540. https://doi.org/10.1007/s10980-015-0210-8

Skarin, A., Niebuhr, B. B., Sandström, P., and Tømmervik, H. (2022). Den ekologiska bevisföringen i Fosenmålet - Analys av renens användning av vinterbetesmarkerna och konsekvenser av vindkraftutbyggnad. TIDSSKRIFT FOR UTMARKSFORSKNING 2022–1: 19–27. https://hdl.handle.net/11250/2995433

Skarin, A., Sandström, P. and Alam, M. (2018) Out of sight of wind turbines – reindeer response to wind farms in operation, *Ecology and Evolution* 8(19), 9906–9919. https://doi.org/10.1002/ece3.4476

Skarin, A., Sandström, P., Brandão Niebuhr Dos Santos, B., Alam, M. and Adler, S. (2021) *Renar, Renskötsel och Vindkraft: Vinter- och Barmarksbete* [Reindeer, Reindeer Husbandry and Wind Power: Winter and Snow-Free Grazing Lands]. Rapport 7011, Naturvårdsverket, Stockholm.

Swedish Environmental Protection Agency. (2013) Levande skogar [Sustainable forests], http://www.miljomal.se/Miljomalen/12-Levande-skogar/ (Accessed November 22, 2022).

Swedish Environmental Protection Agency. (2014) *Förslag till en Strategi för Miljömålet Storslagen Fjällmiljö.* [Proposed Strategy for the Environmental Goal Magnificant Mountain Landscapes]. Skrivelse 2014-06-05, Redovisning av ett Regeringsuppdrag, Stockholm.

Swedish Forest Agency. (2011) *Uppföljning av Hänsyn till Rennäringen*, [Follow-Up of the Attention to Reindeer Husbandry]. Jönköping, Sweden: Skogsstyrelsen. http://shop.skogsstyrelsen.se/shop/9098/art30/10527330-5ac9fc-1579.pdf (Accessed November 22, 2022).

Swedish Forest Agency. (2013) *Dialog och Samverkan Mellan skogsbruk och Rennäring.* Jönköping, [Dialogue and Collaboration between Forestry and Reindeer Husbandry]. Sweden: Skogsstyrelsen. http://shop. skogsstyrelsen.se/shop/9098/art4/16015504-4cc687-1584.pdf (Accessed November 22, 2022).

Swedish Forest Agency. (2022a) *Nationellt Skogsprogram: Förslag till Indikatorer för det Nationella Skogsprogrammet,* [National forest program: proposed indicators for the national forest program]. Jönköping, Sweden: Skogsstyrelsen. https://www.skogsstyrelsen.se/globalassets/om-oss/rapporter/ rapporter-2022/rapport-2022-04-forslag-till-indikatorer-for-det-nationella-skogsprogrammet.pdf (Accessed November 22, 2022).

Swedish Forest Agency. (2022b) *Skogliga grunddata* [Baseline forest data], https://www.skogsstyrelsen. se/skogligagrunddata/ (Accessed November 22, 2022).

TatukGIS. (2014) *TatukGIS.* Gdynia, Poland. https://www.tatukgis.com/Products.aspx (Accessed November 22, 2022).

Vors, L.S. and Boyce, M.S. (2009) Global declines of caribou and reindeer, *Global Change Biology* 15(11), 2626–2633.

Wilson, N.J., Mutter, E., Inkster, J. and Satterfield, T. (2018) Community-based monitoring as the practice of Indigenous governance: a case study of Indigenous-led water quality monitoring in the Yukon River Basin, *Journal of Environmental Management* 210, 290–298. https://doi.org/10.1016/ j.jenvman.2018.01.020

18 Reflections on monitoring: conclusions and ways forward

E. Carina H. Keskitalo, Alan Brown, and Anna Allard

Introduction

Monitoring nature and biodiversity has historically often been the work of individuals or groups of individuals motivated more by curiosity than out of necessity. A more systematic interest in nature was spurred on by increasing population growth, urbanization, and the industrial revolution and, in recent decades, by the realization that what society had once taken for granted – a clean, unchanging environment; abundant fish; and inexhaustible natural resources – now has to be managed, which is not an easy feat (e.g. Wauchope et al. 2022). To manage something, we need to know not only how it has changed and how it will change in the future – just as forest managers have considered for centuries – but also how management can be incentivized and implemented within different systems.

Many of the technologies we now use were developed initially for other purposes, including statistical survey methods, aerial photography, satellite remote sensing, image analysis, and DNA sequencing. Though at first the academic interest in new technologies is more often in how they can be developed, rather than how they can be applied, sooner or later they are taken up by practitioners, using them to answer real, urgent questions, some of which were not considered as part of their original design. All of the technologies discussed in earlier chapters, such as Global Positioning Systems (GPS) and small drones, have been adopted in this way by ecologists and field scientists. Along with the technology comes a new vocabulary and a new set of concepts which enlarge our understanding of the natural world, such as the electromagnetic spectrum or probability functions. For example, we are now so used to associating living plants with absorbing red sunlight and reflecting a band of near-infrared light, which can be seen from space by satellite instruments thanks partly to a military need to find camouflaged vehicles.

All of these changes bring with them not only new approaches to monitoring but also new issues of integrating new types of data and evidence. Monitoring is thus a field in constant development, and it is crucial to consider how we can continue managing this change and adapt changing technologies, methods, and instruments.

The chapters throughout the book have shown the tremendous breadth and rate of change not only in the field of monitoring itself but in the methods and technologies that can be used to collect and analyze the results of monitoring. The book illustrates the need to be open to new technologies and to encourage new data types to be included within monitoring parameters and objectives. However, it it hoped that it also illustrates the need to pay attention to the passage of time. This includes taking into account how to design or adapt existing monitoring programmes to include new data or objectives.

DOI: 10.4324/9781003179245-18

Given the ever-changing context, it is important to be already aware of the need to include new factors over time when designing programmes. We also need to understand how historical uses have formed the practices we see today. Monitoring should not only be concerned with today's landscape or that of the future (for which it is explicitly designed) but should also take into account those traces of the past that exist in the landscape. In this way, for instance, anthropology or broader social science data can help us to better understand how yesterday's land use has helped form today's land use and how both can potentially shape the future landscape. Careful design of monitoring may here include both understanding how cases differ from each other, where bias and errors may be found, and how monitoring may need to increasingly relate to the social context – the major influence on nature in the Anthropocene. All of these factors are discussed in this chapter. The following sections also include a discussion of what this book has been able to cover in relation to a broader monitoring and social field.

Learning from case studies and examples: continuity and innovation

The book has shown that comprehensive coverage in monitoring systems is not a given but rather something that the monitoring practitioner needs to consider carefully. For instance, to what degree does a specified indicator actually mean the same thing in different countries?

To illustrate this, we have included case studies from multiple areas. We have not been able to cover all European cases relevant to this book or even all land uses (which might almost be an impossibility); instead, the cases we have included show a slant towards Sweden, and sometimes a slant towards forest or grassland systems. We recognize that they cannot be assumed to be typical of all other cases, either in data content (noting that Sweden has a very strong history of data systematization) or in the way the governance system works (for instance, sectorially), which cannot always be generalized to other countries. However, though understanding something about the forest sector may not mean that we automatically understand the river sector – far from – it may mean that we can understand the common design features and make better-informed decisions. Understanding the detail of design features in one case supports our understanding of what design features and data types could be applicable to other cases.

The book shows that monitoring change means understanding both older and newer methods of data collection and the complex relationship between continuity and innovation. Some of the methods introduced are so specialized as to require their own expertise, yet for the individual student or reader it is nevertheless crucial to understand enough to select methods and models and to take expert advice. Alongside these new and innovative methods, students also have to understand the methods used for past observations and to create legacy datasets, including thematic maps and even (showing the pace of change!) archival remote sensing imagery. Understanding both old and new methods is essential for making comparisons that can take into account different sources of uncertainty – real change is not, for example, simply the absolute difference between the older and more recent map!

There is a crucial balance not only between different methods but also practically in combining continuity skills and new expertise. At first glance, *continuity* means continuing with the original protocols (the standards, methods, and detailed instructions) for sampling, data collection, and analysis. Innovation (e.g. the use of GPS) can maintain continuity while making the same work more efficient, as well as adding metadata and supplementary observations that improve the statistical power of the same primary data collection by modelling and post-stratification. On the other hand, innovations such as remote sensing and citizen science can also look like attractive ways to reduce or replace monitoring, potentially breaking continuity. Innovation can also add wholly new types of evidence and context, including existing social data in databases that have so far not been used for monitoring purposes. A more sophisticated interpretation tells us that continuity is not about doing everything the same way forever – and monitoring schemes can come to a natural end – but about how schemes adapt. Monitoring should be about making the best of new innovations to maintain the ability to make sound statistical and thematic inferences as we compare new observations with those observations already collected.

To be able to demonstrate the considerable complexity of biodiversity monitoring and how design choices are made, the cases in the book have often covered the national level, although case studies on a more local level are included in some chapters. This focus on largely national monitoring covering extended areas has been preferred because these schemes can strongly influence the ability to monitor progress and setbacks at the local level, such as restoration projects (which by most definitions are local and, less seldom, regional). In addition, the examples covering both environmental and social data suggest that a better understanding of why and when restoration works, or does not work, can be best realized in relation to social use. The very success or failure of any environmental endeavour (outside the protected areas) will ultimately fail, or at least be under-achieved, if the persons involved in managing the everyday land do not understand how biodiversity is described and change measured, or do not agree with the goal of restoration.

However, whilst having attempted to cover a large range of methods and data sources, we have not covered all of them. The developing field of monitoring, in concert with the relevant social science approaches that have been discussed here but are not commonly included in monitoring, is broader than we have room to show. For instance, we have not included questions of social media or other technologies that are novel in social science. Though we are aware that there are numerous types of studies that could potentially be used, to keep the book manageable and understandable to a broad range of readers, the important issue was to cover major established social methodologies to illustrate their logic, use, and application.

The general logic in this book is that if we can understand the issues considered in tailoring monitoring to the requirements of one context well enough, we should be able to make similar assessments (or at least well-grounded assessments) in other contexts, by understanding the demands and difficulties at the different levels or scales involved. We hope this will make the student well placed to understand the complexity of cases that differ from the ones illustrated here – including the requirement for both continuity and innovation as monitoring schemes adapt over time.

Understanding, avoiding, and gathering information on bias

Bias is an important issue in monitoring, and the more traditional understanding of bias has been discussed in several chapters, including statistical bias, observer bias, and the need to avoid bias in, for instance, questionnaires during qualitative research. There seem to be few completely successful strategies for avoiding bias. Survey sampling designs can be paired with unbiased estimators of population parameters, but these designs may not be able to avoid observer bias, which is not statistically modelled, and instead assume that all of the observations or measurements are accurate and reproducible between observers. Reducing observer bias can lead to adopting measures that are much less variable between subjects but only because they have much less information content. Getting this balance right needs to be solved case by case.

Because monitoring is geared toward taking actions (or being certain enough that no new actions are needed) such as practical management and policy changes, any evidence that changes our belief in the current state of biodiversity – related to our topic of interest – can be legitimate. This might come from quantitative evidence (statistical estimators), maps, models suggesting a high enough risk of unwanted future states, historical research, unexpected and compelling observations, changes in public preference that move trigger points for action, or counterfactuals. If this is the case, it suggests the need for dialogue between experts used to handling different types of evidence from different sources spanning biodiversity monitoring and the social context.

This means that *interpretation* – understanding and putting data in context – is crucial in quantitative as well as qualitative work. Both in situ field monitoring and remote sensing image acquisition need some pre-processing and validation before the results can be compared, such as atmospheric compensation, calibration (offset [bias] and gain), removing observer bias, and filling in missing data (see chapters 5 and 7 for details). Though we do not see the design stage of remote sensing instruments, platforms, and orbits, this is equivalent to sample survey design in the sense that it dictates how the observations can be interpreted and what errors or uncertainties are possible. Notice the difference between knowing about sources of uncertainty and compensating for or minimizing them. In both cases, metadata are critical (instrument specification; the previous experience of the in situ observer), including the ability to re-visit a location for a repeat observation, which in the case of remote sensing depends on pixel size (or ground sampling distance, GSD). In some cases, the same calibration steps can inadvertently introduce and hide new sources of uncertainty. The details matter.

In qualitative research, it is equally important to understand our own role as instruments making observations – and the bias we may bring, as well as how to limit this. Bias exists not just in sampling but in the attitudes we carry, remembering that, for instance, ecologists and statisticians also have their own social context and sets of beliefs (points well made, for instance, by Lindenmeyer and Likens 2018).

Understanding this type of bias is crucial not only to limit our own bias but to make sure that these same processes amongst stakeholders are made a topic of investigation, in order to understand how their values and preferences influence management.

We may, for instance, mistakenly assume that stakeholders or managers understand an issue in the same way we do or that if managers are provided with updated information, they will change the way they conduct management in the way we expect. However, in social science, it has long been established that people do not always change what they do when confronted with new information (e.g. Keskitalo 2022). Land managers or land users may also hold very different conceptions of drivers or important factors in the landscape than a researcher in a specific area does, or they may be incentivized in specific ways (e.g. Mansourian et al. 2005). For these reasons, understanding social data becomes increasingly important: land use management is more often than not one of the most important drivers of environmental change. So monitoring should – depending on the area – not only try to eliminate bias and recognize the assumptions, beliefs, and biases that are unavoidable in the inferential process – but also understand and be aware of the bias that is part of how we all see the world or even make it a topic of research. In this respect, qualitative methods can be helpful: here it is possible to research those *biases* – the views of the general world or of professionals– that any particular stakeholders or sectors may include.

Partial data, errors, uncertainty, expertise: hard choices and the promise of new methods

Designing or developing monitoring can be about coping with data uncertainty, bias, and errors by being aware of these and, in some cases, seeing them as a topic for investigation. However, though it is crucial to be aware of sources of errors and how to limit our own bias, it is also crucial to be aware that not everything can be done "by the book". Even when the analyst is conscious of the large number of factors important to monitoring, often we need to work with ongoing systems developed for earlier reasons, data, or resource constraints. It may not always be possible to include all relevant data or undertake large enough surveys so that the statistical basis for drawing conclusions is sufficient. Chapters in this book vary somewhat in how they treat this aspect. Early chapters in this book are more about outlining the process and best practice, whereas later chapters consider the use of additional (auxiliary) data or qualify what can be done, either by discussing this in general or by illustrating the choices that were made in specific examples.

As an example, monitoring surveys can include at least three sources of error: legacy data such as historical records that have unknown biases, citizen science projects where participants choose where and when to make observations, and schemes of partial data collection where only some population units are safe or convenient to include in a sample. As chapter 4 discusses, none of these would be considered to be "probability samples" with all units available for selection and unbiased estimators, but they may also be the only data sources available. In other cases, a probability sampling design may be perfectly good (and design unbiased) but the affordable sample size is too small to give enough precision in the estimates. To solve these problems of bias and lack of precision, some model-based sample survey methods claim that a sufficiently strongly associated auxiliary variable can both fill holes in the data by prediction and give unbiased estimates of population parameters. This is becoming an increasingly important question, with remote sensing having only partial validation/calibration data points and with the growth of citizen science and modelling.

One of the general tensions in surveying and monitoring is that between using classification systems (ontologies) very well matched to the local range of habitats and land cover or universal, more general systems designed to give comparable data across regions and countries. As some of the chapters have shown, this type of concern also extends to the social systems: in a comparative survey across countries, how do we ensure that the questions asked both match local conditions and allow comparability? (see chapter 13). In traditional monitoring, the boundaries between fewer, broader classes are often locally ambiguous and increase observer differences (only some of which can be managed statistically as observer error). In social survey design, broad descriptions would mean ambiguities in how respondents understand questions and uncertainties as to what exactly they consider in their response, and combining data may mean that we combine areas where, as researchers, we may not even have perfect understanding of the local specificities that may influence response. For such reasons, undertaking interviews (chapter 14) or working with groups in several different areas can be supportive.

It is easy to become overwhelmed by this range of possible technologies, methods, objectives, and competing ideas of what is most important to monitor, and even more so when we have added social considerations. Actually, we are often sufficiently overwhelmed when trying to record a small piece of mire full of different *Sphagnum* species (bog mosses, which can be hard to identify)! However, different chapters in the book show not only the diversity of methods but provide a common set of steps intended to support making these choices. The hope is that the book illustrates some of the progression from a policy statement, to a simple conceptual model where we decide what to monitor, to survey sampling design choices, and then to how to record data in a range of environments and making choices about what factors to include. We also hope that it includes some illustrations of how auxiliary data from technologies such as remote sensing can help make the correct inferences and draw the right conclusions from a manageable set of observations and how supplementary information can link biodiversity monitoring to land use, as can working with people who study land use and land use assessment from a social perspective.

To be a success, this means avoiding the temptation to observe everything, everywhere: we need to discard as well as include. All monitoring involves finding the right context and asking the right questions, simplifying the question into something that can be measured, making inferences from uncertain data, and using these to make the best decisions, while bearing in mind the utility (value) of different outcomes – what if we are wrong? – balanced against the opportunity cost of waiting until we have more certain data – but now it is too late to act.

As noted above, one of the large challenges of monitoring is how to compare data and observations collected in the past, using methods then available, with current data collected with new technologies. We can also use new technologies to make new observations about the past; for example, from peat and ice cores or archaeological digs. This means understanding how both new and traditional methods work, and part of the art of monitoring is getting the best balance between continuity of traditional (typically field) methods and innovative technologies and methods.

Even with knowledge of best practice and what doing something by the book means, the monitoring student and practitioner will be faced with hard choices. Any dataset may not be complete enough to answer the questions set by politicians. Having some wall-to-wall data may not guard against biased estimators from related partial datasets. What the monitoring student and practitioner needs to do is to be aware of these limitations and say only what the data can support (but not more) – but possibly go further to indicate what could be said with improved assessment or additional data. This goes beyond the much more cautious approach that is taken to interpreting the results of experiments, where the better control of design means that in many cases no observed difference can typically be interpreted as no real difference.

For instance, the monitoring student or practitioner needs to be aware of some specific data requirements for different methods, some of which are a particular partial sample. Expert systems use domain knowledge (e.g. expertise in habitat mapping from aerial photo interpretation) to specify rules in a classification tree (CART), and object-based image analysis (OBIA) captures existing knowledge. In this case, the outputs can then be used as training data in machine learning to create more efficient rule bases: training data go in, rules come out. However, in general, the problem with "black box" sets of rules from machine learning is that they cannot easily be adjusted to new situations – instead, the rules can only be fine-tuned by introducing new training data. In the case of machine learning using support vector machines, for instance, we may not need a set of training data spanning the full feature space but rather only need to know the values surrounding, so to speak, the "snakes in a plane" (see Domingos 2015 for the example). Similar arguments have been made about partial data collection along the relevant parts of gradients in monitoring small protected sites (see the case study in chapter 16).

All of these are examples of using partial datasets and samples in new ways, not only through the robust statistical designs introduced in chapter 4. Monitoring is not just about learning to work with a range of methods, finding innovation across disciplines, but fundamentally about learning to work with a range of experts, valuing how to combine concepts and insights. Thus, the case study in chapter 16 shows how small, partial samples can be used to draw conclusions about habitat condition. Chapter 17 exemplifies how the right information can be identified and collected using a community-based monitoring programme, or *co-production* – combining the knowledge and resources of researchers, public bodies, and, perhaps most important, the local people.

A future of multiple disciplines and partnerships

What is evident in concluding this book is that monitoring of biodiversity is truly work for a community, because it is nearly impossible for an individual to keep track of all disciplines involved in the performance of modern monitoring. To invite and include scientists from other disciplines is often rewarding and opens up insights into new ways of thinking and new solutions. The same methodologies can be used for many purposes – such as drone-derived lidar data over Angkor Wat showing that the dense forest area was once a busy urban place; or drones used to estimate fish abundance at sea. Environmental DNA is another example, used to investigate organisms in the same way but for different purposes, such as deciding contemporary biodiversity in ocean sediments or the biodiversity of a piece of landscape millennia ago!

This difficulty of combining data – and the need to do this, not to obscure but to bring forward the role of different disciplines – has been a key focus in the book. We have tried to illustrate a breadth of different methods and data available, including up-to-date references, with the aim of allowing the monitoring practitioner to gain a better understand what is out there and to make better-focused decisions relevant to the organization or context within which they work. Rather than apply or assume transdisciplinarity (beyond seeing it as a way of treating knowledge reflexively; e.g. Popa et al. 2015), we thus want to illustrate the large variety of knowledge and methods that are needed to combine ecological and social study. Developing these complex combinations can perhaps be done most efficiently by inviting cooperation. In fact, in our own work, we have often noted that finding experts in different areas and then working towards common research problems or monitoring issues, with each drawing on their own knowledge to provide answers, may be the most practicable solution. That said, the practical organization of multidisciplinary work is not the focus here. Discussions on, for instance, adaptive management, co-management, or how different knowledge systems (traditional and local) can be included have not been the general focus of the book, although some of these factors are taken up in specific chapters. Rather, the aim has been to provide the conceptual foundation and knowledge of a range of methods suitable to make decisions situated in a context of monitoring. Knowing this – and knowing what different disciplines may be able to supply – can be a first step towards working together in a problem-based setting where each different area of expertise can supply their specialised knowledge, methods, and tools.

Making it work

In practice, then, moving towards utilizing multiple sources of knowledge is about *reflexivity*: understanding the limitations of knowledge, the existence of bias, the need to minimize and/or be aware of it in relation to the method and data utilized, and, finally, what can be said on the basis of this (e.g. Koot et al. 2020; Gregg et al. 2022). This means that an important issue as a student of biodiversity monitoring is at least to understand the range of new methods and new types of expertise, including knowing enough specialized language and concepts to be able to talk to experts in other relevant fields.

Perhaps we can agree with Kant that all of our knowledge begins with experience (Kant, as originally published in 1781) and agree with Shakespeare's (1594) Moth that experience is purchased with observation! As Kant went on to say, though all of our knowledge begins with experience, it does not follow that it all arises out of experience. We can add what we observe to what we already know and also question what we think we know. What are the potential biases and the areas we need to include to really support stakeholders? What are the groups we should be involved with, and what are the different concerns we need to take into account?

Learning to observe whilst recognizing all of our varying expertise and knowledge gives us a common language flexible enough to accommodate a diversity of understanding about monitoring. What follows is how we can best join up our new observations to our existing understanding. The best monitoring projects are those designed to combine any available methods to make the most efficient set of observations and the best use of data. The result may not be perfect – limitations in funding and time often prevent this – but at least we will be able to assess what our observations and data can say and what it cannot. Knowing what we can say, how we can be wrong, and what our limitations in any given case are – as well as how we could address these – is often as important as getting the perfect dataset. If we know how our data are limited, we can also delimit what we can say on the basis of these data. And if we know what is limiting, perhaps it is also possible for monitoring to reach out to establish new cooperations and gain support for new research – adding new types of data and funding – that can support the quality of both the monitoring and the conclusions drawn from it.

This can be complex and hard to accomplish or inspiring for the curious individual (which is most often the case). But perhaps this is a good place to say, "don't panic" (as a much later author than Kant and Shakespeare noted; Adams 1979): this richness is what makes monitoring an exciting and rewarding career, solving real-world problems!

Key messages

- Monitoring has become increasingly complex, reflecting the large requirements placed on it by concerns over global change and the need to connect biodiversity with land use and other management pressures and to take up the opportunities of new technologies. This means that it is increasingly important to understand design choices as well as what different data, disciplines, and methods can supply.

• Practical monitoring always means making hard choices, including coping with partial knowledge, bias, and uncertainty and meeting the need to reflect a community of interests extending beyond scientists. These choices can be made easier by working in cooperation with experts from different disciplines, including those with local knowledge and insight.

Study questions

1 Why is it not always possible to do things "by the book"?
2 Why is it important to be knowledgeable about what different disciplines, knowledge, data, and methods can provide?
3 Why can't you do it all yourself; that is, cover all relevant disciplines?

Further reading

We urge the student to search for the latest books, websites, or handbooks to keep abreast of what is going on; handbooks can get a bit old rather quickly in these times of fast development. Although, as stressed at several places in this book, the foundations often remain, even when the ways of doing monitoring are constantly changing. Learn from others, and keep a focus on what you need to address and answer!

References

Adams, D. (1979) *The Hitchhiker's Guide to the Galaxy*. London: Pan Books.

Domingos, P. (2015) *The Master Algorithm: How the Quest for Ultimate Learning Machines Will Remake Our World*. London: Penguin.

Gregg, E.A., Kidd, L.R., Bekessy, S.A., Martin, J.K., Robinson, J.A. and Garrard, G.E. (2022) Ethical considerations for conservation messaging research and practice, *People and Nature* 4(5), 1098–1112.

Kant, E. (1781) *Critik der Reinen Vernunft* [The Critique of Pure Reason]. Riga, Latvia: Johann Friedrich Hartknoch.

Keskitalo, E.C.H. (2022) *The Social Aspects of Environmental and Climate Change. Institutional Dynamics beyond the Linear Model*. London and New York: Routledge.

Koot, S., Hebinck, P. and Sullivan, S. (2020) Science for success—a conflict of interest? Researcher position and reflexivity in socio-ecological research for CBNRM in Namibia, *Society & Natural Resources*. doi: 10.1080/08941920.2020.1762953

Lindenmayer, D.B. and Likens, G.E. (2018) *Effective Ecological Monitoring*. Clayton, Australia: CSIRO Publishing.

Mansourian, S., Vallauri, D. and Dudley, N. (2005) *Forest Restoration in Landscapes. Beyond Planting Trees*. New York: Springer. doi: 10.1007/0-387-29112-1

Popa, F., Guillermin, M. and Dedeurwaerdere, T. (2015) A pragmatist approach to transdisciplinarity in sustainability research: from complex systems theory to reflexive science, *Futures* 65, 45–56.

Shakespeare, W. (1594) Love's Labour's Lost, Act III scene 1, https://www.opensourceshakespeare. org/views/plays/play_view.php?WorkID=loveslabours&Act=3&Scene=1&Scope=scene (Accessed November 22, 2022).

Wauchope, H.S., Jones, J.P.G., Geldmann, J., Simmons, B.I., Amano, T., Blanco, D.E., Fuller, R.A., Johnston, A., Langendoen, T., Mundkur, T., et al. (2022). Protected areas have a mixed impact on waterbirds, but management helps, *Nature* 605, 103–107.

Appendix 1

Horvitz-Thompson estimators

Alan Brown

Introduction

Mathematical equations can be several things at once: an elegant statement of a process of thought, a visual pattern that prompts instant recognition of something we already know or tells us there is something new to learn, and an unambiguous set of instructions on how to carry out an analysis of monitoring data. They can seem daunting because, stepping outside the comfort of the linear text, they show us everything at once. But with time they become familiar old friends, allowing us to scan a paper for new concepts and solutions even when we have very little proficiency in the written language.

For the student of monitoring, critical skills include not only being able to read an equation but how to translate equations into flowcharts, spreadsheets, lines of code in computer algorithms, and graphs. Even though in practice we use computer packages, at the learning stage this can be done at first with small datasets, varying the data and assumptions to see what effect these have on the estimators and other outputs. Graphical methods are helpful for illustrating statistical power calculations,[1] and alternative Bayesian methods[2] show how equations, algorithms, and graphs can be combined to reverse the back-to-front logic forced on 20th-century "statistical tests", making statistics more intuitive in the age of computers.

Monitoring often relies on sampling from a real or more abstract *population* of units. Sample counts or measurements are made to estimate the population statistics (sometimes known as *parameters*), drawing on assumptions and statistical theory about the way in which the sample was selected. As the examples throughout the book show, this may be a real population or a more abstract population of potential measurements; for example, of timber volume in a woodland. Though we are most interested in the sample statistics as a way of getting an unbiased estimation of the population statistics, the process of sampling can also be very informative about the context within which these can be interpreted, including sources of inaccuracy and variation not captured within the statistical design.

Chapter 4 introduces sampling designs, each of which has its own set of *design-unbiased estimators*. One of the aims of this appendix is explaining how "unbiased" is a property of the design, not of any individual sample.

Though basic designs of sampling – such as simple random sampling (SRS) – are useful for explaining the principles of estimation and statistical analysis, these are not so often used in practice, especially for very large and expensive schemes where they can be inefficient and risk actual bias in an individual sampling scheme, due to chance, even though with repetition the same scheme is designed to be unbiased in the long run.

One feature of the most basic schemes is that all available sample units have an equal probability of being selected. This *probability sampling* underpins the unbiased design. However, equal probability sampling is often not possible or desirable. Natural sample units may not be of equal size (think of a group of small islands) or have the same amount of habitat in which a scarce plant might be found. The probability of an individual tree being the nearest to a random point increases when the local tree density decreases. There can also be easy-to-observe auxiliary variables that are correlated with the less-easy-to-observe variables being monitored, in which case the sampling scheme can benefit from allowing those units with higher values of this indicator to have a higher chance of selection. In all of these cases, we need estimators of population statistics that allow *unequal probability sampling* but are nevertheless unbiased. This is one of the properties of Horvitz-Thompson (H-T) estimators, which were devised in the 1950s (Horvitz and Thompson 1952) and are now very widely used. Introducing these is a good way of introducing a range of methods which are close to practical monitoring schemes.

Inclusion probabilities and weights allow unequal probability designs

In any monitoring scheme, the whole population of sample units should be available to be selected as part of the sample. The chance of being included is called the *inclusion probability*. In an equal probability scheme this is simply the number of samples divided by the total number of units – though the exact calculation depends on whether units are with or without *replacement*. Replacement is the same as a lottery in which the same person can win more than one prize if there is more than one draw. Though it does not make sense for a real set of field observations to be repeated just because they happen to have been selected more than once, the mathematics of sampling with replacement are a bit easier to follow.

The inclusion probability for unequal probability sampling is proportional to some auxiliary variable that is thought to be predictive of what is being monitored, such as the size of the island, the cliff space available for nesting birds, the age of a woodland, the patch of suitable wetland habitat, or local tree density.

With equal inclusion probabilities, the relative chance of any unit being included in the sample does not increase with the number of draws (sample size) that make up the sample.[3] For unequal probability sampling, this is not the case. Units starting with a higher inclusion probability have a better chance of being selected (or re-selected) in each draw, so their advantage is compounded with increasing sample size. For individual units, this has to be worked out according to the sample size, at least for the units, which are actually included in the sample.

To come up with an unbiased estimator, the values in each unit – the counts or measurements made in the field – are then *weighted* inversely proportional to the inclusion probability, removing any bias that would come from high probability units being more likely to be in the sample. At the same time, the sample statistics can take more advantage of the greater information content and precision in higher counts or bigger measures, rather than risking having a lot of sample units with no species or features or interest, low counts, or only marginal measurements. This will be illustrated using two examples, below.

It is important to consider the ecological drivers which might create patterns, such as different densities of individuals in sample units of different sizes, alongside the implied statistical model. In a set of ponds on the Welsh border, great crested newts (*Triturus cristatus*) breed in both smaller and larger ponds but, in the absence of predatory fish, will tend to have more reliable and consistent densities in the larger ponds, simply by chance. It might also be the case that larger ponds are more suitable or breeding is more successful in those with good shading canopy and cover from vegetation near the pond margins. So there may be a combination of statistical patterns (larger densities are less variable and larger ponds can accommodate more individuals) as well as some measurable feature that makes different sizes better. The key point here is that, by accommodating unequal inclusion probabilities geared to some predictive variable, H–T estimators can be used to support sampling designs that take advantage of these patterns.

The inclusion probabilities and derived weights for individual units enables a design-unbiased estimator for the total population number of individuals, or the true measurement across the area of interest. These are called *first-order inclusion probabilities*. The variance of this estimator depends on not only the range of values in the sample units, weighted according to their inclusion probabilities, but an estimate of how likely it is, for example, that several units with high counts appear in the same sample. This means working out the joint inclusion probability of all combinations of pairs of units in the sample, known as the *second-order inclusion probabilities*. Notice that this measure of covariance accounts for all the higher-order joint probabilities: three units together, four units together, and so on.

This gives H–T estimators two possible components for the variance estimator, reflected in the two parts of the equation, by contrast with equal probability designs where the variance calculation is straightforward and equally weighted and there is no need to calculate covariance at all.

Calculating the first-order inclusion probability (π_i)

It is helpful to think of a sample being selected one at a time in a sequence of random draws until we have the required sample size. We know the probability of any unit being included in the first draw (p_i) because this is in the design of the sampling scheme. For example, it might be proportional to the size of the pond. It will always be a number between 0 and 1, with the total probability for the population of units adding up to 1. Suppose it is a large pond, with an inclusion probability in each draw of 0.2. For a sample of six ponds (i.e. six random draws), the probability of inclusion in the sample can be calculated. If there is replacement, an individual unit might be selected more than once, so it is easiest to calculate the probability of a unit *not* being included first. If p_i is the inclusion probability in a single draw, $(1 - p_i)$ is the probability of not being included:

$$\pi_i = \text{the probability of inclusion of the } i\text{th unit}$$
$$= 1 - \text{Probability } (i\text{th unit not included})$$
$$= 1 - (1 - p_i)^n.$$

For our example pond and six random draws:

$$\pi_i = 1 - (1 - 0.2)^6 = 0.738.$$

Though $(1 - p_i)$ is typically a much larger number than p_i, it is still less than 1, so the repeated multiplication in raising to the power of n (= the number of draws) makes this smaller and smaller and π_i larger.

For the reader familiar with probability, there are some subtleties here. Because the total probability of a unit being selected in each draw is 1, at first sight the total probability after n draws should be n. Already we can see from the above example that the total inclusion probability for every pond available to be selected will be greater than 1. However, these are the probabilities of ponds being included *one or more times* in the sample of n draws: because we are allowing each one to be available for selection in every draw (that is "with replacement"), the same pond can be selected more than once. Because these duplicates are not being counted in the calculation of inclusion probabilities, the total inclusion probability of all ponds after n draws will be greater than 1 but less than n.

For the same reason, the number of *unique* samples (ponds) selected can be less than the total number of draws, so though n is the sample size, the number of sample ponds that we will use in the calculations may be fewer.

Unbiased estimator for the total count or measure

Now that we know how to calculate first-order inclusion probabilities, we can use the weighed H–T estimator for the total count or measurement:

$$\widehat{T}_\pi = \sum_{i=1}^{v} \frac{y_i}{\pi_i}.$$

Here, y_i is the score or measurement from unit i, π_i is the inclusion probability, and v is the number of *unique* sample units, so a unit is counted only once even if it is drawn more than once. Notice that this estimator is exactly the same whether or not there is replacement. It is obvious how the inclusion probabilities are used to weight the influence that each observation makes towards the total. The hat over the T is to remind us that this is an estimate.

Second-order inclusion probability (π_{ij})

To work out the variance of our estimator, we also need the second-order inclusion probabilities. This is a little harder to work out because it combines probabilities for two units. We need to start by understanding the logic of different combinations of two units in just three draws, shown in Table A1.1 as unit A and unit B.

Of course, these draws are not all equally likely, and what is most striking is the number of combinations even with a sample size of 3! We need a better way of working this out for larger samples. We can use Boolean logic:

$$\begin{aligned} P(A \cap B) &= P(A) + P(B) - P(A \cup B) \\ &= P(A) + P(B) - [1 - P(A^c \cap B^c)], \end{aligned}$$

where $P(A \cap B)$ is the probability of both A and B being selected, π_{ij}, and A^c is the complement, the probability of A *not* being selected. This is much easier to work with,

Table A1.1 Calculating second-order inclusion probabilities.

					Draws				
					1	2	3		
1	–	–	–	Not A or B	0.7	0.7	0.7	0.343	
1	A	–	–	Just A	0.1	0.7	0.7	0.049	
2	–	A	–		0.7	0.1	0.7	0.049	
3	–	–	A		0.7	0.7	0.1	0.049	
4	A	A	–		0.1	0.1	0.7	0.007	
5	A	–	A		0.1	0.7	0.1	0.007	
6	–	A	A		0.7	0.1	0.1	0.007	
7	A	A	A		0.1	0.1	0.1	0.001	
1	B	–	–	Just B	0.2	0.7	0.7	0.098	
2	–	B	–		0.7	0.2	0.7	0.098	
3	–	–	B		0.7	0.7	0.2	0.098	
4	B	B	–		0.2	0.2	0.7	0.028	
5	B	–	B		0.2	0.7	0.2	0.028	
6	–	B	B		0.7	0.2	0.2	0.028	
7	B	B	B		0.2	0.2	0.2	0.008	
1	A	B	–	A & B	0.1	0.2	0.7	0.014	0.014
2	A	–	B		0.1	0.7	0.2	0.014	0.014
3	–	A	B		0.7	0.1	0.2	0.014	0.014
4	B	A	–		0.2	0.1	0.7	0.014	0.014
5	B	–	A		0.2	0.7	0.1	0.014	0.014
6	–	B	A		0.7	0.2	0.1	0.014	0.014
11	A	B	A		0.1	0.2	0.1	0.002	0.002
12	A	A	B		0.1	0.1	0.2	0.002	0.002
13	A	B	B		0.1	0.2	0.2	0.004	0.004
14	B	A	B		0.2	0.1	0.2	0.004	0.004
15	B	A	A		0.2	0.1	0.1	0.002	0.002
16	B	B	A		0.2	0.2	0.1	0.004	0.004
								Total	**Total**
								1	**0.102**

The probabilities on the right are calculated from a starting probability of $p(A) = 0.1$ and $p(B) = 0.2$ on each draw, with the probability of any other unit of 0.7. There are 12 combinations of three draws. Those that include both units, A and B, are shaded. The probability of each combination is shown – for example, $0.7 \times 0.7 \times 0.7 = 0.343$ – with (as expected) a total probability of 1 for all combinations. The sum of probabilities of combinations with both A and B is 0.102.

because $P(A^c \cap B^c)$ is always just one combination, the top, empty row in Table A1.1 and we already know $P(A)$ and $P(B)$ as the first-order inclusion probabilities, shown earlier as π_i and π_j. So we can put our first-order probabilities into the equation

$$\pi_{ij} = \pi_i + \pi_j - \left[1 - \left(1 - p_i - p_j \right)^n \right],$$

where n is the sample size.[4]

$$\pi_A = \pi_i = 1 - (1 - 0.1)^3 = 0.271$$

$$\pi_B = \pi_j = 1 - (1 - 0.2)^3 = 0.488$$

$\pi_{AB} = \pi_{ij} = 0.271 + 0.488 - [1 - (1 - 0.1 - 0.2)^3] = \mathbf{0.102}$, the same result as in Table A1.1.

Unbiased estimator for variance

Variance is a statistical measure of the amount of variation in counts or measurements between sample units that allows the analysis to model the uncertainty in the estimate of the mean or total. In SRS, sample variance around the mean (average) is generally understood as the sum of differences between each sample measurement and the mean, calculated (to avoid all of the positive and negative differences cancelling each other out) by taking the square root of the sum of the squared counts or measurements.

It is helpful to reflect on the practical, ecological interpretation of variance. Where there are natural sample units — for example, counts of nesting birds (measure) or trees (unit) — the variance can have obvious ecological meaning. Where the units are more artificial, such as 10m-diameter circular vegetation plots, both the mean and variance are tied to the sample unit, and the upper and lower limits of estimates of plant presence, abundance, and cover will change if the plot size changes.

Different estimators also have their own formula for variance, which may not have an intuitive ecological meaning but nevertheless gives an unbiased estimate of the range of uncertainty in other estimators which are more obviously meaningful, such as the population total count or measurement. For our H–T estimator, the estimated variance is given by:

$$\widehat{V}ar(\widehat{T}_\pi) = \sum_{i=1}^{v}\left(\frac{1-\pi_i}{\pi_i^2}\right)y_i^2 + \sum_{i=1}^{v}\sum_{j\neq1}\left(\frac{\pi_{ij}-\pi_i\pi_j}{\pi_i\pi_j}\right)\frac{1}{\pi_{ij}}y_iy_j,$$

which at first sight looks very little like the familiar equation for variance under SRS. However, at least we know all of the components, so it should be straightforward to follow the instructions in the equation and run the calculation. Lots and lots of calculation!

The equation comes in two parts added together. The left-hand part has only first-order inclusion probabilities and the squared measure for single units. The right-hand part also has second-order inclusion probabilities and two sample units. After a while, it is easy to see that these two parts are similar, only the second part has to cope with multiplying two different probabilities and two different counts, rather than squaring one of each.

However, the most obvious difference between the two parts is having one or two summations, shown by the sigma (Σ) symbols. On the left, summation is over, at the most, *n* instances (recall that $v \leq n$, and $v = n$ where there is no replacement). However, the right-hand component is summed over $(v * v) - v$ repetitions, roughly proportional to the square of the sample size, though the actual number of unique calculations is only half this. Nevertheless, it is easy to see why these estimators are only practical using computers to do the work. Even then, adding up this very larger number of often very small differences can be vulnerable to the propagation of rounding errors in the digital storage of numbers in floating point arithmetic. There is more of a risk than in SRS

i

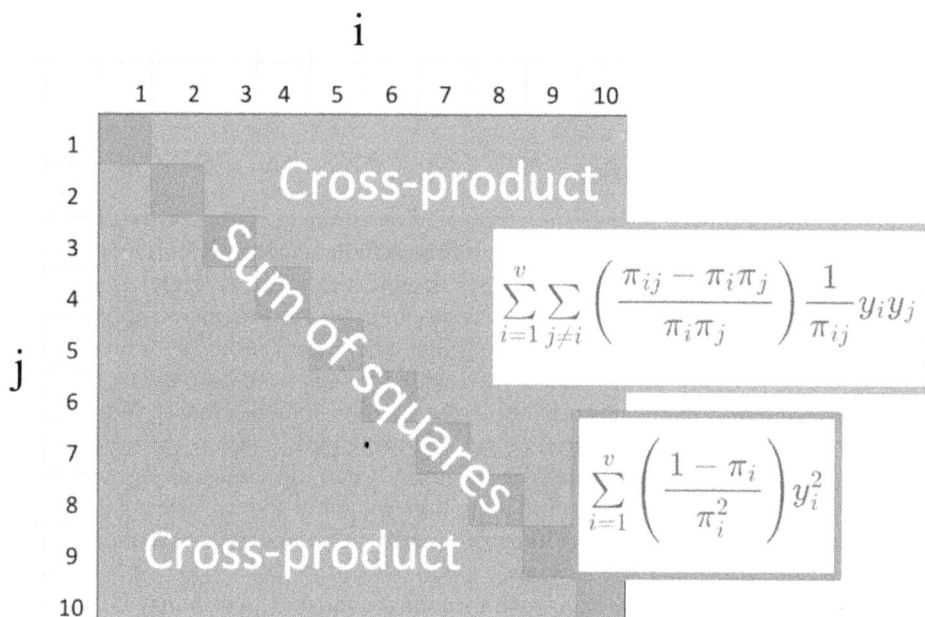

Figure A1.1 The relationship between products (the sum of squares) and cross-products in the calculation of variance, shown for a sample size of 10.

because the H–T variance estimator has inclusion probabilities, which are typically very small numbers.

The scope of calculations can be visualized as a two-dimensional matrix on a spreadsheet, in which the first part of the equation is the diagonal elements and the second part is the two (mirrored) off-diagonals, in effect combining the sum of squares ($i = j$, diagonals) and cross-products or covariance ($i \neq j$, off-diagonals), shown here for a sample size of $n = v = 10$. Notice the role of first- and second-order inclusion probabilities (Figure A1.1).

Example 1

A population of 435 bushes is scattered across a wetland divided into 99 unequal-sized islands separated by deep gullies (Figure A1.2). Notice that this is a finite population, all of which have a probability greater than zero of being included in a sample, so it is design based (see chapter 4). A H–T estimator is used to estimate the total number of trees from a sample of just ten islands. Though this example is entirely artificial,[5] it is typical of the situation where the natural sampling units are unequal in size and inconvenient to access and there is no obvious process generating a density pattern or gradient. For the purposes of the example, we assume the bushes cannot easily be identified and counted from the air – perhaps they are covered by the canopy of taller trees – though we might have some information about the extent of the most suitable habitat for them.

The only candidate for p_i, the inclusion probability in each draw, is the relative size of the island unit within the rectangular study area of 6,331,946 (pixels). As a

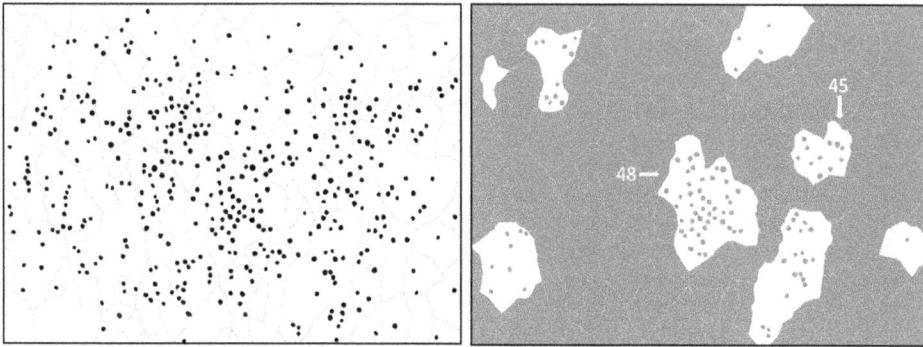

Figure A1.2 An artificial dataset representing a set of bushes and sample units with unequal areas (left) and a sample of ten randomly drawn units (right), with selection probability proportional to the unit size (two pairs of samples are neighbours and appear as one white shape).

Table A1.2 The probability (p_i) and cumulative probability for the first five units selected.

i	Unit area	p_i	Cumulative p_i
1	28,520	0.00450414	0.00450414
2	7886	0.0124543	0.00574958
3	33,763	0.00533217	0.01108174
4	31,086	0.00490939	0.01599113
5	24,036	0.00379599	0.01978712

demonstration, selection will be done with replacement. The random selection of ten samples is shown on the right.

Table A1.2 shows the probabilities for the first five units. For example, $p_1 = 28,520/6,331,946$. The fourth column is the cumulative p_i, adding up $p_1 + p_2 + \ldots$. To draw units in proportion to p_i, we simply have to generate a random number between 0 and 1 and select the unit with the closest cumulative p_i value above this. For example, if the random number generated is 0.00592, unit 3 would be selected, noting that the larger islands have larger intervals in the cumulative probability column. This is an example of PPS (probability proportional to size) selection with replacement. It is also possible to use PPS selection without replacement, but this is a little more complicated than can easily be shown in a simple example.[6]

In the following sample of ten units, the sample observations are in column y_i (Figure A1.3a).

The light (white) columns contain formulas with the contributions towards the estimator in the next to last (blue) column labelled y_i/π_i. Though the total of 459 happens to be close to the actual figure, this could be by chance, and how much we should believe it will also depend on the sample variance. The biggest unit here is number 48 – note the large value of p_i – which has a count of 46 bushes. Notice, however, that the contribution of the much smaller unit 45, which has only 5 bushes, is nearly half that of unit 48 because it is given a much bigger relative weight. The final column of ratios is much more evenly

(a)

Unit	pi	pi^2	yi	yi^2	πi	π^2	yi/πi	((1-πi)/πi^2)yi^2
11	0.0262025	0.00068657	4	16	0.2332	0.0544	17.1532	294.23
22	0.0170098	0.00028933	10	100	0.1577	0.0249	63.4312	4023.51
28	0.0046082	2.1236E-05	0	0	0.0451	0.0020	0.0000	0.00
45	0.0075876	5.7571E-05	5	25	0.0733	0.0054	68.1788	4648.35
47	0.0090293	8.1528E-05	6	36	0.0867	0.0075	69.1953	4787.99
48	0.0508506	0.00258578	46	2116	0.4066	0.1653	113.1327	12799.01
64	0.0417906	0.00174646	19	361	0.3475	0.1207	54.6818	2990.10
66	0.0109631	0.00012019	1	1	0.1044	0.0109	9.5806	91.79
67	0.023097	0.00053347	7	49	0.2084	0.0434	33.5918	1128.41
76	0.0033748	1.1389E-05	1	1	0.0332	0.0011	30.0842	905.06
						Total	**459.0296**	**31668.45**

(b)

Second-order inclusion probabilities

πij		1	2	3	4	5
		0.2332	0.1577	0.0451	0.0733	0.0867
1	0.2332	-	0.033761867	0.009614309	0.015641073	0.018505299
2	0.1577	0.033761867	-	0.00647198	0.010529518	0.012458022
3	0.0451	0.009614309	0.00647198	-	0.002997143	0.003546199
4	0.0451	0.009614309	0.00647198	0.001842064	-	0.003546199
5	0.0867	0.018505299	0.012458022	0.003546199	0.005769717	-

Measured variables (counts)

yi*yj		1	2	3	4	5
		4	10	0	5	6
1	4	-	40	0	20	24
2	10	40	-	0	50	60
3	0	0	0	-	0	0
4	0	0	0	0	-	0
5	6	24	60	0	30	-

Cross-products

		1	2	3	4	5
1		-	-96.71963141	0	-109.1991044	-110.0041301
2		-96.71963141	-	0	-423.8948925	-427.03609
3		0	0	-	0	0
4		0	0	0	-	0
5		-110.0041301	-427.03609	0	-481.9121241	-

Figure A1.3 (a) The sample of ten units, showing the probability (p_i), count (y_i), inclusion probability (π_i), and weighted count (y_i/π_i), with the H-T estimator in the final column. (b) Part of the matrix of second-order inclusion probabilities, counts, and cross-products, shown for the first five units.

balanced than the column of counts (y_i). Even though the units are of different sizes, these different weightings make it possible to combine them in an unbiased estimator. The blue column of ratios (y_i/π_i) is much more evenly balanced than the column of counts (y_i).

It is worth underlining why this possibility is important: it is because it allows us to select a bigger proportion of those sample units that are most likely to have high counts (in this example, the larger islands). Again, in the real world, when we have sample units with no observations, we may not know whether this unit is actually part of the population; for example, whether the habitat is suitable and what to make of absence data. In these cases, stratification or a better definition of the area of interest might help.

Calculating the variance

The first part of the variance is shown in the red (last) column. The second part, the covariance or cross-product, can be structured as a stack of symmetrical matrices (part of which are shown in Figure A1.3b). The most important intermediate variables are the second-order inclusion probabilities (π_{ij}), the counts ($y_i * y_j$), and the final cross-product, shown in green in the bottom matrix. Notice that these are all zero or negative.

The two components (rounded up) are 31,668.5 and −13,003, giving an estimated variance of 18,665.5.

There are several other points worth noticing here. As expected, all of the tables are symmetrical either side of the diagonal, so really we only need to make half the calculations and double the result. Though the result is still 10 + 60 cross-products, of course these calculations only need to be made for the units actually sample, not for the remaining units – here only 10 out of 99. Even with only 99 units, the probabilities are very small compared with the counts.

The second cross-product or covariance component reduces the total estimated variance. Because the cross-product is zero if either of the units has a count (y_i or y_j) of zero, including units where nothing is observed does not reduce the variance, so relatively speaking these zero observations increase the uncertainty of the H-T estimator.

Confidence limits and estimator bias

For this example so far we have a single sample, giving a single estimate of the population total and population variance. To fit *confidence limits* and carry out associated tests such as *t* tests, we need to be able to make an assumption about the distribution of estimates from repeated draws of random samples from the same population: this sampling distribution must be a *normal distribution*. Generally, we can make this assumption in two situations:

1   If the parent distribution (here, the distribution of bushes) is normally distributed.
2   If the sample size is big enough (conventionally $n \geq 30$), the sampling distribution is normal, *regardless of the shape of the parent distribution*.

The second case is one of the most important results in statistics: the *central limit theorem*.

The parent distribution here is very over-dispersed and strongly skewed with a lot of very low counts and a few very high counts. Of course, we do not actually know this when carry out our sampling but would either have to make assumptions from our ecological knowledge or field observations or look at the distribution of the sample itself. If this is obviously *over-dispersed* (very asymmetrical when plotted as a frequency

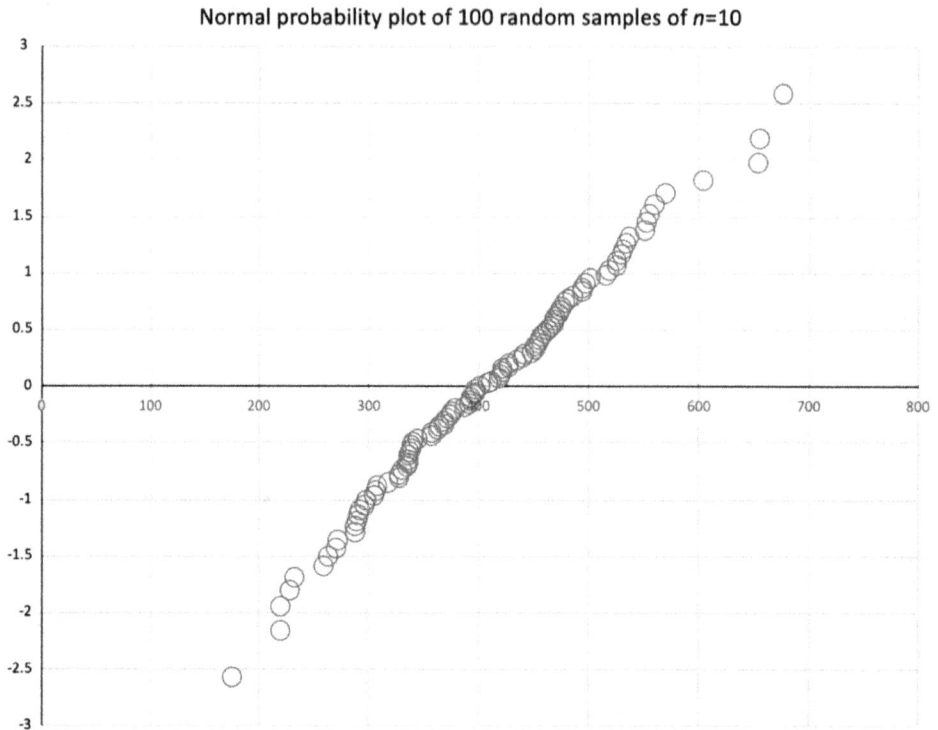

Figure A1.4 Normal probability plot for 100 sets of random samples, showing probability on the vertical axis and estimates of T on the horizontal axis. Normally distributed data should show as a straight line.

distribution), it is a good idea to *transform* the data before fitting confidence limits and then back-transform them, allowing the limits to be asymmetrical as well.[7]

In this artificial example, exceptionally, we know the entire population and can generate the sampling distribution of the estimator, $\widehat{T}$, by randomly drawing 100 samples and looking at the distribution. To check for normality, these can be graphed as a *normal probability plot*, shown in Figure A1.4.

The plot shows that, even with the small sample size, $\widehat{T}$ is close to normally distributed – though once again it is worth emphasizing that in the real world we do not have the data for this type of plot. Notice that the line cuts the axis a little below the true value, 435.

Assuming this is close enough to a normal distribution, the confidence limits are

$$\widehat{T} \pm t \sqrt{\frac{\widehat{V}(\widehat{T})}{\nu}} = \text{approximately } 380 \text{ to } 538$$

for total $\widehat{T} = 459$, estimated variance $\widehat{V}(\widehat{T}) = 18{,}665$,

where ν is the sample size (here, $\nu = 10$) and t is taken from *Student's t distribution* for $\nu - 1$ degrees of freedom, $p = 0.05$. We need the t distribution, which is slightly more spread out than the normal distribution, to take into account the additional uncertainty from the variance estimated from the data; that is, $\widehat{V}$ rather than V.

The distribution of estimates is a good reminder of what having an *unbiased estimator* means. Taking a large number of random samples, the mean converges towards the true value, here 411 coming close to 435. However, there is a great deal of spread in the range of sample values, and even though these crowd towards the mode, there is still a significant chance that an individual sample might be over a standard deviation above or below the true value. The design is unbiased, but individual samples may be biased.

Example 2

In the first example, sample unit size was used as an auxiliary variable to make sampling more efficient. Example 2 shows how auxiliary information can be also be used in stratification. Here, an advantage of H–T estimators using unequal probabilities is allowing the existing sample units to be divided up in any way, no matter how irregular.

We might suspect that the suitable habitat for our target species is lost towards the edges of the area of interest (the sampling frame), suggesting that the peripheral units – no matter how large or small – might not be part of the same population. Suppose we have remote sensing imagery describing the limits of the most suitable habitat and reduce our sampling frame to correspond with this boundary (Figure A1.5).

Now there are 312 bushes in 56 units. Using exactly the same sampling as in Example 1, Figure A1.6 shows the normal probability plot.

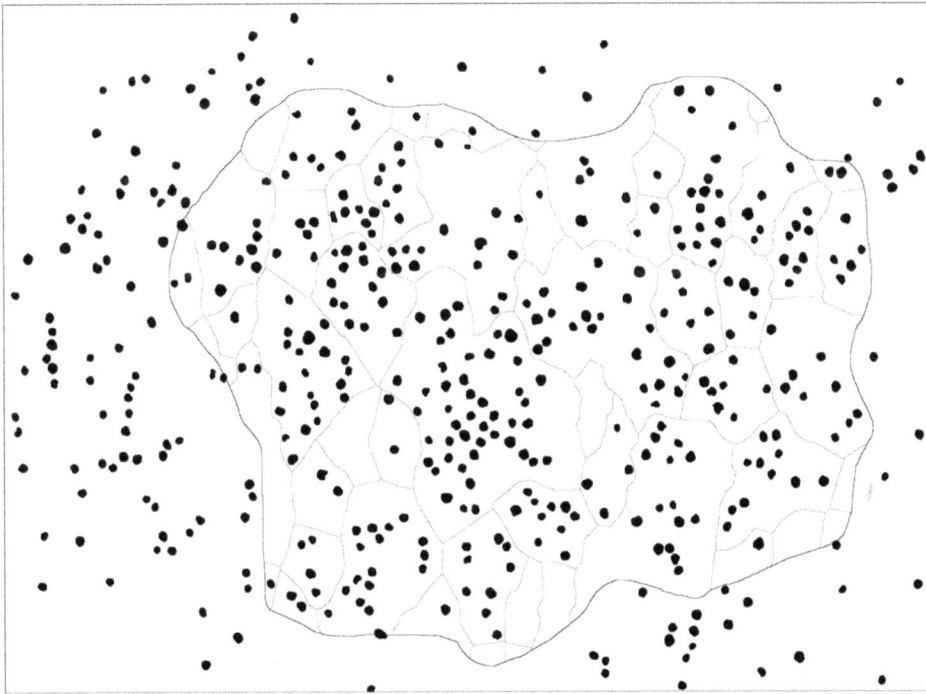

Figure A1.5 The sampling frame reduced to cover only the most suitable habitat.

(a)

(b)

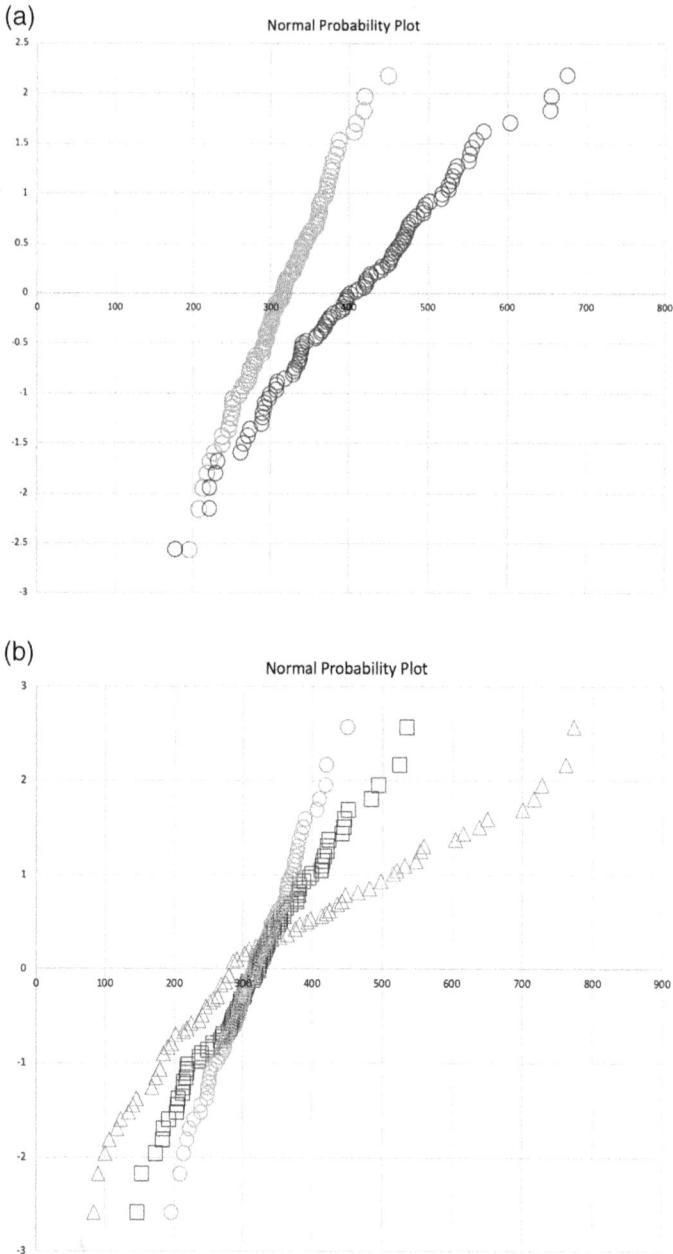

Figure A1.6 (a) Normal probability plot for 100 sets of random samples from Example 2 (left, red) versus Example 1 (right, blue).[8] Notice that the mean values – where the plotted circles cut the *Y* axis – are different in Examples 1 and 2. The increased steepness of the line of points shows greater precision. (b) Normal probability plot for 100 sets of random samples from Example 2 with simple random sampling (triangles), simple random sampling converting counts to ratios (squares), and the H-T estimator (circles). All three methods are shown to be design-unbiased – they all cross the *X* axis at almost the same point, but the H-T estimator is the most precise and closest to a normal distribution, shown by the steep near-straight line of circles.

Here the mean value of 100 samples is 316, very close to the true value. The normal plot, shown in red, has a greater density of values towards the mode and is much steeper, showing less spread of variance. In fact, 74% of confidence limits include population total, with the lower confidence limits including the true value inside 89% of the lower limits and 85% of the upper limits, showing that the sampling distribution is symmetrical and near-normal.

This demonstrates how well H-T estimators can give unbiased results, even with a small sample size – but that an unbiased result may still not be very precise. We may need to increase our sample size not only to ensure that we can fit confidence limits and use *t* tests but to increase precision and statistical power.

Practical H-T estimators

The large number of calculations involved, especially in the cross-products, means that H-T estimators are not practical for large datasets in a spreadsheet. However, there are plenty of excellent free open-source software packages that will run calculations, many of them with example datasets. For example, the R package (R Core Team 2021) "sampling" (Tillé and Matei 2016; see also Tillé 2020) and Thomas Lumley's "survey" package (Lumley 2020).

Some conclusions and observations

We could use a naïve version of simple random sampling to select units and divide the count by unit area to get a comparable index of bush density, carrying out the analysis using the SRS estimators of mean and variance around the mean. However, as the examples show, this would tend, on average, to include a greater proportion of small units with few or no counts, inflating the variance of the estimator. Horvitz-Thompson estimators can be more efficient because they weight sample selection towards the most informative samples – those that we predict will have the higher counts – using inclusion probabilities that are not simply proportional to the unit size but take into account the number of successive draws that make up the sample size.

Though artificial, these examples are typical of situations where it is impractical to divide the sample frame into equal-sized units. Often it is impossible; for example, when selecting scattered trees if they are closest to random co-ordinates: trees in less densely wooded areas will have more chance of selection.

Unequal probability sampling opens the door to a range of methods that can take advantage of the ability of auxiliary variables to predict the observed values of our target species or variable and provides a common, unifying design that can be adapted to stratified sampling, systematic sampling, and simple random sampling by adjusting the rules around the first- and second-order inclusion probabilities (see Overton and Stehman 1995). The use of auxiliary variables of any sort is increasingly common in model-based and model-assisted monitoring, taking advantage especially of new wall-to-wall data from remote sensing.

As a more general observation, we might expect those samples that happen to include units more evenly spread across the sampling frame to be more representative and perhaps give estimates closer to the true population value. Especially when very few samples are selected, we might want to prefer samples that are spatially balanced and include some sort of stratification across the sampling frame or include auxiliary spatial variables to achieve this, while keeping the estimators unbiased.

Finally, the examples show how we can understand sampling designs by using a real or artificial dataset and generating multiple samples.

Notes

1 The G*Power package (Faul et al. 2007) shows this well.
2 Outside the scope of this book, but see, for example, Sivia and Skilling (2006) for an excellent introduction.
3 In a finite population of units selected without replacement, the inclusion probability in the next draw increases slightly as units are removed, but all units have an equal chance of being selected sooner or later in the whole sequence of draws.
4 In Horvitz and Thompson's (1952) paper, the first section on "sampling with arbitrary probabilities" develops the equivalent formula for sampling without replacement, summarized in equations (4) and (5) on p. 666.
5 In fact, the scattered trees were generated by tipping lentils onto a white sheet and separating any that overlapped; the "islands" were generated from a Voronoi partition of a photograph of part of a flock of starlings. All processing and counts were carried out using open-source software Fiji (Schindelin et al. 2012).
6 Also, different selection methods have different formulas for inclusion probabilities.
7 Alternatively, we can use Bayesian statistics and plot asymmetrical credibility limits on a posterior distribution.
8 In Figure A1.6 (a) some bushes on the boundary are left out for comparability, slightly reducing the spread on the vertical axis.

References

Faul, F., Erdfelder, E., Lang, A.-G. and Buchner, A. (2007) G*Power 3: a flexible statistical power analysis program for the social, behavioral, and biomedical sciences, *Behavior Research Methods* 39, 175–191.

Horvitz, D.G. and Thompson, D.J. (1952) A generalization of sampling without replacement from a finite universe, *Journal of American Statistical Association* 47(260), 663–685.

Lumley, T. (2020) survey: analysis of complex survey samples, R package version 4.0, https://CRAN.R-project.org/package=survey

Overton, W.S. and Stehman, S.V. (1995) The Horvitz-Thompson theorem as a unifying perspective for probability sampling: with examples from natural resource sampling, *The American Statistician* 49(3), 261–268.

R Core Team. (2021) *R: A Language and Environment for Statistical Computing.* Vienna, Austria: R Foundation for Statistical Computing. https://www.R-project.org/

Schindelin, J., Arganda-Carreras, I., Frise, E., Kaynig, V., Longair, M., Pietzsch, T., Preibisch, S., Rueden, C., Saalfeld, S. and Schmid, B. (2012) Fiji: an open-source platform for biological-image analysis, *Nature Methods* 9(7), 676–682. doi: 10.1038/nmeth.2019

Sivia, D. and Skilling, J. (2006) *Data Analysis: A Bayesian Tutorial.* 2nd edn. Oxford University Press.

Tillé, Y. (2020) *Sampling and Estimation from Finite Populations.* John Wiley & Sons.

Tillé, Y. and Matei, A. (2016) sampling: survey sampling. R package version 2.8. https://CRAN.R-project.org/package=sampling

Appendix 2

Summary table of survey sample designs

Alan Brown

Sampling design	Analysis	Notes
Non-probability sampling		
1 Purposive/judgemental sampling (including complete population census)	No formal statistical analysis, so cannot guarantee unbiased estimators of population statistics – unless this is a complete census	Can be effective when the area of interest or population is very small, for making structured observations where no unbiased estimator is needed; can be the model in citizen science observations
2 Snowball sampling (used in qualitative surveys to select new subjects for interviews using contact information from previous subjects)	Qualitative – the aim is to get a complete coverage of the full range of opinion and preference as efficiently as possible	Can be seen as a type of cluster sampling
Basic designs – probability* sampling		
3 Simple random sampling (SRS), equal inclusion probability	Statistical analysis: estimators for population mean and variance, *t* tests, power analysis, non-parametric methods, Bayesian analysis	Small samples (<30) need testing for normal distribution and transformation or use Bayesian methods; larger samples can use standard tests assuming errors are normally distributed (central limit theorem).
4 Systematic sampling (with random start)		Systematic sampling used for eDNA surveys
5 Cluster sampling		
Advanced designs – complete coverage without sampling		
6 Wall-to-wall imaging	Image analysis, including expert systems, machine learning, and deep learning; expert visual interpretation and thematic classification, including stereo viewing	Satellite remote sensing. Uncertainty is still present but is not handled as a sampling design. Variables from remote sensing are important for model-based estimators (11 and 12).

(Continued)

Sampling design	Analysis	Notes
Advanced designs – probability* sampling		
7 Simple random sampling, unequal inclusion probability	Statistical analysis: Horvitz–Thompson estimators	Typically more efficient than SRS. Unequal probability sampling and cluster sampling often made necessary by the practicalities of field observations
8 Stratified random sampling, unequal inclusion probability		
9 PPS (probability proportional to size) sampling	Statistical analysis as a form of unequal probability sampling	Allows the most informative sample to be collected while maintaining unbiased estimators of population statistics
10 Spatially balanced sampling	Statistical analysis	R package: BalancedSampling
Model-based and model-assisted methods		
11 Geostatistical designs	Kriging, indicator kriging: geostatistical methods used to predict spatial patterns of variables based on a limited set of field observations	Model-based sampling methods developed by soil scientists and mining prospectors; semi-variance used to express spatial correlation
12 Ratio/regression estimators	Auxiliary variables used to predict observations to reduce the estimated variance	Model-based monitoring becoming increasingly important with the availability of auxiliary variables from satellite and airborne remote sensing. Model-assisted methods may be more reliably unbiased.
13 Post-stratification	Auxiliary variables used to refine the stratification and reduce the estimated variance	
Advanced designs – adapted to field methods (examples)		
14 Bitterlich sampling	Statistical analysis as a form of unequal probability sampling	Forest inventory methods using a "relascope", which has a variable sample area depending on tree diameter
15 Distance methods	Statistical modelling of detection probabilities (refer to e.g. Thomas et al. 2010)	Field methods for bird and animal sightings away from a fixed transect
16 Mark–release–recapture	Jolly-Seber models	Field method where animals are captured, marked, and recaptured; can also be used to measure repeatability in plant counts

(*Continued*)

Sampling design	Analysis	Notes
Stages in sampling design		
I Single-stage design	Statistical power analysis	Only one round of selecting sample units risks having too few samples, with too little statistical power to detect change
II Two-stage and multi-stage sampling (as a feature of other designs)		Pilot studies are used to adjust sample size; spatial and temporal designs for the main data collection
Comparing field observations		
A Re-randomization	Statistical analysis compares the estimated population statistics (e.g. mean, variance) from each round of observation.	The basic method of comparison, often necessary where exact re-visits cannot be made; for example, benthic grab sampling
B Paired comparisons	Direct comparison of pairs (or multiples) of observations made at the same point; statistical analysis of whether the change in the samples is significant as a change in the population	Allows spatial expression of observed, direct changes but vulnerable to the adopted sample units changing in a way that is gradually no longer representative of changes in the population
C Paired comparisons with partial replacement	Paired comparisons (as B) but with gradual replacement of a proportion of observed units – usually by re-randomization	Useful for habitats sensitive to disturbance during recording (e.g. mire systems), designed to keep the sample representative of the population

*Note that *probability sampling design* refers only to the *spatial* design. The *temporal* aspect – timings – is very rarely random and much more likely to be determined by the practicalities of field methods and expert knowledge of seasonality: when it is safe to visit and when the species or measurements under observation are most likely to be easy to find and observe. The timing of visits may introduce bias in estimators and can be modelled, so some schemes combine design based (spatial) and model based (temporal). Restrictions on timings apply much less to satellite remote sensing, which, increasingly, acquires imagery every few days – or even every day – throughout the year. Airborne remote, especially using drones, is similar to other field methods with seasonal restrictions and more limited opportunities for acquisitions.

Reference

Thomas, L., Buckland, S.T., Rexstad, E.A., Laake, J.L., Strindberg, S., Hedley, S.L., Bishop, J.R., Marques, T.A. and Burnham, K.P. (2010) Distance software: design and analysis of distance sampling surveys for estimating population size, *Journal of Applied Ecology* 47(1), 5–14. doi: 10.1111/j.1365-2664.2009.01737.x

Appendix 3

Definitions of terms: micro, aggregated, and register databases

Urban Lindgren and Einar Holm

Micro database

The concept *micro database* refers to collections of microdata items. One such collection within the database is given different names, often contingent on data source and use; that is, a row, a case, a record, an instance, an object, an individual, a decision-making unit. Each microdata item is selected based on some common defining property like being a member of the species *Homo sapiens*. Such a microdata item – that is, representing an individual human – in turn contains a set of attributes and sometimes also events. An example micro database might be labelled "Population in Sweden 2000–2200" and contain some 10 million individuals, each one with attributes like identity, origin, sex, mother, year, partner, age, living place, X–Y coordinates, education, profession, and income. In addition, it can contain events like give birth, mate, move, get a job, etc. With 13 such attributes and four potential events for each person each year, the example database will contain $10^6 \times 22 \times 17$ cells or fields, each with a particular value (374 million cells). Similar micro databases might contain all (or a sample of) examples of a certain animal or plant in a county. If both databases contain a reference to location (place, coordinates) for each instance, the social and biological data can be combined for analysis in full detail.

Aggregated database

In an aggregated database (or table), the values of one or several or all instance fields are replaced by counts or means in cells defined by classifying fields into more coerce dimensions. An aggregated table with, for example, four dimensions – year (in ten-decade groups), age (in 20 five-year groups), sex (in two groups), place (in 290 municipalities) – gives a selected small subset of the information of the micro database from which it was obtained. But that descriptive information might sometimes be enough and precisely what is demanded as a result or as an input for further analysis. A more in-depth causal analysis, however, usually requires the use of more variables than can be contained in a descriptive multi-dimensional table.

Register database

A register database is defined not only by target content but also by a current agency (SCB) producing and/or storing the data. Access to certain register data is controlled by the agency in charge of it as reflecting internal policy or legal rules.

A register database can contain a micro database or an aggregated database or something else – any information that can be stored and presented in table format.

There is an additional, more hidden, characteristic/requirement of the micro unit in a micro database: it has a certain complexity (own adaption, metabolism, etc.) making each instance easy to distinguish from each other and from their environment. One common limit might be between living and dead organisms or matter. One could treat each cubic centimetre of water in a sea as a separate object like humans in a population, but there is really no point in that. In the case of water, it does not matter how the borders of the water "objects" are defined; that is, mean value of temperature or nitrate concentration. The water molecules do not care much about the human-imposed borders between some of them for total mean frequency or interaction between them. Consequently, a large amount of register data tables of dead matter does not need an intermediate level of objects to be useful and maintained as a register database.

Regarding the content, many or most available register-based micro databases are created for administrative purposes; that is, income, taxation, work, location of population, etc. Although this might sometimes create a collection bias (e.g. under-reported income sources in order to avoid taxes), the advantage compared to survey-based data collection is close to 100% coverage of the target population giving small amounts of selection errors, year by year, in total population registers. Maintenance transparency in data definitions and collection methods is also superior compared to comparing surveys with different selection biases, created for different purposes different years. The main bias of extensively using register data (in Sweden) is that the selection of available attributes gravitates towards demography, mobility, family formation, work, education, income, taxes, etc., whereas attitudes, opinions, and other non-economic behaviours are almost completely absent in Swedish total population registers. Instead, some survey-based repeated data collections exist, like the Labour Force Surveys (LFS), describing the labour market developments for the Swedish population aged 15 to 74.

Appendix 4

Additional examples of potential microsimulation applications

Urban Lindgren and Einar Holm

The importance of labour mobility for firm performance and regional development

There is extensive literature on this research theme within economic geography. Together with colleagues at the Department of Geography in Umeå, we have published several journal articles on this topic (e.g. Boschma et al. 2009, 2014; Eriksson and Lindgren 2009; Eriksson 2011; Östbring and Lindgren 2013; Borggren et al. 2016; Östbring et al. 2017, 2018). The research questions posed in all of these papers stem from some generic questions of economic geography: how is firm competitiveness affected by the region where they operate? And what characteristics make a region economically prosperous?

Most people would acknowledge that different parts of a country – or regions – are variously successful. In comparison to regions in the periphery, metropolitan regions tend to have a more diversified labour market with many different types of jobs, higher average income, lower unemployment rates, and a wide variety of services and amenities. Today's differences between regions are partly due to historical reasons often related to the existence of natural geographical advantages such as proximity to natural harbours, waterways, streaming water, etc. More recent theorization on regional development stresses the importance of co-location of economic activities. Co-location has two types of advantages: those providing lower costs and those emerging from social interaction between people. Co-location in a city increases the likelihood of people knowing each other (or at least having heard of), which is an important piece of information regarding whom to trust or not. Being able to skip costly inquiries about potential business partners reduces costs. The second type of advantage refers to factors such as business climate and untraded interdependencies; that is, place-based routines, norms, and values. These are links between firms that are not directly connected to business transactions. Informal norms and behaviours that influence decisions of firms are regarded as important factors for the competitiveness of the local trade and industry.

Learning, creativity, and innovations also belong to the second type of advantage. It is widely recognized that related co-located firms create a local milieu that provides a breeding ground for knowledge dissemination and learning (e.g. Malmberg and Maskell 2002). It is in this theoretical context the referred studies on labour mobility should be viewed. Because the matching of skill portfolios within firms influences the potential for knowledge dissemination and learning – which in turn affects innovations, firm productivity, and ability to compete – there will be regional implications in the sense that the distribution of successful firms will vary across regions. Empirical studies show such patterns indicating a correlation between regional population density and successful regional trade and industry.

Register data are key for the advancement of this research. Briefly described, the papers mentioned above make use of comprehensive longitudinal population registers that combine individual demographic and socioeconomic characteristics and firm attributes. By analyzing skill portfolios (e.g. formal education or work experience) of in-house staff or comparing differences between in-house staff and inflow of new staff, cognitive distance within the entire workforce can be calculated. These calculations (entropy measures) are usually used to determine whether the firm skill portfolio is unrelated, related, or similar. A general conclusion made in this literature is that a related skill portfolio tends to be more beneficial for innovation and learning, which make firms more competitive and successful. The theoretical explanation for this result is that long cognitive distances (unrelated), due to very different educational backgrounds, imply difficulties in communication and sharing professional ideas. On the other hand, if the educational backgrounds are more or less identical (similar), individuals know the same things, which may be detrimental to work efforts aiming at novelty. The sweet spot seems to be a cognitive distance somewhere in between (*lagom kognitivt avstånd*), which enables new related ideas to thrive in an interactive communicative environment.

The processes put forward in this research have a strong bearing on regional development. The results hint at the importance of labour market matching, which is related to factors such as spatial distribution of population, migration, and gravitational shifts in the urban system. In particular, the representation of migration destination choice is a difficult task in microsimulation modelling.

In a specific application, some – but not all – aspects of the broad above-discussed set of potential relations between conditions and effects have to be operationalized, estimated, and implemented in the model. To explore "what if" questions related to labour mobility using simulation, what type of labour mobility will be used to explore effects on firm performance and regional development has to be decided and how this is covered by variables/indicators present and available in the model must be known.

A core driver of several relations is the impact of agglomeration economies on firm and regional performance (via co-location, access to skills, knowledge, networks, etc.). This might be indicated by a labour supply potential (access to labour via a production-constrained interaction model) measured from each firm's current location (in addition to controlling for other confounding covariates). A simulation experiment could be to change the home–work distance decay sensitivity in the labour potential calculation, run the model with and without the change, and compare short- and long-term impacts on firm and regional performance (employment, productivity, etc.). This would provide some information about the extent to which distance and density matter for performance.

Another example would be to explore the potential impacts of similarity or relatedness between old and new recruitments to firms. Changing the proportions can be implemented on different arenas "before" recruitment. This can be done by going back in history and changing proportions between individuals' education, skills, etc., used as relatedness indicators or by just changing how the allocation between people with given supplied skills is made to firms in demand. Is it possible to construct a match giving much more related recruitments and, in turn, does this influence firm and regional performance?

Changing attitudes to the balance between work and leisure – downshifting as a way of leading a "simpler" life

Work is important in people's lives – it brings livelihood, and for some it is part of their identity, too. From a societal point of view, life is commonly subdivided into three parts: childhood, adulthood, and old age. Work is strongly linked to adulthood, not least because adults should provide for those who cannot work. Children are in need of care and the elderly have a right to a pension after working throughout their lives. A brief look at the social insurance system of any welfare society shows the reliefs of these phases of life. Because working is a strong social norm, the growing interest in trends towards voluntary simplicity and downshifting that strive for a reduction of consumption and time spent on work (e.g. Etzioni 1998) appears to be at odds with dominating notions such as full employment, full-time jobs, and career making. In a recently published journal article (Eimermann et al. 2021), we investigated to what extent downshifting shows any footprints in Swedish register data.

The ideas of voluntary simplicity and downshifting can be traced back several decades. Many early studies were conducted in the UK, North America, and Australia (e.g. Saltzman 1991; Etzioni 1998; Schor 1999; Tan 2000; Hamilton 2003; Hampton 2008). Alexander and Ussher (2002) argued that as many as 200 million people in Western countries may have tried various ways of leading a "simpler life", including less consumption and fewer working hours. In a similar vein, Hamilton (2003) reported that 25% of British adults aged 30 to 60 had downshifted and that this shift was equally distributed across sociodemographic groups. There are many different ways to obtain such a goal. Saltzman (1991) suggested several types of downshifting, such as avoiding promotions, stepping down the organizational ladder, leaving jobs in competitive organizations for self-employment doing similar tasks, and escaping large cities by setting up viable businesses in small towns in the countryside. In a recent study by Sandow and Lundholm (2020) analyzing counterurbanization in Sweden, they showed that there was a persistent outflow (during their study period 2003–2013) of families moving from medium-sized towns and small towns to more sparsely populated parts of the country. These couples are commonly highly educated and work as public-sector professionals.

In the Swedish context, the popular debate – not least manifested in blogs and other channels of social media – on downshifting has been going on for many years. In some blog communities, discussions centre on ways of escaping the rat race; that is, quit working, reduce consumption, and take control over their own time, occasionally by moving to smaller towns or the countryside (e.g. *Onkel Toms stuga / Tid och pengar* and *Farbror Fri / Enkel boning*). From a work point of view, some bloggers such as Miljonär innan 30 och pensionär innan 40 have pushed the argument even further by aiming for retirement sometime during their 30s and 40s. The implementation of such plans varies, but a common theme is to reduce consumption and increase savings of the take-home pay as much as possible. According to calculations by the well-known U.S. blogger Peter Adeney, the time it takes to become financially independent (defined as returns on investments exceeding monthly household expenditures) is a function of savings rate. Provided a 5% inflation-adjusted return on investment and a 50% savings rate, it takes roughly 16 years to reach the goal. This is a vital source of inspiration for bloggers aspiring to early retirement, but it goes without saying that these views also give rise to harsh criticism by opponents who point out many different concerns, mostly with regard to consequences for welfare provision of reduced income tax revenues (e.g. Tjernström Lustig 2021).

Based on experiences of quite extensive numbers of downshifting in other countries and a presumed upswing of people interested in voluntary simplicity, we set out to investigate whether these trends can be observed in recent register data (Eimermann et al. 2021). To do so, we needed a definition of *downshifting* that includes a theoretical content that can be observed in register data. The work by Amitai Etzioni (1998), which is widely recognized in this field of research, offers a theorization of voluntary simplicity that is helpful for our purpose. In particular, the description of the type of voluntary simplicity denoted *holistic simplifiers* fits the bill. Holistic simplifiers "adjust their whole life patterns according to the ethos of voluntary simplicity. They often move from affluent suburbs or gentrified parts of major cities to smaller towns, the countryside, farms and less affluent or urbanised parts of the country ... with the explicit goal of leading a 'simpler' life" (Etzioni 1998). This definition includes observable factors such as income changes, neighbourhood characteristics, and migration.

The results of our analyses showed that during the mid-2010s, downshifting in the style of holistic simplifiers was rare – only 3,188 individuals met the criteria of significantly reducing work income (50% drop or more) and moving away from affluent neighbourhoods in large cities (parishes having mean work income above the parish distribution average). On an annual basis, this corresponds to 0.33% of all movers crossing a municipality border. Moreover, the results showed that only a few of the 3,188 identified holistic simplifiers reduced their consumption (relating to cars and housing) in accordance with what could be expected from previous studies.

According to Eimermann et al. (2021), downshifting à la holistic simplifiers does not seem to be a widely spread phenomenon in Sweden. There may be groups of people who consider such a change and have executed softer types of downshifting, but it is unclear how common such patterns are in general. In regard to microsimulation modelling, it can be concluded that downshifting is not a factor that will likely have large impacts on modules in a microsimulation model. Modules and their behavioural equations relating to, for example, labour supply, migration, and residential mobility do not appear to need downshifting-related variables to increase their explanatory power.

Even if this may be the case, it would be possible to explore certain aspects of downshifting; for example, as a component of size of the living place locally and regionally. If the downshifter is moving to a place that is smaller both locally – for example, population within 1km – and regionally – for example, population within 50km (or measured across administrative and functional regions such as RegSo and local labour market regions), this would be a considerably softer move but still a move to a smaller vicinity, with regard to both neighbours and employment. From this line of reasoning, a number of model-related research questions can be posed: How many are they? And how is this influencing individual income over time (given other covariates)? Moreover, if there is a remaining influence on income, this could be added to the general income equation in the model, and experiments with different levels of local and regional density could be explored by simulation. This would generate additional questions such as: for how many individuals is this move associated with decreased income? And who are the movers?

Exploring social frontiers as an additional dimension of residential segregation – patterns and consequences

In an ongoing project called "Life at the frontier: researching the impact of social frontiers on the social mobility and integration of migrants" led by Director Prof. Gwilym Pryce of the University of Sheffield and funded by NordForsk, the aim is to analyze the educational, employment, and earnings outcomes of people living in highly segregated parts of large cities. Research on residential segregation – or the spatial separation of groups along the lines of ethnicity, socioeconomic differences, and demographic characteristics – has been attracting attention among scholars for more than 50 years. In the seminal work by Massey and Denton (1988), five dimensions of residential segregation were identified, but most of these widely applied measures are aspatial (Piekut et al. 2019) and are calculated for comparatively large spatial units. None of the five dimensions pay attention to spatial transitions between communities. It has been suggested that these transitions can be materialized as social boundaries that may be linked to social mobility and integration. Dean et al. (2019) argued that social boundaries may turn into social frontiers when there are distinct spatial divisions in the residential ethnic and socioeconomic composition of adjacent communities. Theoretically, residing close to such social frontiers might produce many different outcomes related to social mobility and integration. It has been suggested that these neighbourhoods isolate ethnic minorities and poor people from the networks of information and employment opportunities, increase risk for prejudice and misunderstandings leading to conflicts, and increase recognition of lower socioeconomic status among different vulnerable groups.

Dean et al. (2019) developed an algorithm for detecting social frontiers using British data. Within the "Life at the frontier"-project, we have applied this methodology and run the algorithm on Swedish register data for Stockholm, Göteborg, and Malmö. The results show that there are significant differences between neighbourhoods (DeSO) sharing a social frontier defined as neighbourhood differences between the share of foreign-born people. Briefly, descriptive statistics indicate that there are differences between immigrant groups in regard to living on the high immigrant proportion side or the low immigrant proportion side of a social frontier.

For example, immigrants born in OECD countries outside Europe more frequently reside on the low immigrant proportion side of the social frontier, whereas the opposite is true for immigrants from the Global South and Eastern Europe. Regarding labour market and income factors, employment rates are higher on the low proportion side and so are wage incomes. On the other hand, income from one's own business is higher on the high proportion side. This might be related to necessity-driven entrepreneurship (e.g. Rataj 2020) triggered by recognized difficulties in obtaining permanent employment paying a minimum wage for immigrants with short formal education (e.g. Bornhäll et al. 2019). The high immigrant proportion sides of social frontiers have more grocery stores, pharmacies, and gyms, which is an interesting finding given the supposed shortage of service facilities in vulnerable areas characterized by low purchasing power and social unrest.

The overall focus of the ongoing studies within the project is to analyze causal effects of living close to social frontiers during childhood and adolescence. What long-term consequences for labour market outcomes (employment, career development, earnings, etc.), economic mobility (i.e. inter-generational changes in earnings and occupational status), and economic integration (i.e. diminishing economic differences between, for example, natives and immigrants) does this type of residential segregation bring to people? In this research, we try to separate between, on the one hand, neighbourhood effects triggered by the composition of and the mobility flows of people living in the vicinity and, on the other hand, place-based factors related to the specific geography of the neighbourhood – for example, physical characteristics of the built environment that may have impact on people's well-being and safety. We intend to do this by research designs that use natural experiments appearing as a consequence of policy changes – for example, the Swedish settlement dispersal strategy of refugee immigrants during the 1980s and 1990s (e.g. Edin et al. 2004). Another approach is to use quasi-experimental designs shedding light on mechanisms that generate outcomes for children via neighbourhoods (Chetty and Hendren 2018). In this study, they compare outcomes of siblings who shared the same family context but through family migration were exposed to different neighbourhood contexts during their childhood.

Residential segregation influences people in many different ways. It is likely that the childhood residential environment will have an impact on future labour market outcomes and the chances of economic betterment in comparison to the parents' generation. Microsimulation models that take these aspects into account will make more accurate representations concerning factors such as labour market participation, employment, unemployment, work income, dependence on social benefits, and population redistribution. However, contrafactual simulation cannot by itself replace the quest to control for effects of selection on future education and earnings by comparing siblings who grew up in different kinds of neighbourhoods. However, it can facilitate studying the long-term effects of different types of neighbourhood exposure.

References

Alexander, S. and Ussher, S. (2002) The voluntary simplicity movement: A multi-national survey analysis in theoretical context, *Journal of Consumer Culture* 12, 66–86.

Borggren, J., Eriksson, R.H. and Lindgren, U. (2016) Knowledge flows in high-impact firms: How does relatedness influence survival, acquisition and exit? *Journal of Economic Geography* 16(3), 637–665.

Bornhäll, A., Daunfeldt, S.-O. and Seerar Westerberg, H. (2019) Less than 30 percent of non-Western immigrants earn a monthly wage that exceeds 2000 euro after nine years in Sweden, *HFI Notes* No. 1.

Boschma, R., Eriksson, R. and Lindgren, U. (2009) How does labour mobility affect the performance of plants? – the importance of relatedness and geographical proximity, *Journal of Economic Geography* 9(2), 169–190.

Boschma, R., Eriksson, R. and Lindgren, U. (2014) Labour market externalities and regional growth in Sweden – the importance of labour mobility between skill-related industries, *Regional Studies* 48(10), 1669–1690.

Chetty, R. and Hendren, N. (2018) The impacts of neighborhoods on intergenerational mobility I: childhood exposure effects, *The Quarterly Journal of Economics* 133(3), 1107–1162.

Dean, N., Dong, G., Piekut, A. and Pryce, G. (2019) Frontiers in residential segregation: understanding neighbourhood boundaries and their impacts, *Tijdschrift voor Economische en Sociale Geografie* 110(3), 271–288.

Edin, P.-E., Fredriksson, P. and Åslund, O. (2004) Settlement policies and the economic success of immigrants, *Journal of Population Economics* 17, 133–155.

Eimermann, M., Lindgren, U. and Lundmark, L. (2021) Nuancing holistic simplicity in Sweden: a statistical exploration of consumption, age and gender, *Sustainability* 13, 8340.

Eriksson, R. (2011) Localized spillovers and knowledge flows: how does proximity influence the performance of plants? *Economic Geography* 87(2), 127–152.

Eriksson, R. and Lindgren, U. (2009) Localised mobility clusters: impacts of labour market externalities on firm performance, *Journal of Economic Geography* 9(1), 33–53.

Etzioni, A. (1998) Voluntary simplicity: characterization, select psychological implications, and societal consequences, *Journal of Economic Psychology* 19, 619–643.

Hamilton, C. (2003) *Downshifting in Britain: A Sea-Change in the Pursuit of Happiness*. Discussion Paper 58. Canberra: The Australia Institute.

Hampton, R.S. (2008) *Downshifting, Leisure Meanings and Transformation in Leisure.* PhD Thesis, Pennsylvania State University.

Malmberg, A. and Maskell, P. (2002) The elusive concept of localization economies: towards a knowledge-based theory of spatial clustering, *Environment and Planning A* 34, 429–449.

Massey, D.S. and Denton, N.A. (1988) The dimensions of residential segregation, *Social Forces* 67(2), 281–315.

Östbring, L., Eriksson, R.H. and Lindgren, U. (2017) Labor mobility and organizational proximity: routines as supporting mechanisms for variety, skill integration and productivity, *Industry and Innovation* 24(8), 775–794.

Östbring, L., Eriksson, R.H. and Lindgren, U. (2018) Relatedness through experience: on the importance of collected worker experiences for plant performance, *Papers in Regional Science* 97(3), 501–518.

Östbring, L. and Lindgren, U. (2013) Labour mobility and plant performance: on the (dis)similarity between labour- and capital-intensive sectors for knowledge diffusion and productivity, *Geografiska Annaler B* 95(4), 287–305.

Piekut, A., Pryce, G. and van Gent, W. (2019) Segregation in the twenty first century: processes, complexities and future directions, *Tijdschrift voor Economische en Sociale Geografie* 110(3), 225–234.

Rataj, M. (2020) *The Geography of Entrepreneurship: Regional and individual Determinants of New Firm Formation in Sweden.* PhD Thesis, Umeå University.

Saltzman, A. (1991) *Downshifting – Reinventing Success on a Slower Track.* New York: HarperCollins.

Sandow, E. and Lundholm, E. (2020) Which families move out from metropolitan areas? Counterurban migration and professions in Sweden, *European Urban and Regional Studies* 27(3), 276–289.

Schor, J.B. (1999) *The Overworked American.* New York: Basic Books.

Tan, P. (2000) *Leaving the Rat Race to Get a Life: A Study of Midlife Career Downshifting.* PhD Thesis, Swinburne University.

Tjernström Lustig, A. (2021) Ekorrhjulsavhopparna snyltar på de som jobbar och betalar skatt, https://www.gp.se/debatt/ekorrhjulsavhopparna-snyltar-p%C3%A5-de-som-jobbar-och-betalar-skatt-1.40121181

Index

Note: page numbers in italic type refer to Figures; those in bold type refer to Tables.